深入浅出 Windows Phone 8 应用开发

Windows Phone 8: Developing and Deploying

林政 著
Lin Zheng

清華大學出版社
北京

内容简介

本书系统论述了 Windows Phone 8 操作系统的基本架构、开发方法与开发实践。全书内容共分三篇：开发基础篇、开发技术篇和开发实例篇。开发基础篇包括第 1 章～第 3 章，介绍了 Windows Phone 8 的技术架构及基本特性、开发环境的搭建、创建项目的方法，以及 XAML 语法基础；开发技术篇包括第 4 章～第 21 章，介绍了常用控件、布局管理、数据存储、图形动画、多媒体、启动器与选择器、手机感应编程、MVVM 模式、网络编程、异步编程与并行编程、联系人和日程安排、手机文件读取、Socket 编程、墓碑机制与后台任务、蓝牙通信和近场通信、响应模式，以及 C++ 编程；开发实例篇包括第 22 章～第 24 章，分别介绍了普通应用实例（时钟、日历、统计表、记事本、快速邮件）、网络应用实例（RSS 阅读器、博客园主页、网络留言板、快递 100）和记账本应用等。

本书配套光盘提供了书中实例源代码及开发实例的讲解视频，最大限度满足读者高效学习和快速动手实践的需要。

本书内容覆盖面广、实例丰富、注重理论学习与实践开发的配合，非常适合于 Windows Phone 8 开发入门的读者，也适合于从其他智能手机平台转向 Windows Phone 8 平台的读者；对于有 Windows Phone 开发经验的读者，也极具参考价值。

图书在版编目(CIP)数据

深入浅出：Windows Phone 8 应用开发/林政著. --北京：清华大学出版社，2013.1(2013.11 重印)
ISBN 978-7-302-30836-2

Ⅰ. ①深… Ⅱ. ①林… Ⅲ. ①移动电话机－应用程序－程序设计 Ⅳ. ①TN929.53

中国版本图书馆 CIP 数据核字(2012)第 284671 号

责任编辑：盛东亮
封面设计：李召霞
责任校对：时翠兰
责任印制：王静怡

出版发行：清华大学出版社
网　　址：http://www.tup.com.cn，http://www.wqbook.com
地　　址：北京清华大学学研大厦 A 座　　**邮　　编**：100084
社 总 机：010-62770175　　**邮　　购**：010-62786544
投稿与读者服务：010-62776969，c-service@tup.tsinghua.edu.cn
质 量 反 馈：010-62772015，zhiliang@tup.tsinghua.edu.cn
课 件 下 载：http://www.tup.com.cn，010-62795954
印 装 者：清华大学印刷厂
经　　销：全国新华书店
开　　本：186mm×240mm　　**印　　张**：39.5　　**字　　数**：904 千字
版　　次：2013 年 1 月第 1 版　　**印　　次**：2013 年 11 月第 4 次印刷
印　　数：7501～9500
定　　价：79.00 元

产品编号：050834-01

推荐序

PREFACE

开发者编写代码、创造奇迹——堪称IT行业的魔术师。随着微软Windows Phone和Windows 8设备的快速普及，移动互联网开发者的黄金时代正在到来。微软的开发工具和应用平台在进一步降低开发者的应用开发成本，新的应用分发模式和巨大的客户群让这个时代充满机会。

微软为开发者提供专业、全面和便捷的开发工具，支持从云到端应用的开发、设计和测试环境的高一致性。Windows Phone开发者可以免费下载所有的开发、设计和测试工具。

微软一直致力于从平台层面为开发者提供最好的开发体验。Windows Phone 8和Windows 8公共内核使微软的应用开发平台具备更好的一致性。开发者同时为Windows 8和Windows Phone构建应用时，将极大地降低开发成本。开发者可以利用一套开发工具，使用一致的开发模型，复用大量代码，在Windows Phone和Windows 8上构建用户体验一致的应用和游戏。共享内核意味着：①工具是相同的；②语言是相同的；③XAML的用户界面定义背后的大多数概念是类似的；④在某些情况下，开发者可能在两种平台中共享大部分相同代码，来设计自己的应用和游戏。Windows Phone 8还为开发者提供了很多新的功能，例如：C和C++原生代码，遵循微软设计风格的开始屏幕；最新的IE 10；中文语音识别及其文语转换；NFC技术；企业级应用部署方案。

微软的Windows Phone和Windows 8平台和应用生态系统处于快速发展周期，每天都有大量的开发人员加入并贡献他们开发的应用。微软应用商店的管理规则严谨而透明，高质量应用可以用相对较低的成本获取用户的关注。随着移动互联网的飞速发展，开发者在移动领域面临前所未有的机遇。微软不断带来的云和终端技术创新，使开发者可以充分发挥创意，并为消费者带来新体验。

本书为国内第一部较为详细地阐述Windows Phone 8核心技术与最佳实践的图书，可以为广大读者提供开发入门的捷径，值得认真阅读。愿本书助力中国开发者抓住移动互联网的发展机遇，共享成功！

赵立威

技术顾问总监

微软(中国)有限公司

2012年12月1日

赞 誉

REVIEW

移动终端的发展日新月异，所有全平台的移动互联网产品都在支持 Windows Phone 8，而对于开发者而言，通过本书可以了解到 Windows Phone 8 体系的方方面面，是一本非常实用的开发教程，该书既包含了平台核心技术，又提供了真实的项目案例，本书适合希望从事 Windows Phone 8 开发的工程师，也可作为高校开展相关专业的教材，不容错过。

——沈大海 工信部移动互联网人才培养办公室

非常荣幸第一时间阅读了本书的样章，通读之后可以看出作者在内容编排以及示例代码编写上下了足够的功夫。伴随着 Windows Phone 生态系统的快速成长，越来越多的开发者加入到了 Windows Phone 阵营并赚到了自己的第一桶金。如果你想加入 Windows Phone 8 开发者阵营，或者对 Windows Phone 8 开发有所了解并期望提高自己的开发水平和实战经验，本书都可以助你一臂之力。

——陈啸天 诺基亚开发者生态系统技术支持经理
兼诺基亚体验创新中心项目总监

移动互联是未来趋势，终端为王的今日，Windows Phone 8 在微软的推动下一定会成为不可忽视的一个移动平台。作者用平实直观的方式全面阐述了 Windows Phone 8 的应用开发技术，是学习 Windows Phone 8 开发的一本优秀参考书籍。让我们现在就加入微软移动开发阵营吧。

——张凌华 移动开发专家、麦可网创始人

Windows Phone 8 有可能是未来移动操作系统中最重要的系统之一，里边蕴藏着巨大的机会。这本 Windows Phone 8 的图书，浅显易懂，深入浅出，是一本不错的入门级书籍，相信对学习 Windows Phone 8 开发的人有很大的作用，推荐 eoe 移动开发者社区中对 Windows Phone 8 感兴趣的人认真阅读此书。

——靳岩 eoe 联合创始人、CEO

感谢各位业界资深人士给予点评；点评按给予时间先后顺序，无区分。——作者、编辑注

Windows Phone 作为 Android 和 iOS 最强有力的竞争对手，经过微软两年多时间的精雕细琢，已经逐渐成熟；急需大量应用程序来进行武装；因为 Windows 本身机制的改变，所以其应用程序开发方式也和传统的 Windows Mobile 相差甚远，本书不仅从最基本的原理机制出发，逐步深入 Windows Phone 应用开发的各个方面，还提供了大量实战案例，是学习 Windows Phone 应用开发的绝佳选择。感谢作者给我们带来如此好书！

——杨丰盛 MORE-TOP 团队创始人、畅销 IT 图书作者

iOS 的出现引起了智能手机的革命，Android 的诞生则推动这场革命走向高潮，Windows Phone 的出现给了这场革命增加了更多的选择和可能性。在这场智能手机操作系统之间的竞争中，开发者扮演了一个非常关键的角色，非常有幸在 Android 诞生时我写了中国大陆第一本 Android 技术书籍让很多开发者开始进入 Android 开发阵营并成为先行者，现在摆在大家眼前的是第一本 Windows Phone 8 的技术书籍，而你正有机会成为新一批的先行者。

——姚尚朗 eoe 联合创始人、CTO

2011 年，诺基亚与微软"牵手成功"，这一业内普遍看好的"软硬结合"把 Windows Phone 推到了业界焦点，并与 iPhone、Android 将移动平台变成了三国鼎立之势，从此这个年轻的手机操作系统映入开发者的视野。本书从 Windows Phone 的发展讲起，涵盖其开发的各个方面，其中还不乏实例，既可阅读学习，还可随时调试和运行，是一本不可多得的好教程。如果你是一位 Windows Phone 8 应用初学者，或者你想转行做 Windows Phone 8 开发，强烈推荐你细读这本书，定当获益匪浅！

——曹亚莉 51CTO 技术社区博客总编

欢迎加入 Windows Phone 阵营。2010 年至今，我们一步步看着 Windows Phone 在中国从无到有，从小众到赢得关注，我们也亲身体会到诺基亚和微软在国内开发者生态扶持与协助方面所做的努力，越来越多的开发商也开始重视在此领域的扩张。Windows Phone 8 相对之前的版本是一次飞跃，内核与桌面系统统一，开放了更多更实用的系统接口，请尽快掌握本书内容，一定会让您在此领域拥有更多施展的机会。

——黄斌 智机网站长

前言

FOREWORD

创新与革命一直都是IT行业的灵魂，苹果的iPhone是一个颠覆式的革命者，它重新定义了手机的含义，给予人们一种独一无二的体验，并且打造出了一种前所未有的商业模式，让其iPhone产品在推向市场后大受欢迎。接下来，谷歌公司收购了Android操作系统，把这场智能手机领域的革命推向了另一个高潮，谷歌的开源策略让Android手机遍地开花，大受追捧。然而，革命总是有人欢喜有人忧，昔日的王者诺基亚，已经失去了当年在手机领域呼风唤雨的地位了，Symbian系统的臃肿和落后让诺基亚已经力不从心，微软的Windows Mobile手机操作系统的市场占有率也日渐下降。创新和革命一直都没有停止过，面对着严峻的形势，微软重新审视了手机操作系统的研发，果断地抛弃了落后的Windows Mobile操作系统，研发出了Windows Phone系列手机操作系统，从2010年的第一个版本Windows Phone 7开始到2012年的Windows Phone 8面世，微软一路上不停地开拓创新、精雕细琢，打造出一个强大的手机操作系统和完善的Windows Phone生态圈。2011年4月，诺基亚和微软正式结盟，诺基亚渐渐地放弃Symbian操作系统而转向Windows Phone操作系统，2012年9月，三星、诺基亚、HTC相继发布了搭载Windows Phone 8的旗舰智能手机，标志着Windows Phone操作系统的又一个新的里程碑，也展现了Windows Phone 8无限的发展潜力。

Windows Phone是一个年轻的手机操作系统，它是微软在面对着iPhone和Android的激烈竞争，综合地考虑了iPhone和Android的优点以及缺点的基础上诞生的，具有不可比拟的优势和发展潜力。在未来的智能手机操作系统的领域中，Windows Phone将会起着举足轻重的作用。2011年年初微软与诺基亚达成战略合作协议，共同发展Windows Phone手机操作系统和打造Windows Phone操作系统的生态圈，这对于Windows Phone系统的发展是一个极大的利好消息，同时微软和诺基亚的结盟使得Windows Phone成为了Android系统和iPhone系统的一个强劲对手。

本书包含哪些内容

本书内容涵盖Windows Phone 8手机应用开发的各方面的知识，比如控件、数据存储、图形动画、多媒体、MVVM模式、互联网编程、蓝牙、近场通信、支持C++编程等。全书内容涉及面广，实例丰富，深入浅出地介绍了Windows Phone 8应用开发的方方面面。本书的一些实例应用已经发布到Windows Phone Marketplace，可以直接用Windows Phone 8手机下载使用；当然，书中给出了这些实例的源代码。

光盘包含哪些内容

配套光盘涵盖了两部分内容：书中实例源代码及书中实例的开发视频文件。源代码为第2章及第4章～第24章等共22章内容涉及的实例源代码；开发视频文件为RSS阅读器、博客园主页、记事本、快递100、快递邮件、日历、时钟、统计图表、网络留言表等9个综合实例的开发过程视频录像。

如何高效阅读本书

由于本书的实例代码主要使用的是C#语言开发的(C++编程章节使用的是C++编程语言)，所以需要读者有一定的C#编程基础。本书的各章节之间有一定的知识关联，由浅至深地渐进式叙述，建议初学者按照章节的顺序来阅读和学习本书；对于有一定Windows Phone 8编程经验的读者，可以略过一些章节，直接阅读自己感兴趣的内容。

如何快速动手实践

本书每个知识点都配有相应的实例，读者可以直接用Microsoft Visual Studio 2012 Express for Windows Phone开发工具打开工程文件进行调试和运行。由于微软的开发工具和Windows Phone SDK更新较频繁，所以不能保证最新的开发环境和本书中描述的内容完全一致，要获取最新的开发工具和Windows Phone SDK请关注微软的Windows Phone开发的中文网站(https://dev.windowsphone.com/zh-cn)的动态。

本书适合哪些读者

本书适合于Windows Phone 8应用开发初学者，也适合其他手机平台的开发者快速地转入Windows Phone 8的开发平台，同时对于有一定的Windows Phone 8开发经验的读者也有很好的参考学习价值。

由于作者水平有限，Windows Phone 8开发知识极其广泛，书中难免存在疏漏和不妥之处，敬请广大读者批评指正。

作者联系方式：zheng-lin@foxmail.com

编辑联系方式：shengdl@tup.tsinghua.edu.cn

作　者

2012年12月5日

目录

CONTENTS

开发技术篇

开发实例篇

开发基础篇

万丈高楼平地起，本篇将带领读者快速地了解 Windows Phone 8 操作系统，并且动手开发出第一个 Windows Phone 8 的应用程序。

本篇是对 Windows Phone 8 的一个概括性介绍，读者可以快速阅读本篇，了解 Windows Phone 8 的一些基础的知识和语法的介绍。Windows Phone 8 是微软的一个全新的智能手机操作系统，其设计理念和技术结构都和微软过去的智能手机操作系统有很大的差异，所以通过第 1 章了解 Windows Phone 8 的发展脉络和 Windows Phone 8 的技术结构是学习 Windows Phone 8 应用开发的第一步，也是进入这个领域必须要了解的基本知识。开发环境的搭建也是做手机软件开发不可缺少的一个环节，第 2 章按照搭建开发环境的步骤一步步地搭建好 Windows Phone 8 的应用开发环境，同时快速地开发出第一个 Windows Phone 8 的应用程序。在第一个应用程序中，读者可能会对一些语法结构不太明白，不过，这只是一个大概的了解和介绍，在接下来的学习中读者会慢慢地掌握其中的原理。本篇还对 Windows Phone 8 的基本语法进行了简单的介绍，可以先学习和适应这种语法的结构和编程的方式，在以后的 Windows Phone 8 应用开发的过程中，会进一步与这种 XAML 格式的语法打交道，这里只是从总体上来了解这种语法。

本篇包括以下章节：

第 1 章　概述

介绍 Windows Phone 8 的发展情况并概括性地总结 Windows Phone 8 的技术架构，帮助读者快速了解 Windows Phone 8 手机操作系统。

第 2 章　开发环境

介绍开发环境的搭建步骤，详细地叙述第一个 Windows Phone 8 应用程序的开发以及 Windows Phone 8 项目工程的结构。

第 3 章　XAML 简介

介绍 Windows Phone 8 Silverlight 应用程序开发的基本语法，深刻地诠释 XAML 页面文件的设计和控件的表示。

通过本篇的学习，读者可以了解微软的智能手机操作系统发展历程，Windows Phone 智能手机的发展以及 Windows Phone 8 的技术架构；学会 Windows Phone 8 的环境搭建，创建出自己的 Windows Phone 8 应用程序并了解 Windows Phone 8 应用程序的工程结构、语法结构；初步掌握 XAML 语法知识，适应这个新的语法结构及其编程方式。

第1章 概述

Windows Phone 是一个诞生于移动互联网以及智能手机爆发期间的操作系统，是微软绝地反击苹果 iPhone 和谷歌 Android 的利器。Windows Phone 是微软这位巨人在移动领域的一次冲击，是一次风险与机遇共存的挑战。Windows Phone 8 是 Windows Phone 系列的目前最新的版本，它的诞生意味着 Windows Phone 手机操作系统对 iPhone 和 Android 系统新一轮的反攻。2012 年 9 月诺基亚发布 Windows Phone 8 旗舰手机 Lumia 920，其流畅优美的用户体验，重量级的硬件技术，优秀的操作系统功能展示出了 Windows Phone 8 智能手机的优越和强大。

Windows Phone 8 的操作系统在兼容 Windows Phone 7 的基础上，实现了一次换心手术，把 Windows CE 内核换成了 Windows NT 内核，运行在 Windows 运行时的架构上与 Windows 8 系统形成了统一的编程模式。

1.1 Windows Phone 的技术特点

Windows Phone 是微软公司设计的手机操作系统，因为微软公司之前发布的手机操作系统 Windows Mobile 6.5 是最后的一款 Windows Mobile 系统，所以新的操作系统命名为 Windows Phone 并以 Windows Phone 7 作为 Windows Phone 系列的第一个版本号。目前 Windows Phone 系列最新的操作系统为 Windows Phone 8 操作系统，Windows Phone 8 操作系统向下兼容所有 Windows Phone 7 的应用程序，不过不支持 Windows Phone 7 的硬件设备升级到 Windows Phone 8 操作系统。Windows Phone 8 和 Windows 8 都是运行在 Windows 运行时的架构上，使用了 Windows NT 的内核，两个操作系统可以共用大部分的相同的基于 Windows 运行时的 API。

1.1.1 Windows Phone 的发展

微软的手机操作系统加起来都有十几年历史了，在这十几年的时间里微软向世人树立了自己的手机智能系统的标杆，同时也划开了一个时代的帷幕。微软的一路走来，前半段是一路高歌，后半段是跌跌撞撞，直到现在的 Windows Phone。微软的手机操作系统发展历程如图 1.1 所示。

Windows Phone 是一个在危机中诞生的产品，虽然微软在手机操作系统研发领域已有十几年的历史，但面对 iPhone 和 Android 这些更加易用和极具创新性的产品，Windows Mobile 系统所占的市场份额陡然下降。鲍尔默曾经在 All Things Digital 大会上说："我们曾在这场游戏里处于领先地位，现在我们发现自己只名列第五，我们错过了一整轮。"意识到自己需要亟待追赶之后，微软最终决定按下 Ctrl＋Alt＋Del 组合键，重启自己止步不前的移动操作系统，迎来新的开始。

手机操作系统领域的竞争异常激烈，如果不变革就只有等待着被淘汰。面对这样的形势，微软采取了主动出击的策略。巨人并没有修补 Windows Mobile 这艘漏船，而是精心设计了一个全新的智能手机平台，以应对 iPhone 和 Android 带来的挑战，于是，Windows Phone 以一种崭新的面貌出现在用户的面前，如图 1.2 所示。不要错误地认为微软开发 Windows Phone 的主要目的就是为了赚点授权费，其真正的动机是保卫微软的核心业务：

图 1.1 微软手机发展历程

图 1.2 Windows Phone 8 手机的主屏幕

Windows 和 Office 产品线。移动需求以及智能手机已经变得无处不在，微软必须有一个令人信服的手机系统，以防止越来越多的用户陷进苹果和谷歌的生态圈。

目前，iPhone 和 Android 手机随处可见。智能手机未来的发展趋势非常明显，iPhone 和 Android 很可能成为最主要的两大平台。不过，微软的 Windows Phone 也不可小觑，Windows Phone 这个系统代表着软件巨人的一次冲击，微软在智能机市场发展的早期击败了 Palm 和其他竞争者，却眼睁睁看着动作更快、创新更多的苹果带着出人意料的猛将“iPhone”闯入市场，提高了行业门槛，也提升了人们对手机行业的期望。进行了一些深层次的研究以后，微软走出了正确的一步，从零开始开发了一个全新的独具特色的手机平台——Windows Phone。跟今天的竞争者比起来，它正如 3 年前的 iPhone 一样充满了创新性和差异性。

虽然 Windows Phone 系统推出的时间比较晚，但是该系统应用程序数量增长并不缓慢，Windows Phone 应用程序数量突破 15 000 的时间为 26 周，比苹果当年达到同数量应用程序时间还提前 1 周，这足以证明这一平台从一开始就受到了开发者的追捧。

美国互联网数据中心预计，到 2015 年，Windows Phone 的市场份额将达到 20.9%，超过苹果目前所占份额的 15%。发生变化的主要原因是诺基亚的战略转变，诺基亚已经宣布与微软合作，将从 Symbian 平台转向 Windows Phone 平台。

图 1.3 诺基亚 CEO 史蒂芬·艾洛普(左)与微软 CEO 史蒂夫·鲍尔默

诺基亚于 2011 年 2 月 11 日宣布与微软达成战略合作关系，诺基亚手机将采用 Windows Phone 系统，并且将参与该系统的开发。如图 1.3 所示为诺基亚 CEO 和微软 CEO 的握手合作。双方将达成广泛的战略合作，诺基亚将把 Windows Phone 作为智能手机的主要操作系统，并融合部分微软的互联网服务。两家公司建立一个全新的“移动生态圈”，诺基亚的内容和应用商店将与微软的 Microsoft Marketplace 整合，诺基亚将向微软提供硬件设计和语言支持方面的专业技术，并提供营销支持，协助 Windows Phone 手机丰富价格定位，获得更多市场份额，并进军更多地区市场。微软拿出一套工具，让开发者能更容易地开发出 Nokia Windows Phone 的 App；微软也将 Bing 服务和 adCenter 广告服务整合进诺基亚手机，诺基亚地图将成为 Bing 地图的一部分。

在微软和诺基亚两大巨头的联合之下，Windows Phone 成长的潜力不可估量。Pyramid Research 公司在 2011 年 5 月份发出一份研究报告，根据今年第一季度的数据预测出手机操作系统市场到 2015 年的情况。随着诺基亚加入到 Windows Phone 的大军，Pyramid 认为微软的这款操作系统最早将在 2013 年击败 iPhone、BlackBerry OS，甚至包括 Android 在内的其他操作系统，其分析图表如图 1.4 所示。这一结论是基于该公司对全球所有市场的研究得出的，而诺基亚擅长的低端手机在许多中小市场非常受欢迎。

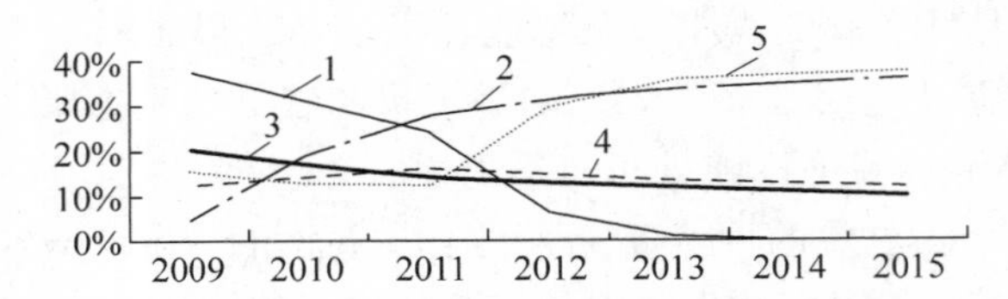

图 1.4 Pyramid Research 公司对手机系统市场占有率的预测走势

1—Symbian；2—Android；3—BlackBerry OS；4—Apple iOS；5—Windows

金字塔的高级分析师 Stela Bokun 解释说，Windows Phone 的市场份额将有望在更早的时间点上——2013 年实现对 Android 的超越。Bokun 认为，事实上从 2011 年开始，Windows Phone 的普及速度远远超过 2008 年谷歌发布 Android 系统后的普及速度。分析师将此趋势归功于微软与诺基亚合作后的光明前景，诺基亚将降低 Windows Phone 给终端用户带来的成本压力，从而加速这一系统的普及。

1.1.2 Windows Phone 8 的出现

Windows Phone 8 是微软在 2012 年 6 月 21 日发布的最新的 Windows Phone 系列的操作系统，搭载 Windows Phone 8 的智能手机也在 2012 年陆续地上市。Windows Phone 8 是 Windows Phone 系统的下一个版本，也是目前 Windows Phone 的第三个大型版本。由于内核变更，所有 Windows Phone 7.5 系统手机无法升级到 Windows Phone 8。

Windows Phone 8 将与即将发布的 Windows 8 操作系统共享核心代码，这意味着 Windows Phone 手机用户将可使用更多的设备和应用，表明微软朝着一体化 Windows 产品组合的方向迈出了新的一步，将给计算行业带来彻底的变革。Windows Phone 8 采用和 Windows 8 相同的针对移动平台精简优化 NT 内核，这标志着移动版 Windows Phone 将提前与 Windows 系统(ARM)同步，部分 Windows 8(ARM)应用可以更方便的移植到手机上，例如不需要重写代码等。

Windows Phone 8 系统也是第一个支持双核 CPU 的 Windows Phone 版本，宣布 Windows Phone 进入双核时代，同时宣告着 Windows Phone 7 退出历史舞台。Windows Phone 8 兼容所有 Windows Phone 7.5 的应用程序，但 Windows Phone 8 的所有原生程序无法在 Windows Phone 7.5 上运行，属于单向兼容。

1.1.3 Windows Phone 8 的新特性

Windows Phone 8 是 Windows Phone 系列操作系统一次重大的升级，它添加了很多新的特性，给 Windows Phone 8 的手机提供了更加强大完善的功能。

1. 硬件提升

此次 Windows Phone 8 系统首次在硬件上获得了较大的提升，处理器方面 Windows Phone 8 将支持双核或多核处理器，理论上最高可支持 64 核，而 Windows Phone 7.5 时代只能支持单核处理器。Windows Phone 8 支持三种分辨率：800×480(15：9)、1280×720(16：9)

和 1280×768(15∶9),Windows Phone 8 屏幕支持 720P 或者 WXGA。Windows Phone 8 将支持 MicroSD 卡扩展,用户可以将软件安装在数据卡上。同时所有 Windows Phone 7.5 的应用将全部兼容 Windows Phone 8。

2. 浏览器改进

Windows Phone 8 内置的浏览器升级到了 IE10 移动版。相比 Windows Phone 7.5 时代,JavaScript 性能提升 4 倍,HTML 5 性能提升 2 倍。

3. 游戏移植更方便

换上新内核的 Windows Phone 8 开始向所有开发者开放原生代码(C 和 C++),应用的性能将得到提升,游戏更是基于 DirectX,方便移植。由于采用 Windows 8 内核,Windows Phone 8 手机将可以支持更多 Windows 8 上的应用,而软件开发者只需要对这些软件做一些小的调整。除此以外,Windows Phone 8 首次支持 ARM 构架下的 Direct3D 硬件加速,同时由于基于相同的核心机制,因此 Windows 8(ARM)平台向 Windows Phone 8 平台移植程序将成为一件轻松的事情。

4. 支持 NFC 技术

Windows Phone 8 将支持 NFC 移动传输技术,这项功能在之前 Windows Phone 7 时代是没有的。而通过 NFC 技术,Windows Phone 8 可以更好地在手机、笔记本、平板之间将实现互操作,共享资源变得更加简单。

5. 实现移动支付等功能

由于 NFC 技术的引进,移动钱包也出现在 Windows Phone 8 中了,支持信用卡、贷记卡,以及会员卡等,也支持 NFC 接触支付。微软称之为"最完整的移动钱包体验"。同时微软为 Windows Phone 8 开发了程序内购买服务,也可以通过移动钱包来支付。Windows Phone 8 中将直接内置 Wallet Hub(钱包中心),这是一项结合了可让移动运营商参与的安全 NFC 支付以及信用卡、会员卡信息存储的功能,同时也有点类似苹果 iOS 6 中的 Passbook 功能。

6. 内置诺基亚地图

Windows Phone 8 将用诺基亚地图来替代 Bing 地图,地图数据将由 NAVTEQ 提供,微软 Windows Phone 8 内置的地图服务全部具备 3D 导航与硬件加速功能。同时,所有机型都将内置原来诺基亚独占的语音导航功能,而诺基亚的 WP8[①] 手机地图支持离线查看、Turn By Turn 导航等功能。诺基亚与微软的合作正在逐步加深。

7. 商务与企业功能

由于 Windows Phone 7.5 对于商业的支持不够全面,因此 Windows Phone 8 对于移动商业服务大幅改进,Windows Phone 8 支持 BitLocker 加密、安全启动、LOB 应用程序部署、设备管理,以及移动 Office 办公等。企业功能可以算是 Windows Phone 8 的重头戏,新增 BitLocker 加密、更强的安全性,管理方面,可使用类似于管理 Windows PC 的工具对手机

① 为了方便,习惯上将 Windows Phone 8 简记为 WP8,后文不再声明。

和应用进行管理，而且支持远程，企业也可以拥有应用的私有分发渠道。

8. 新的待机界面

Windows Phone 8 拥有了新的动态磁贴界面，磁贴可以分为大中小三种，并且每一小方块的颜色可以自定义。需要注意的是，按住瓷片原来只可以调整位置或者删除，而 Windows Phone 8 中可以通过右下角的剪头调整瓷片大小，甚至可以横向拉宽到整个屏幕。同时 Windows Phone 8 上实时的地图导航可以在主界面的磁贴块中直接显示。

1.2 Windows Phone 的技术架构

从 Windows Phone 7 操作系统到 Windows Phone 8 操作系统最大的改变就是把 Windows CE 内核更换了 Windows NT 内核，并且底层的架构使用了 Windows 运行时的架构。在该平台上的编程语言支持 C#、VB. NET 和 C++，在 XAML 普通应用程序开发框架里面可以使用 C#或者 VB. NET 语言，使用 C++编程需要通过 Windows 运行时组件来调用，不能够直接与 XAML 页面进行交互。

1.2.1 Windows 运行时

Windows 运行时，英文名称 Windows Runtime，或 WinRT，是 Windows 8 和 Windows Phone 8 中的一种跨平台应用程序架构。Windows 运行时支持的开发语言包括 C++（一般包括 C++/CX）和托管语言 C# 和 VB，还有 JavaScript。Windows 运行时应用程序同时原生支持 x86 架构和 ARM 架构，同时为了更好的安全性和稳定性，也支持运行在沙盒环境中。Windows Phone 8 使用的 Windows 运行时是一个精简版本的 Windows 运行时和 Windows 8 上还是有着一定的差异性，比如 Windows Phone 8 版本的 Windows 运行时不支持 JavaScript 的编程等。

由于依赖于一些增强 COM 组件，Windows 运行时本质上是一基于 COM 的 API。正因为其 COM 风格的基础，Windows 运行时可以像 COM 那样轻松地实现多种语言代码之间的交互联系，不过本质上是非托管的本地 API。API 的定义存储在以“. winmd”为后缀的元数据文件中，格式编码遵循 ECMA 335 的定义，和. Net 使用的文件格式一样，不过稍有改进。使用统一的元数据格式相比于 P/Invoke，可以大幅减少 Windows 运行时调用. NET 程序时的开销，同时拥有更简单的语法。全新的 C++/CX（组件扩展）语言，借用了一些 C++/CLI 语法，允许授权和使用 Windows 运行时组件，但相比传统的 C++下 COM 编程，对于程序员来说，有更少的粘合可见性，同时对于混合类型的限制相比 C++/CLI 也更少。在新的称为 Windows Runtime C++ Template Library（WRL）的模板类库的帮助下，也一样可以在 Windows 运行时组件里面使用标准 C++的代码。

在 Windows 运行时上任何耗时超过 50ms 的事情都应该通过使用了 Async 关键字的异步调用来完成，以确保流畅、快速的应用体验。由于即便当异步调用的情况存在时，许多开发者仍倾向于使用同步 API 调用，因此在 Windows 运行时深处建立了使用 Async 关键字的异步方法从而迫使开发者进行异步调用。

1.2.2 Windows Phone 8 应用程序模型

Windows Phone 8 平台支持多种应用程序模型，各种应用程序模型都各有自己的开发规则和使用场景，下面来看一下 Windows Phone 8 的各种应用程序模型。

1. 托管应用

托管应用程序是指普通的使用 XAML 作为界面的 Windows Phone 应用程序，可以使用 C#或者 VB.NET 作为托管应用程序的编程语言。在托管的 Windows Phone 8 应用程序里面兼容现在的 Windows Phone 7 应用程序，Windows Phone 7 版本的 SDK 的 API 都可以继续使用。Windows Phone 8 上的托管 API 在 Windows Phone 7 版本的基础上添加了一些新的 API，如新的诺基亚地图控件、电子钱包 API 等。

2. 托管应用＋Windows 运行时组件

在 Windows Phone 8 里面不支持直接使用 C++语言来编写 XAML 应用程序（在 Windows 8 里面可以），如果要在 XAML 应用程序里面使用 C++来进行编程需要通过 Windows 运行时组件来调用基于 C++的 API 或者标准的 C++代码。Windows Phone 8 新增加了 Windows 运行时的 API，Windows 运行时 API 支持 C#，C++和 VB.NET 编程语言，包括大量的 Windows 8 SDK 的子集，使开发者能够在 Windows 8 和 Windows Phone 8 中共享代码，可以通过少量的修改就可以将应用程序兼容两个平台。

3. Direct3D 游戏

Windows Phone 8 新增支持了使用 C++进行编码的 Direct3D 游戏的应用程序。这意味着，一个基于 DirectX 的 PC 游戏可以和 Windows Phone 8 手机版本的游戏共享代码，共用相关的组件和引擎，也极大地方便了将 PC 的 DirectX 游戏移植到 Windows Phone 8 手机上。同时，Direct3D 游戏也为 Windows Phone 8 平台的高质量高性能大型游戏提供了强大的开发框架。

4. 托管应用＋ DirectX

托管应用＋ DirectX 的应用程序模式主要是为了那些既需要使用 Direct3D 图形处理能力有需要使用相关 XAML 应用程序的功能的应用程序。比如在游戏中要使用 XAML 的相关控件等。

5. XNA 游戏

Windows Phone 8 的 SDK 将不支持 XNA 游戏的开发，但是 Windows Phone 8 的手机将继续兼容 XNA 游戏。如果在 Windows Phone 8 中要开发 XNA 框架的游戏，可以选择 Windows Phone 7.1 的 SDK 来创建 XNA 游戏，游戏依然可以流畅地在 Windows Phone 8 中运行，建议对性能要求不高的游戏可以选择 XNA 框架来开发，对性能要求高的 3D 游戏选择 Direct3D 框架开发。

6. 托管应用＋ JavaScript

在 Windows Phone 8 里面不支持 JavaScript 的应用程序，因为 Windows Phone 8 版本

的 Windows 运行时并没有提供 JavaScript 的相关 API，不可以使用 JavaScript 直接调用系统的 API 比如打电话等。然而，开发人员可以创建一个托管应用程序使用 XAML 的前端，使用嵌入式浏览器控件来显示本地 HTML 内容，并有可以使用 InvokeScript 方法和 ScriptNotify 的事件来访问电话的 API。此外，在 Windows Phone 8 在手机的浏览器已经升级到 IE10，IE10 提供了强大的 HTML5/CSS3 的新功能，如可伸缩矢量图形（SVG）、ES5、索引型数据库、手势事件，以及高性能的积分脚本引擎等，也可以为 Windows Phone 8 创造出有趣的新型的应用程序。

开发环境

工欲善其事，必先利其器。本章将介绍如何搭建 Windows Phone 8 的手机应用开发的环境并开发第一个 Windows Phone 8 的手机应用程序。

2.1 搭建开发环境

本节介绍 Windows Phone 8 开发环境的要求和开发工具的安装。

2.1.1 开发环境的要求

进行 Windows Phone 8 的开发，计算机配置应满足以下要求：

(1) 操作系统为 Windows 8 64 位(x64)版本；

(2) 系统盘需要至少 8G 的剩余硬盘空间；

(3) 内存空间达到 4GB 或者以上；

(4) Windows Phone 8 模拟器基于 Hyper-V，需要 CPU 支持二级地址转换技术。

注意，部分电脑会默认关闭主板 BIOS 的虚拟化技术，这时候需要进入主板 BIOS 设置页面开启虚拟化技术，然后在启动或者关闭 Windows 功能界面启动 Hyper-V 服务。

2.1.2 开发工具的安装

微软将 Windows Phone 8 的开发工具免费提供给开发者使用，可以到微软的 Windows Phone 8 的官方网站下面下载需要的开发工具。工具安装可以在线安装或者下载完整的 ISO 安装包进行安装，不过安装的过程都很简单，只需要按照提示单击下一步就可以很方便地安装了。

Windows Phone Developer Tools 是 Windows Phone 8 开发的主工具包，里面包含了程序的 SDK、运行模拟器和编程工具。Windows Phone Developer Tools 包含的工具集合详细信息如下：

1) Visual Studio 2012 Express for Windows Phone

Visual Studio 2012 Express for Windows Phone 是 Windows Phone 的集成开发环境

(IDE)，其包括了 C# 和 XAML 代码编辑功能、简单界面的布局与设计功能、编译程序、连接 Windows Phone 模拟器、部署程序，以及调试程序等功能。

2) Windows Phone Emulator

Windows Phone Emulator 是 Windows Phone 的模拟器，开发者可以在没有真实设备的情况下继续开发 Windows Phone 的应用，本书讲述的内容都是基于 Windows Phone 模拟器的。但是当前版本的模拟器不是什么都可以做的，具有一定的限制性。没有电话模拟器(cellar emulator)，不能打出和接受电话，也不能发送和接收短信。没有 GPS 模拟器，不能自动产生 GPS 的模拟数据。重力加速器(Accelerometer)模拟器的模拟数据不会更新，一直保留为矩阵(0,0,-1)，表示模拟器一直没有移动过。不能模拟内置镜头。当然如果有已经解锁好的 Windows Phone 8 手机，可以直接使用手机来调试和运行编写的程序。

3) Microsoft Expression Blend for Windows Phone

Microsoft Expression Blend for Windows Phone 是强大的 XAML UI 设计工具，使用 Expression Blend 可以弥补 Visual Studio 2012 Express 所缺乏的 UI 设计功能，例如 Animation 等。当开发 Windows Phone 程序时可以使用 Visual Studio 2012 Express 与 Expression Blend 相互协作，无缝结合。

2.2 创建 Windows Phone 8 应用

开发工具安装完毕之后，那么接下来的事情就是创建一个 Windows Phone 8 的应用。本节介绍如何利用 Visual Studio 2012 Express for Windows Phone 开发工具来创建一个 Windows Phone 8 的应用和详细地解析一个 Windows Phone 8 的工程项目的结构。

2.2.1 创建 Hello Windows Phone 项目

1. 新建一个 Windows Phone 的应用程序

打开 Visual Studio 2012 Express for Windows Phone 开发工具，选择 File 菜单，选择新建一个工程 New Project，在这里新建一个 Windows Phone Application 工程，如图 2.1 所示。

2. 选择 Windows Phone 版本号

如果同时安装了多个版本的 SDK 那么则需要你选择项目工程的 API 版本号，目前有 7.1 版本和 8.0 版本，选择了 8.0 版本就可以使用 Windows Phone 8 的最新 API。选择版本号如图 2.2 所示。

3. 编写程序代码

新建的 Windows Phone 应用程序已经是一个可以运行的完整的 Windows Phone 应用了，所以只需要在上面修改一下，就可以完成第一个 Hello Windows Phone 程序的开发。创建好的 Windows Phone 8 的项目工程如图 2.3 所示。

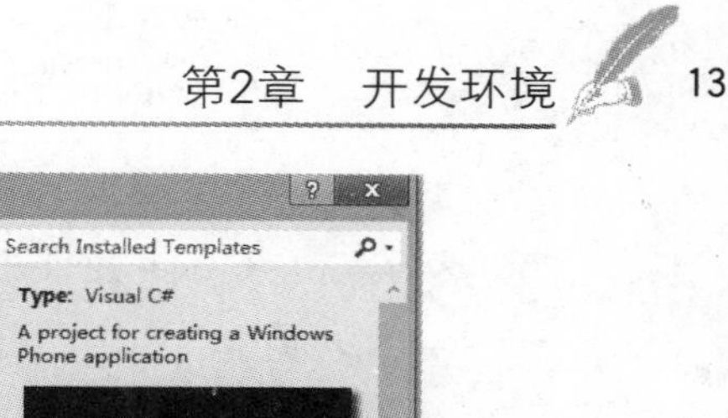

图 2.1　新建一个项目

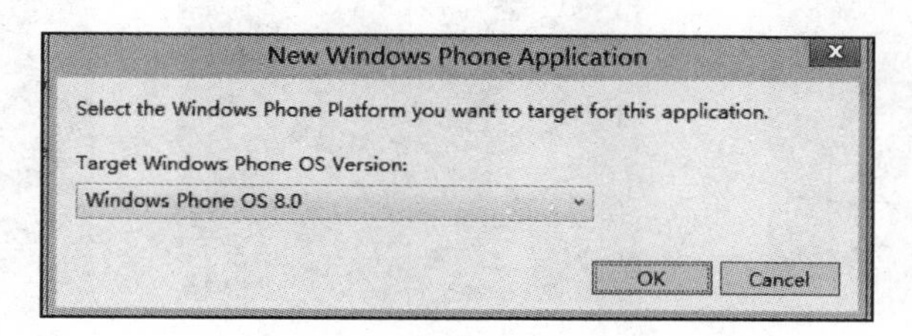

图 2.2　选择程序的版本号

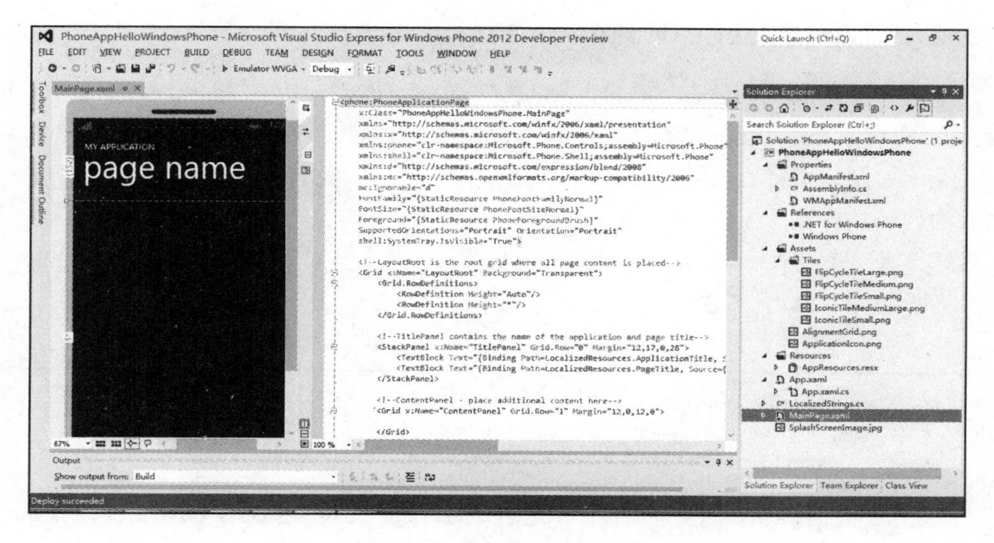

图 2.3　新建好的项目的界面

将左边的工具箱的 Button 控件和 TextBlock 控件拖放到可视化编辑界面如图 2.4 所示，然后双击 Button 控件添加以下代码：

```
private void button1_Click(object sender, RoutedEventArgs e)
{
    textBlock1.Text = "Hello Windows Phone ";
}
```

右键解决方案，选择 Deploy Solution 在模拟器上运行应用程序，运行的效果如图 2.5 和图 2.6 所示。

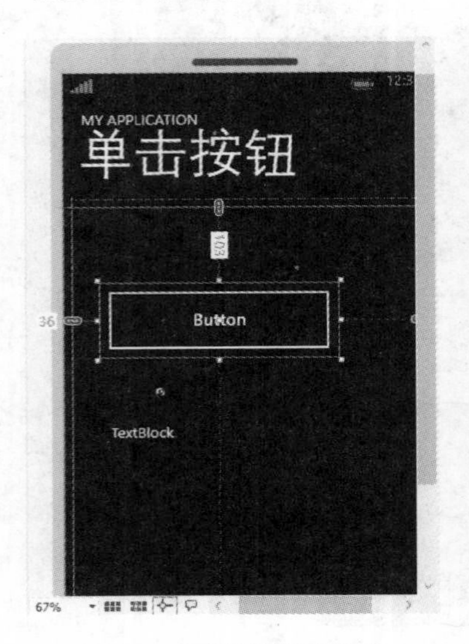

图 2.4 可视化编辑界面

图 2.5 程序列表的界面

图 2.6 应用程序运行的效果

2.2.2 解析 Hello Windows Phone 应用

在 Hello Windows Phone 项目工程里面包含了 MainPage. xaml 文件、MainPage. xaml. cs 文件、App. xaml 文件、App. xaml. cs 文件、WMAppManifest. xml 文件、AppManifest. xml 文件、AssemblyInfo. cs 文件和一些图片文件。下面来详细地解析一下每个文件的代码和作用。

代码清单 2-1：Hello Windows Phone(源代码：第 2 章\Examples_2_1)

1. MainPage. xaml 文件

MainPage. xaml 文件代码

```
<phone:PhoneApplicationPage
    x:Class = "PhoneAppHelloWindowsPhone.MainPage"
    xmlns = "http://schemas.microsoft.com/winfx/2006/xaml/presentation"
    xmlns:x = "http://schemas.microsoft.com/winfx/2006/xaml"
    xmlns:phone = "clr - namespace:Microsoft.Phone.Controls;assembly = Microsoft.Phone"
    xmlns:shell = "clr - namespace:Microsoft.Phone.Shell;assembly = Microsoft.Phone"
```

```
    xmlns:d="http://schemas.microsoft.com/expression/blend/2008"
    xmlns:mc="http://schemas.openxmlformats.org/markup-compatibility/2006"
    mc:Ignorable="d" d:DesignWidth="480" d:DesignHeight="768"
    FontFamily="{StaticResource PhoneFontFamilyNormal}"
    FontSize="{StaticResource PhoneFontSizeNormal}"
    Foreground="{StaticResource PhoneForegroundBrush}"
    SupportedOrientations="Portrait" Orientation="Portrait"
    shell:SystemTray.IsVisible="True">
    <!--默认的XAML页面使用了网格Grid控件来进行布局-->
    <Grid x:Name="LayoutRoot" Background="Transparent">
        <Grid.RowDefinitions>
            <RowDefinition Height="Auto"/>
            <RowDefinition Height="*"/>
        </Grid.RowDefinitions>
        <!--StackPanel控件里定义了是程序的名称和当前页面的名称-->
        <StackPanel x:Name="TitlePanel" Grid.Row="0" Margin="12,17,0,28">
            <TextBlock x:Name="ApplicationTitle" Text="Hello Windows Phone" Style=
"{StaticResource PhoneTextNormalStyle}"/>
            <TextBlock x:Name="PageTitle" Text="单击按钮" Margin="9,-7,0,0" Style=
"{StaticResource PhoneTextTitle1Style}"/>
        </StackPanel>
        <!--该Grid控件里面的是按钮控件和显示Hello Windows Phone的文本控件-->
        <Grid x:Name="ContentPanel" Grid.Row="1" Margin="12,0,12,0">
            <Button Content="Button" Height="101" HorizontalAlignment="Left" Margin=
"36,103,0,0" Name="button1" VerticalAlignment="Top" Width="313" Click="button1_Click" />
            <TextBlock Height="83" HorizontalAlignment="Left" Margin="50,295,0,0" Name=
"textBlock1" Text="TextBlock" VerticalAlignment="Top" Width="299" />
        </Grid>
    </Grid>
</phone:PhoneApplicationPage>
```

LayoutRoot是PhoneApplicationPage中的根Grid,所有页面内容全部位于LayoutRoot中,TitlePanel是拥有两个TextBlock控件的StackPanel。这两个控件分别是ApplicationTitle和PageTitle。ApplicationTitle：默认情况下,它的“Text”属性被设为“MY APPLICATION”,可以修改为你自己的应用程序标题名字,如实例中修改为Hello Windows Phone。PageTitle：默认情况下,它的“Text”属性被设为“page name”,如果应用程序有多个页面,可以使用这个TextBlock指定一个真实的页面,如果应用程序只有一个页面需要控件,这个TextBlock就会占用不必要的空间,如果你删除它,StackPanel的高度值会自动调整,因此,当需要放置更多的控件时,可以移除PageTitle。<Button …/>和<TextBlock … />就是刚才通过拖拉工具箱的Button控件和TextBlock控件产生的XAML代码。

在MainPage.xaml文件里面有若干个命名空间,这些命名空间的含义如下：

(1) xmlns代表的是默认的空间,如果在UI里面控件没有前缀则代表它属于默认的名字空间。例如,MainPage.xaml文件里面的Grid标签。

(2) xmlns:x代表专属的名字空间,比如一个控件里面有一个属性叫Name,那么x:Name则代表这个xaml的名字空间。

（3）xmlns:phone 包含在 Miscroft. phone 的引用 DLL。

（4）xmlns:shell 包含在 Miscroft. sell 的引用 DLL，此文件可以帮助管理生命周期。

（5）xmlns:d 呈现一些设计时的数据，而应用真正运行起来时会帮我们忽略掉这些运行时的数据。

（6）xmlns:mc 布局的兼容性，这里主要配合 xmlns:d 使用，它包含 Ignorable 属性，可以在运行时忽略掉这些设计时的数据。

2. MainPage. xaml. cs 文件

MainPage. xaml. cs 文件代码

```
using System;
using System.Collections.Generic;
using System.Linq;
using System.Net;
using System.Windows;
using System.Windows.Controls;
using System.Windows.Documents;
using System.Windows.Input;
using System.Windows.Media;
using System.Windows.Media.Animation;
using System.Windows.Shapes;
using Microsoft.Phone.Controls;
namespace PhoneAppHelloWindowsPhone
{
    public partial class MainPage : PhoneApplicationPage
    {
        ///< summary >
        ///应用程序的初始化方法
        ///</summary >
        public MainPage()
        {
            //初始化页面组件
            InitializeComponent();
        }
        ///< summary >
        ///Button 按钮的单击处理事件
        ///</summary >
        ///< param name = "sender">触发该事件的对象</param >
        ///< param name = "e">触发的事件</param >
        private void button1_Click(object sender, RoutedEventArgs e)
        {
            //给文本框控件 textBlock1 的 Text 属性赋值
            textBlock1.Text = "Hello Windows Phone";
        }
    }
}
```

MainPage. xaml. cs 文件是 MainPage. xaml 文件对应的后台代码的处理，在 MainPage. xaml. cs 文件会完成程序页面的控件的初始化工作和处理控件的触发的事件，如 button1_Click 方法就是对应的 MainPage. xaml 中 Button 的单击事件。

3. App. xaml 文件

App. xaml 文件代码

```
<Application
    x:Class = "PhoneAppHelloWindowsPhone.App"
    xmlns = "http://schemas.microsoft.com/winfx/2006/xaml/presentation"
    xmlns:x = "http://schemas.microsoft.com/winfx/2006/xaml"
    xmlns:phone = "clr-namespace:Microsoft.Phone.Controls;assembly = Microsoft.Phone"
    xmlns:shell = "clr-namespace:Microsoft.Phone.Shell;assembly = Microsoft.Phone">
    <!-- 应用程序的资源 -->
    <Application.Resources>
        <local:LocalizedStrings xmlns:local = "clr-namespace:PhoneAppHelloWindowsPhone" x:
Key = "LocalizedStrings"/>
    </Application.Resources>    <Application.ApplicationLifetimeObjects>
        <!-- 应用程序生命周期的事件处理 -->
        <shell:PhoneApplicationService
            Launching = "Application_Launching" Closing = "Application_Closing"
            Activated = "Application_Activated" Deactivated = "Application_Deactivated"/>
    </Application.ApplicationLifetimeObjects>
</Application>
```

App. xaml 文件中的<Application. Resources></Application. Resources>节点的作用是用来加载整个应用程序的资源的，如果你需要在应用程序中加载同样样式的话，则需要在这个节点下添加共用的样式。其中 Application. ApplicationLifetimeObjects 标签内定义了应用程序的启动过程（Launching）、程序的关闭过程（Closing）、程序的重新激活（Activated）、失去激活（Deactivated），这些事件都在 App. xaml. cs 文件中进行了定义。

4. App. xaml. cs 文件

App. xaml. cs 文件代码

```
using System;
using System.Collections.Generic;
using System.Linq;
using System.Net;
using System.Windows;
using System.Windows.Controls;
using System.Windows.Documents;
using System.Windows.Input;
using System.Windows.Media;
using System.Windows.Media.Animation;
using System.Windows.Navigation;
using System.Windows.Shapes;
using Microsoft.Phone.Controls;
```

```
using Microsoft.Phone.Shell;
namespace PhoneAppHelloWindowsPhone
{
    public partial class App : Application
    {
        /// <summary>
        /// 提供了应用程序 UI 底层的框架的入口
        /// </summary>
        /// <returns>应用程序 UI 的底层框架</returns>
        public PhoneApplicationFrame RootFrame { get; private set; }
        /// <summary>
        ///构建一个应用程序
        ///</summary>
        public App()
        {
            //全局监控未捕获的异常信息
            UnhandledException += Application_UnhandledException;
            //标准的 Silverlight 程序初始化
            InitializeComponent();
            //手机特有功能的初始化
            InitializePhoneApplication();
            //如果程序正在调试,这里设置调试的图形信息
            if (System.Diagnostics.Debugger.IsAttached)
            {
                //展示当前框架的速度的计算
                Application.Current.Host.Settings.EnableFrameRateCounter = true;
                //展现出应用程序被刷新的区域
                Application.Current.Host.Settings.EnableRedrawRegions = true;
                //当手机被闲置时,设置程序的服务被禁用
                //注意：这只是调试模式下使用,检测到手机被闲置时禁止应用程序将继续运行,
//以至于不再多消耗电池电量
                    PhoneApplicationService.Current.UserIdleDetectionMode = IdleDetectionMode.
Disabled;
            }
        }
        //当应用程序启动的时候会执行这个方法体的代码
        //应用程序被激活的时候这个方法体的代码不会被执行
        private void Application_Launching(object sender, LaunchingEventArgs e)
        {
        }
        //当应用程序被激活的时候会执行这个方法体的代码
        //应用程序第一次启动的时候这个方法体的代码不会被执行
        private void Application_Activated(object sender, ActivatedEventArgs e)
        {
        }
        //当应用程序被禁止的时候会执行这个方法体的代码
        //应用程序关闭的时候这个方法体的代码不会被执行
    private void Application_Deactivated(object sender, DeactivatedEventArgs e)
        {
        }
```

```
//当应用程序关闭的时候会执行这个方法体的代码
//应用程序被禁止的时候这个方法体的代码不会被执行
private void Application_Closing(object sender, ClosingEventArgs e)
{
}
//当应用程序导航失败时会执行这个方法体的代码
private void RootFrame_NavigationFailed(object sender, NavigationFailedEventArgs e)
{
    if (System.Diagnostics.Debugger.IsAttached)
    {
        //一个导航失败中断程序的执行
        System.Diagnostics.Debugger.Break();
    }
}
//当应用程序发生未处理异常的时候会执行这个方法体的代码
private void Application_UnhandledException(object sender, ApplicationUnhandledExceptionEventArgs e)
{
    if (System.Diagnostics.Debugger.IsAttached)
    {
        //一个未处理的异常中断程序的执行
        System.Diagnostics.Debugger.Break();
    }
}
#region Phone application initialization
//避免重复初始化程序的标示符
private bool phoneApplicationInitialized = false;
//不要在这个方法体添加任何的代码
private void InitializePhoneApplication()
{
    if (phoneApplicationInitialized)
        return;
    //创建应用程序的框架
    //导航到的应用程序初始化完成后,程序才会做出反应
    RootFrame = new PhoneApplicationFrame();
    RootFrame.Navigated += CompleteInitializePhoneApplication;
    //注册导航失败的事件
    RootFrame.NavigationFailed += RootFrame_NavigationFailed;
    //设置初始化标示符为true,表示已进行了初始化
    phoneApplicationInitialized = true;
}
//不要在这个方法体添加任何的代码
private void CompleteInitializePhoneApplication(object sender, NavigationEventArgs e)
{
    //设置应用程序可视的UI为RootFrame
    if (RootVisual != RootFrame)
        RootVisual = RootFrame;
    //移除手机应用程序初始化完成的事件
    RootFrame.Navigated -= CompleteInitializePhoneApplication;
}
#endregion
```

```
    }
}
```

App. xaml. cs 文件是一个控制着整个应用程序的全局文件，整个应用程序的生命周期都在该文件中进行定义和处理。下面来解析一下 Windows Phone 8 应用程序的生命周期。

1）启动(Launching)

当用户单击了手机上应用程序安装列表里的某一应用程序，或者单击了开始屏幕上的代表某一应用程序的小方块图标，此时一个 Windows Phone 应用程序就被启动了。无论用户使用哪种方式启动一个应用程序，该应用程序的实例已经被创建了。当应用程序被启动了，也就是一个启动事件被触发了。处理这个启动事件时，应用程序应该从一个独立的存储中读取所有必要的数据来为用户创建一个新的应用程序的会话进程。应用程序不应该试图从以前的应用程序实例中恢复瞬时状态。当用户启动一个应用程序，就出现了一个新的应用程序实例。需要注意的是：启动和激活事件是互斥的。

2）运行(Running)

当启动事件被触发了，一个应用程序就开始运行了。应用程序处于运行状态时，用户进行浏览该应用程序的页面等相关操作，此时应用程序会自己管理自己的状态。如果应用程序处于运行状态，那么与执行模型相关的唯一操作就是逐步的保存设置以及其他应用程序持久化数据(Persisting data)，这样做的目的是为了避免当应用程序的状态发生改变时需要保存大量的数据。这是可选的，因为当应用程序只有少量的持久化数据，这个操作就不是必需的。

3）关闭 (Closing)

应用程序处于运行状态之后的状态是取决于用户采取了哪种操作。一个可能的操作是用户按下手机上的回退(Back)键从而回退到应用程序的前一页面，甚至翻过了应用程序的第一个页面。当这种情况发生时，关闭事件将会被触发，此时应用程序被终止了。处理关闭事件，应用程序应该把所有的持久化数据保存到独立的存储中。此时没有必要保存瞬时状态数据，即那些只和前应用程序实例相关的数据。因为用户如果要返回一个已经被终止的应用程序，唯一的方式就是重新启动它，打开它的首页。当用户启动应用程序，它将会以一个全新的实例出现。

4）禁止 (Deactivating)

如果一个应用程序正在运行，随后在操作系统前台被另一个应用程序或体验替代，例如，锁屏或者启动一个 Chooser，这时第一个应用程序将会被禁止。有好多种方法能够实现应用程序的禁止状态。当用户单击手机开始键或者手机由于超时导致设备自动处于锁屏状态都会使当前应用程序处于禁止状态，在这种状态下应用程序被逻辑删除了。当用户调用一个 Launcher 或者 Chooser 时，当前应用程序同样会被禁止，辅助应用程序允许用户执行拍照或者发送电子邮件等常见任务。无论以上哪种情况，当前运行的应用程序将会被禁止，禁止事件被触发了。并不像应用程序被终止一样，一个被禁止的应用程序可能会被逻辑删除。这就意味着应用程序不再运行。这个应用程序的进程已经被挂起或者终止，操作系统会保存能够代表应用程序的记录以及其一系列状态数据。这就使用户返回一个被禁止的应用程序成为可能性，应用程序才能被再次激活并把上次用户浏览的页面呈现出来。

在禁止状态事件处理过程中，一个应用程序应该存储其当前状态信息，状态信息是从PhoneApplicationService类中的State属性公开的状态信息词典中获得的。在状态信息词典中存储的数据是瞬时状态数据或者是能帮助应用程序在被禁止时恢复其状态的数据。由于并不能保证一个被禁止的应用程序会被重新激活，所以在此事件的处理中应用程序需要一直把持久化数据保存到一个独立的存储空间。

禁止事件处理程序所进行的所有操作必须在10s内完成，否则操作系统将会直接终止应用程序而不是逻辑删除他。正是出于这个原因，当应用程序有大量的持久化数据需要保存，应用程序会在运行过程中对数据逐步地进行保存。

5）激活（Activating）

当一个应用程序被禁止后，有可能这个应用程序永远不会被再次激活。用户可能会从头重新启动该应用程序，从而得到一个新的应用程序实例。或者用户可以启动其他几个应用程序，这样就会把处在应用程序堆栈最后的即使利用回退按键也不可能到达的欺骗性程序关闭掉。

当然用户也有继续要使用原应用程序的可能性。这种情况可能发生在用户不停地敲击回退键直到指定的应用程序。或者，一个Launcher或Chooser操作导致了当前应用程序被禁止，用户可以完成这个操作任务或取消这个新的操作任务。当用户返回一个处于逻辑删除状态的应用程序，该程序将会被重新激活，激活事件将会被触发。在此事件中，用户的应用程序将会从一独立存储中取回应用程序的持久化数据。这个应用程序同样也会从PhoneApplicationService类的状态信息词典中读取状态信息从而恢复到被禁止的应用程序之前的状态。

5. WMAppManifest.xml 文件

WMAppManifest.xml 文件代码

```
<?xml version="1.0" encoding="utf-8"?>
<Deployment xmlns="http://schemas.microsoft.com/windowsphone/2012/deployment" AppPlatformVersion=
"8.0">
  <DefaultLanguage xmlns="" code="en-US"/>
  <App xmlns="" ProductID="{86519139-0c70-43bd-801a-c01d9bfb7d97}" Title=
"PhoneAppHelloWindowsPhone" RuntimeType="Silverlight" Version="1.0.0.0" Genre="apps.
normal" Author="PhoneAppHelloWindowsPhone author" Description="Sample description"
Publisher="PhoneAppHelloWindowsPhone" PublisherID="{7026ca60-c0cf-4cc9-b67b-
c1f23979f84f}">
    <IconPath IsRelative="true" IsResource="false">Assets\ApplicationIcon.png</IconPath>
    <Capabilities>
      <Capability Name="ID_CAP_NETWORKING"/>
      <Capability Name="ID_CAP_MEDIALIB_AUDIO"/>
      <Capability Name="ID_CAP_SENSORS"/>
      <Capability Name="ID_CAP_WEBBROWSERCOMPONENT"/>
    </Capabilities>
    <Tasks>
```

```
        <DefaultTask  Name ="_default" NavigationPage="MainPage.xaml"/>
      </Tasks>
      <Tokens>
        <PrimaryToken TokenID="PhoneAppHelloWindowsPhoneToken" TaskName="_default">
          <TemplateFlip>
          <SmallImageURI IsRelative = " true " IsResource = " false " > Assets \ Tiles \
FlipCycleTileSmall.png</SmallImageURI>
            <Count>0</Count>
            <BackgroundImageURI IsRelative = " true " IsResource = " false"> Assets \ Tiles \
FlipCycleTileMedium.png</BackgroundImageURI>
            <Title>PhoneAppHelloWindowsPhone</Title>
            <BackContent></BackContent>
            <BackBackgroundImageURI></BackBackgroundImageURI>
            <BackTitle></BackTitle>
            <LargeBackgroundImageURI></LargeBackgroundImageURI>
            <LargeBackContent></LargeBackContent>
            <LargeBackBackgroundImageURI></LargeBackBackgroundImageURI>
            <DeviceLockImageURI></DeviceLockImageURI>
            <HasLarge></HasLarge>
          </TemplateFlip>
        </PrimaryToken>
      </Tokens>
      <ScreenResolutions>
        <ScreenResolution Name="ID_RESOLUTION_WVGA"/>
        <ScreenResolution Name="ID_RESOLUTION_WXGA"/>
        <ScreenResolution Name="ID_RESOLUTION_HD720P"/>
      </ScreenResolutions>
    </App>
  </Deployment>
```

WMAppManifest.xml 是一个包含与 Windows Phone Silverlight 应用程序相关的特定元数据的清单文件，记录了应用程式的相关属性描述且包含了用于 Windows Phone 的 Silverlight 所具有的特定功能。App 节点的一些属性的含义如表 2.1 所示。<Capabilities>相关的区块，在这个区块中则是描述了应用能够使用的功能性，例如能不能使用网路的功能

表 2.1 WMAppManifest.xml 文件 App 节点属性的含义

属性名称	说　明
ProductID	代表应用程式的 GUID 字串列表
RuntimeType	设定应用程式是 Silverlight 或是 XNA 的类
Title	专案的预设名称，这里的文字也会显示在应用程式清单
Version	应用程式的版本编号
Genre	当应用程式为 Silverlight 时会为 apps.normal，apps.game 则为 XNA
Author	作者名称
Description	应用程式的描述（说明）
Publisher	这个值预设会是专案的名称，当你的应用程式有使用到 Push 的相关功能，这个值是一定要有的

或是存取媒体库(Media library)的内容。在一般的情形下,我们是不需要去修改到这个部分的,假设当你移除了某些功能,例如说移除了 WebBrowser 的部分,那么当你在代码中有使用到 WebBrowser 相关的功能时,程式便会出错了,而 unhandle exception 在 Silverlight for Windows Phone 中是会直接关闭应用程式的。DefaultTask 属性表示程序默认的启动主页,Tokens 节点包含了程序相关的图标和磁贴的设置,ScreenResolutions 节点包含 3 种屏幕的适配。

6. AppManifest.xml 文件

AppManifest.xml 文件代码

```
<Deployment xmlns="http://schemas.microsoft.com/client/2007/deployment"
        xmlns:x="http://schemas.microsoft.com/winfx/2006/xaml">
    <Deployment.Parts>
    </Deployment.Parts>
</Deployment>
```

AppManifest.xml 是一个生成应用程序包所必需的应用程序清单文件。

7. AssemblyInfo.cs 文件的解析

AssemblyInfo.cs 文件代码

```
using System.Reflection;
using System.Runtime.CompilerServices;
using System.Runtime.InteropServices;
//有关程序集的常规信息是通过下面的一组属性来控制
//更改这些属性值可修改与程序集关联的信息
[assembly: AssemblyTitle("PhoneAppHelloWindowsPhone")]
[assembly: AssemblyDescription("")]
[assembly: AssemblyConfiguration("")]
[assembly: AssemblyCompany("")]
[assembly: AssemblyProduct("PhoneAppHelloWindowsPhone")]
[assembly: AssemblyCopyright("Copyright © 2011")]
[assembly: AssemblyTrademark("")]
[assembly: AssemblyCulture("")]
//将 ComVisible 设置为 false 使程序集对 COM 组件是不可见的类型
//如果需要在程序集里面访问 COM 组件,那么就要将 ComVisible 属性设置为 true
[assembly: ComVisible(false)]
//应用的 Guid 字符串,是机器自动生成的一个唯一的字符串
[assembly: Guid("ff94ed8f-b021-4463-ace2-378e63cd4d0a")]
//关联的版本号
[assembly: AssemblyVersion("1.0.0.0")]
[assembly: AssemblyFileVersion("1.0.0.0")]
```

AssemblyInfo.cs 包含名称和版本的元数据,这些元数据将被嵌入到生成的程序集。

XAML简介

Windows Phone 8 普通应用程序开发使用的是 Silverlight 的框架，Windows Phone 8 应用程序中的界面都是由 XAML 文件组成的，和这些 XAML 文件一一对应起来的是 XAML.CS 文件，这就是微软典型的 Code-Behind 模式的编程方式。XAML 文件的语法类似于 XML 和 HTML 的结合体，这是 Silverlight 程序特有的语法结构，本章将介绍有关 XAML 方面的知识和语法。

3.1 什么是 XAML

XAML 是一种声明性标记语言，如同应用于.NET Framework 编程模型一样，XAML 简化了为.NET Framework 应用程序创建 UI 的过程。在声明性 XAML 标记中可以创建可见的 UI 元素，然后使用代码隐藏文件（通过分部类定义与标记相连接）将 UI 定义与运行时逻辑相分离。XAML 直接以程序集中定义的一组特定后备类型表示对象的实例化，就如同其他的基于 XML 的标记语言一样，XAML 大体上也遵循 XML 的语法规则。例如，每个 XAML 元素包含一个名称以及一个或多个属性。在 XAML 中，每个属性都是和某个 Windows Phone 类的属性相对应的，而且所有的元素名称都和 Windows Phone 类库中定义的类名称相匹配。例如，<Button/>元素就和 System.Windows.Controls.Button 类对应。

XAML 是一个纯粹的标记语言，这也就意味着某个元素要实现一个事件的处理时，需要在该元素中通过特定的属性来指定相应的事件处理方法名，而真正的事件处理逻辑可以通过 C# 或 VB.NET 语言实现，用户是无法通过 XAML 来编写相应的事件处理逻辑的。如果对 ASP.NET 技术比较了解的话，那么应该对代码后置这个概念不会陌生。对于一个 Windows Phone 程序来说，也可以像 ASP.NET 那样采用代码后置模型，将页面和相应的逻辑代码分别存放在不同的文件中，也可以以一种内联的方式将页面和逻辑代码都存放在同一个文件中。

XAML 开发人员应注意，声明一个 XAML 元素时，最好用 Name 属性为该元素指定一个名称，这样应用程序逻辑开发人员才可以通过代码来访问此元素。这是因为某种类型的元素可能在 XAML 页面上声明多次，但是如果不显式地指明各个元素的 Name 属性，则无法区分哪个是想要操作的元素，也就无法通过 C＃或 VB. NET 来操作该元素和其中的属性。

下面是声明一个 XAML 元素必须遵循的 4 大原则：

(1) XAML 是区分大小写的，元素和属性的名称必须严格区分大小写，例如，对于 Button 元素来说，其在 XAML 中的声明应该为<Button>，而不是<button>；

(2) 所有的属性值，无论它是什么数据类型，都必须包含在双引号中；

(3) 所有的元素都必须是封闭的，也就是说，一个元素必须是自我封闭的，例如<Button.../>，或者是有一个起始标记和一个结束标记，例如<Button>...</Button>；

(4) 最终的 XAML 文件也必须是合适的 XML 文档。

在 Windows Phone 应用程序开发过程中，XAML 发挥着以下的作用：

(1) XAML 是用于声明 UI 及该 UI 中元素的主要格式。通常，项目中至少有一个 XAML 文件表示应用程序中用于最初显示的 UI 页面。其他 XAML 文件可能声明其他用于导航 UI 或模式替换 UI 页面。另外一些 XAML 文件可以声明资源，如模板或其他可以重用或替换的应用程序元素。

(2) XAML 是用于声明样式和模板的格式，这些样式和模板应用于控件和 UI 的逻辑基础。可以执行此操作来模板化现有控件，或作为为控件提供默认模板的控件来执行此操作。

(3) XAML 是用于为创建 UI 和在不同设计器应用程序之间交换 UI 设计提供设计器支持的常见格式。最值得注意的是，应用程序的 XAML 可在 Expression Blend 产品与 Visual Studio 之间互换。

(4) XAML 定义 UI 的可视外观，而关联的代码隐藏文件定义逻辑。可以对 UI 设计进行调整，而不必更改代码隐藏中的逻辑。就此作用而言，XAML 简化了负责可视化设计的人员与负责应用程序逻辑和信息设计的人员之间的工作交流。

(5) 由于支持可视化设计器和设计图面，因此，XAML 支持在早期开发阶段快速构造 UI 原型，并在整个开发过程中使设计的组成元素更可能保留为代码访问点，即使可视化设计发生了较大变化也不例外。

3.2 XAML 语法概述

编写 XAML 文件时，必须严格遵守 XAML 的语法，下面将介绍 XAML 的一些重要的语法。

3.2.1 XAML 命名空间

按照针对编程的广泛定义，命名空间确定如何解释引用编程实体的字符串标记。如果重复使用字符串标记，命名空间还可以解决多义性。命名空间概念的存在使得编程框架能够区分用户声明的标记与框架声明的标记，并通过命名空间限定来消除可能的标记冲突，等等。XAML 命名空间是为 XAML 语言提供此用途的命名空间概念。就 XAML 的常规作用及其面向 Windows Phone 的应用程序而言，XAML 用于声明对象、这些对象的属性和对象-属性关系(表示为层次结构)。声明的对象由类型库提供支持，相关的库可以是以下任意一项：

(1) Windows Phone 核心库；

(2) 分布式库，它们是在包中再分发的 SDK 的一部分(可能带有应用程序库缓存选项)；

(3) 表示应用程序中融入的和应用程序包再分发的第三方控件的定义的库；

(4) 用户自己的库，这是用户通过 Windows Phone 项目创建，用于容纳某些或所有应用程序的用户代码的库；

(5) 其他库，即用户在单独的项目中定义，通过应用程序模型进行引用的库。

XAML 命名空间概念使用标记中提供的 XML 样式命名空间声明(xmlns)，并将以 CLR 命名空间格式表示的后备类型信息和程序集信息与特定的 XAML 命名空间关联。这使得读取 XAML 文件的 XAML 处理器能够区分标记(markup)中的标记(token)，并且在创建运行时对象表示形式时，该处理器能够从与该 XAML 命名空间关联的后备程序集中查找类型和成员。

XAML 文件几乎始终在其根元素中声明一个默认的 XAML 命名空间。默认 XAML 命名空间定义可以声明哪些元素，而无需通过前缀进一步进行限定。例如，用户声明一个元素<Balloon/>，则该元素 Balloon 应存在且在默认 XAML 命名空间中有效。相反，如果 Balloon 不在所定义的默认 XAML 命名空间中，则必须转而使用一个前缀来限定该引用。例如，<party:Balloon/>，该前缀指示此实体存在于与默认命名空间不同的 XAML 命名空间中，尤其是，用户已将某个 XAML 命名空间映射到前缀 party 以便于使用。

XAML 命名空间应用于声明它们的特定元素，同时应用于 XAML 结构中该元素所包含的任何元素。因此，XAML 命名空间几乎始终在根元素上声明，以充分利用此继承概念。

来自除核心库之外的其他库的类型将要求用户使用前缀声明和映射 XAML 命名空间，然后才能从该库中引用类型。针对默认命名空间之外的其他 XAML 命名空间的 XAML 命名空间声明提供了 3 项信息：

(1) 一个前缀，该前缀会作为后续 XAML 标记中引用到该 XAML 命名空间的标记(markup)；

(2) 在该 XAML 命名空间中定义元素的后备类型的程序集，XAML 处理器必须访问此程序集才能基于 XAML 声明创建对象；

(3) 该程序集中的一个 CLR 命名空间。

SDK 库具有 CLR 特性，以便加载程序集的设计器可以建议使用特定的前缀。在 Visual Studio 中，对于已由某个项目引用的任何程序集，都可以使用自动完成功能从所引用的程序中读取 CLR 特性。这一 Visual Studio 功能要么将所有可能的 XAML 命名空间显示为下拉列表，要么使用建议的前缀作为提示以帮助建议特定的映射选择。

在几乎每个 XAML 文件中声明的一个特定的 XAML 命名空间是针对由 XAML 语言定义的元素的 XAML 命名空间。根据约定，XAML 语言 XAML 命名空间映射到前缀 x:。Windows Phone 项目的默认项目和文件模板始终同时将默认的 XAML 命名空间(无前缀，只有 xmlns=)和 XAML 语言命名空间(映射到前缀 x:)定义为根元素的一部分。例如：

```
< phone:PhoneApplicationPage
  xmlns = "http://schemas.microsoft.com/winfx/2006/xaml/presentation"
  xmlns:x = "http://schemas.microsoft.com/winfx/2006/xaml"
...>
```

“x:前缀”类型的命名空间包含多个将在 XAML 中频繁使用的编程构造。下面列出了最常见的“x:前缀”类型的命名空间构造：

(1) x:Key：为 ResourceDictionary 中的每个资源设置一个唯一键。

(2) x:Class：指定为 XAML 页提供代码隐藏的类的 CLR 命名空间和类名称，并命名由标记编译器在应用程序模型中创建的类。必须具有一个这样的类才能支持代码隐藏或支持初始化为 RootVisual。

(3) x:Name：处理 XAML 中定义的对象元素后，为运行时代码中存在的实例指定运行时对象名称。对于不支持 FrameworkElement.Name 属性的情形，可以将 x:Name 用于元素命名方案。默认情况下，通过处理对象元素而创建的对象实例没有可供在代码中使用的唯一标识符或固有的对象引用。在代码中调用构造函数时，几乎总是使用构造函数结果为构造的实例设置一个变量，以便以后在代码中引用该实例。为了对通过标记定义创建的对象进行标准化访问，XAML 定义了 x:Name 属性。可以在任何对象元素上设置 x:Name 属性的值。在代码隐藏文件中，所选择的标识符等效于引用构造实例的实例变量。在任何方面，命名元素都像它们是对象实例一样工作(此名称只是引用该实例)，并且代码隐藏文件可以引用该命名元素来处理应用程序内的运行时交互。

3.2.2　声明对象

一个 XAML 文件始终只有一个元素作为其根，该元素声明的一个对象将作为某些编程结构(如页面)的概念根，或者是应用程序的整个运行时定义的对象图。

根据 XAML 语法，可以通过 3 种方法在 XAML 中声明对象：

1) 直接使用对象元素语法

直接使用对象元素语法是使用开始标记和结束标记将对象实例化为 XML 格式的元素。可以使用此语法声明根对象或创建用于设置属性值的嵌套对象。

2) 间接使用属性语法

间接使用属性语法是使用内联字符串值声明对象。在概念上，这可能用于实例化除根之外的任何对象。可以使用此语法设置属性值。这是一个针对 XAML 处理器的间接操作，因为必须要通过某个过程在了解如何设置属性、该属性的类型系统特性和所提供的字符串值的基础上创建新对象。通常，这表明相关类型或属性要么支持可处理字符串输入的类型转换器，要么 XAML 分析器支持进行本机转换。

3）使用标记扩展

这并不意味着始终可以选择使用任何语法以给定的 XAML 词汇创建对象。词汇中的某些对象只能使用对象元素语法创建。少量对象只能通过初始设置为属性值来创建。事实上，在 Windows Phone 中，可以使用对象元素或属性语法创建的对象比较少。即使这两种语法格式都是可能的，也只有其中一种语法格式占主流或是最适合方案使用的格式。除了以等同于实例化对象的方式声明对象之外，XAML 中还提供了一些可用来引用现有对象的方法。这些对象可能在 XAML 的其他区域中定义，或者通过平台及其应用程序或编程模型的某种行为隐式存在。

若要使用对象元素语法声明对象，需要使用以下模式编写标记，其中，objectName 是要实例化的类型的名称。在本文中，经常出现术语“对象元素用法”，这是用于用对象元素语法创建对象的特定标记的简称。

```
<objectName>
</objectName>
```

下面的示例是用于声明 Canvas 对象的对象元素用法。

```
<Canvas></Canvas>
```

许多 Windows Phone XAML 对象（例如 Canvas）可以包含其他对象。

```
<Canvas>
  <Rectangle>
  </Rectangle>
</Canvas>
```

为方便起见（且作为 XAML 与 XML 的一般关系的一部分），如果对象不包含其他对象，则可以使用一个自结束标记（而不是开始/结束标记对）来声明对象元素，如下面示例中的 <Rectangle /> 标记所示。

```
<Canvas>
  <Rectangle />
</Canvas>
```

在某些情况下，属性值并不只是语言基元（如字符串），此时可以使用属性语法来实例化设置该属性的对象，并初始化用于定义新对象的键属性。由于此行为绑定到属性设置，请参见后面有关如何使用属性语法在一个语法步骤中声明对象并设置其属性的信息。

3.2.3 设置属性

可以设置使用对象元素语法声明的对象的属性。可以通过多种方法使用 XAML 设置属性：

(1) 使用属性语法；

(2) 使用属性元素语法；

(3) 使用内容元素语法；

(4) 使用集合语法(通常是隐式集合语法)。

对于对象声明,用于在 XAML 中设置对象属性的此方法列表并不表示可以使用这些方法中的任何一种来设置给定的属性。某些属性只支持其中一种方法,某些属性可能支持组合,例如,支持内容元素语法的属性可能还通过属性元素语法或备选属性语法支持更详细的格式。这取决于属性和属性使用的对象类型。Windwos Phone 中的对象还有一些无论使用任何方式都无法使用 XAML 设置的属性,只能使用代码来设置这些属性。

无论使用任何方式(包括 XAML 或代码)都无法设置只读属性,除非有其他机制适用。该机制可能是调用一个设置为属性的内部表示形式的构造函数重载、一个并非是严格意义上的属性访问器的帮助器方法或一个计算属性。计算属性依赖于其他可设置属性的值,以及服务或行为对该属性值的影响,而这些功能在依赖项属性系统中提供。

1. 使用属性语法设置属性

设置属性使用以下语法。其中 objectName 是要实例化的对象,propertyName 是要对该对象设置的属性的名称,propertyValue 是要设置的值。

```
<objectName propertyName = "propertyValue" .../>
```

或者

```
<objectName propertyName = "propertyValue">
...
</objectName>
```

使用上述任何一种语法都可以声明对象并设置该对象的属性。虽然第一个示例是标记中的单一元素,实际上这里有一些与 XAML 处理器如何分析此标记有关的分离步骤。首先,对象元素的存在表明必须实例化新的 objectName 对象。只有存在这样的实例后,才可以对它设置实例属性 propertyName。

下面的示例使用 4 个属性的属性语法来设置 Rectangle 对象的 Name、Width、Height 和 Fill 属性。

```
<Rectangle Name = "rectangle1" Width = "100" Height = "100" Fill = "Blue" />
```

如果清楚地了解 XAML 分析器如何解释此标记和定义对象树,则等效的代码可能类似以下伪代码:

```
Rectangle rectangle1 = new Rectangle();
rectangle1.Width = 100.0;
rectangle1.Height = 100.0;
rectangle1.Fill = new SolidColorBrush(Colors.Blue);
```

许多属性可以使用属性元素语法来设置。若要使用属性元素语法,必须能够指定对象元素的新实例才能“填充”属性元素值。

若要使用属性元素语法,需要为要设置的属性创建 XAML 元素。这些元素的形式为 <object. property>。在标准的 XML 中,此元素只被视为在名称中有一个点的元素。但是使用 XAML 时，元素名称中的点将该元素标识为属性元素，且 property 是 object 的属性。

在下面的语法中，property 是要设置属性的名称，propertyValueAsObjectElement 是声明新对象的新对象元素，其值类型是该属性期望的值。

```
<object>
 <object.property>
propertyValueAsObjectElement
 </object.property>
</object>
```

下面的示例使用属性元素语法通过 SolidColorBrush 对象元素来设置 Rectangle 的填充。（在 SolidColorBrush 中，Color 使用属性语法来设置。）此 XAML 的呈现结果等同于前面使用属性语法设置 Fill 的 XAML 示例：

```
<Rectangle
  Name="rectangle1"
  Width="100"
  Height="100"
>
  <Rectangle.Fill>
    <SolidColorBrush Color="Blue"/>
  </Rectangle.Fill>
</Rectangle>
```

2. 使用 XAML 内容语法设置属性

一些 Windows Phone 类型定义了一个启用 XAML 内容元素语法的属性。在 XAML 内容元素语法中，可以忽略该属性的属性元素，并可以通过提供所属类型的对象元素标记中的内容来设置该属性，该内容通常为一个或多个对象元素，这称为 XAML 内容语法。

例如，Border 的 Child 属性页显示了 XAML 内容语法（而非属性元素语法），以设置 Border 的单一对象 Child 值。下面的示例与这一用法类似：

```
<Border>
  <Button .../>
</Border>
```

如果声明为 XAML 内容属性的属性也支持"松散"对象模型（在此模型中，属性类型为 Object，或具体而言为类型 String），则可以使用 XAML 内容语法将纯字符串作为内容放入开始对象标记与结束对象标记之间。例如，TextBlock 的 Text 属性页显示了另一种 XAML 语法，该语法使用 XAML 内容语法（而不是属性语法）来为 Text 设置一个字符串值。下面的示例说明该用法并设置 TextBlock 的 Text 属性，而不显式指定 Text 属性。Text 使用将 XML 视为内容或"内部文本"的内容进行设置，而不是通过使用属性或声明对象元素来设置。

```
<TextBlock>Hello!</TextBlock>
```

3. 使用集合语法设置属性

在 XAML 中，有几个集合语法的变体，这看上去似乎允许"设置"只读集合属性，而实际上,XAML 允许的操作是向集合中添加项。实现 XAML 支持的 XAML 语言和 XAML 处理器依赖于后备集合类型中的约定来启用此语法。

通常,XAML 语法中不存在保留集合项的集合类型的属性(如索引器或 Items 属性)。对于集合而言,XAML 中的集合实际所需的未必是属性,而是方法——Add 方法。调用 Add 方法就是上述约定。当 XAML 处理器遇到 XAML 集合语法中的一个或多个对象元素时,首先通过使用其对象标记创建每个对象,然后通过调用集合的 Add 方法以声明顺序将每个新对象添加到集合中。

下面的示例演示了一个使用可构造集合类型的集合属性(可以定义实际的集合并将其实例化为 XAML 中的一个对象元素):

```
<LinerGradientBrush>
  <LinearGradientBrush.GradientStops>
    <GradientStopCollection>
      <GradientStop Offset="0.0" Color="Red" />
      <GradientStop Offset="1.0" Color="Blue" />
    </GradientStopCollection>
  </LinearGradientBrush.GradientStops>
</LinearGradientBrush>
```

不过,对于采用集合的 XAML 属性而言,XAML 分析器可根据集合所属的属性隐式知道集合的后备类型。因此,可以省略集合本身的对象元素,如下所示:

```
<LinearGradientBrush>
  <LinearGradientBrush.GradientStops>
    <GradientStop Offset="0.0" Color="Red" />
    <GradientStop Offset="1.0" Color="Blue" />
  </LinearGradientBrush.GradientStops>
</LinearGradientBrush>
```

另外,有一些属性不但是集合属性,还标识为类的 XAML 内容属性。前面示例中以及许多其他 XAML 属性中使用的 GradientStops 属性就是这种情况。在这些语法中,也可以省略属性元素,如下所示:

```
<LinearGradientBrush>
  <GradientStop Offset="0.0" Color="Red" />
  <GradientStop Offset="1.0" Color="Blue" />
</LinearGradientBrush>
```

在广泛用于控件合成的类(如面板、视图或项控件)中,此集合和内容语法的组合是最常见的。例如,下面的示例演示将两个 UI 元素合成到一个 StackPanel 中的显式 XAML 以及最简单的 XAML:

```
<StackPanel>
  <StackPanel.Children>
    <!--UIElementCollection-->
    <TextBlock>Hello</TextBlock>
    <TextBlock>World</TextBlock>
```

```
      <!--/UIElementCollection-->
    </StackPanel.Children>
  </StackPanel>
  <StackPanel>
    <TextBlock>Hello</TextBlock>
    <TextBlock>World</TextBlock>
  </StackPanel>
```

请注意显式语法中注释掉的UIElementCollection。将其注释掉是因为即使将在对象树中创建相关集合，也无法在XAML中显式指定它。这是因为UIElementCollection不是可构造的类。在运行时对象树中获取的值是所属类中的一个默认初始化值，在初始化之后无法更改此值。在某些情况下，标记中会特意且显式包含集合类（例如，赋予集合一个x:Name，以便可以在代码中更方便地引用该集合）。但是，注意不要显式声明由于其后备类型的特征而无法由XAML分析器构造的集合类。

4. 何时使用属性语法或属性元素语法来设置属性

所有支持使用XAML设置的属性都支持用于直接值设置的属性语法或属性元素语法，但可能不会互换支持每种语法。某些属性支持上述两种语法，某些属性还支持其他语法选项（如前面所示的Text的内容元素语法）。属性支持的XAML语法的类型在某种程度上取决于该属性用作其属性类型的对象的类型。如果该属性类型为基元类型（如双精度、整型或字符串），则该属性始终支持属性语法。

下面的示例使用属性语法设置Rectangle的宽度。Width属性支持属性语法，这是因为属性值是双精度值。

```
<Rectangle Width="100" />
```

如果可以通过对字符串进行类型转换来创建用于设置某属性的对象类型，也可以使用属性语法来设置该属性。对于基元，始终是这种情况。但是，某些其他对象类型也可以使用指定为属性值的字符串（而不是需要对象元素语法）来创建。此方法使用该特定属性或该属性类型通常所支持的基本类型转换。属性的字符串值经过分析后，字符串信息用于设置对新对象的初始化非常重要的属性。特定类型转换器还可能创建公共属性类型的不同子类，这取决于它处理字符串中的信息的独特方式。

下面的示例使用属性语法设置Rectangle的填充。当使用SolidColorBrush设置Fill属性时，该属性支持属性语法。这是因为支持Fill属性的Brush抽象类型支持类型转换语法，该语法可以创建一个通过将属性指定的字符串作为其Color来初始化的SolidColorBrush。

```
<Rectangle Width="100" Height="100" Fill="Blue" />
```

如果用于设置某属性的对象支持对象元素语法，则可以使用属性元素语法来设置该属性。如果该对象支持对象元素语法，该属性也支持属性元素语法。下面的示例使用属性元素语法设置Rectangle的填充。当使用SolidColorBrush设置Fill属性时，该属性支持属性元素语法，这是因为SolidColorBrush支持对象元素语法并满足该属性的使用Brush类型设

置其值的要求。（SolidColorBrush 也使用属性语法设置了其 Color 属性，此 XAML 的呈现结果等同于前面使用属性语法设置 Fill 的 XAML 示例。）

```
<Rectangle Width="100" Height="100">
  <Rectangle.Fill>
    <SolidColorBrush Color="Blue"/>
  </Rectangle.Fill>
</Rectangle>
```

3.2.4 标记扩展

标记扩展是一个在 Windows Phone XAML 实现中广泛使用的 XAML 语言概念。在 XAML 属性语法中，花括号"{"和"}"表示标记扩展用法。此用法指示 XAML 处理不要像通常那样将属性值视为文本字符串或者视为可直接转换为文本字符串的值。相反，分析器通常应调用支持该特定标记扩展的代码，该标记扩展可帮助从标记中构造对象树。

Windows Phone 支持在其默认的 XAML 命名空间下定义且其 XAML 分析器可以理解的以下标记扩展：

(1) Binding：支持数据绑定，此绑定将延迟属性值，直至在数据上下文中解释此值。

(2) StaticResource：支持引用在 ResourceDictionary 中定义的资源值。

(3) TemplateBinding：支持 XAML 中可与模板化对象的代码属性交互的控件模板。

(4) RelativeSource：启用特定形式的模板绑定。

采用引用类型值(类型没有转换器)的属性需要属性元素语法(该语法始终创建新实例)或通过标记扩展的对象引用。XAML 标记扩展通常返回一个现有实例或将值延迟到运行时。通过使用标记扩展，每个可使用 XAML 设置的属性都可能在属性语法中设置。即使属性不支持对直接对象实例化使用属性语法，也可以使用属性语法为属性提供引用值；或者可以使特定行为能够符合用值类型或实时创建的引用类型填充 XAML 属性这一常规要求。

例如，下面的 XAML 使用属性语法设置 Border 的 Style 属性的值。Style 属性采用了 Style 类的实例，这是默认情况下无法使用属性语法字符串创建的引用类型。但在本例中，属性引用了特定的标记扩展 StaticResource。当处理该标记扩展时，它返回对以前在资源字典中定义为键控资源的某个样式的引用。

```
<Canvas.Resources>
  <Style TargetType="Border" x:Key="PageBackground">
    <Setter Property="BorderBrush" Value="Blue"/>
    <Setter Property="BorderThickness" Value="5"/>
  </Style>
</Canvas.Resources>
...
<Border Style="{StaticResource PageBackground}">
  ...
</Border>
```

在许多情况下，可以使用标记扩展来提供一个作为对现有对象的引用的值，或者标记扩展可以提供一个可以以属性格式设置属性的对象。文本"{"值因为左花括号符号"{"是标记扩展序列的开始标记，所以必须使用转义符序列，以便指定以"{"开头的文本字符值。转义序列是"{}"。例如，若要指定作为单个左花括号的字符值，请将属性值指定为"{"。还可以在某些情况下使用替代引号（如""分隔的属性值内的'），以便将"{"值作为字符串提供。

3.2.5 事件

XAML 是用于对象及其属性的声明性语言，但它也可以包含用于将事件处理程序附加到标记中的对象的语法。接着，可以通过特定的技术扩展 XAML 事件语法约定，这会通过编程模型集成 XAML 声明的事件。可以将相关事件的名称指定为处理该事件的对象的属性名称。对于属性值，可以指定在代码中定义的事件处理程序函数的名称。XAML 处理器使用此名称在加载的对象树中创建一个委托表示形式，并将指定的处理程序添加到内部处理程序列表中。

大多数基于 Windows Phone 的应用程序都是由标记和代码隐藏源生成的。在一个项目中，XAML 被编写为.xaml 文件，而使用 CLR 语言（如 VisualBasic 或 C#）编写代码隐藏文件。编译 XAML 文件时，通过将一个命名空间和类指定为 XAML 页的根元素的 x:Class 属性来确定每个 XAML 页的 XAML 代码隐藏文件的位置。

开发技术篇

对 Windows Phone 8 有一定的了解之后，开始进入本篇的学习，本篇的知识具有一定的独立性，同时也会有一些关联性，涉及了 Windows Phone 8 的应用程序开发的大部分常用的技术，每个知识点一般都会配有对应的实例，读者可以边学习边做实际编程的练习，这样会有更好的学习效果，同时也锻炼了自己开发程序的动手能力，让自己更加深刻地掌握这些知识。

本篇对 Windows Phone 8 的开发技术进行了全面而详细的讲解，从简单的技术开始，然后再循序渐进地深入，各章节之间会有一定的联系，后面的章节会包含前面章节的知识。本篇会涉及 Windows Phone 8 的控件、数据存储、设计模式、网络编程、蓝牙、近场通信等方面的知识。本篇共包含了 18 章的篇幅，有一定基础的读者可以挑选自己感兴趣的技术的章节阅读。

开发技术篇包括了以下的章节：

第 4 章　常用控件

介绍默认的常用控件的使用。

第 5 章　布局管理

介绍应用程序的整理布局设计和管理，有常用的绝对布局、相对布局等，也有 Windows Phone 8 特有的全景视图布局和枢轴布局。

第 6 章　数据存储

介绍 Windows Phone 8 的独立存储和 SQL Service CE 数据库存储。

第 7 章　图形动画

介绍图形和动画的编程。

第 8 章　多媒体

介绍在 Windows Phone 8 上的使用音频和视频的编程。

第 9 章　启动器与选择器

介绍 Windows Phone 8 上用系统提供的选择器和启动器来调用系统

自带的一些功能。

第 10 章　手机感应编程

介绍手机的重力感应编程、触摸感应编程、指南针、陀螺仪和语音控制。

第 11 章　MVVM 模式

介绍 Windows Phone 8 手机上的 MVVM 模式的编程，也包含一些数据绑定各方面的内容。

第 12 章　Silverlight Toolkit 组件

介绍 Silverlight ToolKit 这套组件的控件的使用，可以当做 Windows Phone 8 默认控件的一个扩充控件库来使用。

第 13 章　网络编程

介绍 Windows Phone 8 手机上的网络编程技术，涉及 Http 协议、Web Service、WCF Service 等多方面的网络编程的知识。

第 14 章　异步编程与并行编程

介绍了 Windows Phone 8 里面新的异步编程语法和新增加的并行编程方式。

第 15 章　联系人和日程安排

介绍联系人日程安排的信息获取，联系人的增删改操作。

第 16 章　手机文件数据读写

介绍了存储卡数据的读取、手机应用程序的文件读写，以及音频图片数据的读写。

第 17 章　Socket 编程

介绍 Socket 编程的相关知识，包括 Windows Phone 7.1 版本的和 Windows Phone 8 新版本的 Socket 编程知识。

第 18 章　墓碑机制与后台任务

介绍 Windows Phone 里面的墓碑机制的原理和如何去使用，以及 Windows Phone 8 支持的一些后台任务的编程。

第 19 章　蓝牙和近场通信

介绍如何使用蓝牙和近场通信的 API 去与近距离的设备进行无线通信。

第 20 章　响应式编程

介绍响应式编程的框架，如何在应用程序去使用响应式编程框架。

第 21 章　C++编程

介绍 Windows Phone 8 支持的 C++/CX 语法以及如何在 Windows Phone 8 中使用 C++进行编程。

通过本篇的学习，读者将会对 Windows Phone 8 的编程技术有一个深入的了解，通过运用这些知识可以很顺利地开发出一些简单的手机应用程序，同时也对 Windows Phone 8 的知识有一个较为全面的把握。

常用控件

Windows Phone 8 系统默认提供了丰富的可视化控件，有些控件跟 Silverlight 中的控件大同小异，比如 Button 控件、TextBlock 控件等；有些是 Windows Phone 8 特有的控件，比如菜单栏控件等。熟练地掌握这些控件的使用是开发 Windows Phone 应用程序的必要条件，也是给用户提供良好的用户体验保证。

Windows Phone 8 平台的用户界面上引入了 Metro 的概念，可以称之为 Windows Phone 8 UI 设计的一门语言，只不过这门语言是由一些文字、界面和版式构成。Windows Phone 8 的 UI 在设计方面给人的感觉就是简约，无论是图片、文字还是一些色块都要具备简单和形象的风格，给用户一种清爽和舒适的感觉，Windows Phone 8 系统里面默认的一些控件都保持着这样的一种风格。

常用控件这部分都是普通应用程序中非常常见的一些控件，在 Visual Studio 2012 编程工具里面可以通过控件面板拖拉这些控件到程序界面上，很方便地生成一个最原始的控件。

4.1 控件的基类

大部分的 Windows Phone 控件都间接或者直接继承了 System. Windows. UIElement、System. Windows. FrameworkElement 和 System. Windows. Controls. Control 这 3 个基类。这 3 个基类封装了 Windows Phone 的 Silverlight 程序的界面元素的一些共同的特性和通用的方法，Windows Phone 的控件的实现都是通过直接或者间接继承这些基类来扩展的，然后再根据控件的特性来定义和实现控件自身的属性和方法，创建自定义的控件和界面元素同样是这样的原理。下面简单地介绍这 3 个基类的含义和常用的属性方法，更加详细的介绍请参考微软的 API 文档。

这 3 个基类的继承层次结构如下：

```
System.Windows.UIElement
  System.Windows.FrameworkElement
    System.Windows.Controls.Control
```

1. System. Windows. UIElement

UIElement 是 Silverlight 中具有可视外观并可以处理基本输入的大多数对象的基类。在 UI 用户界面中，大多可视元素的输入行为都是在 UIElement 类中定义的。其中包括键盘、鼠标和触笔输入事件以及焦点事件。UIElement 类的常用属性如表 4.1 所示，常用事件如表 4.2 所示。

表 4.1 UIElement 类常用的属性

名　称	说　明
CacheMode	获取或设置一个值，该值指示应在可能时高速缓存已呈现内容
Clip	获取或设置用于定义 UIElement 的内容边框的 Geometry
DesiredSize	获取此 UIElement 在布局过程的测量处理过程中计算的大小
Dispatcher	获取与此对象关联的 Dispatcher（继承自 DependencyObject）
Effect	获取或设置用于呈现此 UIElement 的像素着色器效果
IsHitTestVisible	获取或设置此 UIElement 的包含区域是否可为命中测试返回 true 值
Opacity	获取或设置对象的不透明度
OpacityMask	获取或设置用于改变此对象区域的不透明度的画笔
Projection	获取或设置在呈现此 UIElement 时要应用的透视投影（三维效果）
RenderSize	获取 UIElement 的最终呈现大小
RenderTransform	获取或设置影响 UIElement 的呈现位置的变换信息
RenderTransformOrigin	获取或设置由 RenderTransform 声明的任何可能呈现变换的原点，相对于 UIElement 的边界
UseLayoutRounding	获取或设置一个值，该值确定对象及其可视化子树的呈现是否应使用将呈现与整像素对齐的舍入行为
Visibility	获取或设置 UIElement 的可见性；不可见的 UIElement 不呈现，也不将其所需大小告知布局

表 4.2 UIElement 类常用的事件

名　称	说　明
GotFocus	当 UIElement 收到焦点时发生
KeyDown	在 UIElement 具有焦点的情况下按下键盘上的某个键时发生
KeyUp	在 UIElement 具有焦点的情况下释放键盘上的某个键时发生
LostFocus	当 UIElement 失去焦点时发生
LostMouseCapture	在 UIElement 失去鼠标捕获时发生
ManipulationCompleted	对于 UIElement 的操作和延时完毕时发生
ManipulationDelta	当输入设备在操作期间更改位置时发生
ManipulationStarted	当输入设备对 UIElement 开始操作时发生
MouseEnter	当鼠标（或触笔）进入 UIElement 的边界区域时发生
MouseLeave	当鼠标（或触笔）离开 UIElement 的边界区域时发生
MouseLeftButtonDown	当按下鼠标左键（或触笔的笔尖接触 Tablet）并且鼠标指针悬停在 UIElement 上时发生
MouseLeftButtonUp	当鼠标（或触笔）悬停在 UIElement 上（或 UIElement 具有鼠标捕获）并且用户松开鼠标左键（或从 Tablet 上移开触笔的笔尖）时发生
MouseMove	当鼠标（或触笔）的坐标位置更改并且悬停在 UIElement 上（或 UIElement 具有鼠标捕获）时发生

2. System. Windows. FrameworkElement

FrameworkElement 为 Silverlight 布局中涉及的对象提供公共 API 的框架。FrameworkElement 还定义在 Silverlight 中与数据绑定、对象树和对象生存期功能区域相关的 API。FrameworkElement 扩展了 UIElement 并添加了布局和数据绑定两大重要的功能。

FrameworkElement 引入的主要策略是关于应用程序布局。FrameworkElement 在 UIElement 引入的基本布局协议之上生成，并增加了布局“插槽”的概念，使布局制作者可以方便地拥有一组面向属性的一致的布局语义。HorizontalAlignment、VerticalAlignment、MinWidth 和 Margin 等属性使得从 FrameworkElement 派生的所有组件在布局容器内具有一致的行为。

FrameworkElement 引入的两个最关键的内容是数据绑定和样式。直接或间接继承 FrameworkElement 基类的控件将可以使用控件的数据绑定功能和自定义样式功能。

使用 FrameworkElement 的数据绑定可通过一种简单的方式来表达，将给定元素中的一个或多个属性绑定到一个数据片断。数据绑定的最值得关注的功能之一是引入了数据模板。利用数据模板，可以声明性地指定某个数据片断的可视化方式；可以将问题换个方向，让数据来确定将要创建的显示内容，而无须创建可绑定到数据的自定义用户界面。

样式实际上是轻量级的数据绑定。使用样式，可以将共享定义的一组属性绑定到元素的一个或多个实例。通过显式引用（设置 Style 属性）或通过将样式与元素的 CLR 类型隐式关联，便可以将样式应用到元素。FrameworkElement 类在 UIElement 类的基础上新增的属性如表 4.3 所示，新增的事件如表 4.4 所示。

表 4.3　FrameworkElement 类新增的属性

名　称	说　明
ActualHeight	获取 FrameworkElement 的呈现高度
ActualWidth	获取 FrameworkElement 的呈现宽度
Cursor	获取或设置鼠标指针悬停在 FrameworkElement 上时所显示的光标
DataContext	获取或设置 FrameworkElement 参与数据绑定时的数据上下文
Height	获取或设置 FrameworkElement 的建议高度
HorizontalAlignment	获取或设置在布局父级（如面板或项控件）中构成 FrameworkElement 时应用于此元素的水平对齐特征
Language	获取或设置应用于 FrameworkElement 的本地化/全球化语言信息
Margin	获取或设置 FrameworkElement 的外边距
MaxHeight	获取或设置 FrameworkElement 的最大高度约束
MaxWidth	获取或设置 FrameworkElement 的最大宽度约束
MinHeight	获取或设置 FrameworkElement 的最小高度约束
MinWidth	获取或设置 FrameworkElement 的最小宽度约束
Name	获取或设置对象的标识名称，该名称提供最初已编译标记的引用；在 XAML 处理器从标记创建对象树后，运行时代码可以使用此名称引用标记元素

续表

名　称	说　明
Parent	获取对象树中此 FrameworkElement 的父对象
Resources	获取本地定义的资源字典，在 XAML 中，可以通过 XAML 隐式集合语法，将资源项建立为 frameworkElement. Resources 属性元素的子对象元素
Style	获取或设置呈现过程中应用于此对象的实例 Style
Tag	获取或设置一个可用于存储有关此对象的自定义信息的任意对象值
Triggers	获取为 FrameworkElement 定义的动画的触发器的集合
VerticalAlignment	获取或设置在父对象(如面板或项控件)中构成 FrameworkElement 时应用于此元素的垂直对齐特征
Width	获取或设置 FrameworkElement 的宽度

表 4.4　FrameworkElement 类新增的事件

名　称	说　明
BindingValidationError	在绑定源报告数据验证错误时发生
LayoutUpdated	当 Silverlight 可视化树的布局更改时发生
Loaded	当已构造 FrameworkElement 并将其添加到对象树中时发生
SizeChanged	当 FrameworkElement 上的 ActualHeight 或 ActualWidth 属性的值发生更改时发生

3. System. Windows. Controls. Control

Control 表示用户界面元素的基类，这些元素使用 ControlTemplate 来定义其外观。Control 类是添加到应用程序中的许多控件的基类。Control 类很少定义行为，但可以将 Control 添加到应用程序中，添加继承自 Control 的控件(如 Button 或 ListBox)的情况更为常见。

Template 属性是一个 ControlTemplate，可以指定 Control 的外观。Silverlight 附带的许多控件使用 ControlTemplate 并继承自 Control。如果要更改控件的外观并保留其功能，应考虑创建新的 ControlTemplate 而不是创建新的类。

如果要创建具有自定义行为的控件并允许其他人自定义控件的外观，则所创建的控件可以从 Control 类继承并定义一个 ControlTemplate。如果要扩展现有控件的行为，可以从继承自 Control 的类继承。

控件最重要的功能是模板化。如果将控件的组合系统视为一个保留模式呈现系统，则控件可通过模板化以一种参数化的声明性方式描述其呈现。ControlTemplate 实际上不过是一个用于创建一组子元素的脚本，同时绑定到由控件提供的属性。

Control 提供一组常用属性，如 Foreground、Background、Padding 等。模板创作者可以使用这些常用属性来自定义控件的显示。控件的实现提供了数据模型和交互模型。交互模型定义了一组命令，以及到输入笔势的绑定。数据模型提供了一组属性，用于自定义交互模型或自定义显示(由模板确定)。Control 类在 FrameworkElement 类和 UIElement 类的基础上新增的属性如表 4.5 所示，新增的事件如表 4.6 所示。

表 4.5 Control 类新增的属性

名称	说明
Background	获取或设置一个用于提供控件背景的画笔
BorderBrush	获取或设置一个用于描述控件的边框背景的画笔
BorderThickness	获取或设置控件的边框宽度
DefaultStyleKey	获取或设置一个键，该键引用控件的默认样式
FontFamily	获取或设置用于在控件中显示文本的字体
FontSize	获取或设置此控件中文本的大小
FontStretch	获取或设置字体在屏幕上的压缩或扩展程度
FontStyle	获取或设置呈现文本时使用的样式
FontWeight	获取或设置指定字体的粗细
Foreground	获取或设置一个用于描述前景色的画笔
HorizontalContentAlignment	获取或设置控件内容的水平对齐方式
IsEnabled	获取或设置一个值，该值指示用户是否可以与控件交互
IsTabStop	获取或设置一个值，该值指示是否将某个控件包含在 Tab 导航中
Padding	获取或设置控件内的边距
TabIndex	获取或设置一个值，该值决定在用户使用 Tab 键在控件中导航时元素接收焦点的顺序
TabNavigation	获取或设置一个值，该值修改 Tab 键和 TabIndex 针对此控件的工作方式
Template	获取或设置控件模板
VerticalContentAlignment	获取或设置控件内容的垂直对齐方式

表 4.6 Control 类新增的事件

名称	说明
IsEnabledChanged	当 IsEnabled 属性更改时发生

4.2 按钮(Button)

按钮(Button)控件是一个表示按钮的控件，是单击之后要触发事件的控件，代表 Button 的类是 System. Windows. Controls. Button。控件的 XAML 语法如下：

```
<Button .../>
<Button>属性元素
</Button>
<Button ...>内容</Button>
```

按钮控件最主要的作用是实现其单击的操作，单击 Button 按钮的时候，将引发一个

Click 事件，通过设置 Click 属性来处理按钮的单击事件，如设置为 Click="button1_Click"，那么在 cs 页面上会自动生成一个 button1_Click 事件的处理方法，在这个方法下面就可以处理按钮的单击事件了。按钮控件有 3 个状态，分别是按下状态(Press)、悬停状态(Hover)和释放状态(Release)。通过设置按钮控件的 ClickMode 属性值来控制按钮在哪种状态下才执行 Click 事件。例如，设置 ClickMode="Release"则表示直到手指释放了按钮的时候才开始执行 Click 按钮的单击事件。

下面给出设置按钮触发事件的示例：设置按钮尺寸、样式及文本颜色。

代码清单 4-1：按钮设置(源代码：第 4 章\Examples_4_1)

MainPage.xaml 文件主要代码

```
<Grid x:Name = "ContentPanel" Grid.Row = "1" Margin = "12,0,12,0">
            <! - 添加按钮单击事件 -->
            <Button Content = "按钮 1" Height = "72" HorizontalAlignment = "Left" Margin =
"46,47,0,0" Name = "button1" VerticalAlignment = "Top" Width = "364" Click = "button1_Click" />
            <! - 设置按钮的样式 -->
            <Button Content = "按钮 2"
            FontSize = "48"
            FontStyle = "Italic"
            Foreground = "Red"
            Background = "Blue"
            BorderThickness = "10"
            BorderBrush = "Yellow"
            Padding = "20"
            HorizontalAlignment = "Center"
VerticalAlignment = "Center" Margin = "63,117,185,342" />
            <! - 图片按钮 -->
             <Button HorizontalAlignment = "Center" VerticalAlignment = "Center" Margin =
"105,237,186,237" Width = "165">
                <StackPanel>
                    <Image Source = "ApplicationIcon.png"
                  Stretch = "None" Height = "61" Width = "94" />
                </StackPanel>
            </Button>
        </Grid>
```

MainPage.xaml.cs 文件主要代码

```
public partial class MainPage : PhoneApplicationPage
{
    //Constructor
    public MainPage()
    {
        InitializeComponent();
    }
    //处理按钮单击事件，更新按钮的显示文本内容
```

```
    private void button1_Click(object sender, RoutedEventArgs e)
    {
        button1.Content = "你点击了我啦!";
    }
}
```

运行的效果如图 4.1～图 4.4 所示。

图 4.1　按钮 1

图 4.2　你点击我了

图 4.3　按钮 2

图 4.4　图片按钮

4.3　文本块(TextBlock)

文本块(TextBlock)控件是用于显示少量文本的轻量控件，可以通过 TextBlock 呈现只读的文本。控件的 XAML 语法如下：

```
<TextBlock .../>
<TextBlock ...>内容</TextBlock>
<TextBlock> 属性元素</TextBlock>
```

TextBlock 控件在 Windows Phone 应用中非常普遍，它就相当于一个只是用于呈现文本的标签一样。写过 HTML 页面的用户都知道，在 HTML 语法中，可以直接将文本写在各种 HTML 的标签外面，但是在 XAML 语法中，不能直接把文本写在 XAML 的各种控件之外，如果要这样做就必须将文本写到 TextBlock 控件中才能展示出来。在 TextBlock 控件中需要经常用到的属性如下：

(1) FontFamily 字体名称，可以设置为“Courier New”、“Times New Roman”、“Verdana”等字体；

(2) FontSize 文字大小，以像素为单位，可以赋值为 1、2 等数字；

(3) FontStyle 字体样式，可以赋值为 Arial，verdana 等；

(4) FontWeight 文字的粗细，可设置为 Thin、ExtraLight、Light、Normal、Medium、SemiBold、Bold、ExtraBold、Black、ExtraBlack，这些值是否起作用还要取决于所选择的字体；

(5) Foreground 字体前景颜色，可以赋值为 Red、Green 等颜色；

(6) Width 文字区块宽度，可以赋值数字，如 400、500 等；

(7) Height 文字区块高度，可以赋值数字，如 400、500 等；

(8) Opacity 文字透明度，可以赋值为 0～1.0 的数字(0 表示全透明，1 表示不透明)；

(9) Text 字体正文，可以赋值各种文字；

(10) TextWrapping 字体换行，可以赋值 Wrap，NoWrap，WrapWithOverflow，可以通过设置 TextWrapping 属性的值为“Wrap”来强制其换行，默认情况下文字会被截断而不自动换行。

下面给出设置文本块样式设置的示例：演示如何在一个 TextBlock 控件里面的文字定义多种样式、文字断行、内容自动折行和 cs 文件生成 TextBlock 控件。

代码清单 4-2：文本块控件演示(源代码：第 4 章\Examples_4_2)

MainPage.xaml 文件主要代码

```
<Grid x:Name = "ContentPanel" Grid.Row = "1" Margin = "12,0,12,0">
    <!-创建一个简单的 TextBlock 控件 -->
    <TextBlock x:Name = "TextBlock2" Height = "30" Text = "你好,我是 TextBlock 控件" Foreground
= "Red" Margin = "25,84,177,493">
     </TextBlock>
<!-给同一 TextBlock 控件的文字内容设置多种不同的样式 -->
     <TextBlock Margin = "25,171,19,384">
     <TextBlock.Inlines>
          <Run  FontWeight = "Bold" FontSize = "14" Text = "TextBlock. " />
          <Run FontStyle = "Italic" Foreground = "Red" Text = "red text. " />
          <Run FontStyle = "Italic" FontSize = "18" Text = "linear gradient text. ">
               <Run.Foreground>
                    <LinearGradientBrush>
                         <GradientStop Color = "Green" Offset = "0.0" />
                         <GradientStop Color = "Purple" Offset = "0.25" />
                         <GradientStop Color = "Orange" Offset = "0.5" />
                         <GradientStop Color = "Blue" Offset = "0.75" />
                    </LinearGradientBrush>
               </Run.Foreground>
          </Run>
          <Run FontStyle = "Italic" Foreground = "Green" Text = " green " />
     </TextBlock.Inlines>
   </TextBlock>
     <!-自定义断行 -->
     <TextBlock Margin = "9,344,230,130">
     你好!
     <LineBreak/>
     我是 TextBlock
     <LineBreak/>
     再见
     <LineBreak/>
     --2011 年 6 月 8 日
     </TextBlock>
     <!-设置 TextBlock 控件自动折行 -->
     <TextBlock TextWrapping = "Wrap" Margin = "279,344,19,154">
好像内容太长长长长长长长长长长长长长长长长长长了
     </TextBlock>
```

```
    <! - 不设置自动折行 -- >
    < TextBlock   Margin = "248,477,50,77">
好像内容太长长长长长长长长长长长长长长长长长长了
    </TextBlock >
    <! - 设置 TextBlock 控件内容的颜色渐变 -- >
    < TextBlock Text = "颜色变变变变变变"    Margin = "25,519,109,21">
    < TextBlock.Foreground >
        < LinearGradientBrush >
            < GradientStop Color = " # FF0000FF" Offset = "0.0" />
            < GradientStop Color = " # FFEEEEEE" Offset = "1.0" />
        </LinearGradientBrush >
    </TextBlock.Foreground >
    </TextBlock >
</Grid >
```

MainPage.xam.cs 文件主要代码

```
public MainPage()
{
    InitializeComponent();
    //在 cs 页面动态生成 TextBlock 控件
    TextBlock txtBlock = new TextBlock();
    txtBlock.Height = 50;
    txtBlock.Width = 200;
    txtBlock.Text = "在 CS 页面生成的 TextBlock";
    txtBlock.Foreground = new SolidColorBrush(Colors.Blue);
    ContentPanel.Children.Add(txtBlock);
}
```

程序运行的效果如图 4.5～图 4.10 所示。

图 4.5　红色字体

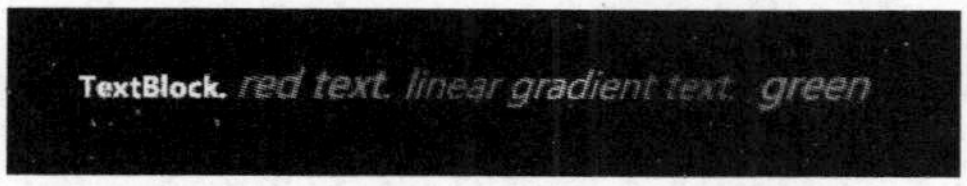

图 4.6　同一 TextBlock 控件的文字内容不同的样式显示效果

图 4.7　CS 页面生成的 TextBlock 控件的效果

图 4.8　设置内容文字的颜色渐变效果

图 4.9　自定义折行的文本内容

图 4.10　设置自动折行和不设置自动折行的效果对比

4.4 文本框(TextBox)

文本框(TextBox)控件是表示一个可用于显示和编辑单格式、多行文本的控件。TextBox 常用于在表单中编辑非格式化文本。例如,如果一个表单要求输入用户姓名、电话号码等,则可以使用 TextBox 控件来进行文本输入。控件的 XAML 语法如下：

```
<TextBox .../>
```

TextBox 的高度可以是一行,也可以包含多行。对于输入少量纯文本(如表单中的“姓名”、“电话号码”等)而言,单行 TextBox 是最好的选择。还可以创建一个使用户可以输入多行文本的 TextBox。例如,表单要求输入较多的文字,可能需要使用支持多行文本的 TextBox。设置多行文本的方法很简单,将 TextWrapping 特性设置为 Wrap 会使文本在到达 TextBox 控件的边缘时换至新行,必要时会自动扩展 TextBox 控件以便为新行留出空间。将 AcceptsReturn 特性设置为 true 会导致在按换行键时插入新行,必要时会再次自动扩展 TextBox 以便为新行留出空间。文本框的大小固定时,内容过多就需要添加滚动条来显示全部的内容了,这时候需要设置 HorizontalScrollBarVisibility 或 VerticalScrollBarVisibility 属性以启用水平滚动条或垂直滚动条,通过这两个属性中的一个可以向 TextBox 添加一个滚动条,以便在 TextBox 超出包含它的框架或窗口的大小时,可以滚动 TextBox 的内容。更多有关 TextBox 控件的一些属性和事件请参考表 4.7 和表 4.8。

表 4.7 TextBox 控件增加的属性

名　称	说　明
AcceptsReturn	获取或设置一个值,该值确定文本框是否允许和显示换行符或回车符
CaretBrush	获取或设置用于呈现指示插入点的竖线的画笔
FontSource	获取或设置应用于 TextBox 以呈现内容的字体源
HorizontalScrollBarVisibility	获取或设置水平滚动条的可见性
InputScope	获取或设置此 TextBox 使用的输入的上下文
IsReadOnly	获取或设置一个值,该值确定用户是否能够在文本框中更改文本
MaxLength	获取或设置一个值,该值确定用户输入所允许的最大字符数
SelectedText	获取或设置文本框中当前选择的内容
SelectionBackground	获取或设置填充选定文本的背景的画笔
SelectionForeground	获取或设置用于文本框中选定文本的画笔
SelectionLength	获取或设置文本框中当前选定内容的字符数
SelectionStart	获取或设置文本框中选定文本的起始位置
Text	获取或设置文本框的文本内容
TextAlignment	获取或设置文本应在文本框中进行对齐的方式
TextWrapping	获取或设置当一行文本超过文本框的可用宽度后如何进行换行
VerticalScrollBarVisibility	获取或设置垂直滚动条的可见性

表 4.8　TextBox 控件增加的事件

名　　称	说　　明
SelectionChanged	在文本选定内容更改后发生
TextChanged	在文本框中的内容更改时发生

下面给出文本框的示例：创建一个自定义样式的 TextBox 控件，可以自动折行，并且捕获 TextBox 内容变化事件。

代码清单 4-3：文本框控件演示(源代码：第 4 章\Examples_4_3)

MainPage. xaml 文件主要代码

```
<phone:PhoneApplicationPage.Resources>
    <ResourceDictionary>
        <!-- 设置 TextBox 的样式 -->
        <Style x:Key = "TextBoxStyle1" TargetType = "TextBox">
            <Setter Property = "FontFamily" Value = "{StaticResource PhoneFontFamilyNormal}"/>
            <Setter Property = "FontSize" Value = "{StaticResource PhoneFontSizeMediumLarge}"/>
            <Setter Property = "Background" Value = "{StaticResource PhoneTextBoxBrush}"/>
            <Setter Property = "Foreground" Value = "Blue"/>
            <Setter Property = "BorderBrush" Value = "{StaticResource PhoneTextBoxBrush}"/>
            <Setter Property = "SelectionBackground" Value = "{StaticResource PhoneAccentBrush}"/>
            <Setter Property = " SelectionForeground " Value = " { StaticResource
PhoneTextBoxSelectionForegroundBrush}"/>
            <Setter Property = "BorderThickness" Value = "{StaticResource PhoneBorderThickness}"/>
            <Setter Property = "Padding" Value = "2"/>
        </Style>
    </ResourceDictionary>
</phone:PhoneApplicationPage.Resources>
<Grid x:Name = "LayoutRoot" Background = "Transparent">
…
    <Grid x:Name = "ContentPanel" Grid.Row = "1" Margin = "12,0,12,0" >
        <!-- 创建一个自动折行并使用上面定义的 TextBoxStyle1 样式资源的控件，AcceptsReturn =
"True" 表示换行 -->
        <TextBox Style = "{StaticResource TextBoxStyle1}" Margin = "12,31,22,473" Name =
"TextBox1" TextWrapping = " Wrap" HorizontalScrollBarVisibility = " Disabled" AcceptsReturn =
"true"  TextChanged = "TextBox1_TextChanged"/>
        <TextBlock Height = "60" HorizontalAlignment = "Left" Margin = "12,221,0,0" Name =
"textBlock1" Text = "" VerticalAlignment = "Top" Width = "364" />
        <TextBlock Height = "43" HorizontalAlignment = "Left" Margin = "12,172,0,0" Name =
"textBlock2" Text = "获取 TextBox 的输入:" VerticalAlignment = "Top" Width = "170" />
    </Grid>
</Grid>
```

MainPage. xam. cs 文件主要代码

```
//处理 TextBox 变化的事件
private void TextBox1_TextChanged(object sender, TextChangedEventArgs e)
{
    textBlock1.Text = TextBox1.Text;
}
```

程序运行的效果如图 4.11 所示。

图 4.11 获取文本输入

4.5 边框(Border)

边框(Border)控件是指在另一个对象的周围绘制边框、背景或同时绘制二者的控件。控件的 XAML 语法如下：

```
<Border>
子控件对象
</Border>
```

Border 控件通常会是其他控件的一个外观显示的辅助控件，它很少单独使用，一般都是配合其他控件一起来使用，从而展示出其他控件的边框效果。Border 只能包含一个子对象。如果要在多个对象周围放置一个边框，应将这些对象包装到一个容器对象中，如 StackPanel。可以通过设置 Border 控件的属性来展现出各种各样的边框效果，比如可以通过设置 CornerRadius 属性将边框的各角改为圆角，并且可以通过设置 Padding 属性以在 Border 中定位对象等。Border 控件类的一些新增的属性如表 4.9 所示。

表 4.9 Border 类新增加的属性

名称	说明
BorderBrush	获取或设置用于创建边框的 Brush
BorderThickness	获取或设置边框的粗细
Child	获取或设置要在其周围绘制边框的子元素
CornerRadius	获取或设置边框的角的半径
Padding	获取或设置边框与其子对象之间的距离

下面给出边框控件的示例：演示了各种 Border 样式的使用。

代码清单 4-4：边框样式演示(源代码：第 4 章\Examples_4_4)

MainPage.xaml 文件主要代码

```
<Grid x:Name = "ContentPanel" Grid.Row = "1" Margin = "12,0">
<!--
```

```
BorderThickness - 边框的宽度(像素值: 上下左右; 左右,上下; 左,上,右,下)
BorderBrush - 边框的颜色
CornerRadius - 边框角的半径
-->
<Border Background="Coral" Margin="0,12,241,489" Padding="10" CornerRadius="30,38,150,29"
    BorderThickness="8 15 10 2" BorderBrush="Azure"></Border>
<Border BorderThickness="1,3,5,7" BorderBrush="Blue" CornerRadius="10" Width="120"
Margin="221,0,115,537">

    <TextBlock Text="蓝色的 Border" ToolTipService.ToolTip="这是蓝色的 Border 吗?"
TextAlignment="Center" />
</Border>
<!--单击后将显示边框 -->
<Border x:Name="TextBorder" BorderThickness="10"  Margin="-6,143,6,356" >
    <Border.BorderBrush>
        <SolidColorBrush Color="Red" Opacity="0" />
    </Border.BorderBrush>
    <TextBlock MouseLeftButtonDown="TextBlock_MouseLeftButtonDown" Text="请单击一下我!" />
</Border>
<!--颜色渐变的边框 -->
<Border x:Name="brdTest" BorderThickness="4" Width="200" Height="150" Margin="98,257,
158,200">
    <Border.BorderBrush>
        <LinearGradientBrush x:Name="borderLinearGradientBrush" MappingMode="RelativeToBoundingBox"
StartPoint="0.5,0" EndPoint="0.5,1">
            <LinearGradientBrush.GradientStops>
                <GradientStop Color="Yellow" Offset="0" />
                <GradientStop Color="Blue" Offset="1" />
            </LinearGradientBrush.GradientStops>
        </LinearGradientBrush>
    </Border.BorderBrush>
</Border>
</Grid>
```

MainPage.xaml.cs 文件主要代码

```
public partial class MainPage : PhoneApplicationPage
{
    public MainPage()
    {
        InitializeComponent();
        //动态填充 brdTest 里面的子元素
        Rectangle rectBlue = new Rectangle();
        rectBlue.Width = 1000;
        rectBlue.Height = 1000;
        SolidColorBrush scBrush = new SolidColorBrush(Colors.Blue);
        rectBlue.Fill = scBrush;
        this.brdTest.Child = rectBlue;
    }
    //单击事件,通过修改 Opacity 来实现,当用户单击文本时,出现一个文本的边框.
    private void TextBlock_MouseLeftButtonDown(object sender, System.Windows.Input.
```

```
MouseButtonEventArgs e)
    {
        //0 表示完全透明的,1 表示完全显示出来
        TextBorder.BorderBrush.Opacity = 1;
    }
}
```

程序运行的效果如图 4.12 以及图 4.13 所示。

图 4.12　BorderDemo 的运行效果

图 4.13　单击了"请单击一下我!"之后的显示效果

4.6　超链接(HyperlinkButton)

超链接(HyperlinkButton)控件是表示显示超链接的按钮控件。控件的 XAML 语法如下：

```
<HyperlinkButton .../>
<HyperlinkButton>
属性元素
</HyperlinkButton>
<HyperlinkButton>内容</HyperlinkButton>
```

HyperlinkButton 也继承了 ButtonBase 基类，大部分的特性和 Button 控件一样，不过 HyperlinkButton 控件多了 NavigateUri 属性，通过将 URI 的绝对地址或者相对地址赋值给 NavigateUri 属性，可以实现单击 HyperlinkButton 时要导航到用户设置的 URI。单击 HyperlinkButton 后，用户可以导航到同一应用程序内的外部网页或内容。可以使用 NavigateUri 属性为 HyperlinkButton 设置 URI。如果此 URI 设置为外部网页，则可以使用 TargetName 属性指定应在其中打开页面的目标窗口或框架。如果 URI 设置为同一个应用程序中的内容，则可以使用 TargetName 属性指定要导航到的对象的名称。不必处理 HyperlinkButton 的单击事件，即可自动导航到为 NavigateUri 指定的值。

下面给出超链接导航的示例：使用 HyperlinkButton 超链接按钮，导航打开网页和打

开应用程序的其他页面。

代码清单 4-5：超链接导航演示(源代码：第 4 章\Examples_4_5)

MainPage. xaml 文件主要代码

```
<Grid x:Name = "ContentPanel" Grid.Row = "1" Margin = "12,0,12,0">
    <!-- 创建一个 HyperlinkButton 按钮 -->
    <HyperlinkButton Width = "200" Height = "30"
         Content = "链接按钮"
         Background = "Blue" Foreground = "Orange"
         FontWeight = "Bold" Margin = " - 12,19,268,558">
    </HyperlinkButton>
    <!-- 单击 Google 按钮,将会跳转到 ie 并打开网页 http://google.com -->
     <HyperlinkButton Content = "Google" NavigateUri = "http://google.com" TargetName =
"_blank" Margin = "12,72,342,486" />
    <!-- 单击 Page1 页面按钮,将会跳转到 Page1.xaml 页面 -->
    <HyperlinkButton Width = "142" Height = "47" x:Name = "HomeLink"
            Content = "Page1 页面" Foreground = "Yellow"
             FontWeight = "Bold" HorizontalAlignment = "Left" Margin = "6,127,0,0"
             VerticalAlignment = "Top" NavigateUri = "/Page1.xaml" />
</Grid>
```

程序运行的效果如图 4.14 所示。

图 4.14 链接控件的运行效果

4.7 单选按钮(RadioButton)

单选按钮(RadioButton)控件是表示一个单选按钮的控件。使用该按钮控件,用户可从一组选项中选择一个选项。控件的 XAML 语法如下：

```
<RadioButton .../>
<RadioButton ...>
内容
</RadioButton>
```

可以通过将 RadioButton 控件放到父控件内或者为每个 RadioButton 设置 GroupName 属性来对 RadioButton 进行分组。当 RadioButton 元素分在一组中时,按钮之间会互相排斥,用户一次只能选择 RadioButton 组中的一项。RadioButton 有两种状态：选定(选中)或清除(未选中)。是否选中了 RadioButton 由其 IsChecked 属性的状态决定。选定 RadioButton 后,IsChecked 属性为 true; 清除 RadioButton 后,IsChecked 属性为 false。要清除一个 RadioButton,可以单击该组中的另一个 RadioButton,但不能通过再次单击该单选按钮自身来清除。但是可以通过编程方式来清除 RadioButton,方法是将其 IsChecked 属性设置为 false。

下面给出单项选择的示例:创建一组 RadioButton 按钮,并捕获你选择的按钮的内容。

代码清单 4-6:单项选择(源代码:第 4 章\Examples_4_6)

MainPage.xaml 文件主要代码

```
<Grid x:Name="ContentPanel" Grid.Row="1" Margin="12,0,12,0">
    <RadioButton Margin="6,92,4,433" GroupName="MCSites" Background="Yellow"
     Foreground="Blue" Content="A、诺基亚"  Click="RadioButton_Click" Name="a">
    </RadioButton>
    <RadioButton Margin="6,160,4,360" GroupName="MCSites" Background="Yellow"
     Foreground="Orange" Content="B、苹果"  Click="RadioButton_Click" Name="b" >
    </RadioButton>
    <RadioButton Margin="6,236,4,298" GroupName="MCSites" Background="Yellow"
     Foreground="Green" Content="C、HTC"  Click="RadioButton_Click" Name="c">
    </RadioButton>
    <RadioButton Margin="9,296,1,227" GroupName="MCSites" Background="Yellow"
     Foreground="Purple" Content="D、其他的"  Click="RadioButton_Click" Name="d">
    </RadioButton>
    <TextBlock FontSize="34" Height="57" HorizontalAlignment="Left" Margin="0,29,0,0"
Name="textBlock1" Text="你喜欢哪一个品牌的手机?" VerticalAlignment="Top" Width="446" />
    <TextBlock FontSize="34" Height="57" HorizontalAlignment="Left" Margin="0,386,0,0"
Name="textBlock2" Text="你选择的答案是:" VerticalAlignment="Top" Width="446" />
    <TextBlock FontSize="34" Height="57" HorizontalAlignment="Left" Margin="4,449,0,0"
Name="answer" Text="" VerticalAlignment="Top" Width="446" />
    </Grid>
</Grid>
```

MainPage.xaml.cs 文件主要代码

```
//RadioButton 按钮单击事件
private void RadioButton_Click(object sender, RoutedEventArgs e)
{
    if (a.IsChecked == true)
        answer.Text = a.Content.ToString();
    else if (b.IsChecked == true)
        answer.Text = b.Content.ToString();
    else if (c.IsChecked == true)
        answer.Text = c.Content.ToString();
    else
        answer.Text = d.Content.ToString();
}
```

程序运行的效果如图 4.15 所示。

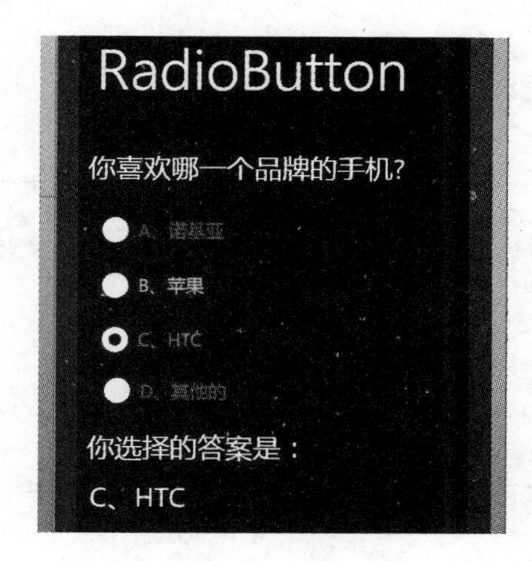

图 4.15　RadioButton 的运行效果

4.8　复选框(CheckBox)

复选框(CheckBox)控件,表示用户可以选中或不选中的控件。控件的 XAML 语法如下：

```
<CheckBox .../>
<CheckBox>内容</CheckBox>
```

CheckBox 和 RadioButton 控件都允许用户从选项列表中进行选择。CheckBox 控件允许用户选择一组选项。而 RadioButton 控件则允许用户从互相排斥的选项中进行选择。CheckBox 控件继承自 ToggleButton,可具有 3 种状态：选中、未选中和不确定。那么 CheckBox 控件可以通过 IsThreeState 属性来获取或设置指示控件是支持两种状态还是 3 种状态的值和通过 IsChecked 属性获取或设置是否选中了复选框控件。需要注意的是,如果将 IsThreeState 属性设置为 true,则 IsChecked 属性将为已选中或不确定状态返回 true。

下面给出复选框的示例：判断复选框是否选中。

代码清单 4-7：复选框(源代码：第 4 章\Examples_4_7)

MainPage.xaml 文件主要代码

```
<Grid x:Name="ContentPanel" Grid.Row="1" Margin="12,0,12,0">
    <!-- 一个 CheckBox 控件 -->
    <CheckBox Name="McCheckBox1" Foreground="Orange"
     Content="Check Me"
      FontFamily="Georgia" FontSize="20" FontWeight="Bold" Margin="0,6,0,527">
    </CheckBox>
    <!-- 添加了选中和不选中的处理事件 -->
    <CheckBox Name="McCheckBox3"
        Content="Check Me"
        IsChecked="True"
        Checked="McCheckBox_Checked"
```

```
            Unchecked = "McCheckBox_Unchecked" Margin = "0,75,0,441">
        </CheckBox>
    </Grid>
```

MainPage.xaml.cs 文件主要代码

```
private void McCheckBox_Checked(object sender, RoutedEventArgs e)
{//第一次进入页面的时候 McCheckBox3 还没有初始化,所以必须先判断一下 McCheckBox3 是否
//为 null
    if (McCheckBox3 != null)
    {
        McCheckBox3.Content = "Checked";
    }
}
private void McCheckBox_Unchecked(object sender, RoutedEventArgs e)
{
    if (McCheckBox3 != null)
    {
        McCheckBox3.Content = "Unchecked";
    }
}
```

程序运行的效果如图 4.16 所示。

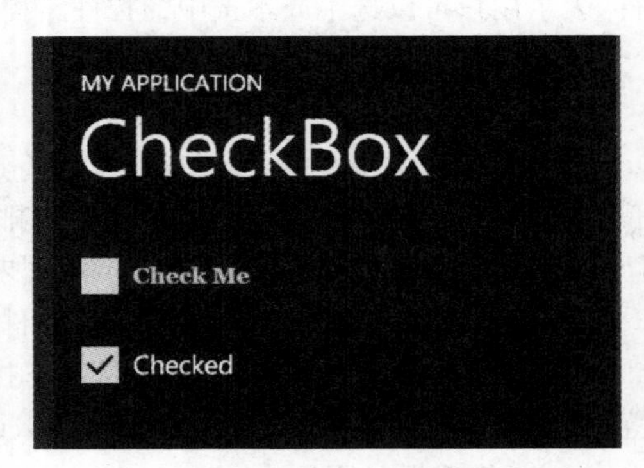

图 4.16　CheckBox 的运行效果

4.9　进度条(ProgressBar)

进度条(ProgressBar)控件表示一个指示操作进度的控件。控件的 XAML 语法如下：

```
<ProgressBar .../>
```

ProgressBar 控件的两种形式：重复模式和非重复模式。这两种类型的 ProgressBar 控件是由 IsIndeterminate 属性来确定的，下面来阐述这两种模式下的进度条的特点。

1. 重复模式进度条

ProgressBar 的属性 IsIndeterminate 设置为 true 时为重复模式，也是 ProgressBar 的默认值。当无法确定需要等待的时间或者无法计算等待的进度情况时，适合使用重复模式，进

度条会一直产生等待的效果，直到关掉进度条或者跳转到其他的页面。重复模式进度条的运行效果如图 4.17 所示。

2. 非重复模式进度条

ProgressBar 的属性 IsIndeterminate 设置为 false 时为非重复模式，也是根据值来填充的进度条。当可以确定执行任务的时间或者可以跟踪完成任务的进度情况时，适合使用非重复模式的进度条，从而可以根据完成任务的情况而改变进度条的值从而产生动态的运行效果。当 IsIndeterminate 为 false 时，可以使用 Minimum 和 Maximum 属性指定范围。默认情况下，Minimum 为 0，而 Maximum 为 100。通过 ValueChanged 事件可以监控到进度条控件的值的变化，若要指定进度值，那么就需要通过 Value 属性来设置。非重复模式进度条的运行效果如图 4.18 所示。

图 4.17　重复模式的进度条的运行效果

图 4.18　非重复模式的进度条的运行效果

使用非重复模式的进度条时，通常需要使用 System. ComponentModel. BackgroundWorker 类来配合 ProgressBar 控件来跟踪程序完成任务的进度。BackgroundWorker 类允许在单独的专用线程上运行操作。耗时的操作(如下载和数据库事务)在长时间运行时可能会导致用户界面(UI)似乎处于停止响应状态。如果需要能进行响应的用户界面，而且面临与这类操作相关的长时间延迟，则可以使用 BackgroundWorker 类方便地解决问题。若要在后台执行耗时的操作，可以创建一个 BackgroundWorker，侦听那些报告操作进度并在操作完成时发出信号的事件。若要设置后台操作，可以为 DoWork 事件添加一个事件处理程序。在此事件处理程序中调用耗时的操作。若要启动该操作，可以调用 RunWorkerAsync。若要收到进度更新通知，可以对 ProgressChanged 事件进行处理。若要在操作完成时收到通知，可以对 RunWorkerCompleted 事件进行处理。需要注意的是，必须确保在 DoWork 事件处理程序中不操作任何用户界面对象，而应该通过 ProgressChanged 和 RunWorkerCompleted 事件与用户界面进行通信。BackgroundWorker 事件不跨 AppDomain 边界进行封送处理。不能使用 BackgroundWorker 组件在多个 AppDomain 中执行多线程操作。如果后台操作需要参数，可以在调用 RunWorkerAsync 时给出参数。在 DoWork 事件处理程序内部，可以从 DoWorkEventArgs. Argument 属性中提取该参数。BackgroundWorker 类的常用属性和常用事件分别如表 4.10 和表 4.11 所示。

表 4.10　BackgroundWorker 类常用的属性

名　称	说　明
CancellationPending	获取一个值，该值指示应用程序是否已请求取消后台操作；该属性通常会被用于 DoWork 事件中，用于标识目前正在执行的后台是否有被取消

续表

名　称	说　明
IsBusy	获取一个值，该值指示 BackgroundWorker 是否正在运行后台操作
WorkerReportsProgress	获取或设置一个值，该值指示 BackgroundWorker 能否报告进度更新；如果要使用 ReportProgress 方法和 ProgressChanged 事件时需要设置为 true
WorkerSupportsCancellation	获取或设置一个值，该值指示 BackgroundWorker 是否支持异步取消

表 4.11　BackgroundWorker 类常用的事件

名　称	说　明
DoWork	调用 RunWorkerAsync 时发生
ProgressChanged	调用 ReportProgress 时发生
RunWorkerCompleted	当后台操作已完成、被取消或引发异常时发生

下面给出两种模式的进度条的示例：创建重复模式的进度条和创建非重复模式的进度条。

代码清单 4-8：两种模式的进度条(源代码：第 4 章\Examples_4_8)

MainPage.xaml 文件主要代码

```
<Grid x:Name = "ContentPanel" Grid.Row = "1" Margin = "12,0,12,0">
    <TextBlock Text = "选择 ProgressBar 的类型:" />
    <RadioButton Content = "Determinate 类型" Height = "71" Name = "radioButton1" GroupName =
"Type" Margin = "0,45,0,491" />
    <RadioButton Content = "Indeterminate 类型" Height = "71" Name = "radioButton2" GroupName
= "Type" IsChecked = "True" Margin = "0,122,0,414" />
    <Button Content = "启动 ProgressBar" Height = "72" HorizontalAlignment = "Left" Margin =
"3,247,0,0" Name = "begin" VerticalAlignment = "Top" Width = "386" Click = "begin_Click" />
    <Button Content = "取消 ProgressBar" Height = "72" HorizontalAlignment = "Left" Margin =
"6,338,0,0" Name = "cancel" VerticalAlignment = "Top" Width = "383" Click = "cancel_Click" />
     < ProgressBar Height = " 56 " HorizontalAlignment = " Left " Name = " progressBar1
" VerticalAlignment = "Top" Width = "462" IsIndeterminate = "true" Margin = "0,443,0,0"  />
</Grid>
```

MainPage.xaml.cs 文件代码

```
using System.Windows;
using Microsoft.Phone.Controls;
using System.Threading;
using System.ComponentModel;
namespace ProgressBarDemo
{
    public partial class MainPage : PhoneApplicationPage
    {
        //定义一个后台处理类
        private BackgroundWorker backgroundWorker;
        public MainPage()
```

```
{
    InitializeComponent();
    //第一次进入页面,设置进度条为不可见
    progressBar1.Visibility = System.Windows.Visibility.Collapsed;
}
private void begin_Click(object sender, RoutedEventArgs e)
{
    //设置进度条为可见
    progressBar1.Visibility = System.Windows.Visibility.Visible;
    if (radioButton1.IsChecked == true)
    {
        //设置进度条为不可重复模式
        progressBar1.IsIndeterminate = false;
        //创建一个后台处理类的对象
        backgroundWorker = new BackgroundWorker();
        //调用 RunWorkerAsync 后台操作时引发此事件,即后台要处理的事情写在这个
        //事件里面
        backgroundWorker.DoWork += new DoWorkEventHandler(backgroundWorker_DoWork);
        //当后台操作完成事件
        backgroundWorker.RunWorkerCompleted += new RunWorkerCompletedEventHandler
(backgroundWorker_RunWorkerCompleted);
        //当处理进度(ReportProgress)被激活后,进度的改变将触发 ProgressChanged 事件
        backgroundWorker.ProgressChanged    +=    new    ProgressChangedEventHandler
(backgroundWorker_ProgressChanged);
        //设置为可报告进度情况的后台处理
        backgroundWorker.WorkerReportsProgress = true;
        backgroundWorker.RunWorkerAsync();
    }
    else
    {
        //设置进度条的值为 0
        progressBar1.Value = 0;
        //设置进度条为重复模式
        progressBar1.IsIndeterminate = true;
    }
}
private void cancel_Click(object sender, RoutedEventArgs e)
{
    //隐藏进度条
    progressBar1.Visibility = System.Windows.Visibility.Collapsed;
}
//进度改变处理
void backgroundWorker_ProgressChanged(object sender, ProgressChangedEventArgs e)
{
    Dispatcher.BeginInvoke(() =>
    {
        //把进度改变的值赋值给 progressBar1 进度条的值
        progressBar1.Value = e.ProgressPercentage;
    }
 );
```

```
            }
            //后台操作完成
            void backgroundWorker_RunWorkerCompleted(object sender, RunWorkerCompletedEventArgs e)
            {
                Dispatcher.BeginInvoke(() =>
                {
                    //隐藏进度条
                    progressBar1.Visibility = System.Windows.Visibility.Visible;
                });
            }
            //后台操作处理
            void backgroundWorker_DoWork(object sender, DoWorkEventArgs e)
            {
                for (int i = 0; i < 100; i++)
                {
                    i += 20;
                    //赋值当前进度的值,同时会触发进度改变事件
                    backgroundWorker.ReportProgress(i);
                    //为了能看到进度条的变化效果,这里用进程暂停了 1 秒
                    Thread.Sleep(1000);
                }
            }
        }
    }
```

程序运行的效果如图 4.19 所示。

图 4.19　进度条的运行效果

4.10　滚动区域(ScrollViewer)

滚动区域(ScrollViewer)控件是表示可包含其他可见元素的可滚动区域。程序界面中的内容通常比手机屏幕的显示区域大。利用ScrollViewer控件可以方便地使应用程序中的内容具备滚动功能。控件的XAML语法如下:

```
<ScrollViewer .../>
<ScrollViewer ...>内容</ScrollViewer>
```

ScrollViewer元素封装一个内容元素和若干ScrollBar控件(最多两个)。范围包括ScrollViewer的所有内容。内容的可见区域称为视区。HorizontalScrollBarVisibility和VerticalScrollBarVisibility属性分别控制垂直和水平ScrollBar控件出现的条件。如果它们设置为Hidden,则可以在代码中使用ComputedHorizontalScrollBarVisibilityProperty和ComputedVerticalScrollBarVisibilityProperty,以便发现其在运行时的实际状态。ScrollViewer控件是针对大内容控件的布局控件。由于该控件内仅能支持一个子控件,所以在多数情况下,ScrollViewer控件都会和Stackpanel、Canvas和Grid相互配合使用。如果遇到内容较长的子控件,ScrollViewer会生成滚动条,提供对内容的滚动支持。ScrollViewer控件的常用属性如表4.12所示。

表4.12　ScrollViewer类常用的属性

名　称	说　明
ComputedHorizontalScrollBarVisibility	获取一个值,该值指示水平ScrollBar是否可见
ComputedVerticalScrollBarVisibility	获取一个值,该值表示垂直ScrollBar是否可见
ExtentHeight	获取ScrollViewer中显示的所有内容的垂直大小
ExtentWidth	获取ScrollViewer中显示的所有内容的水平大小
HorizontalOffset	获取一个值,该值包含滚动内容的水平偏移量
HorizontalScrollBarVisibility	获取或设置一个值,该值指示是否应显示水平ScrollBar
ScrollableHeight	获取一个表示可滚动区域的垂直大小的值,或一个表示范围宽度和视区宽度差别的值
ScrollableWidth	获取一个表示可滚动区域的水平大小的值,或一个表示范围宽度和视区宽度差别的值
VerticalContentAlignment	获取或设置控件内容的垂直对齐方式
VerticalOffset	获取一个值,该值包含滚动内容的垂直偏移量
VerticalScrollBarVisibility	获取或设置一个值,该值指示是否应显示垂直ScrollBar
ViewportHeight	获取一个值,该值包含可见内容的垂直大小
ViewportWidth	获取一个值,该值包含可见内容的水平大小

下面给出查看大图片的示例:创建一个ScrollViewer控件来存放一张大图片,然后在空间内可以左右上下地拖动来查看图片。

代码清单 4-9：查看大图片(源代码：第 4 章\Examples_4_9)

MainPage.xaml 文件主要代码

```
<ScrollViewer Height="300" Width="300"
    VerticalScrollBarVisibility="Auto"
    HorizontalScrollBarVisibility="Auto">
    <ScrollViewer.Content>
        <StackPanel>
        <Image Source="/ScrollViewerPicture;component/cat.jpg"></Image>
    </StackPanel>
    </ScrollViewer.Content>
</ScrollViewer>
```

程序运行的效果如图 4.20 所示。

图 4.20　ScrollViewer 控件的运行效果

下面给出滚动图片的示例：创建一组可以向上或者向下滚动的图片。

代码清单 4-10：滚动图片(源代码：第 4 章\Examples_4_10)

MainPage.xaml 文件主要代码

```
<ScrollViewer Name="scrollViewer1" VerticalScrollBarVisibility="Hidden" VerticalAlignment=
"Top" HorizontalAlignment="Left" Height="600" Width="332">
    <StackPanel Name="stkpnlImage" Width="327" />
</ScrollViewer>
<Button Content="往上" Height="111" HorizontalAlignment="Left" Margin="338,162,0,0"
Name="btnUp" VerticalAlignment="Top" Width="112" FontSize="30"  Click="btnUp_
Click" />
<Button Content="往下" FontSize="30" Height="116" HorizontalAlignment="Left" Margin=
"338,279,0,0" Name="btnDown" VerticalAlignment="Top" Width="112" Click="btnDown_
Click"  />
<Button Content="停止" FontSize="30" Height="116" HorizontalAlignment="Left" Margin=
"338,419,0,0" Name="stop" VerticalAlignment="Top" Width="112" Click="stop_Click" />
```

MainPage.xaml.cs 文件主要代码

```
//往下滚动的定时触发器
private DispatcherTimer tmrDown;
```

```
//往上滚动的定时触发器
private DispatcherTimer tmrUp;
public MainPage()
{
    InitializeComponent();
    //添加图片到 ScrollViewer 里面的 StackPanel 中
    for (int i = 0; i <= 30; i++)
    {
        Image imgItem = new Image();
        imgItem.Width = 200;
        imgItem.Height = 200;
        //4 张图片循环添加到 StackPanel 的子节点上
        if (i % 4 == 0)
        {
            imgItem.Source = (new BitmapImage(new Uri("a.jpg", UriKind.Relative)));
        }
        else if (i % 4 == 1)
        {
            imgItem.Source = (new BitmapImage(new Uri("b.jpg", UriKind.Relative)));
        }
        else if (i % 4 == 2)
        {
            imgItem.Source = (new BitmapImage(new Uri("c.jpg", UriKind.Relative)));
        }
        else
        {
            imgItem.Source = (new BitmapImage(new Uri("d.jpg", UriKind.Relative)));
        }
        this.stkpnlImage.Children.Add(imgItem);
    }
    //初始化 tmrDown 定时触发器
    tmrDown = new DispatcherTimer();
    //每 500 毫秒跑一次
    tmrDown.Interval = new TimeSpan(500);
    //加入每次 tick 的事件
    tmrDown.Tick += tmrDown_Tick;
    //初始化 tmrUp 定时触发器
    tmrUp = new DispatcherTimer();
    tmrUp.Interval = new TimeSpan(500);
    tmrUp.Tick += tmrUp_Tick;
}
void tmrUp_Tick(object sender, EventArgs e)
{
    //将 VerticalOffset - 10,将出现图片将往上滚动的效果
    scrollViewer1.ScrollToVerticalOffset(scrollViewer1.VerticalOffset - 10);
}
void tmrDown_Tick(object sender, EventArgs e)
{
    //先停止往上的定时触发器
     tmrUp.Stop();
```

```
        //将 VerticalOffset + 10,将出现图片将往下滚动的效果
        scrollViewer1.ScrollToVerticalOffset(scrollViewer1.VerticalOffset + 10);
    }
    //往上按钮事件
    private void btnUp_Click(object sender, RoutedEventArgs e)
    {
        //先停止往下的定时触发器
         tmrDown.Stop();
        //tmrUp 定时触发器开始
         tmrUp.Start();
    }
    //往下按钮事件
    private void btnDown_Click(object sender, RoutedEventArgs e)
    {
        //tmrDown 定时触发器开始
        tmrDown.Start();
    }
    //停止按钮事件
    private void stop_Click(object sender, RoutedEventArgs e)
    {
        //停止定时触发器
        tmrUp.Stop();
        tmrDown.Stop();
    }
```

程序运行的效果如图 4.21 所示。

图 4.21　滚动的图片

4.11　列表框(ListBox)

列表框(ListBox)控件是指包含可选项列表。控件的 XAML 语法如下：

<ListBox .../>

ListBox 是一个显示项集合的控件。一次可以显示 ListBox 中的多个项。可以使用 SelectionMode 属性指定 ListBox 是否允许多重选择。ListBox 是 ItemsControl，可以使用 Items 或 ItemsSource 属性设置其内容，可以直接填充 ListBox 控件，也可以将该控件绑定到项集合。ListBox 控件的常用属性如表 4.13 所示。

如何在 ListBox 上显示数据，有两种方式：

(1) 直接在 ListBox 控件 XAML 页面上添加子控件元素。ListBox 控件支持在控件内添加多个控件来展示其内容。

(2) 通过数据绑定的方式来使用 ListBox 控件。将 ListBox 控件的 ItemsSource 属性设置为一个集合，再设置 ItemTemplate 以自定义每个 ListBoxItem 的显示方式。

表 4.13　ListBox 类常用的属性

名　称	说　明
SelectedItems	获取 ListBox 控件的当前选定项的列表
SelectionMode	获取或设置 ListBox 控件的选择行为
DataContext	获取或设置 FrameworkElement 参与数据绑定时的数据上下文（继承自 FrameworkElement）
Items	获取用于生成控件内容的集合（继承自 ItemsControl）
ItemsPanel	获取或设置模板，它定义了控制项的布局的面板(继承自 ItemsControl)
ItemsSource	获取或设置用于生成 ItemsControl 的内容的集合（继承自 ItemsControl）
ItemTemplate	获取或设置用于显示每个项的 DataTemplate（继承自 ItemsControl）

下面给出绑定客户列表的示例：使用 ListBox 控件绑定客户列表。

代码清单 4-11：客户列表(源代码：第 4 章\Examples_4_11)

Customer.cs 文件代码：客户类相当于一个实体类，存储了用户的姓、名和地址

```
public class Customer
{
    public String FirstName { get; set; }
    public String LastName { get; set; }
    public String Address { get; set; }
    public Customer(String firstName, String lastName, String address)
    {
        this.FirstName = firstName;
        this.LastName = lastName;
        this.Address = address;
```

```
    }
}
```

Customers. cs 文件代码：客户信息类存储了客户类对象的信息，继承了 ObservableCollection<T>类

```
public class Customers : ObservableCollection < Customer >
{
    public Customers()
    {
        Add(new Customer("李", "小明","北京市金山路 89 号"));
        Add(new Customer("林", "关关","广东省深圳市深南大道 77 号"));
        Add(new Customer("张", "三三","广东省广州市天河路 80 号"));
        Add(new Customer("李", "思思","上海市中心街 9 好"));
    }
}
```

MainPage. xaml 文件主要代码

```
< phone:PhoneApplicationPage
    …
    xmlns:my = "clr - namespace:ListBoxDemo"
    …>
    <! - 定义客户列表的资源 -->
        < Grid. Resources >
            < my:Customers x:Key = "customers"/>
        </Grid. Resources >
     …
        < Grid x:Name = "ContentPanel" Grid. Row = "1" Margin = "12,0,12,0">
            < ListBox ItemsSource = "{StaticResource customers}" Margin = "0,5, - 12,10">
                < ListBox. ItemTemplate >
                    < DataTemplate >
                        < StackPanel Orientation = "Vertical">
                            < TextBlock Text = "客户姓名: "   Foreground = "Blue"/>
                            < TextBlock Text = "{Binding FirstName}" />
                            < TextBlock Text = "{Binding LastName}" />
                            < TextBlock Text = "客户地址: " Foreground = "Blue" />
                            < TextBlock Text = "{Binding Address}" />
                            < TextBlock Text = " ---------------------------- " />
                        </StackPanel >
                    </DataTemplate >
                </ListBox. ItemTemplate >
            </ListBox >
      …
```

程序运行的效果如图 4.22 所示。

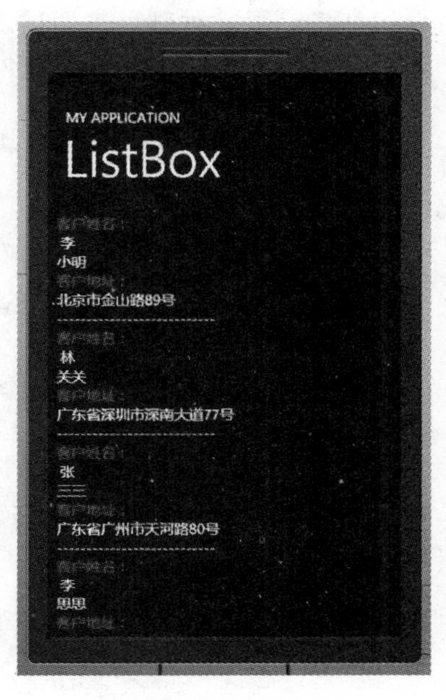

图 4.22　客户列表

4.12　滑动条(Slider)

滑动条(Slider)控件表示一种控件,该控件使用户可以通过沿着一条轨道移动 Thumb 控件来从一个值范围中进行选择。控件的 XAML 语法如下:

```
<Slider .../>
```

Slider 控件可让用户从一个值范围内选择一个值。可以通过设置 Slider 控件的属性来定义该控件。Slider 的方向由 Orientation 属性指定,要么为水平,要么为垂直。Slider 控件的一些常用属性如表 4.14 所示。

表 4.14　Slider 类常用的属性

名　称	说　明
IsDirectionReversed	获取或设置指示增加值方向的值
IsEnabled	获取或设置一个值,该值指示用户是否可以与控件交互
IsFocused	获取一个值,该值指示滑块控件是否有焦点
Orientation	获取或设置 Slider 的方向
SmallChange	获取或设置要从 RangeBase 控件的 Value 加上或减去的 Value
Value	获取或设置范围控件的当前设置,这可能是强制的
Maximum	获取或设置范围元素的最大可能的 Value
LargeChange	获取或设置要从 RangeBase 控件的 Value 加上或减去的值
Minimum	获取或设置范围元素的 Minimum 可能的 Value

下面给出调色板的示例：使用红、蓝、绿三种颜色来调试出综合的色彩。

代码清单 4-12：调色板（源代码：第 4 章\Examples_4_12）

MainPage.xaml 文件主要代码

```
<!-- 设置红色 -->
<TextBlock Grid.Column = "0" Grid.Row = "0" Text = "红色" Foreground = "Red" />
<Slider Name = "redSlider" Grid.Column = "0" Grid.Row = "1" Foreground = "Red" Minimum = "0"
Maximum = "255" ValueChanged = "OnSliderValueChanged" />
<TextBlock Name = "redText" Grid.Column = "0" Grid.Row = "2" Text = "0" Foreground = "Red" />
<!-- 设置绿色 -->
<TextBlock Grid.Column = "1" Grid.Row = "0" Text = "绿色" Foreground = "Green" />
<Slider Name = "greenSlider" Grid.Column = "1" Grid.Row = "1" Foreground = "Green" Minimum = "0"
Maximum = "255" ValueChanged = "OnSliderValueChanged" />
<TextBlock Name = "greenText" Grid.Column = "1" Grid.Row = "2" Text = "0"   Foreground = "Green"   />
<!-- 设置蓝色 -->
<TextBlock Grid.Column = "2" Grid.Row = "0" Text = "蓝色" Foreground = "Blue" />
<Slider Name = "blueSlider" Grid.Column = "2" Grid.Row = "1" Foreground = "Blue" Minimum = "0"
Maximum = "255" ValueChanged = "OnSliderValueChanged" />
<TextBlock Name = "blueText" Grid.Column = "2" Grid.Row = "2" Text = "0" Foreground = "Blue"   />
…
<TextBlock Height = "50" HorizontalAlignment = "Left" Margin = "69,200,0,0" Name = "textBlock1"
Text = "颜色" FontSize = "26" VerticalAlignment = "Top" Width = "318" Grid.Row = "1" />
<Ellipse Grid.Row = "1" Height = "100" HorizontalAlignment = "Left" Margin = "95,74,0,0" Name =
"ellipse1" Stroke = "Black" StrokeThickness = "1" VerticalAlignment = "Top" Width = "224" />
```

MainPage.xaml.cs 文件主要代码

```
public MainPage()
{
    InitializeComponent();
    //三个 slider 控件的初始值
    redSlider.Value = 128;
    greenSlider.Value = 128;
    blueSlider.Value = 128;
}
//slider 控件变化时触发的事件
void OnSliderValueChanged(object sender, RoutedPropertyChangedEventArgs<double> args)
{
    Color clr = Color.FromArgb(255, (byte)redSlider.Value, (byte)greenSlider.Value, (byte)
blueSlider.Value);                                    //获取调剂的颜色
    rect.Fill = new SolidColorBrush(clr);             //用该颜色的笔刷填充大面板
    redText.Text = clr.R.ToString("X2");
    greenText.Text = clr.G.ToString("X2");
    blueText.Text = clr.B.ToString("X2");
}
//手机方向变化触发的事件，重新布局
protected override void OnOrientationChanged(OrientationChangedEventArgs args)
{
    ContentPanel.RowDefinitions.Clear();
    ContentPanel.ColumnDefinitions.Clear();
    //横向位置
    if ((args.Orientation & PageOrientation.Landscape) != 0)
```

```
{
    ColumnDefinition coldef = new ColumnDefinition();
    coldef.Width = new GridLength(1, GridUnitType.Star);
    ContentPanel.ColumnDefinitions.Add(coldef);
    coldef = new ColumnDefinition();
    coldef.Width = new GridLength(1, GridUnitType.Star);
    ContentPanel.ColumnDefinitions.Add(coldef);
    Grid.SetRow(controlGrid, 0);
    Grid.SetColumn(controlGrid, 1);
}
else //竖向位置
{
    RowDefinition rowdef = new RowDefinition();
    rowdef.Height = new GridLength(1, GridUnitType.Star);
    ContentPanel.RowDefinitions.Add(rowdef);
    rowdef = new RowDefinition();
    rowdef.Height = new GridLength(1, GridUnitType.Star);
    ContentPanel.RowDefinitions.Add(rowdef);
    Grid.SetRow(controlGrid, 1);
    Grid.SetColumn(controlGrid, 0);
}
base.OnOrientationChanged(args);
```

程序运行的效果如图 4.23 所示。

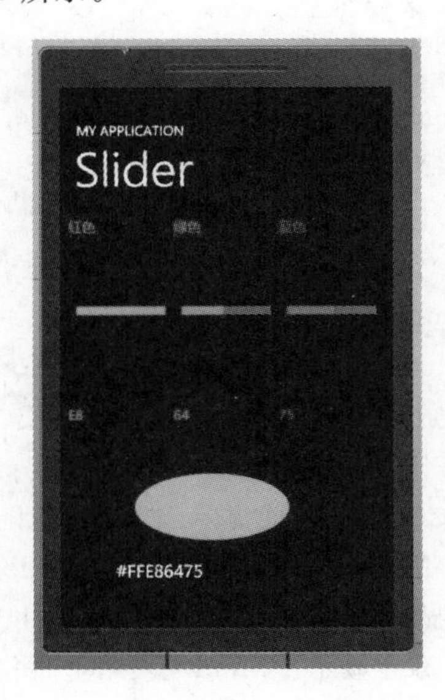

图 4.23　调色板的运行效果

4.13　菜单栏(ApplicationBar)

ApplicationBar 就是所谓的菜单栏控件，包括 Iconbutton 和 Menu 两种类型。Iconbutton 就相当于菜单栏里的工具栏，可以带图标；Menu 就是菜单，但是没有二级菜单。

在创建之前一定要先在 Reference 里加入 Microsoft. Phone. shell，并且在 xaml 页面添加以下代码：

```
xmlns:shell = "clr - namespace:Microsoft.Phone.Shell;assembly = Microsoft.Phone.Shell"
```

WindowsPhone 中的菜单栏最多可以显示 4 个图标按钮。这些图标会自动地被从左向右添加到菜单栏中。如果还有额外的选项可以通过菜单项来添加，这些菜单项默认是不显示的。只有在单击菜单栏右侧的省略号(或省略号下方的区域)时才会显示出来，在电话屏幕的方向改变时，系统会自动处理菜单栏的方向(包括按钮和菜单项)。按钮中的图标应该是 48×48 像素的，其他的尺寸会自动被缩放为 48×48 像素的，不过这通常会导致失真。

下面来看一下下面的一段菜单栏的代码：

```
<phone:PhoneApplicationPage.ApplicationBar>
    <shell:ApplicationBar IsVisible = "True" IsMenuEnabled = "True">
        <shell:ApplicationBarIconButton IconUri = "/Images/appbar_button1.png" Text = "Button 1"/>
        <shell:ApplicationBarIconButton IconUri = "/Images/appbar_button2.png" Text = "Button 2"/>
        <shell:ApplicationBar.MenuItems>
            <shell:ApplicationBarMenuItem Text = "MenuItem 1"/>
            <shell:ApplicationBarMenuItem Text = "MenuItem 2"/>
        </shell:ApplicationBar.MenuItems>
    </shell:ApplicationBar>
</phone:PhoneApplicationPage.ApplicationBar>
```

创建菜单栏时需要注意：

(1) xmlns：shell ＝ " clr-namespace：Microsoft. Phone. Shell；assembly ＝ Microsoft. Phone"为 ApplicationBar 定义字首使用，引入 Microsoft. Phone. Shell 中的主要元素，提供元件、事件的使用。

(2) shell：ApplicationBar：负责设定 application bar 是否要显示、是否具有 menu 选项或按钮的出现。具有 StateChanged 的事件可以监测。

(3) shell：ApplicationBarIconButton：用于 ApplicationBar 集合中，建立按钮选项、icon、Click 事件等。最多可以放 4 个项目。要创建 IconBar，就一定要有 Icon，Windows Phone 对 Icon 的要求比较严——一定是 48×48 像素的. png 格式的图片。注意图片的 Build Action 属性需要设置为 Content 才能显示出来，否则将会出现一个错误的图片。

(4) shell：ApplicationBar. MenuItems：指定 ApplicationBar 要出现的 Menu 项目，内部会包住多个 Item 来显示，纯显示文字，也可以加入 ApplicationBarIconButton 项目，但只能显示其 Text 属性，Icon 内容不会显示。Menus 采用垂直排列。最多可以放 50 个项目。

以上简单说明了 ApplicationBar 常见的 Tag，但是事实上 ApplicationBar 分成以下两种：

1) Local Applibation Bar

Local Applibation Bar 是指特定页面的菜单栏，项目中 MainPage. xaml 中默认的菜单栏就是 Local Application Bar，它属于该. xaml 文件专用。需要注意的是，如果该. xmal 文件在 ＜ phone：PhoneApplicationPage ＞ 中设定了 ApplicationBar ＝ "{ StaticResource globalApplicationBar}"，要记得先删掉，不然会出现错误。

2）Global Application Bar

Global Application Bar 是指全局菜单栏，如果想要支持不同的.xaml 文件也能存在相同的 Application Bar，可以到 App.xaml 文件中，在<Application.Resources>集合中加入<shell:ApplicationBar />，另外要记得设定 x:Key 属性，这样其他的.xaml 文件才有办法找到全局统一的菜单栏。设定好 App.xaml 之后，可以到其他的.xaml 文件中加上如下所提的语法：在<phone:PhoneApplicationPage>中设定 ApplicationBar="{StaticResource globalApplicationBar}"。

下面给出菜单栏的示例：测试菜单栏的各种事件。

代码清单 4-13：菜单栏（源代码：第 4 章\Examples_4_13）

MainPage.xaml 文件主要代码

```
<phone:PhoneApplicationPage.ApplicationBar>
    <shell:ApplicationBar
        IsVisible="True" IsMenuEnabled="True" Opacity="0.5"
        StateChanged="ApplicationBar_StateChanged">
        <shell:ApplicationBarIconButton  IconUri="icon\ie.png" Text="浏览器" Click=
"ApplicationBarIconButton_Click"/>
        <shell:ApplicationBarIconButton IconUri="icon\phone.png" Text="电话" Click=
"ApplicationBarIconButton_Click_1"/>
        <shell:ApplicationBarIconButton IconUri="icon\about.png" Text="关于" Click=
"ApplicationBarIconButton_Click_2"/>
        <shell:ApplicationBar.MenuItems>
            <shell:ApplicationBarMenuItem Text="MenuItem 1" Click="ApplicationBarMenuItem_
Click"/>
            <shell:ApplicationBarMenuItem Text="MenuItem 2" Click="ApplicationBarMenuItem_
Click_1"/>
        </shell:ApplicationBar.MenuItems>
    </shell:ApplicationBar>
</phone:PhoneApplicationPage.ApplicationBar>
```

MainPage.xaml.cs 文件主要代码

```
private void ApplicationBarIconButton_Click(object sender, EventArgs e)
{
    textBlock1.Text = "你单击了浏览器菜单";
}
private void ApplicationBarIconButton_Click_1(object sender, EventArgs e)
{
    textBlock1.Text = "你单击了电话菜单";
}
private void ApplicationBarIconButton_Click_2(object sender, EventArgs e)
{
    textBlock1.Text = "你单击了关于菜单";
}
private void ApplicationBarMenuItem_Click(object sender, EventArgs e)
```

```
{
    textBlock1.Text = "你单击了 MenuItem 1 菜单";
}
private void ApplicationBarMenuItem_Click_1(object sender, EventArgs e)
{
    textBlock1.Text = "你单击了 MenuItem 2 菜单";
}
        private void ApplicationBar_StateChanged(object sender, Microsoft.Phone.Shell.
ApplicationBarStateChangedEventArgs e)
        {
            textBlock1.Text = "你打开了 Menu 菜单列表";
        }
```

程序运行的效果如图 4.24 所示。

图 4.24　菜单栏运行的效果

第5章

布局管理

布局管理是从一个整体的角度去把握手机应用的界面设计，尽管手机应用的界面相对于Web程序和电脑桌面的程序界面小了一些，但是没有一个好的界面布局，也会影响整个应用的用户体验效果。Windows Phone手机应用程序的界面布局容器主要有Grid、Canvas和StackPanel。在一个手机应用里面所有元素的最顶层必须是一个容器（如Grid、Canvas、StackPanel等），然后在容器中摆放元素，容器中也可能包含容器。从布局的角度来说，外层的容器管理里面的容器，而里面的容器也管理它们里面的容器，一层层地递归下去，最后展现出来的就是整个应用程序的布局效果。

每一个布局的容器都会有自己的规则，它需要按照这些规则去管理它里面的控件。这里的容器也一样，容器拥有完全的分配权，不过这里容器不仅仅是分配空间，还决定元素的位置，因为空间总是跟位置相关的。也就是说，容器说想给控件多大空间就只有多大的空间可使用，容器想让控件摆在什么位置，控件就得乖乖呆在什么位置。

在布局方面，Windows Phone还提供了两个非常有用的控件，分别是枢轴控件Pivot和全景视图控件Panorama，使用这两个控件可以很方便地在一个页面上布局出多个视图，扩充了页面的展示范围，并且具有良好的用户体验。

Windows Phone中常见的布局方式有如下几种：

(1) 网格布局（Grid容器）：按照行列方式布局程序的界面。定义一个区域，在此区域内，用户可以使用相对于Canvas区域的坐标显式定位子元素。

(2) 绝对布局（Canvas容器）：按照绝对坐标来布局程序的界面。

(3) 堆放布局（StackPanel容器）：按照垂直或者水平方式布局程序的界面。

(4) 枢轴视图布局（Pivot控件）：通过类似页签的方式在一个页面上展示多个视图。

(5) 全景视图布局（Panorama控件）：通过左右滑动的方式在一个页面显示多个视图。

用户可以根据自己应用的实际情况来选择布局的容器，可以选择一种或者几种结合起来。

5.1 网格布局（Grid）

网格布局（Grid）是一个类似于HTML里面的Table的标签，它定义了一个表格，然后设置表格里面的行和列，在HTML的Table里面是根据tr和td表示列和行的，而Grid是通过附加属性 Grid. Row、Grid. Column、Grid. RowSpan、Grid. ColumnSpan 来决定列和行的大小位置。

例如，在 HTML 中的 Table 布局如下：

```
<table border="1">
    <tr>
        <td>第 1 行第 1 列</td>
        <td>第 1 行第 2 列</td>
        <td>第 1 行第 3 列</td>
    </tr>
    <tr>
        <td>第 2 行第 1 列</td>
        <td>第 2 行第 2 列</td>
        <td>第 2 行第 3 列</td>
    </tr>
</table>
```

用 Wineos Phone 的 Grid 进行布局的时候，也是一样的道理，同样需要制定 Grid 的行和列。不同的是，Grid 是先指定，后使用；而 Table 是边指定，边使用。下面是一个使用 Grid 的例子，使其达到与 Table 同样的布局效果：

```
<Grid x:Name="ContentPanel">
    <Grid.RowDefinitions>
        <RowDefinition Height="Auto"/>
        <RowDefinition Height="Auto"/>
    </Grid.RowDefinitions>
    <Grid.ColumnDefinitions>
        <ColumnDefinition Width="Auto"/>
        <ColumnDefinition Width="Auto"/>
        <ColumnDefinition Width="Auto"/>
    </Grid.ColumnDefinitions>
    <TextBlock Text="第 1 行第 1 列"    Grid.Row="0"   Grid.Column="0"   />
    <TextBlock Text="第 1 行第 2 列"  Grid.Row="0"   Grid.Column="1"   />
    <TextBlock Text="第 1 行第 3 列"   Grid.Row="0"   Grid.Column="2"   />
    <TextBlock Text="第 1 行第 1 列"    Grid.Row="1"   Grid.Column="0"   />
    <TextBlock Text="第 1 行第 2 列"  Grid.Row="1"   Grid.Column="1"   />
    <TextBlock Text="第 1 行第 3 列"   Grid.Row="1"   Grid.Column="2"   />
</Grid>
```

其显示的效果如图 5.1 所示。

下面来看一下 Grid 中几个重要属性的设置：

1）RowDefinitions 和 ColumnDefinitions

这两个属性主要是来指定 Grid 控件的行数和列数。内部嵌套几个 Definition，就代表这个 Grid 有几行几列。

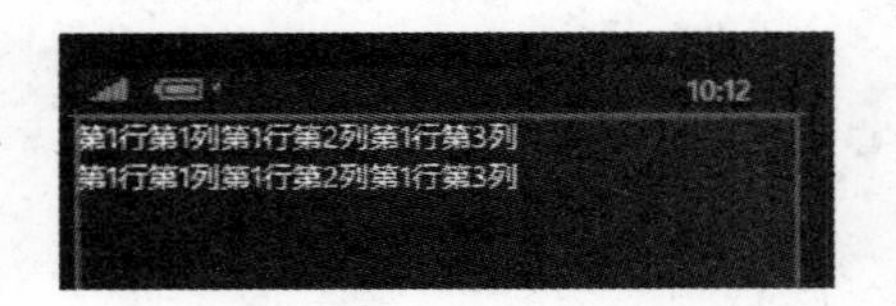

图 5.1　Grid 排版效果

2）Grid. Row 和 Grid. Column

当使用其他控件内置于 Grid 中时，需要使用

Grid.Row 和 Grid.Column 来指定它所在行和列。

3）Height="Auto"和 Width="Auto"

Height="Auto" 表示高度自动适应，Width="Auto"表示宽度自动适应，这种情况下，Grid 的宽度和高度便随着内部内容的大小而自动改变。当然，也可以设置成绝对的宽度和高度。

4）VerticalAlignment 和 HorizontalAlignment

可以根据 VerticalAlignment 和 HorizontalAlignment 属性来设置行和列的位置，这样会自动根据 Grid 容器的大小来自动调整行和列的大小以至于使得整个行和列都铺满了 Grid 的所有的控件。下面是一个使用 VerticalAlignment 和 HorizontalAlignment 属性布局的实例代码。

```
<Grid x:Name = "ContentPanel" ShowGridLines = "True">
    <TextBlock Text = "左上方"
               VerticalAlignment = "Top"
               HorizontalAlignment = "Left" />
    <TextBlock Text = "中上方"
               VerticalAlignment = "Top"
               HorizontalAlignment = "Center" />
    <TextBlock Text = "右上方"
               VerticalAlignment = "Top"
               HorizontalAlignment = "Right" />
    <TextBlock Text = "左中心"
               VerticalAlignment = "Center"
               HorizontalAlignment = "Left" />
    <TextBlock Text = "中心"
               VerticalAlignment = "Center"
               HorizontalAlignment = "Center" />
    <TextBlock Text = "右中心"
               VerticalAlignment = "Center"
               HorizontalAlignment = "Right" />
    <TextBlock Text = "左下方"
               VerticalAlignment = "Bottom"
               HorizontalAlignment = "Left" />
    <TextBlock Text = "中下方"
               VerticalAlignment = "Bottom"
               HorizontalAlignment = "Center"  />
    <TextBlock Text = "右下方"
               VerticalAlignment = "Bottom"
               HorizontalAlignment = "Right" />
</Grid>
```

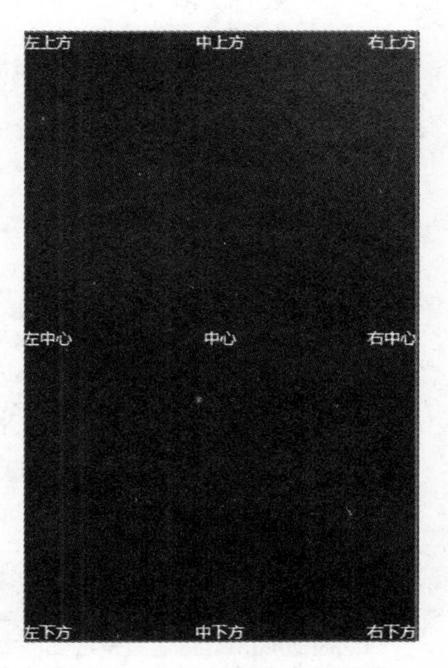

图 5.2　Grid 的整体布局效果

布局的效果如图 5.2 所示。

下面给出简易的计算器的示例：使用 Grid 布局来设计计算器的界面，实现简单的计算功能。

代码清单 5-1：计算器(源代码：第 5 章\Examples_5_1)

MainPage.xaml 文件主要代码

```
<Grid Name = "MyGrid" Height = "704" Width = "440">
    <Grid.ColumnDefinitions>
        <ColumnDefinition />
        <ColumnDefinition />
        <ColumnDefinition />
        <ColumnDefinition />
    </Grid.ColumnDefinitions>
    <Grid.RowDefinitions>
        <RowDefinition Height = "63 * " />
        <RowDefinition Height = "170 * " />
        <RowDefinition Height = "119 * " />
        <RowDefinition Height = "117 * " />
        <RowDefinition Height = "119 * " />
        <RowDefinition Height = "117 * " />
    </Grid.RowDefinitions>
    <!-- 数字按键的布局 -->
    <Button Name = "B7" Click = "DigitBtn_Click" Grid.Column = "0" Grid.Row = "2" Content = "7" />
    <Button Name = "B8" Click = "DigitBtn_Click" Grid.Column = "1" Grid.Row = "2" Content = "8" />
    <Button Name = "B9" Click = "DigitBtn_Click" Grid.Column = "2" Grid.Row = "2" Content = "9" />
    <Button Name = "B4" Click = "DigitBtn_Click" Grid.Column = "0" Grid.Row = "3" Content = "4" />
    <Button Name = "B5" Click = "DigitBtn_Click" Grid.Column = "1" Grid.Row = "3" Content = "5" />
    <Button Name = "B6" Click = "DigitBtn_Click" Grid.Column = "2" Grid.Row = "3" Content = "6" />
    <Button Name = "B1" Click = "DigitBtn_Click" Grid.Column = "0" Grid.Row = "4" Content = "1" />
    <Button Name = "B2" Click = "DigitBtn_Click" Grid.Column = "1" Grid.Row = "4" Content = "2" />
    <Button Name = "B3" Click = "DigitBtn_Click" Grid.Column = "2" Grid.Row = "4" Content = "3" />
    <Button Name = "B0" Click = "DigitBtn_Click" Grid.Column = "0" Grid.Row = "5" Content = "0" />
    <!-- 加减乘除操作按键的布局 -->
    <Button Name = "Plus" Click = "OperationBtn_Click" Grid.Column = "3" Grid.Row = "2" Content = " + " />
    <Button Name = "Minus" Click = "OperationBtn_Click" Grid.Column = "3" Grid.Row = "3"
Content = " - " />
    <Button Name = "Multiply" Click = "OperationBtn_Click" Grid.Column = "3" Grid.Row = "4"
Content = " * " />
    <Button Name = "Divide" Click = "OperationBtn_Click" Grid.Column = "3" Grid.Row = "5"
Content = "/" />
    <!-- 等于和删除按键的布局 -->
    <Button Name = "Del" Grid.Column = "2" Grid.Row = "5" Content = "删除" Click = "Del_Click" />
    <Button Name = "Result" Grid.Column = "1" Grid.Row = "5" Content = " = " Click = "Result_
Click" />
    <!-- OperationResult 显示输入的数字 -->
    <TextBlock Name = "OperationResult" FontSize = "120" Grid.Row = "1" Margin = "6,17,10,17"
Grid.ColumnSpan = "4" HorizontalAlignment = "Right"></TextBlock>
    <!-- InputInformation 显示输入的公式 -->
    <TextBlock Name = "InputInformation" Grid.Row = "0" Margin = "6,12,10,11" Grid.ColumnSpan =
"4" HorizontalAlignment = "Right"></TextBlock>
</Grid>
```

MainPage. xaml. cs 文件代码

```
using System;
using System.Windows;
using System.Windows.Controls;
using Microsoft.Phone.Controls;
namespace Calculator
{
    public partial class MainPage : PhoneApplicationPage
    {
        //之前一次你按下的运算符
        private string Operation = "";
        //结果
        private int num1 = 0;
        public MainPage()
        {
            InitializeComponent();
        }
        //数字按键的事件
        private void DigitBtn_Click(object sender, RoutedEventArgs e)
        {
            if (Operation == "=")
            {
                OperationResult.Text = "";
                InputInformation.Text = "";
                Operation = "";
                num1 = 0;
            }
            //获取你按下的按钮的文本内容,即你按下的数字
            string s = ((Button)sender).Content.ToString();
            OperationResult.Text = OperationResult.Text + s;
            InputInformation.Text = InputInformation.Text + s;
        }
        //加减乘除按键的事件
        private void OperationBtn_Click(object sender, RoutedEventArgs e)
        {
            if (Operation == "=")
            {
                InputInformation.Text = OperationResult.Text;
                Operation = "";
            }
            string s = ((Button)sender).Content.ToString();
            //公式显示
            InputInformation.Text = InputInformation.Text + s;
            //运算
            OperationNum(s);
            //清空你之前输入的数字
            OperationResult.Text = "";
        }
        //按下等于运算符的事件
```

```
private void Result_Click(object sender, RoutedEventArgs e)
{
    OperationNum("=");
    //显示结果
    OperationResult.Text = num1.ToString();
}
//删除之前的所有输入
private void Del_Click(object sender, RoutedEventArgs e)
{
    OperationResult.Text = "";
    InputInformation.Text = "";
    Operation = "";
    num1 = 0;
}
//通过运算符进行计算
private void OperationNum(string s)
{
    //输入的数字不为空,则进行运算,否则只是保存最后一次按下的运算符
    if (OperationResult.Text != "")
    {
        switch (Operation)
        {
            case "":
                num1 = Int32.Parse(OperationResult.Text);
                Operation = s;
                break;
            case "+":
                num1 = num1 + Int32.Parse(OperationResult.Text);
                Operation = s;
                break;
            case "-":
                num1 = num1 - Int32.Parse(OperationResult.Text);
                Operation = s;
                break;
            case "*":
                num1 = num1 * Int32.Parse(OperationResult.Text);
                Operation = s;
                break;
            case "/":
                if (Int32.Parse(OperationResult.Text) != 0)
                    num1 = num1 / Int32.Parse(OperationResult.Text);
                else num1 = 0;
                Operation = s;
                break;
            default: break;
        }
    }
    else
    {
        Operation = s;
```

```
                }
            }
        }
    }
```

程序运行的效果如图 5.3 所示。

图 5.3 计算器

5.2 堆放布局(StackPanel)

堆放布局(StackPanel)的方式是将子元素排列成一行(可沿水平或垂直方向)。StackPanel 的规则是：根据附加属性，让元素横着排列或者竖着排列。

StackPanel 为启用布局的 Panel 元素之一。在特定情形下，比如要将一组对象排列在竖直或水平列表中，StackPanel 就能发挥很好的作用。下面来阐述 StackPanel 中的两个很重要的属性：

1) Orientation 属性

设置 Orientation 属性可确定列表的方向。Orientation ="Horizontal"表示沿水平方向放置 StackPanel 里面的子元素，Orientation =" Vertical"表示沿垂直方向放置 StackPanel 里面的子元素，Orientation 属性的默认值为 Vertical，即默认是垂直方向放置的。

2) VerticalAlignment 属性

VerticalAlignment 属性表示元素垂直方向的位置，VerticalAlignment="Bottom"表示子元素放在 StackPanel 容器的底部，VerticalAlignment =" Center" 表示子元素放在

StackPanel 容器的中间，VerticalAlignment＝"Stretch"表示子元素拉伸铺放在 StackPanel 容器，VerticalAlignment＝" Top "表示子元素放在 StackPanel 容器的上方。VerticalAlignment 属性的默认值为 Stretch。

3）HorizontalAlignment 属性

HorizontalAlignment 属性表示元素水平方向的位置，HorizontalAlignment＝"Left"表示子元素放在 StackPanel 容器的左边，HorizontalAlignment＝"Center"表示子元素放在 StackPanel 容器的中间，HorizontalAlignment ＝" Stretch" 表示子元素拉伸铺放在 StackPanel 容器，HorizontalAlignment＝"Right"表示子元素放在 StackPanel 容器的右边。HorizontalAlignment 属性的默认值为 Stretch。

下面是 StackPanel 两种布局的实现和显示的效果。

1）垂直布局：设置 Orientation＝"Vertical"　默认的布局方式是垂直方式

XAML 代码示例：

```
< StackPanel Orientation = " Vertical" VerticalAlignment = " Stretch" HorizontalAlignment =
"Center">
    < Button  Content = "按钮 1" />
    < Button  Content = "按钮 2" />
    < Button  Content = "按钮 3" />
    < Button  Content = "按钮 4" />
</StackPanel >
```

显示效果如图 5.4 所示。

2）水平布局：设置 Orientation＝"Horizontal"

XAML 代码示例：

```
<StackPanel Orientation = "Horizontal" VerticalAlignment = "Top" HorizontalAlignment = "Left">
    < Button  Content = "按钮 1" />
    < Button  Content = "按钮 2" />
    < Button  Content = "按钮 3" />
    < Button  Content = "按钮 4" />
</StackPanel >
```

显示效果如图 5.5 所示。

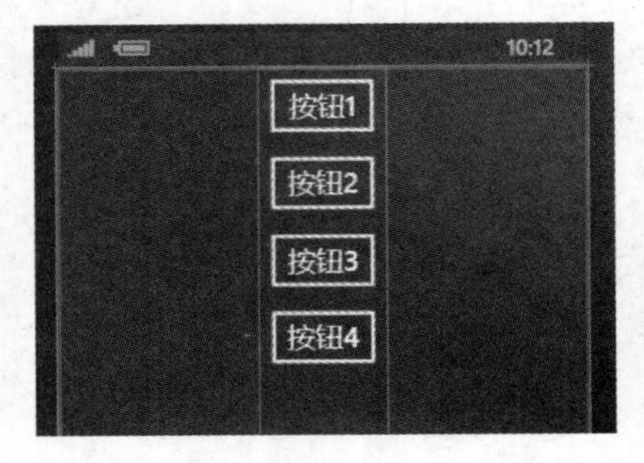

图 5.4　StackPanel 按钮垂直排序

图 5.5　StackPanel 按钮水平排序

下面给出自动折行控件的示例：一个会根据空格自动折行的 TextBlock 控件，这个控件的原理是用了 StackPanel 容器的垂直排列属性来进行折行，先将一个文本根据空格分解成多个文本来创建多个 TextBlock 控件，然后将多个 TextBlock 控件放进 StackPanel 容器里面会产生根据空格自动折行的效果，同时外层使用 ScrollViewer 控件来包住。

代码清单 5-2：自动折行控件（源代码：第 5 章\Examples_5_2）

generic.xaml 文件代码：自动折行控件的样式设置

```
<ResourceDictionary
    xmlns="http://schemas.microsoft.com/winfx/2006/xaml/presentation"
    xmlns:x="http://schemas.microsoft.com/winfx/2006/xaml"
    xmlns:local="clr-namespace:Phone.Controls;assembly=Phone.Controls">
    <!--设置演示的目标对象是 ScrollableTextBlock 控件-->
    <Style TargetType="local:ScrollableTextBlock" >
        <!--设置前景颜色-->
        <Setter Property="Foreground" Value="{StaticResource PhoneForegroundBrush}"/>
        <!--设置背景颜色-->
        <Setter Property="Background" Value="Transparent"/>
        <!--设置字体大小-->
        <Setter Property="FontSize" Value="{StaticResource PhoneFontSizeMedium}"/>
        <Setter Property="Padding" Value="0"/>
        <Setter Property="Width" Value="200" />
        <Setter Property="Height" Value="70" />
        <!--设置模板-->
        <Setter Property="Template">
            <Setter.Value>
                <ControlTemplate TargetType="local:ScrollableTextBlock">
                    <ScrollViewer x:Name = "ScrollViewer" Foreground = "{TemplateBinding
Foreground}"
                                  Background="{TemplateBinding Background}" BorderBrush=
"{TemplateBinding BorderBrush}"
                                  BorderThickness = "{TemplateBinding BorderThickness}"
Padding="{TemplateBinding Padding}">
                        <StackPanel Orientation="Vertical" x:Name="StackPanel" />
                    </ScrollViewer>
                </ControlTemplate>
            </Setter.Value>
        </Setter>
    </Style>
</ResourceDictionary>
```

ScrollableTextBlock.cs 文件代码：ScrollableTextBlock 类通过继承 Control 类来实现自定义控件

```
using System;
using System.Windows;
using System.Windows.Controls;
using System.Text;
namespace Phone.Controls
{
```

```
public class ScrollableTextBlock : Control
{
    private StackPanel stackPanel;
    public ScrollableTextBlock()
    {
        //获取 ScrollableTextBlock 样式,ScrollableTextBlock 样式在 generic.xaml 文件中
        //进行定义
        this.DefaultStyleKey = typeof(ScrollableTextBlock);
    }
    //注册一个 Text 属性
    public static readonly DependencyProperty TextProperty =
        DependencyProperty.Register(
            "Text",
            typeof(string),
            typeof(ScrollableTextBlock),
            new PropertyMetadata("ScrollableTextBlock", OnTextPropertyChanged));
    //定义控件的 Text 属性
    public string Text
    {
        get
        {
            return (string)GetValue(TextProperty);
        }
        set
        {
            SetValue(TextProperty, value);
        }
    }
    //Text 属性改变触发的事件定义
    private static void OnTextPropertyChanged(DependencyObject d, DependencyPropertyChangedEventArgs e)
    {
        ScrollableTextBlock source = (ScrollableTextBlock)d;
        string value = (string)e.NewValue;
        source.ParseText(value);
    }
    //重载 OnApplyTemplate 方法
    public override void OnApplyTemplate()
    {
        base.OnApplyTemplate();
        this.stackPanel = this.GetTemplateChild("StackPanel") as StackPanel;
        this.ParseText(this.Text);
    }
    //解析控件的 Text 属性的赋值
    private void ParseText(string value)
    {
        string[] textBlockTexts = value.Split(' ');
        if (this.stackPanel == null)
        {
            return;
        }
```

```
            //清除之前的 stackPanel 容器的子元素
            this.stackPanel.Children.Clear();
            //重新添加 stackPanel 的子元素
            for (int i = 0; i< textBlockTexts.Length; i++)
            {
                TextBlock textBlock = this.GetTextBlock();
                textBlock.Text = textBlockTexts[i].ToString();
                this.stackPanel.Children.Add(textBlock);
            }
        }
        //创建一个自动折行的 TextBlock
        private TextBlock GetTextBlock()
        {
            TextBlock textBlock = new TextBlock();
            textBlock.TextWrapping = TextWrapping.Wrap;
            textBlock.FontSize = this.FontSize;
            textBlock.FontFamily = this.FontFamily;
            textBlock.FontWeight = this.FontWeight;
            textBlock.Foreground = this.Foreground;
            textBlock.Margin = new Thickness(0, 0, 0, 0);
            return textBlock;
        }
    }
}
```

MainPage.xaml 文件主要代码

```
<phone:PhoneApplicationPage
    …
    xmlns:my = "clr-namespace:Phone.Controls;assembly = Phone.Controls">
    <!-- 应用自定义的控件 -->
    <Grid x:Name = "LayoutRoot" Background = "Transparent">
        <Grid.RowDefinitions>
            <RowDefinition Height = "Auto"/>
            <RowDefinition Height = "*"/>
        </Grid.RowDefinitions>
        <StackPanel x:Name = "TitlePanel" Grid.Row = "0" Margin = "12,17,0,28">
            <TextBlock x:Name = "tt" Text = "会根据空格自动折行的 TextBlock 控件" Style =
"{StaticResource PhoneTextNormalStyle}"/>
            <TextBlock x:Name = "PageTitle" Text = "test page" Margin = "9,-7,0,0" Style =
"{StaticResource PhoneTextTitle1Style}"/>
        </StackPanel>
        <!-- 使用自定义的控件 -->
        <Grid x:Name = "ContentPanel" Grid.Row = "1" Margin = "12,0,12,0">
            <my:ScrollableTextBlock  Text = "ScrollableTextBlock"
                HorizontalAlignment = "Left"  Name = "scrollableTextBlock1"
                  VerticalAlignment = "Top" Height = "618" Width = "427" Margin = "12,-11,0,0" />
        </Grid>
    </Grid>
</phone:PhoneApplicationPage>
```

MainPage.xaml.cs 文件代码

```
using Microsoft.Phone.Controls;
namespace ScrollableTextBoxTest
{
    public partial class MainPage : PhoneApplicationPage
    {
        public MainPage()
        {
            InitializeComponent();
            string text = "购物清单如下：牛奶 咖啡 饼干 苹果 香蕉 苹果 茶叶 蜂蜜 纯净水 猪
肉 鲤鱼 牛肉 橙汁";
            scrollableTextBlock1.Text = text;
            scrollableTextBlock1.FontSize = 30;
        }
    }
}
```

程序运行的效果如图 5.6 所示。

图 5.6 TextBlock 自动排列

5.3 绝对布局(Canvas)

绝对布局(Canvas)画布定义一个区域，在此区域内，用户可以使用相对于 Canvas 区域的坐标显式定位子元素。Canvas 就像一张油布一样，所有的控件都可以堆到这张布上，采用绝对坐标定位，可以使用附加属性(AttachedProperty)对 Canvas 中的元素进行定位，通过附加属性指定控件相对于其直接父容器Canvas控件的上、下、左、右坐标的位置。

Canvas的规则是：读取附加属性Canvas. Left、Canvas. Right、Canvas. Top、Canvas. Bottom，并以此来决定元素的位置。可以嵌套Canvas对象。嵌套对象时，每个对象使用的坐标都是相对于直接包含它们的Canvas而言的。每个子对象都必须是UIElement。在XAML中，将子对象声明为对象元素，这些元素是Canvas对象元素的内部XML。在代码中，可以通过获取由Children属性访问的集合来操作Canvas子对象的集合。由于Canvas为UIElement类型，因此可以嵌套Canvas对象。很多情况下，Canvas仅仅用作其他对象的容器，而没有任何可见属性。如果满足以下任一条件，Canvas即不可见：

(1) Height属性等于0；

(2) Width属性等于0；

(3) Background属性等于null；

(4) Opacity属性等于0；

(5) Visibility属性等于Collapsed。

下面介绍Canvas的一些常用的使用技巧。

1) 添加一个对象到Canvas上

Canvas包含和安置其他对象，要添加一个对象到一个Canvas，将其插入到<Canvas>标签之间。下面举例添加一个Button对象到一个Canvas。

XAML代码如下：

```
<Canvas>
    <Button  Content="按钮 1"  Height="200" Width="200"/>
</Canvas>
```

2) 安置一个对象

在Canvas中安置一个对象，要设置对象的Canvas. Left和Canvas. Top附加属性。附加的Canvas. Left属性标识了对象和其父级Canvas的左边缘距离，附加的Canvas. Top属性标识了子对象和其父级Canvas的上边缘距离。下面的示例使用了前例中的Button，并将其从Canvas的左边移动30像素，从上边移动30像素。

XAML代码如下：

```
<Canvas>
    <Button  Content="按钮 1"  Canvas.Left="30" Canvas.Top="30"
  Height="200" Width="200"/>
</Canvas>
```

如图5.7所示，突出显示了Canvas坐标系统中Button的位置。

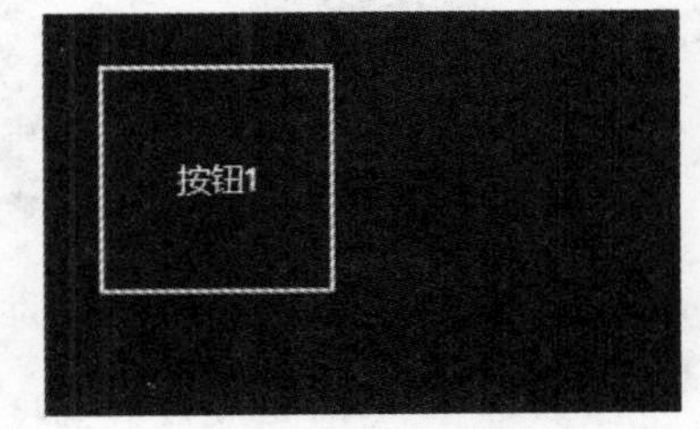

图5.7　Canvas按钮

3) 使用z-order属性产生叠加的效果

默认情况下，Canvas对象的z-order决定于它们声明的顺序。后声明的对象出现在最先声明的对象的前面。下面的示例创建3个Ellipse对象。观察最后声明的 Button(按钮 3)在最前面，位于其他 Ellipse 对象的前面。

XAML 代码如下：

```
<Canvas>
    <Button Content = "按钮 1" Canvas.Left = "30" Canvas.Top = "30" Background = "Blue"
  Height = "200" Width = "200"/>
    <Button Content = "按钮 2" Canvas.Left = "100" Canvas.Top = "100" Background = "Red"
  Height = "200" Width = "200"/>
    <Button Content = "按钮 3" Canvas.Left = "170" Canvas.Top = "170" Background = "Yellow"
  Height = "200" Width = "200"/>
</Canvas>
```

显示的叠加效果如图 5.8 所示。

可以通过设置 Canvas 对象的 Canvas. ZIndex 附加属性来改变这个行为。越高的值越靠近前台；较低的值越远离前台。下面的示例类似于先前的一个，只是 Ellipse 对象的 z-order 颠倒了：最先声明的 Ellipse(银色的那个)现在在前面了。

XAML 代码如下：

```
<Canvas>
    <Button Canvas.ZIndex = "3" Content = "按钮 1" Canvas.Left = "30" Canvas.Top = "30"
Background = "Blue"
  Height = "200" Width = "200"/>
    <Button Canvas.ZIndex = "2" Content = "按钮 2" Canvas.Left = "100" Canvas.Top = "100"
Background = "Red"
  Height = "200" Width = "200"/>
    <Button Canvas.ZIndex = "1" Content = "按钮 3" Canvas.Left = "170" Canvas.Top = "170"
Background = "Yellow"
  Height = "200" Width = "200"/>
</Canvas>
```

显示的叠加效果如图 5.9 所示。

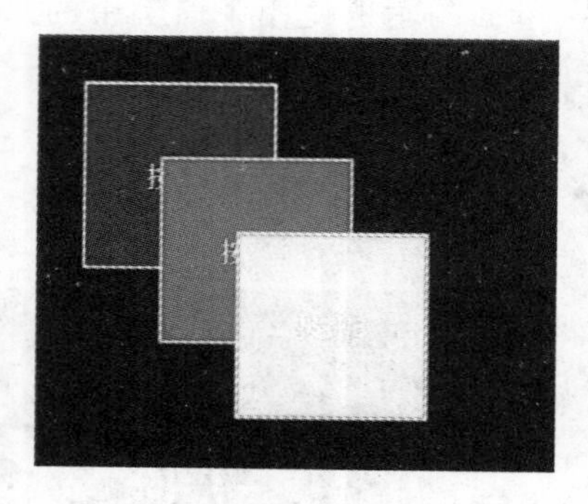

图 5.8　Canvas 叠加按钮布局

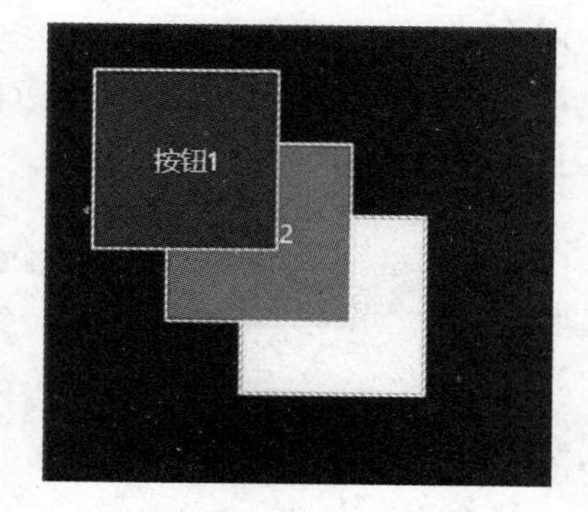

图 5.9　Canvas 叠加按钮布局

4）控制 Canvas 的宽度和高度

Canvas 和许多其他元素都有 Width 和 Height 属性用以标识大小。下面的示例创建了一个 200 像素宽和 200 像素高的 Button。需要注意的是，百分比大小不被 Canvas 所支持。下面的示例展示了设置父级 Canvas 的 Width 和 Height 到 200 像素，并给它一个石灰绿的背景。

XAML 代码如下：

```
<Canvas Width = "200" Height = "200" Background = "LimeGreen">
    <Button  Content = "按钮 1"
             Canvas.Left = "30" Canvas.Top = "30"  Background = "Blue"
             Height = "200" Width = "200"/>
</Canvas>
```

显示的效果如图 5.10 所示，需要注意的是 Button 控件并没有因为超出 Canvas 的范围而被剪切。

5）嵌套 Canvas 对象

Canvas 可以包含其他 Canvas 对象。下面的示例创建了一个 Canvas，它自己包含两个其他 Canvas 对象。

XAML 代码如下：

```
<Canvas Width = "200" Height = "200" Background = "White">
    <Canvas Height = "50" Width = "50" Canvas.Left = "30" Canvas.Top = "30"
            Background = "blue"/>
    <Canvas Height = "50" Width = "50" Canvas.Left = "130" Canvas.Top = "30"
            Background = "red"/>
</Canvas>
```

显示的效果如图 5.11 所示。

图 5.10　Canvas 范围的测试

图 5.11　嵌套 Canvas 对象

下面给出渐变矩形的示例：使用 Canvas 画布渐变的叠加而产生一种立体感的效果的矩形。

代码清单 5-3：渐变矩形(源代码：第 5 章\Examples_5_3)

MainPage.xaml 文件主要代码

```
<Canvas Background = "White">
    <Canvas Height = "400" Width = "400" Canvas.Left = "50" Canvas.Top = "50" Background =
"Gray" Opacity = "0.1" />
```

```
    <Canvas Height = "360" Width = "360" Canvas.Left = "70" Canvas.Top = "70" Background =
"Gray" Opacity = "0.2" />
    <Canvas Height = "320" Width = "320" Canvas.Left = "90" Canvas.Top = "90" Background =
"Gray" Opacity = "0.3" />
    <Canvas Height = "280" Width = "280" Canvas.Left = "110" Canvas.Top = "110" Background =
"Gray" Opacity = "0.4" />
    <Canvas Height = "240" Width = "240" Canvas.Left = "130" Canvas.Top = "130" Background =
"Gray" Opacity = "0.5" />
    <Canvas Height = "200" Width = "200" Canvas.Left = "150" Canvas.Top = "150" Background =
"Gray" Opacity = "0.6" />
    <Canvas Height = "160" Width = "160" Canvas.Left = "170" Canvas.Top = "170" Background =
"Black" Opacity = "0.3" />
    <Canvas Height = "120" Width = "120" Canvas.Left = "190" Canvas.Top = "190" Background =
"Black" Opacity = "0.4" />
    <Canvas Height = "80" Width = "80" Canvas.Left = "210" Canvas.Top = "210" Background =
"Black" Opacity = "0.5" />
    <Canvas Height = "40" Width = "40" Canvas.Left = "230" Canvas.Top = "230" Background =
"Black" Opacity = "0.6" />
</Canvas>
```

程序运行的效果如图 5.12 所示。

图 5.12　Canvas 叠加

5.4　枢轴视图布局(Pivot)

枢轴视图布局(Pivot)提供了一种快捷的方式来管理应用中的视图或页面，通过一种类似于标签的方式来将视图分类，这样就可以在一个界面上通过切换标签来浏览多个数据集，或者切换应用视图。枢轴视图控件水平放置独立的视图，同时处理左侧和右侧的导航，可以通过划动或者平移手势来切换枢轴控件中的视图。

在 Windows Phone 手机上很多系统程序的布局都用到了 Pivot 控件，如图 5.13 所示，手机上的系统设置用了 Pivot 控件。

如系统设置的界面上展示了一个有两个页面的枢轴视图，可以通过划动和平移手势切换页面，向左划动，就由当前页面(如 system)切换到下一个页面(application)，如果切换到最后一个页面，同样操作会回到第一个页面，也就是说，枢轴视图的页面是循环的。另外也可以单击标题来切换，在 system 中，单击其后面并列的灰色 application 就可以切换到相应页面。

图 5.13 Pivot 控件的效果

Pivot 分为两个部分，分别是 Pivot 的 Header 部分和 PivotItem 的 Content 部分。

(1) Header：主要包含 Pivot 的 Title，在图 5.13 中为"SETTINGS"，还有 PivotItem 的 Header，在图 5.13 中为"system application"。可以设置 Pivot 的 Title 属性来改变文字；可以编辑 TitleTemplate 来改变 Title 的样式，比如在 Title 加个小图标等。PivotItem 的 Header 属性与 Pivot 的 Title 属性类似，可以设置它改变文字，但 PivotItem 没有提供 Header 的模板属性，因此目前不能改变它的外观。Header 属性的文字不要设置太长，否则会导致其他 PivotItem 的 Header 无法显示出来。

(2) PivotItem：显示在枢轴视图页面的控件都放到 PivotItem 中，可以把 PivotItem 当作一个容器控件，将要展示的控件都放置在其中。Pivot 中的重要属性和事件分别如表 5.1 以及表 5.2 所示。

表 5.1 Pivot 类常用的属性

名　称	说　明
HeaderTemplate	获取或设置 Pivot 控件的 Header 属性和 PivotItem 子对象的模板
SelectedIndex	获取或设置当前选择的 PivotItem 的索引
SelectedItem	获取或设置当前选择的 PivotItem 的对象
Title	获取或设置 Pivot 的标题
TitleTemplate	获取或设置 Pivot 标题的模板，通过设置它可以改变 Pivot 标题的外观

表 5.2 Pivot 类常用的事件

名　称	说　明
LoadedPivotItem	该事件在加载完成后被触发
LoadingPivotItem	该事件在加载 PivotItem 时被触发，可以用来动态加载数据
SelectionChanged	选择的 PivotItem 改变时该事件被触发
UnloadedPivotItem	该事件在离开当前的 PivotItem 完成时被触发
UnloadingPivotItem	该事件在离开当前的 PivotItem 开始时被触发

下面给出 Pivot 布局的示例：演示 Pivot 控件的事件触发顺序。

代码清单 5-4：Pivot 布局（源代码：第 5 章\Examples_5_4）

MainPage.xaml 文件主要代码

```
<phone:PhoneApplicationPage
xmlns:controls="clr-namespace:Microsoft.Phone.Controls;assembly=Microsoft.Phone.Controls"
    ......>
  <Grid x:Name="LayoutRoot" Background="Transparent">
        <!--Pivot 控件-->
        <controls:Pivot Title="Pivot 的标题" Name="myPivot"
                        LoadedPivotItem="Pivot_LoadedPivotItem"
                        LoadingPivotItem="Pivot_LoadingPivotItem"
                        SelectionChanged="Pivot_SelectionChanged"
                        UnloadedPivotItem="Pivot_UnloadedPivotItem"
                        UnloadingPivotItem="myPivot_UnloadingPivotItem">
            <controls:PivotItem Header="标签一" >
                 <TextBlock Text="这里显示标签一视图的内容." TextWrapping="Wrap"
Style="{StaticResource PhoneTextExtraLargeStyle}"/>
            </controls:PivotItem>
            <controls:PivotItem Header="标签二">
                  <TextBlock Text="这里显示标签二视图的内容." TextWrapping="Wrap"
Style="{StaticResource PhoneTextExtraLargeStyle}"/>
            </controls:PivotItem>
        </controls:Pivot>
    </Grid>
</phone:PhoneApplicationPage>
```

MainPage.xaml.cs 文件主要代码

```
private void Pivot_LoadedPivotItem(object sender, PivotItemEventArgs e)
{
    myPivot.Title = "加载 PivotItem 完成!";
}
private void Pivot_LoadingPivotItem(object sender, PivotItemEventArgs e)
{
    myPivot.Title = "正在加载 PivotItem 中!";
}
private void Pivot_SelectionChanged(object sender, SelectionChangedEventArgs e)
{
    MessageBox.Show("你切换了 Pivot 的标签");
    myPivot.Title = "你切换了 Pivot 的标签";
}
private void Pivot_UnloadedPivotItem(object sender, PivotItemEventArgs e)
{

    MessageBox.Show("你已经离开了 Pivot 的标签");
    myPivot.Title = "你已经离开了 Pivot 的标签";
}
private void myPivot_UnloadingPivotItem(object sender, PivotItemEventArgs e)
{
    MessageBox.Show("你正在离开 Pivot 的标签");
    myPivot.Title = "你正在离开 Pivot 的标签";
}
```

```
        }
    }
```

程序运行的效果如图 5.14 所示。

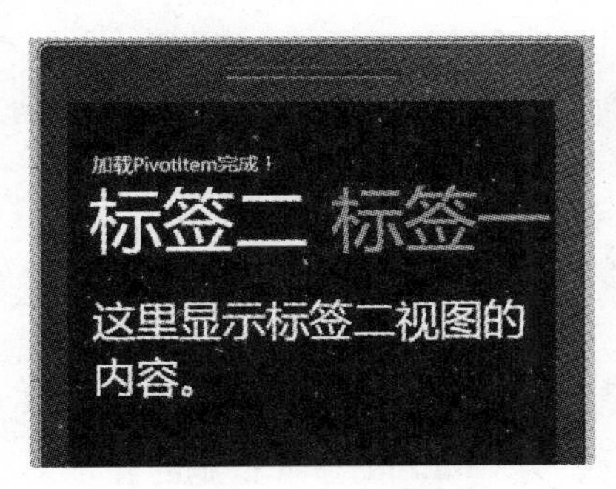

图 5.14 Pivot 控件的演示效果

5.5 全景视图布局(Panorama)

全景视图布局(Panorama)是 Windows Phone 上一个独特的视图控件,给用户提供了一种纵向的拉伸延长的视图布局,它通过使用一个超过屏幕宽度的长水平画布,提供了一种独特显示控件、数据和服务的方式。在 Windows Phone 手机上 Panorama 布局使用很普遍。系统的联系人的界面如图 5.15 所示,就是通过 Panorama 控件来展现的。

图 5.15 Panorama 控件的效果

全景视图布局的用户接口由 3 层类型组成:背景、全景标题、全景区域,它们有各自独立的动作逻辑。

1. 背景(Background)

背景是Panorama控件包装下的整个应用程序的背景，位于全景应用的最底层，这个背景会铺满整个手机屏幕。背景通常是一张全景图，它可能是应用程序最直观的部分，也可以不设置，如联系人应用的背景就是默认的颜色。

整个全景视图的背景是通过Background属性来设置背景。Background是一个Brush类型的属性，可以通过3种画刷来设置：SolidColorBrush——使用一种纯色画刷设置Background；ImageBrush——使用图片做背景；GradientBrush——使用渐变画刷做背景。建议使用图片的高为800像素，宽度大于480像素，但是不要超过2000像素，因为高度不足800像素图片会被拉伸，超过2000像素图片将会被裁减。Panorama的背景图片，其Build Action属性一定要设置为Resource，否则在Panorama第一次显示时不会立刻显示出来。这是因为Build Action属性设置为Content导致资源的异步加载。在其他资源的使用中也应该注意到这点。

背景设计应注意以下要点：

(1) 利用单色的背景，或者是跨度为整个全景的图片。如果决定使用图片，从大小来考虑，可能会使用JPG图片，但是Windows Phone支持的任何UI图片类型都是可以的。

(2) 可以使用多个图片作为背景，但是在任一时刻，只能显示其中一张。

(3) 为了保持良好的程序性能，最少的加载时间，并且无需剪裁，图片大小应该在800×480像素和800×1024像素(高×宽)中选择。

(4) 对于一个具备4个全景区域的应用，使用16×9的屏幕高宽比。

(5) 为了提高文本的易读性，使用一个透明的黑色或者白色过滤器。

(6) 在动态UI元素上，避免使用下拉阴影效果(drop-shadow effects)。

(7) 使用一定比例的与panning手势相关的动作，该panning手势和顶层内容宽度与背景图片的宽度比例有关。

(8) 只有背景艺术出现在应用中时，才使用动画。

(9) 当用户的pan手势超出图片的宽度时，关闭并且返回可见区域。

2. 全景标题(Panorama Title)

全景标题是整个全景应用的标题。其目的是让用户识别该应用，无论是以何种方式进入应用，它都必须是可见的。如联系人应用中的全景标题是people。通过Panorama的HeaderTemplate属性，可以改变它的Title，比如修改字体、颜色，甚至添加一些其他的控件。

全景标题设计应注意以下要点：

(1) 使用简洁的文字或者图片，如将一个logo作为全景标题。使用多个UI元素，如一个logo加文字(或者其他UI元素)也是可以的。

(2) 确保字体或者图片的颜色与整个背景相匹配，而且，标题的可视性不依赖于背景图片。

(3) 为了保持一致性体验，在Start菜单中的应用程序名称和该标题一致。

(4) 避免标题动画，或者动态改变标题的字体。

(5) 使用一定比例的动作，相对于最顶层内容来说较慢，而相对于底层图片来说较快。

(6) 当用户的 pan 手势超出图片的宽度时，关闭并且返回可见区域。

3. 全景区域(Panorama Item)

全景区域是全景应用的组成部分，它封装了其他控件和内容，包含全景视图中的所有 Panorama Item 的集合，如联系人应用中的 recent、all 和 What's New。建议在全景视图中最多不要加入超过 4 个 Item。在 Item 中可以放置比屏幕宽度宽的控件。Item 分为 Header(标题)和 Content(内容)两个部分，标题是全景区域的可选部分，可以不填。Content 就是要展示内容的区域，要展示的控件就会放到这个区域当中。屏幕的宽度比要显示的 Content(内容)区域略宽，下一个 Content 区域的一部分会在当前位置显示出来，就像手机联系人信息页面一样，在当前页面"all"中，会显示出它的后一个页面"what's new"的左侧边界区域，在用户导航操作时，可以预览到下一个页面的一小部分内容，这样做使用户体验更好。

全景区域设计应注意以下要点：

(1) 最大化利用 4 个全景区域，确保全景应用的平滑性能。

(2) 在列表或者网格内使用垂直滚动是可以接受的，但前提是它处于全景区域内，并且不同时与水平滚动出现。

(3) 只要全景区域的宽度小于屏幕的宽度，垂直滚动是可以的。

(4) 支持所有自定义控件和标准控件。

(5) 与手指拖拽的移动比例相同。

(6) 当用户导向到一个新的区域时，开启屏幕动画。

(7) 设计全景区域的布局，使得少量的下一个全景区域可见。提供轻微的重叠，使得用户直觉地利用 Pan 手势来切换应用。

(8) 直到该全景区域有内容要表示时，才显示该全景区域。

下面给出 Panorama 布局的示例：切换 item 时随机切换背景。

代码清单 5-5：Panorama 布局(源代码：第 5 章\Examples_5_5)

MainPage.xaml 文件主要代码

```
<Grid x:Name = "LayoutRoot" Background = "Transparent">
    <!-- Panorama 控件 -->
    <controls:Panorama Title = "my application" SelectionChanged = "Panorama_SelectionChanged"
Name = "myPanorama">
        <controls:Panorama.Background>
            <ImageBrush ImageSource = "PanoramaBackground.png"/>
        </controls:Panorama.Background>
        <!-- Panorama item -->
        <controls:PanoramaItem Header = "first item">
            <StackPanel>
                <TextBlock Text = "第一个 item" FontSize = "50"/>
                <TextBlock Text = "这是第一个 item!" FontSize = "50"/>
```

```
            </StackPanel>
        </controls:PanoramaItem>
        <controls:PanoramaItem Header="second item">
            <StackPanel>
                <TextBlock Text="第二个 item" FontSize="50"/>
                <TextBlock Text="这是第二个 item!" FontSize="50"/>
            </StackPanel>
        </controls:PanoramaItem>
        <controls:PanoramaItem  Header="third item">
            <StackPanel>
                <TextBlock Text="第三个 item" FontSize="50"/>
                <TextBlock Text="这是第三个 item!" FontSize="50"/>
            </StackPanel>
        </controls:PanoramaItem>
    </controls:Panorama>
</Grid>
```

MainPage.xaml.cs 文件代码

```
using System;
using System.Windows.Controls;
using System.Windows.Media;
using Microsoft.Phone.Controls;
using System.Windows.Media.Imaging;
namespace PanoramaDemo
{
    public partial class MainPage : PhoneApplicationPage
    {
        public MainPage()
        {
            InitializeComponent();
        }
        //获取 Panorama 控件 item 改变的事件
        private void Panorama_SelectionChanged(object sender, SelectionChangedEventArgs e)
        {
            //根据当前的时间的秒数来随机取 image1.jpg 或者 image2.jpg 图片为背景
            BitmapImage bitmapImage;
            int tem = DateTime.Now.Second;
            if (tem % 2 == 0)
            {
                bitmapImage = new BitmapImage(new Uri("Images/image1.jpg", UriKind.Relative));
            }
            else
            {
                bitmapImage = new BitmapImage(new Uri("Images/image2.jpg", UriKind.Relative));
            }
            //创建一个图片画刷
            ImageBrush imageBrush = new ImageBrush();
            imageBrush.ImageSource = bitmapImage;
            //设置 Panorama 控件的背景
            this.myPanorama.Background = imageBrush;
```

```
            }
        }
}
```

程序运行的效果如图 5.16 所示。

图 5.16　Panorama 控件的演示效果

数 据 存 储

数据的输入、保存、查询等数据的操作是一个应用程序的常规功能，对于 Windows Phone 8 手机应用程序也是一个基本的功能。Windows Phone 8 手机的数据存储可以分为两种：一种是云存储，另外一种是手机存储。云存储是指将信息存储在云端，也就是存储在互联网上，请数据存储到网络上，这一种情况将会在互联网编程中讲解。另外一种情况就是存储在客户端，所谓的手机存储，这一章将讲解 Windows Phone 8 的数据的手机存储。

6.1 独立存储

独立存储(IsolateStorage)是一种数据存储机制，它在代码与保存的数据之间定义了标准化的关联方式。Windows Phone 8 没有提供直接操作客户端操作系统类文件的手段，但它提供了另一途径来实现在客户端读写数据的需求，这就是 IsolatedStorage。通过使用 IsolatedStorage 数据在虚拟文件系统中隔离，虚拟文件系统可以是根目录中的一个文件，也可以是一个目录和文件树。在使用 IsolatedStorage 时，无须知道 IsolatedStorage 在哪儿或者如何存放数据，它有具备独立存储机制的 API，这些 API 提供了一个虚拟的文件系统和可以访问这个虚拟文件系统的数据流对象，从而提供隔离性和安全性。

6.1.1 独立存储的介绍

独立存储是 Windows Phone 8 操作系统的一个特色的功能，它相当于一个手机内置的数据存储中心，独立的意思就是这个存储相对于一个手机应用程序是完全独立的，两个不同的应用程序是不会在独立存储上冲突、影响以及交互。

独立存储有两种实现的方式：一种方式叫做独立存储设置(Isolated Storage Settings)，通过库中的键/值对，类似于散列表的结构一样，一般程序设置、状态等信息的存储适合使用这种方式；另一种方式叫做独立存储文件(Isolated Storage File)，这是通过创建真实的文件(xml、txt 等)和目录来保存信息的，但是这些文件是在系统内部的，无法在程序外面查看到原始的文件，一般数据量较大较复杂的程序主体信息适合于使用这种方式。Windows Phone 8 系统的这种独特的独立存储结构有一个很重要的特点就是安全，因为用户无法操作系统的存储以及其他应用程序的存储空间。手机的独立文件夹结构图如图 6.1 所示。

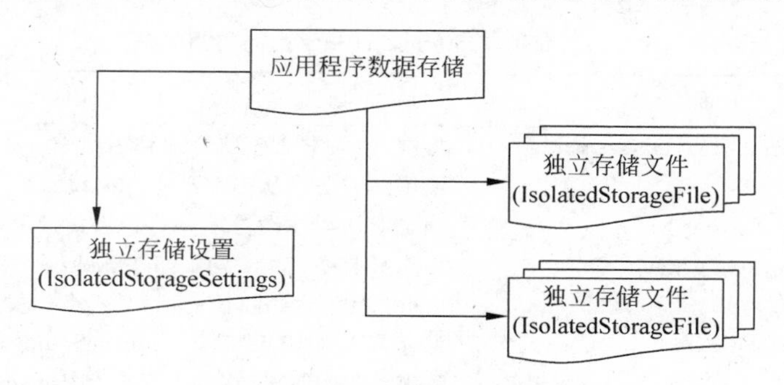

图 6.1　手机的独立文件夹结构图

IsolatedStorage 这个命名空间里有几个重要的类是操作独立存储用到的，下面将作简单的说明：

1）System. IO. IsolatedStorage. IsolatedStorageException（mscorlib. dll 中）

IsolatedStorageException 类专门处理当使用独立存储发生失败时，所引发的异常内容。它使用 ISS_E_ISOSTORE，错误代码为 0x80131450。一般会发生的错误主要有两类：一类是遗失证据（Miss evidence），evidence 代表的是该应用程序的 assembly 与 origin 的资讯，IsolatedStorage 需要透过 evidence 来验证 code 的识别并连接正确的 Isolated Storage 控件，这个信息不见了就无法运作；另外一类是错误的操作（Invalid operations），有一些的 FileStream 的功能，不支持 Isolated Storage 将会引发这个异常。

2）System. IO. IsolatedStorage. IsolatedStorageFile（mscorlib. dll 中）

IsolatedStorageFile 类提供取得隔离区中的文件，目录的方法与属性，如 CreateDirectory()、CreateFile()等。通过该类取得控制权后，配合 IsolatedFileStream 存取区域内的资料。

3）System. IO. IsolatedStorage. IsolatedStorageSettings（System. Windows. dll 中）

IsolatedStorageSettings 类提供可以在隔离储存区中储存索引键值组的 Dictionary ＜TKey, TValue＞。在应用程序里面，可以自动建立一个 Dictionary 类型的设定值，透过 key/value 的方式来取得设定值。通常用于程序共用变量，程序的设置。

4）System. IO. IsolatedStorage. IsolatedFileStream（mscorlib. dll 中）

IsolatedFileStream 类主要提供操作隔离区中的文件与目录，它与 FileStream 相似，扩充了 FileStream 的功能，通常会配合的 StreamWriter 与 StreamReader 的操作隔离区的内容。

6.1.2　使用独立存储设置（IsolatedStorageSettings）

独立存储设置（IsolatedStorageSettings）允许用户在一个字典中存储键/值对（无须任何设定），然后再读取出来。这些数据会一直保存着，无论应用程序停止/启动，或者关机等。除非用户删除它，或者卸载该应用程序，否则它一直存在。IsolatedStorageSettings 类的一些常用的方法和属性分别如表 6.1 和表 6.2 所示。

表 6.1 IsolatedStorageSettings 类的常用方法

名　称	说　明
public void Add(string key,Object value)	将键/值对的对应项添加到字典中
public void Clear()	使用此方法可从当前独立存储设置范围中移除所有值
public bool Contains(string key)	确定应用程序设置字典是否包含指定键
public bool Remove(string key)	移除具有指定键的项，如果 key 不存在，此方法将返回 false；将不会删除任何项，也不会引发任何异常
public void Save()	保存写入到当前 IsolatedStorageSettings 对象的数据，应用程序关闭的时候，将保存写入到 IsolatedStorageSettings 对象中的数据；如果希望应用程序立即写入独立存储，则可以在应用程序代码中调用 Save 方法
public bool TryGetValue<T>(string key, out T value)	获取指定键的值；如果找到指定的键，则为 true；否则为 false

表 6.2 IsolatedStorageSettings 类的常用属性

名　称	说　明
ApplicationSettings	获取一个 IsolatedStorageSettings 实例，如果不存在这样的实例，则创建一个用于在独立存储中存储键/值对的字典的新实例，这个实例是特定于你当前的应用程序的
Count	获取存储在字典中的键/值对的数目，Int 类型
Item	获取或设置与指定的键相关联的值
Keys	获取包含词典中的键的集合
Values	获取包含词典中的值的集合

在 Windows Phone 应用程序中使用独立存储设置的步骤如下：

(1) 添加 System. IO. IsolatedStorage 空间的引用，如 using System. IO. IsolatedStorage;

(2) 创建独立存储设置类的对象实例，如 IsolatedStorageSettings settings = IsolatedStorageSettings. ApplicationSettings;

(3) 使用 IsolatedStorageSettings 类对象进行存储数据，如 settings. Add("state", false), settings["state "] = false，判断特定键是否存在 settings. Contains("state ")，取值 Bool state=(Bool) settings["state "]。

下面给出独立存储设置使用的示例：该示例演示了保存、删除、查找、修改和清空独立存储设置。

代码清单 6-1：独立存储设置的使用(源代码：第 6 章\Examples_6_1)

MainPage. xaml 文件主要代码

```
<Grid x:Name = "ContentGrid" Grid.Row = "1">
    <TextBlock Height = "39" HorizontalAlignment = "Left" Margin = "30,42,0,0"
```

```
            Name = "textBlock1" Text = "Key" VerticalAlignment = "Top" />
    < TextBox Height = "72" HorizontalAlignment = "Left"
            Margin = "115,24,0,0" Name = "txtKey" Text = ""
            VerticalAlignment = "Top" Width = "212" />
    < TextBox Height = "74" HorizontalAlignment = "Left" Margin = "115,109,0,0"
            Name = "txtValue" Text = "" VerticalAlignment = "Top" Width = "212" />
    < TextBlock Height = "39" HorizontalAlignment = "Left" Margin = "12,131,0,0"
            Name = "textBlock2" Text = "Value" VerticalAlignment = "Top" />
    < Button Content = "保存" Height = "70"
            HorizontalAlignment = "Left" Margin = "0,255,0,0"
            Name = "btnSave" VerticalAlignment = "Top" Width = "160"
            Click = "btnSave_Click" />
    < Button Content = "删除" Height = "70" HorizontalAlignment = "Left"
            Margin = "150,255,0,0" Name = "btnDelete" VerticalAlignment = "Top"
            Width = "160" Click = "btnDelete_Click" />
    < TextBlock Height = "39" HorizontalAlignment = "Left" Margin = "94,347,0,0"
            Name = "textBlock3" Text = "Keys 列表" VerticalAlignment = "Top" />
    < ListBox Height = "168" HorizontalAlignment = "Left" Margin = "94,392,0,0"
            Name = "lstKeys" VerticalAlignment = "Top" Width = "274"
            BorderThickness = "1" SelectionChanged = "lstKeys_SelectionChanged" />
    < Button Content = "清空所有" Height = "72" HorizontalAlignment = "Left" Margin = "308,255,
0,0" Name = "deleteall" VerticalAlignment = "Top" Width = "160" Click = "deleteall_Click" />
</Grid >
```

MainPage.xaml.cs 文件主要代码

```
//声明 IsolatedStorageSettings 实例
private IsolatedStorageSettings _appSettings;
public MainPage()
{
    InitializeComponent();
    //获取当前应用程序的 IsolatedStorageSettings 存储
    _appSettings = IsolatedStorageSettings.ApplicationSettings;
    //绑定到 list 控件
    BindKeyList();
}
//保存 IsolatedStorageSettings 键值
private void btnSave_Click(object sender, RoutedEventArgs e)
{
    //检查 key 输入框不为空
    if (!String.IsNullOrEmpty(txtKey.Text))
    {
        if (_appSettings.Contains(txtKey.Text))
        {
            //如果 key 值已经存在则修改它的值
            _appSettings[txtKey.Text] = txtValue.Text;
        }
        else
        {
            //不存在则新增一个 key 值
            _appSettings.Add(txtKey.Text, txtValue.Text);
```

```
        }
        //保存独立存储设置
        _appSettings.Save();
        BindKeyList();
    }
    else
    {
        MessageBox.Show("请输入 key 值");
    }
}
//删除在 List 中选中的独立存储设置的键
private void btnDelete_Click(object sender, RoutedEventArgs e)
{
    //如果选中了 List 中的某项
    if (lstKeys.SelectedIndex > -1)
    {
        //移除这个键的独立存储设置
        _appSettings.Remove(lstKeys.SelectedItem.ToString());
        _appSettings.Save();
        BindKeyList();
    }
}
//List 控件选中项的事件,将选中的键和值显示在上面的文本框中
private void lstKeys_SelectionChanged(object sender, SelectionChangedEventArgs e)
{
    if (e.AddedItems.Count > 0)
    {
        //获取在 List 中选择的 key
        string key = e.AddedItems[0].ToString();
        //检查独立存储设置是否存在这个 key
        if (_appSettings.Contains(key))
        {
            txtKey.Text = key;
            //获取 key 的值并且显示在文本框上
            txtValue.Text = _appSettings[key].ToString();
        }
    }
}
//将当前程序中所有的 key 值绑定到 List 上
private void BindKeyList()
{
    //先清空 List 控件的绑定值
    lstKeys.Items.Clear();
    //获取当前应用程序的所有的 key
    foreach (string key in _appSettings.Keys)
```

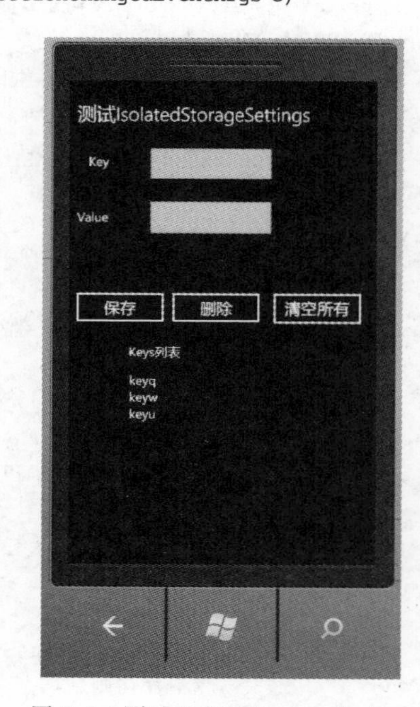

图 6.2 测试 IsolatedStorageSettings

```
    {
        //添加到 List 控件上
        lstKeys.Items.Add(key);
    }
    txtKey.Text = "";
    txtValue.Text = "";
}
private void deleteall_Click(object sender, RoutedEventArgs e)
{
    //清空所有的 key
    _appSettings.Clear();
    BindKeyList();
}
```

程序运行的效果如图 6.2 所示。

6.1.3 使用独立存储文件(IsolatedStorageFile)

独立存储文件(IsolatedStorageFile)表示包含文件和目录的独立存储区。使用IsolatedStorageFile 是一种可以在用户的设备中存储真实文件的机制,它使独立存储的虚拟文件系统抽象化。IsolatedStorageFile 对象对应于特定的独立存储范围,在该范围中存在由 IsolatedStorageFileStream 对象表示的文件。应用程序可以使用独立存储将数据保存在文件系统中这些数据自己的独立部分,而不必在文件系统中指定特定的路径。IsolatedStorageFile 类的一些常用的方法如表 6.3 所示。

表 6.3 IsolatedStorageFile 类的常用方法

名　称	说　明
public IsolatedStorageFileStream CreateFile (string path)	在独立存储区中创建文件,path 是要在独立存储区中创建的文件的相对路径,返回一个独立存储文件流
public void DeleteFile(string file)	删除独立存储区中的文件;删除的文件一经删除便无法恢复
public bool FileExists(string path)	确定指定的路径是否指的是独立存储区中的现有文件
public IsolatedStorageFileStream OpenFile (string path,FileMode mode)	在指定的模式中打开文件,mode 取值以下几种: FileMode.CreateNew 指定操作系统应创建新文件 FileMode.Create 指定操作系统应创建新文件 FileMode.Open 指定操作系统应打开现有文件 FileMode.OpenOrCreate 指定操作系统应打开文件(如果文件存在);否则,应创建新文件 FileMode.Truncate 指定操作系统应打开现有文件,文件一旦打开,就将被截断为零字节大小 FileMode.Append 打开现有文件并查找到文件尾,或创建新文件

续表

名　称	说　明
public IsolatedStorageFileStream OpenFile(string path,FileMode mode,FileAccess access)	Access 取值有以下几种： FileAccess.Read 对文件的读访问 FileAccess.Write 文件的写访问 FileAccess.ReadWrite 对文件的读访问和写访问
public string[] GetFileNames()	获取独立存储区根处的文件的名称，返回独立存储区根处文件的相对路径的数组，零长度数组指定根处没有任何文件
public string[] GetFileNames(string searchPattern)	枚举独立存储范围中与给定模式匹配的文件
public void CreateDirectory(string dir)	在独立存储范围中创建目录，创建的目录最初不包含任何文件
public void DeleteDirectory(string dir)	删除独立存储范围中的目录
public bool DirectoryExists(string path)	确定指定的路径是否指的是独立存储区中的现有目录
public string[] GetDirectoryNames()	枚举独立存储区根处的目录，返回独立存储区根处目录的相对路径的数组，零长度数组指定根处没有任何目录
public string[] GetDirectoryNames(string searchPattern)	枚举独立存储范围中与给定模式匹配的目录
public static IsolatedStorageFile GetUserStoreForApplication()	获取从手机调用的应用程序所使用的用户范围的独立存储
public void Dispose()	释放由 IsolatedStorageFile 使用的所有资源
public void Remove()	移除独立存储区范围及其所有内容

使用 IsolatedStorageFile 对象来实现手机信息的存储，有以下 3 个主要步骤：

1）调用手机的独立存储

例如：

```
IsolatedStorageFile storage = IsolatedStorageFile.GetUserStoreForApplication()
```

2）创建独立存储文件流

例如：

```
IsolatedStorageFileStream location = new IsolatedStorageFileStream(nameTxt.Text + ".item",
System.IO.FileMode.Create, storage)
```

3）读写该文件流

例如，将独立存储文件流转化为可写流：

```
System.IO.StreamWriter file = new System.IO.StreamWriter(location)
```

将 XML 文件保存到流 file 上，即已经写入到手机独立存储文件上，_doc 是用户创建的文件：

```
_doc.Save(file)
```

转化为可读流：

```
System.IO.StreamReader file = new System.IO.StreamReader(location)
```

解析流，转化为 XML：

```
_xml = XElement.Parse(file.ReadToEnd())
```

下面给出购物清单应用的示例：该示例演示了使用 XML 的文件格式来存储信息。

代码清单 6-2：购物清单（源代码：第 6 章\Examples_6_2）

MainPage.xaml 文件主要代码：清单列表界面

```
<Grid x:Name = "LayoutRoot" Background = "Transparent">
    …
    <! -- 应用的标题 -->
    <StackPanel x:Name = "TitlePanel" Grid.Row = "0" Margin = "24,24,0,12">
        <TextBlock x:Name = "PageTitle" Text = "购物清单" Margin = " - 3, - 8,0,0" Style =
"{StaticResource PhoneTextTitle1Style}"/>
    </StackPanel>
    <! -- 使用 ListBox 控件来绑定购物清单的数据 -->
    <Grid x:Name = "ContentGrid" Grid.Row = "1">
        <ListBox Grid.Row = "0" Margin = "10" FontSize = "48" Name = "Files">
        </ListBox>
    </Grid>
</Grid>
<! -- 菜单栏,放了新增一个账目的按钮 -->
<phone:PhoneApplicationPage.ApplicationBar>
    <shell:ApplicationBar IsVisible = "True" IsMenuEnabled = "True">
        <shell:ApplicationBar.MenuItems>
            <shell:ApplicationBarMenuItem Text = "新增" Click = "New_Click"/>
        </shell:ApplicationBar.MenuItems>
    </shell:ApplicationBar>
</phone:PhoneApplicationPage.ApplicationBar>
```

MainPage.xaml.cs 文件代码：清单列表的后台处理

```
using System;
using System.Windows;
using System.Windows.Controls;
using Microsoft.Phone.Controls;
using System.IO.IsolatedStorage;
namespace ShoppingList_Demo
{
    public partial class MainPage : PhoneApplicationPage
    {
        public MainPage()
        {
            InitializeComponent();
            //加载页面触发 Loaded 事件
            Loaded += (object sender, RoutedEventArgs e) =>
            {
                Files.Items.Clear();//先清空一下 ListBox 的数据
                //获取应用程序的独立存储文件
                using (IsolatedStorageFile storage = IsolatedStorageFile.GetUserStoreForApplication())
```

```
                {
                    //获取并循环 *.item 的存储文件
                    foreach (string filename in storage.GetFileNames("*.item"))
                    {
                        //动态构建一个 Grid
                        Grid a = new Grid();
                        //定义第一列
                        ColumnDefinition col = new ColumnDefinition();
                        GridLength gl = new GridLength(200);
                        col.Width = gl;
                        a.ColumnDefinitions.Add(col);
                        //定义第二列
                        ColumnDefinition col2 = new ColumnDefinition();
                        GridLength gl2 = new GridLength(200);
                        col2.Width = gl;
                        a.ColumnDefinitions.Add(col2);
                        //添加一个 TextBlock 显示文件名到第一列
                        TextBlock txbx = new TextBlock();
                        txbx.Text = filename;
                        Grid.SetColumn(txbx, 0);
                        //添加一个 HyperlinkButton 链接到购物详细清单页面,这是第二列
                        HyperlinkButton btn = new HyperlinkButton();
                        btn.Width = 200;
                        btn.Content = "查看详细";
                        btn.Name = filename;
                        btn.NavigateUri = new Uri("/DisplayPage.xaml?item=" + filename,
UriKind.Relative);//传递文件名到商品详细页面
                        Grid.SetColumn(btn, 1);
                        a.Children.Add(txbx);
                        a.Children.Add(btn);
                        Files.Items.Add(a);
                    }
                }
            }
        }
        private void New_Click(object sender, EventArgs e)
        {
            NavigationService.Navigate(new Uri("/AddItem.xaml", UriKind.Relative));
        }
    }
}
```

AddItem.xaml 文件主要代码：添加商品界面

```
<TextBlock Grid.Column="0" Grid.Row="0" Text="名称:" HorizontalAlignment="Center"
VerticalAlignment="Center" />
<TextBox Name="nameTxt"  Grid.Column="1" Margin="8,8,8,0" Padding="2" />
<TextBlock Grid.Column="0" Grid.Row="1" Text="价格:"  HorizontalAlignment="Center"
VerticalAlignment="Center" />
<TextBox x:Name="priceTxt"  Grid.Column="1" Margin="8,8,8,0" Padding="2" Grid.Row="1"  />
```

```
<TextBlock Grid.Column = "0" Grid.Row = "2" Text = "数量:" HorizontalAlignment = "Center"
VerticalAlignment = "Center" />
<TextBox Name = "quanTxt" Grid.Column = "1" Margin = "8,8,8,375" Padding = "2" Grid.Row = "2"
Grid.RowSpan = "2" />
</Grid>
<Button x:Name = "BtnSave" Content = "保存" HorizontalAlignment = "Right" Margin = "0,0,17,0"
Grid.Row = "1" VerticalAlignment = "Bottom" Click = "BtnSave_Click" />
```

AddItem.xaml.cs 文件代码：添加商品后台处理

```
using System;
using System.Windows;
using Microsoft.Phone.Controls;
using System.IO.IsolatedStorage;
using System.Xml.Linq;
namespace ShoppingList_Demo
{
    public partial class AddItem : PhoneApplicationPage
    {
        public AddItem()
        {
            InitializeComponent();
        }
        private void BtnSave_Click(object sender, RoutedEventArgs e)
        {
            using (IsolatedStorageFile storage = IsolatedStorageFile.GetUserStoreForApplication())
            {
                XDocument _doc = new XDocument();
                XElement _item = new XElement(nameTxt.Text);            //创建一个 XML 元素
                XAttribute price = new XAttribute("price", priceTxt.Text);
                                                                        //创建一个 XML 属性
                XAttribute quantity = new XAttribute("quantity", quanTxt.Text);
                _item.Add(price, quantity);                  //将这两个属性添加到 XML 元素上
                //用_item 新建一个 XML 的 Linq 文档
                _doc = new XDocument(new XDeclaration("1.0", "utf-8", "yes"), _item);
                //创建一个独立存储的文件流
                IsolatedStorageFileStream    location    =    new    IsolatedStorageFileStream
(nameTxt.Text + ".item",System.IO.FileMode.Create, storage);
                //将独立存储文件流转化为可写流
                System.IO.StreamWriter file = new System.IO.StreamWriter(location);
                //将 XML 文件保存到流 file 上,即已经写入到手机独立存储文件上
                _doc.Save(file);
                file.Dispose();                     //关闭可写流
                location.Dispose();                 //关闭手机独立存储流
                //调回清单主页
                NavigationService.Navigate(new Uri("/MainPage.xaml", UriKind.Relative));
            }
        }
    }
}
```

DisplayPage. xaml 文件主要代码：商品详细的界面

```
< TextBlock Grid. Column = "0" Grid. Row = "0" Text = "名称:" HorizontalAlignment = "Center"
VerticalAlignment = "Center" />
< TextBlock Name = "nameTxt"  Grid. Column = "1" Margin = "8" Padding = "2" Height = "59"  />
< TextBlock Grid. Row = "1" Text = "价格:" HorizontalAlignment = "Center" VerticalAlignment =
"Center" />
< TextBlock x: Name = "priceTxt"  Grid. Column = "1" Margin = "8" Padding = "2" Height = "59"
Grid. Row = "1"  />
< TextBlock Grid. Column = "0" Grid. Row = "2" Text = "数量:" HorizontalAlignment = "Center"
VerticalAlignment = "Center" />
< TextBlock Name = "quanTxt" Grid. Column = "1" Margin = "8" Padding = "2" Height = "59" Grid.
Row = "2" />
…
< Button x:Name = "BtnBack" Content = "回到购物清单" HorizontalAlignment = "Right" Margin = "0,
0,17,0" Grid. Row = "1" VerticalAlignment = "Bottom" Click = "BtnBack_Click" />
```

DisplayPage. xaml. cs 文件代码：商品详细的后台处理

```
using System;
using System. Windows;
using Microsoft. Phone. Controls;
using System. IO. IsolatedStorage;
using System. Xml. Linq;
using System. Windows. Navigation;
namespace ShoppingList_Demo
{
    public partial class DisplayPage : PhoneApplicationPage
    {
        public DisplayPage()
        {
            InitializeComponent();
        }
        //OnNavigatedTo 事件是当跳转到当前的页面的时候触发的,现在在这里处理详细商品
        //信息的显示
        protected override void OnNavigatedTo(NavigationEventArgs e)
        {
            String itemName = "";
            base. OnNavigatedTo(e);
            //获取上一页面传递过来的 item 值
            bool itemExists = NavigationContext. QueryString. TryGetValue("item", out itemName);
            if (itemExists)
            {
                PageTitle. Text = itemName;
            }
            using (IsolatedStorageFile storage = IsolatedStorageFile. GetUserStoreForApplication())
            {
                XElement _xml;                    //定义 Linq 的 XML 元素
                //打开独立存储文件
                IsolatedStorageFileStream  location  =  new  IsolatedStorageFileStream
(itemName, System. IO. FileMode. Open, storage);
```

```
                //转化为可读流
                System.IO.StreamReader file = new System.IO.StreamReader(location);
                //解析流转化为 XML
                _xml = XElement.Parse(file.ReadToEnd());
                if (_xml.Name.LocalName != null)
                {
                    XAttribute priceTemp = _xml.Attribute("price");          //获取价格
                    priceTxt.Text = priceTemp.Value.ToLower();
                    XAttribute quanTemp = _xml.Attribute("quantity");        //获取数量
                    quanTxt.Text = quanTemp.Value.ToLower();
                    nameTxt.Text = itemName;
                }
                file.Dispose();
                location.Dispose();
            }
        }
        //返回购物清单事件处理
        private void BtnBack_Click(object sender, RoutedEventArgs e)
        {
            NavigationService.Navigate(new Uri("/MainPage.xaml", UriKind.Relative));
        }
    }
}
```

程序运行的效果如图 6.3～图 6.5 所示。

图 6.3　购物单的列表

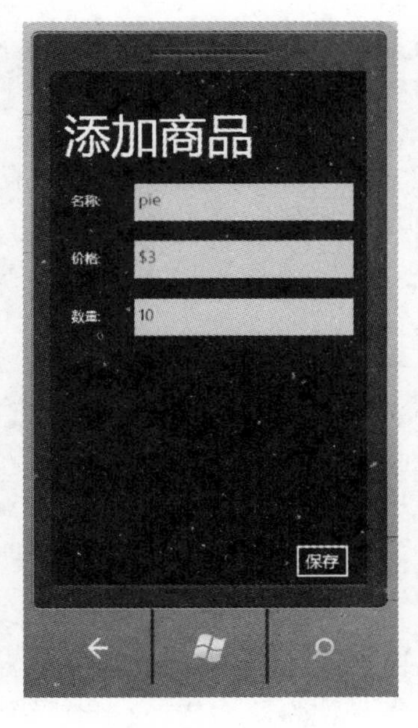

图 6.4　添加商品

图 6.5　商品详细

6.2 SQL Server CE 数据库

SQL Server CE 全名为 SQL Server Compact Edition，是微软推出的一个适用于嵌入到移动应用的精简数据库产品，它与关系型数据库 SQL Server 是完全不同的，在 Windows Phone 中使用 SQL Server CE 数据库时需要使用面向对象的方式来操作数据库，以前在关系型数据库里面使用的 SQL 语句，那么在 Windows Phone 的 SQL Server 数据库里面将要使用 Linq 语句来代替 SQL 语句。SQL Server CE 数据库也是将数据库的存储文件放在应用程序的独立存储空间里面，跟其他的第三方数据库的形式差不多，只是各自的存储文件规则和形式不一样。

6.2.1 创建数据表

创建一个数据表类需要继承 INotifyPropertyChanging 和 INotifyPropertyChanged 接口，主要是为了监控表的数据的变化。在数据表类里面你需要定义表的字段，设置字段的属性类型等，使用的是 Linq 的语法。

创建数据表类的语法格式如下：

```
[Table( Name = "Tasks")]
public partial class Task : INotifyPropertyChanging, INotifyPropertyChanged, ITodoTable
{
    [Column(Storage = "_EstimateTime", DbType = "DateTime")]
    public System.Nullable<System.DateTime> EstimateTime
    {
        get
        {
            return this._EstimateTime;
        }
        set
        {
            if ((this._EstimateTime != value))
            {
                this.OnEstimateTimeChanging(value);
                this.SendPropertyChanging();
                this._EstimateTime = value;
                this.SendPropertyChanged("EstimateTime");
                this.OnEstimateTimeChanged();
            }
        }
    }
    [Association(Name = "Attachment_Items", Storage = "_Attachment",
        ThisKey = "ItemID", OtherKey = "ItemID", DeleteRule = "NO ACTION")]
    public EntitySet<Attachment> Attachment
    {
        get
        {
```

```
            return this._Attachment;
        }
        set
        {
            this._Attachment.Assign(value);
        }
    }
}
```

创建一个数据库对应的 DataContext 类。

使用 System.Data.Linq.DataContext 类来映射每一个数据库实体对象,那么每对应一个数据库你需要创建继承一个类来继承 System.Data.Linq.DataContext 类,用来表示该数据库的上下文数据对象。

创建数据库对应的 DataContext 类的语法格式如下:

```
public partial class DataContextBase : System.Data.Linq.DataContext
{

    private static System.Data.Linq.Mapping.MappingSource mappingSource =
        new AttributeMappingSource();

    public DataContextBase(string connection) :
        base(connection, mappingSource)
    {
        OnCreated();
    }
    public DataContextBase(string connection,
        System.Data.Linq.Mapping.MappingSource mappingSource) :
        base(connection, mappingSource)
    {
        OnCreated();
    }
    public System.Data.Linq.Table<Task> Items
    {
        get
        {
            return this.GetTable<Task>();
        }
    }
}
```

6.2.2 创建数据库

创建一个数据库可以直接调用继承的 System.Data.Linq.DataContext 类的 CreateDatabase 方法,但是需要传入数据库连接字符串(即数据库的位置、名称、密码等)。

例如:

```
TodoDC = new DataContextBase ( " Data Source = ' " + DatabaseFilename + " ' " +
```

```
(DatabasePassword.Length == 0 ? "" : ";Password='" + DatabasePassword + "'"));
```

部署一个数据库即引入数据库的文件来创建一个数据库。

例如：

```
private void CreateDatabaseFromResourceFile(string inFilename, string outFilename)
{
    //第一步
    Stream str = Application.GetResourceStream(new Uri(inFilename, UriKind.Relative)).Stream;
    //第二步
    IsolatedStorageFileStream outFile = iso.CreateFile(outFilename);
    //第三步
    outFile.Write(ReadToEnd(str), 0, (int)str.Length);
    str.Close();
    outFile.Close();
}
```

6.2.3 增删改操作

增加一条数据的方法使用 Table<T>的 InsertOnSubmit 方法。删除一条数据的方法使用 Table<T>的 DeleteOnSubmit 方法。修改一条数据的方法时直接获取了需要修改的数据对象，修改后直接调用 SubmitChanges 方法就可以了。查询数据需要使用 Linq 语法来操作，如 var tasks = from t in todoDC.Items select t;查询表里面的所有数据。

6.2.4 实例：员工信息操作

下面给出员工信息应用的示例：该示例演示了使用 SQL Server CE 数据库的增删改操作的实现。

代码清单 6-3：员工信息应用(源代码：第 6 章\Examples_6_3)

EmployeeTable.cs 文件代码：创建一个员工信息表，
用于保存员工的名字和简介，员工表有一个自增的 ID

```
using System.Data.Linq.Mapping;
using System.ComponentModel;
namespace SQLServerDemo
{
    [Table]
    public class EmployeeTable : INotifyPropertyChanged, INotifyPropertyChanging
    {
        //定义员工表独立增长 ID,设置为主键
        private int _employeeId;
        [Column(IsPrimaryKey = true, IsDbGenerated = true, DbType = " INT NOT NULL
Identity", CanBeNull = false, AutoSync = AutoSync.OnInsert)]
        public int EmployeeID
        {
            get
            {
```

```
            return _employeeId;
        }
        set
        {
            if (_employeeId != value)
            {
                NotifyPropertyChanging("EmployeeID");
                _employeeId = value;
                NotifyPropertyChanged("EmployeeID");
            }
        }
    }
    //定义员工名字字段
    private string _employeeName;
    [Column]
    public string EmployeeName
    {
        get
        {
            return _employeeName;
        }
        set
        {
            if (_employeeName != value)
            {
                NotifyPropertyChanging("EmployeeName");
                _employeeName = value;
                NotifyPropertyChanged("EmployeeName");
            }
        }
    }
    //定义员工简介字段
    private string _employeeDesc;
    [Column]
    public string EmployeeDesc
    {
        get
        {
            return _employeeDesc;
        }
        set
        {
            if (_employeeDesc != value)
            {
                NotifyPropertyChanging("EmployeeDesc");
                _employeeDesc = value;
                NotifyPropertyChanged("EmployeeDesc");
            }
        }
    }
```

```
        #region INotifyPropertyChanged Members
        public event PropertyChangedEventHandler PropertyChanged;
        //用来通知页面表的字段数据产生了改变
        private void NotifyPropertyChanged(string propertyName)
        {
            if (PropertyChanged != null)
            {
                PropertyChanged(this, new PropertyChangedEventArgs(propertyName));
            }
        }
        #endregion
        #region INotifyPropertyChanging Members
        public event PropertyChangingEventHandler PropertyChanging;
        //用来通知数据上下文表的字段数据将要产生改变
        private void NotifyPropertyChanging(string propertyName)
        {
            if (PropertyChanging != null)
            {
                PropertyChanging(this, new PropertyChangingEventArgs(propertyName));
            }
        }
        # endregion
    }
}
```

EmployeeDataContext.cs 文件代码：创建数据库的 DataContent，定义一个 EmployeeDataContext 类来继承 DataContext，在 EmployeeDataContext 中定义数据库连接字符串，以及员工信息表

```
using System.Data.Linq;
namespace SQLServerDemo
{
    public class EmployeeDataContext : DataContext
    {
        //数据库链接字符串
        public static string DBConnectionString = "Data Source = isostore:/Employee.sdf";
        //传递数据库连接字符串到 DataContext 基类
        public EmployeeDataContext(string connectionString)
            : base(connectionString){ }
        //定义一个员工信息表
        public Table<EmployeeTable> Employees;
    }
}
```

EmployeeDataContext.cs 文件代码：创建页面数据绑定的集合

```
using System.ComponentModel;
using System.Collections.ObjectModel;
namespace SQLServerDemo
{
    //EmployeeCollection 用于跟页面的数据绑定
```

```
    public class EmployeeCollection : INotifyPropertyChanged
    {
        //定义 ObservableCollection 来绑定页面的数据
        private ObservableCollection<EmployeeTable> _employeeTables;
        public ObservableCollection<EmployeeTable> EmployeeTables
        {
            get
            {
                return _employeeTables;
            }
            set
            {
                if (_employeeTables != value)
                {
                    _employeeTables = value;
                    NotifyPropertyChanged("EmployeeTables");
                }
            }
        }
        public event PropertyChangedEventHandler PropertyChanged;
        //用于通知属性的改变
        private void NotifyPropertyChanged(string propertyName)
        {
            if (PropertyChanged != null)
            {
                PropertyChanged(this, new PropertyChangedEventArgs(propertyName));
            }
        }
    }
}
```

App.xaml.cs 文件主要代码：在程序加载过程中进行创建数据库

```
private void Application_Launching(object sender, LaunchingEventArgs e)
{
    //如果数据库不存在则创建一个数据库
    using (EmployeeDataContext db = new EmployeeDataContext (EmployeeDataContext.
DBConnectionString))
    {
        if (db.DatabaseExists() == false)
        {
            //创建一个数据库
            db.CreateDatabase();
        }
    }
}
```

MainPage. xaml 文件主要代码：员工信息增删改的页面

```
<Grid x:Name = "ContentPanel" Grid.Row = "1" Margin = "12,0,12,0">
    <Grid Margin = "0,0,0,385">
        ...
        <! -- 员工信息编辑文本框 -->
        <TextBlock FontSize = "30" Height = "37" HorizontalAlignment = "Left" Margin = "12,18,
0,0" Name = "textBlock1" Text = "员工名字:" VerticalAlignment = "Top" />
        <TextBox Name = "name" Text = ""  Margin = "145,0,6,144" />
        <TextBlock FontSize = "30" Height = "52" HorizontalAlignment = "Left" Margin = "18,74,
0,0" Name = "textBlock2" Text = "简介：" VerticalAlignment = "Top" />
        <TextBox Height = "79" HorizontalAlignment = "Left" Margin = "93,65,0,0" Name = "desc"
Text = "" VerticalAlignment = "Top" Width = "357" />
        <Button
            Content = "保存" x:Name = "addButton"
            Click = "addButton_Click" Margin = "219,132,6,6" />
    </Grid>
    <! -- 员工信息 ListBox 绑定 -->
    <ListBox x:Name = "toDoItemsListBox" ItemsSource = "{Binding EmployeeTables}" Margin =
"12,241,12,0" Width = "440">
        <ListBox.ItemTemplate>
            <DataTemplate>
                <Grid HorizontalAlignment = "Stretch" Width = "440">
                    <Grid.ColumnDefinitions>
                        <ColumnDefinition Width = "50" />
                        <ColumnDefinition Width = "*" />
                        <ColumnDefinition Width = "100" />
                    </Grid.ColumnDefinitions>
                    <TextBlock Text = "{Binding EmployeeName}" FontSize = "{StaticResource
PhoneFontSizeLarge}" Grid.Column = "1" VerticalAlignment = "Center"/>
                    <Button Grid.Column = "2" x:Name = "deleteButton" BorderThickness = "0"
 Margin = "0" Click = "deleteButton_Click" Content = "删除">
                    </Button>
                    <Button Grid.Column = "1" x:Name = "editButton" BorderThickness = "0"
Margin = "209,0,81,0" Click = "editButton_Click" Content = "编辑" Grid.ColumnSpan = "2">
                    </Button>
                </Grid>
            </DataTemplate>
        </ListBox.ItemTemplate>
    </ListBox>
</Grid>
```

MainPage. xaml. cs 文件代码：实现员工信息的增删改操作

```
using System.Linq;
using System.Windows;
using System.Windows.Controls;
using Microsoft.Phone.Controls;
using System.Collections.ObjectModel;
namespace SQLServerDemo
{
```

```
public partial class MainPage : PhoneApplicationPage
{
    //创建 DataContext 实例用于操作独立的数据库
    private EmployeeDataContext employeeDB;
    private EmployeeCollection employeeCol = new EmployeeCollection();
    public MainPage()
    {
        InitializeComponent();
        //连接数据库并初始化 DataContext 实例
        employeeDB = new EmployeeDataContext(EmployeeDataContext.DBConnectionString);
        //使用 Linq 查询语句查询 EmployeeTable 表的所有数据
        var employeesInDB = from EmployeeTable employee in employeeDB.Employees
                            select employee;
        //将查询的结果返回到页面数据绑定的集合里面
        employeeCol. EmployeeTables = new ObservableCollection < EmployeeTable >
(employeesInDB);
        //赋值给当前页面的 DataContext 用于数据绑定
        this.DataContext = employeeCol;
    }
    ///<summary>
    ///删除操作
    ///</summary>
    private void deleteButton_Click(object sender, RoutedEventArgs e)
    {
        //获取单击的按钮实例
        var button = sender as Button;
        if (button != null)
        {
            //获取当前按钮绑定的 DataContext,即当前的删除的 EmployeeTable 实例
            EmployeeTable employeeForDelete = button.DataContext as EmployeeTable;
            //移除绑定集合里面要删除的 EmployeeTable 记录
            employeeCol.EmployeeTables.Remove(employeeForDelete);
            //移除数据库里面要删除的 EmployeeTable 记录
            employeeDB.Employees.DeleteOnSubmit(employeeForDelete);
            //保存数据库的改变
            employeeDB.SubmitChanges();
        }
    }
    ///<summary>
    ///保存操作,处理新增和编辑员工信息
    ///</summary>
    private void addButton_Click(object sender, RoutedEventArgs e)
    {
        //控制员工名字和简介不能为空
        if (name.Text != "" && desc.Text != "")
        {
            if (State.Count > 0 && State["employee"] != null )    //编辑状态
            {
                //获取编辑的 EmployeeTable 对象
                EmployeeTable employee = (EmployeeTable)State["employee"];
```

```
                    employee.EmployeeName = name.Text;
                    employee.EmployeeDesc = desc.Text;
                    //保存数据库的改变
                    employeeDB.SubmitChanges();
                    //添加绑定集合的数据,因为在单击编辑的时候移除了
                    employeeCol.EmployeeTables.Add(employee);
                    State["employee"] = null;
                }
                else//新增状态
                {
                    //创建一条表的数据
                    EmployeeTable newEmployee = new EmployeeTable { EmployeeName = name.
Text, EmployeeDesc = desc.Text };
                    //添加绑定集合的数据
                    employeeCol.EmployeeTables.Add(newEmployee);
                    //插入数据库
                    employeeDB.Employees.InsertOnSubmit(newEmployee);
                    //保存数据库的改变
                    employeeDB.SubmitChanges();
                }
                name.Text = "";
                desc.Text = "";
            }
            else
            {
                MessageBox.Show("姓名和简介不能为空!");
            }
        }
        ///<summary>
        ///编辑操作
        ///</summary>
        private void editButton_Click(object sender, RoutedEventArgs e)
        {
            //获取单击的按钮实例
            var button = sender as Button;
            if (button != null)
            {
                //获取当前按钮绑定的 DataContext,即当前的编辑的 EmployeeTable 实例
                EmployeeTable employeeForEdit = button.DataContext as EmployeeTable;
                name.Text = employeeForEdit.EmployeeName;
                desc.Text = employeeForEdit.EmployeeDesc;
                //将需要编辑的表实例存储在 State 里面
                State["employee"] = employeeForEdit;
```

```
                employeeCol.EmployeeTables.Remove(employeeForEdit);
            }
        }
    }
}
```

程序运行的效果如图 6.6 所示。

图 6.6　员工列表的显示效果

图形动画

Windows Phone 系统的图形处理能力非常强大，对于 Silverlight 框架下的图形图像处理的效果很出色。Windows Phone 的图形图像处理技术是 Windows Phone 编程的重要基础知识，掌握好 Windows Phone 的图形图像处理技术是创建和设计一个漂亮的应用的首要条件。

Windows Phone 系统中的图形图像处理可以分为两类：一类是静态图像处理，包括各种各样的多边形、图片位图以及绘图等；另一类是动画，就是动态图形图画的处理，使用动画可以实现 Windows Phone 应用中各种动态的效果。

本章将介绍 Windows Phone 图形图像编程的基础。

7.1 基本的图形

在 Windows Phone 中可以使用基本的图形有：Rectangle（矩形）、Ellipse（椭圆）、Line（线）、Polyline（多边线）、Polygon（多边形）和 Path（有弧线的多边形）。你可以结合这些基本的图形创建出更加复杂的图形。这些图形都继承了 Shape 基类 Shape 类的常用属性如表 7.1 所示。

表 7.1 Shape 类常用的属性

名　　称	说　　明
Fill	获取或设置指定形状内部绘制方式的 Brush
GeometryTransform	获取一个表示 Transform 的值，该值在绘制形状之前应用于 Shape 的几何图形
Stretch	获取或设置一个 Stretch 枚举值，该值描述形状如何填充为它分配的空间
Stroke	获取或设置指定 Shape 轮廓绘制方式的 Brush
StrokeDashArray	获取或设置 Double 值的集合，这些值指示用于勾勒形状轮廓的虚线和间隙样式
StrokeDashCap	获取或设置一个 PenLineCap 枚举值，该值指定如何绘制虚线的两端
StrokeDashOffset	获取或设置一个 Double，它指定虚线样式内虚线开始处的距离
StrokeEndLineCap	获取或设置一个 PenLineCap 枚举值，该值描述位于直线末端的 Shape
StrokeLineJoin	获取或设置一个 PenLineJoin 枚举值，该值指定在 Shape 的顶点处使用的联接类型
StrokeMiterLimit	获取或设置对斜接长度与 Shape 元素的 StrokeThickness 的一半之比的限制
StrokeStartLineCap	获取或设置一个 PenLineCap 枚举值，该值描述位于 Stroke 起始处的 Shape
StrokeThickness	获取或设置 Shape 笔画轮廓的宽度

7.1.1 矩形(Rectangle)

矩形(Rectangle)类表示绘制一个矩形形状,该形状可以使用画刷来填充。Rectangle不能支持子对象。如果要绘制一个包含其他对象的矩形区域,可以使用Canvas,也可以使用复合几何图形,但在这种情况下,用户需要使用RectangleGeometry,而不是Rectangle。Rectangle或其他任何具有填充区域形状的填充不一定必须是纯色,它可以是任何Brush。

下面给出设置矩形样式的示例:演示了边框颜色设置、圆角矩形、画刷填充、颜色渐变等。

代码清单7-1:设置矩形样式(源代码:第7章\Examples_7_1)

MainPage.xaml文件主要代码

```
<Canvas >
    <!-- 蓝边填充红色的矩形 -->
    <Rectangle Canvas.Top = "20"
               Canvas.Left = "40"
               Width = "347"
               Height = "80"
               Fill = "Red"
               Stroke = "Blue"
               StrokeThickness = "3">
    </Rectangle >
    <!-- 使用 LinearGradientBrush 画刷填充的圆角矩形 -->
    <Rectangle Canvas.Top = "134" Canvas.Left = "40"
               Width = "347" Height = "100" Stroke = "#000000"
               StrokeThickness = "2" RadiusX = "15" RadiusY = "15">
        <Rectangle.Fill >
            <LinearGradientBrush StartPoint = "0,1">
                <GradientStop Color = "#FFFFFF" Offset = "0.0" />
                <GradientStop Color = "#FF9900" Offset = "1.0" />
            </LinearGradientBrush >
        </Rectangle.Fill >
    </Rectangle >
    <!-- 使用 RadialGradientBrush 画刷的矩形 -->
    <Rectangle Canvas.Top = "280" Canvas.Left = "40"
               Width = "347" Height = "100">
            <Rectangle.Fill >
                <RadialGradientBrush GradientOrigin = "0.5,0.5"
                        Center = "0.5,0.5" RadiusX = "0.5" RadiusY = "0.5">
                    <GradientStop Color = "#0099FF" Offset = "0" />
                    <GradientStop Color = "#FF0000" Offset = "0.25" />
                    <GradientStop Color = "#FCF903" Offset = "0.75" />
                    <GradientStop Color = "#3E9B01" Offset = "1" />
                    </RadialGradientBrush >
                </Rectangle.Fill >
            </Rectangle >
</Canvas >
```

程序运行的效果如图 7.1 所示。

7.1.2 椭圆(Ellipse)

Ellipse 类表示画一个椭圆。通过属性 Height 和 Width 来设置椭圆的高度和宽度，当高度和宽度相等的时候就是一个圆形。

下面给出创建圆形图形的示例：创建一个 Ellipse 图形、创建一个风车形状的图形和一个渐变的圆形。

代码清单 7-2：创建圆形图形(源代码：第 7 章\Examples_7_2)

MainPage.xaml 文件主要代码

```
<!-- 创建一个 Ellipse 图形 -->
<Ellipse Fill="Yellow" Height="200" Width="200" ></Ellipse>
<!-- 利用 Ellipse 图形组成一个风车形状的图形 -->
<Ellipse Fill="Yellow" Height="200" Width="200" Stroke="Blue" StrokeThickness="70"
StrokeDashArray="1"></Ellipse>
<Ellipse Fill="Red" Height="50" Width="50"></Ellipse>
<!-- 创建一个颜色渐变的 Ellipse 图形 -->
<Ellipse Height="200" Width="200" Margin="250,402,6,6">
    <Ellipse.Fill>
        <LinearGradientBrush StartPoint="0,0" EndPoint="0,1">
            <GradientStop Color="White" Offset="0" />
            <GradientStop Color="Yellow" Offset="0.7" />
        </LinearGradientBrush>
    </Ellipse.Fill>
</Ellipse>
```

程序运行的效果如图 7.2 所示。

图 7.1 矩形的样式

图 7.2 EllipseDemo 的运行效果

7.1.3　直线(Line)

直线(Line)类表示在两个点之间绘制一条直线。默认情况下，Line 对象绘制线条的起点和终点都是没有样式的，但可以通过 StrokeStartLineCap、StrokeEndLineCap、StrokeDashCap 属性为直线对象额外增加线帽样式。其中，前两个属性主要用于实线对象，其取值类型为 PenLineCap 枚举。通过 Line 对象绘制虚线效果，需要用到 StrokeDashArray 属性，该属性对应一个 Double 类型的集合。该集合的奇数位表示线段的长度，偶数位表示两个线段之间的间隔长度。如果只是表示普通的虚线，则只需定义一个数值就可以，默认会将该数值作为线段跟间隔的长度。如果想表示一些特殊类型的虚线，那么就需要为 StrokeDashArray 属性设置多个数值。在 Line 对象应用 StrokeDashArray 属性时需要注意的是，其设置的数值并不是线段以及间隔的实际像素值，而是相对于 StrokeThickness 的倍数。

下面给出创建直线图形的示例：创建实线直线、虚线直线和渐变直线。

代码清单 7-3：创建直线图形(源代码：第 7 章\Examples_7_3)

MainPage.xaml 文件主要代码

```
<!-- 创建一条红色的线 -->
<Line X1 = "50" Y1 = "75" X2 = "250" Y2 = "75" Stroke = "Red" StrokeThickness = "10" Margin = "6,
6,27,509"></Line>
<!-- 创建一条红色的虚线 -->
<Line X1 = "50" Y1 = "75" X2 = "250" Y2 = "75" Stroke = "Red" StrokeThickness = "10"
StrokeDashArray = "2" StrokeDashCap = "Round" Margin = "3,104,71,399">
</Line>
<!-- 创建一条颜色渐变的线 -->
<Line X1 = "50" Y1 = "75" X2 = "250" Y2 = "75" StrokeThickness = "10" Margin = "6,214,18,266">
    <Line.Stroke>
        <LinearGradientBrush StartPoint = "0,0" EndPoint = "1,0">
            <GradientStop Color = "Red" Offset = "0" />
            <GradientStop Color = "Yellow" Offset = "0.25" />
            <GradientStop Color = "Green" Offset = "0.5" />
            <GradientStop Color = "Blue" Offset = "0.75" />
            <GradientStop Color = "Purple" Offset = "1" />
        </LinearGradientBrush>
    </Line.Stroke>
</Line>
```

程序运行的效果如图 7.3 所示。

图 7.3　LineDemo 的运行效果

7.1.4　线形(Polyline)

线形(Polyline)类表示绘制一系列相互连接的直线，表示的图形不需要是闭合的形状。Polyline 类的 Points 属性表示点的集合描述一个或多个点，通过设置 Points 属性来定义多线性的各个节点位置。举一个点设置语法的例子，字符串“0,0 50,100 100,0”将生成一个“V”形折线，锐角放置在“50,100”。线形至少需要两个点以上才能显示出图形的形状，单个点(如 (0,0))是有效值，但不会呈现任何内容。

下面给出创建折线图形的示例：创建实线折线和虚线折线。

代码清单 7-4：创建折线图形(源代码：第 7 章\Examples_7_4)

MainPage. xaml 文件主要代码

```
<! -- 画折线 -->
<Polyline Points = "0,160 25,140 50,160 75,140 100,160 125,140, 150,160 175,140 200,160 225,
140 250,160 275,140 300,160" Stroke = "Blue" StrokeThickness = "5" Margin = "0, - 22,0,446">
</Polyline>
<Polyline Stroke = "Blue" StrokeThickness = "5" Points = "10,150 30,140 50,160 70,130 90,170
110,120 130,180 150,110 170,190 190,100 210,240" Margin = "171,60,6,300">
</Polyline>
<! -- 画虚线 -->
<Polyline Stroke = "Brown" StrokeThickness = "10" StrokeDashArray = "1 1" Canvas. Left = "119"
Canvas. Top = "170" Margin = "12,313,6,212">
   <Polyline. Points>
        <Point X = "0" Y = "0"/>
        <Point X = "30" Y = "30"/>
        <Point X = "100" Y = "20"/>
        <Point X = "200" Y = "50"/>
   </Polyline. Points>
</Polyline>
```

程序运行的效果如图 7.4 所示。

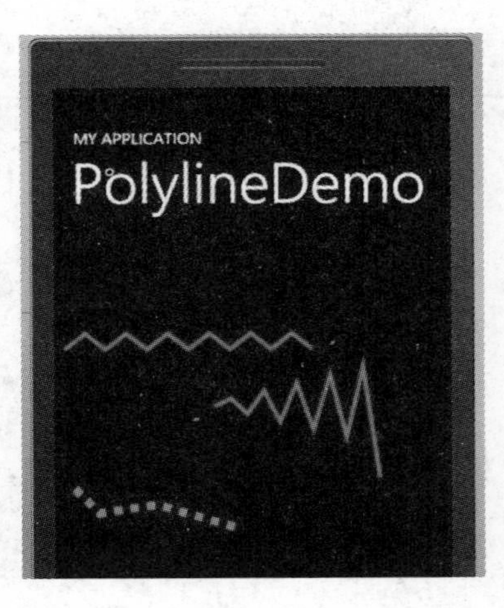

图 7.4 PolyLineDemo 的运行效果

7.1.5 多边形(Polygon)

多边形(Polygon)类是表示绘制一个多边形，与 Polyline 不同的是它是一个闭合的图形。最简单的正多边形是正三角形，在这种情况下，Points 属性具有 3 个项。具有两个 Points 值的 Polygon 对象只要具有一个非零的 StrokeThickness 值和一个非 null 的 Stroke 值，就仍会呈现，但也可使用 Line 对象来实现同样的结果。将不会呈现只有一个点的 Polygon。Polygon 类也是通过 Points 属性来定义多边形的节点的，与 Polyline 的区别是，Polygon 的 Points 属性的点集合第一个点和最后一个点也是用直线串联起来的。

下面给出创建多边形的示例：创建一个红色的三角形、一个金色的鱼形状和一个渐变色彩的飞镖形状。

代码清单 7-5：创建多边形(源代码：第 7 章\Examples_7_5)

MainPage. xaml 文件主要代码

```
<! -- 三角形 -->
<Polygon Fill = "Red" Stroke = "Blue" StrokeThickness = "4"
     Points = "50,50 200,200 350,50" Margin = "0,0,0,394" />
```

```
<! -- 鱼的形状 -- >
< Polygon Points = "40,100 90,140 215,100 265,150 215,200 90,160 40,200"
        Fill = "Orange" Margin = "12,143,57,247"></Polygon >
<! -- 飞镖的形状 -- >
< Polygon Points = "50, 100 200, 100 200, 200 300, 30"
        Stroke = "White" StrokeThickness = "4" Opacity = "0.5" Margin = "95,366,6,0">
    < Polygon. Fill >
        < LinearGradientBrush StartPoint = "0,0" EndPoint = "1,1" >
            < GradientStop Color = "Blue" Offset = "0.25" />
            < GradientStop Color = "Green" Offset = "0.50" />
            < GradientStop Color = "Red" Offset = "0.75" />
        </LinearGradientBrush >
    </Polygon. Fill >
</Polygon >
```

程序运行的效果如图 7.5 所示。

图 7.5 多边形的运行效果

7.1.6 路径(Path)

路径(Path)类表示的是一种比较复杂的图形,利用 Path 绘制一系列相互连接的直线和曲线。直线和曲线维度通过 Data 属性声明,并且可以使用路径特定的 mini-language 或使用对象模型来指定。从根本上讲,Path 是 Shape 对象。但是,可使用 Path 创建比其他 Shape 对象更复杂的二维图形。Path 对象可以绘制闭合或开放的形状、直线和曲线。

Windows Phone 提供了一种称为“mini-language”的属性句法,来描述如何画出轨迹形状,包括 M(移动命令,起始点)、L(直线,结束点)、H(水平线)、V(垂直线)、C(3 次贝塞尔曲线)、Q(2 次贝塞尔曲线)、A(椭圆弧曲线)、Z(结束命令)等。字母 M 代表了移动动作和从给定的点移动到当前点。例如,“M80,200”命令从当前点移动到点(80,200)。

字母 L 绘制一条从当前点线到指定点。例如,“L100,200”命令绘制一条从当前点线,到指定点(100,200)。字母 H 绘制一条从当前点水平线朝 X 轴指定点。字母 V 绘制一条从当前点垂直线走向 Y 轴指定点。字母 C 绘制一条从当前点到第三点的 3 次贝塞尔曲线,两个中间点是作为控制点。字母 S 绘制一条从当前点到第二点的 3 次贝塞尔曲线,第一点是作为控制点。字母 Q 绘制 2 次贝塞尔曲线,从第一点到第二点,第一点是作为控制点。字母 T 绘制平滑的 2 次贝塞尔曲线,从第一点到第二点,第一点是作为控制点。字母 A 绘制一个椭圆弧。它有 5 个参数:size、isLargeAreFlag、rotationAngle、sweepDirectionFlag 和 endpoint。字母 Z 关闭图形,从当前点到出发点的连线。

下面给出创建组合图形的示例:创建一个弧形图形,一个五角星形状和一个组合图形形状。

代码清单 7-6：创建组合图形(源代码：第 7 章\Examples_7_6)

MainPage.xaml 文件主要代码

```
<!-- Path 图形 -->
<Path Stroke="White" StrokeThickness="4"
     Data="M 100,30 A 40,50 50 1 0 100,50" Margin="46,77,233,343" />
<!-- 使用 GeometryGroup 组合图形 -->
<Path Stroke="Red" StrokeThickness="3" Fill="Blue" Margin="6,254,119,130">
    <Path.Data>
        <GeometryGroup >
            <LineGeometry StartPoint="20,200" EndPoint="300,200" />
            <EllipseGeometry Center="80,150" RadiusX="50" RadiusY="50" />
            <RectangleGeometry Rect="80,167 150 30"/>
        </GeometryGroup>
    </Path.Data>
</Path>
<!-- 五角星图形 -->
<Path Data="M 255.30 150.90 L 181.80 179.20 182.00 258.00 132.35
            196.80 57.50 221.30 100.35 155.20 53.90 91.60 130.00 111.90
            176.15 48.10 180.35 126.75 255.30 150.90 "
            HorizontalAlignment="Right" Margin="0,184,6,0"
            Width="203" Stretch="Fill" Height="211"
            VerticalAlignment="Top">
    <Path.Fill>
          <LinearGradientBrush EndPoint="0.5,1" MappingMode="RelativeToBoundingBox"
StartPoint="0.5,0">
            <GradientStop Color="Black"/>
            <!-- 渐变填充 -->
            <GradientStop Color="Red" Offset="0.591"/>
            <GradientStop Color="Green" Offset="0.252"/>
            <GradientStop Color="Blue" Offset="0.957"/>
        </LinearGradientBrush>
    </Path.Fill>
</Path>
```

程序运行的效果如图 7.6 所示。

图 7.6 PathDemo 的运行效果

7.1.7 Geometry 类和 Brush 类

继承 Geometry 基类的几何图形和继承 Brush 基类的画刷都是用来作为其他图形的填充，下面将介绍这两个系列的类。

1. Geometry 类

Geometry 类为用于定义几何形状的对象提供基类。Geometry 对象可用于剪裁区域以及用作将二维图形数据呈现为 Path 的几何图形定义。Geometry 类为基类的几何图形类有 EllipseGeometry 类、GeometryGroup 类、LineGeometry

类、PathGeometry 类和 RectangleGeometry 类。

对于采用 Geometry 的 XAML 语法，需要将 Geometry 的非抽象派生类型指定为对象元素。可以将几何图形概念化为"简单"或"复杂"的几何图形。EllipseGeometry、LineGeometry 和 RectangleGeometry 是简单几何图形，用于将几何形状指定为一个具有基本坐标或维度属性的元素。GeometryGroup 和 PathGeometry 是复杂几何图形。GeometryGroup 将它拥有的其他几何图形组合为子对象。PathGeometry 使用一组嵌套路径图定义元素或简洁的字符串语法来描述几何图形的路径。Geometry 不能完全定义自己的呈现，而是作为 Path 的数据提供。System. Windows. Shapes. Shape 类拥有 Geometry 及其派生类所没有的 Fill、Stroke 和其他呈现属性。Shape 类是一个 FrameworkElement，因而会参与布局系统；其派生类可用作支持 UIElement 子项任何元素的内容。Geometry 类只定义形状的几何图形，无法呈现自身。由于它十分简单，因而用途更加广泛。Geometry 类的常用属性如表 7.2 所示。

表 7.2　Geometry 类常用的属性

名　称	说　明
Bounds	获取一个 Rect，后者指定 Geometry 的与坐标轴对齐的边界框
Empty	获取空的几何图形对象
StandardFlatteningTolerance	获取用于多边形近似的标准公差
Transform	获取或设置应用于 Geometry 的 Transform 对象

2. Brush 类

Brush 类是用于绘制图形对象的对象，从 Brush 派生的类描述了绘制区域的方式。Brush 使用其输出对区域进行绘制，画笔不同，其输出类型也不同。某些画笔使用纯色、渐变或图像绘制区域。下面介绍不同类型的画笔。

1) SolidColorBrush：使用纯色绘制区域

SolidColorBrush 对象可能是最基本的画笔，用于将外观应用到对象。可以通过类型转换语法在 XAML 中将 SolidColorBrush 指定为属性值，该语法使用关于字符串含义的几个约定。其他画笔(如 LinearGradientBrush)需要属性元素语法。可以将 Brush 属性的属性元素语法与对象元素语法 <SolidColorBrush.../> 结合在一起使用；如果要为对象元素提供 Name 值并在以后以其属性作为目标，这可能会很有用。可以使用 ColorAnimation 或 ColorAnimationUsingKeyFrames 对象，对 SolidColorBrush 进行动画处理。若要对 SolidColorBrush 进行动画处理，通常会使用 Fill 等属性(采用 Brush)的间接定向，而不是对 SolidColorBrush 的 Color 属性进行动画处理。

通过名称选择一个预定义的 SolidColorBrush。例如，可以将 Rectangle 对象的 Fill 值设置为 Red 或 MediumBlue。下面的示例使用预定义 SolidColorBrush 的名称来设置 Rectangle 的 Fill。

XAML 示例：

```
<Canvas>
  <Rectangle Width="100" Height="100" Fill="Red" />
</Canvas>
```

2）LinearGradientBrush ：使用线性渐变绘制区域

LinearGradientBrush 对象使用线性渐变绘制区域。线性渐变沿直线定义渐变。该直线的终点由线性渐变的 StartPoint 和 EndPoint 属性定义。LinearGradientBrush 画笔沿此直线绘制其 GradientStops。默认的线性渐变是沿对角方向进行的。默认情况下，线性渐变的 StartPoint 是被绘制区域的左上角（0,0），其 EndPoint 是被绘制区域的右下角（1,1）。所得渐变的颜色是沿着对角方向路径插入的。

XAML 示例：

```
<Canvas>
  <Rectangle Width="200" Height="100">
    <Rectangle.Fill>
      <LinearGradientBrush StartPoint="0,0" EndPoint="1,1">
        <GradientStop Color="Yellow" Offset="0.0" />
        <GradientStop Color="Red" Offset="0.25" />
        <GradientStop Color="Blue" Offset="0.75" />
        <GradientStop Color="LimeGreen" Offset="1.0" />
      </LinearGradientBrush>
    </Rectangle.Fill>
  </Rectangle>
</Canvas>
```

3）RadialGradientBrush：使用径向渐变绘制区域

RadialGradientBrush 对象与 LinearGradientBrush 对象类似。但是，线性渐变有一个起点和一个终点用于定义渐变矢量，而径向渐变有一个椭圆以及一个焦点（GradientOrigin）用于定义渐变行为。该椭圆定义渐变的终点。换言之，1.0 处的渐变停止点定义椭圆圆周处的颜色。焦点定义渐变的中心。0 处的渐变停止点定义焦点处的颜色。

XAML 示例：

```
<Canvas>
  <Rectangle Width="200" Height="100">
    <Rectangle.Fill>
      <RadialGradientBrush GradientOrigin="0.5,0.5" Center="0.5,0.5"
        RadiusX="0.5" RadiusY="0.5">
        <GradientStop Color="Yellow" Offset="0" />
        <GradientStop Color="Red" Offset="0.25" />
        <GradientStop Color="Blue" Offset="0.75" />
        <GradientStop Color="LimeGreen" Offset="1" />
      </RadialGradientBrush>
    </Rectangle.Fill>
  </Rectangle>
</Canvas>
```

4）ImageBrush：使用图像绘制区域

可以使用 ImageBrush 对象为应用程序中的文本创建装饰性效果。例如，TextBlock 对象的 Foreground 属性可以指定 ImageBrush。如果 ImageSource 属性设置为无效格式，或其指定了无法解析的 URI，将引发 ImageFailed 事件。

XAML 示例：

```
<TextBlock FontWeight = "Bold">
  测试
  <TextBlock.Foreground>
        <ImageBrush ImageSource = "forest.jpg"/>
     </TextBlock.Foreground>
   </TextBlock>
```

下面给出画刷的使用示例：创建一个渐变椭圆、一个图片按钮和一个渐变圆形。

代码清单 7-7：画刷（源代码：第 7 章\Examples_7_7）

MainPage.xaml 文件主要代码

```
<Ellipse Height = "82" HorizontalAlignment = "Left" Margin = "24,18,0,0" Name = "ellipse1"
StrokeThickness = "1" VerticalAlignment = "Top" Width = "413" Stroke = "White"></Ellipse>
<TextBlock Height = "96" Margin = "24,132,0,0" Name = "textBlock1" Text = "TextBlock"
VerticalAlignment = "Top" FontSize = "80" HorizontalAlignment = "Left" Width = "444" />
<Rectangle Height = "124" HorizontalAlignment = "Left" Margin = "45,262,0,0" Name =
"rectangle1" Stroke = "Black" StrokeThickness = "1" VerticalAlignment = "Top" Width = "327" />
<Ellipse Height = "161" HorizontalAlignment = "Left" Margin = "105,421,0,0" Name = "ellipse2"
Stroke = "Black" StrokeThickness = "1" VerticalAlignment = "Top" Width = "200" />
```

MainPage.xaml.cs 文件代码

```
using System;
using System.Windows;
using System.Windows.Media;
using Microsoft.Phone.Controls;
namespace BrushDemo
{
    public partial class MainPage : PhoneApplicationPage
    {
        public MainPage()
        {
            InitializeComponent();
            //使用 SolidColorBrush 填充椭圆
            ellipse1.Fill = new SolidColorBrush(Colors.Blue);
            //使用 LinearGradientBrush 来设置文本框的背景
            LinearGradientBrush l = new LinearGradientBrush();
            l.StartPoint = new Point(0.5, 0);
            l.EndPoint = new Point(0.5, 1);
            GradientStop s1 = new GradientStop();
            s1.Color = Colors.Yellow;
            s1.Offset = 0.25;
            l.GradientStops.Add(s1);
            GradientStop s2 = new GradientStop();
```

```
            s2.Color = Colors.Orange;
            s2.Offset = 1.0;
            l.GradientStops.Add(s2);
            textBlock1.Foreground = l;
            //使用 ImageBrush 来填充矩形
            ImageBrush i = new ImageBrush();
            i.Stretch = Stretch.UniformToFill;
            i.ImageSource = new System.Windows.Media.Imaging.BitmapImage(
                new Uri("/ApplicationIcon.png", UriKind.Relative));
            rectangle1.Fill = i;
            //使用 RadialGradientBrush 来设置按钮的背景
            RadialGradientBrush rb = new RadialGradientBrush();
            rb.Center = new Point(0.5, 0.5);
            GradientStop s3 = new GradientStop();
            rb.RadiusX = 0.5;
            rb.RadiusY = 0.5;
            s3.Color = Colors.Yellow;
            s3.Offset = 0.25;
            rb.GradientStops.Add(s3);
          GradientStop s4 = new GradientStop();
            s4.Color = Colors.Orange;
            s4.Offset = 1.0;
            rb.GradientStops.Add(s4);
            ellipse2.Fill = rb;
        }
    }
}
```

运行的效果如图 7.7 所示。

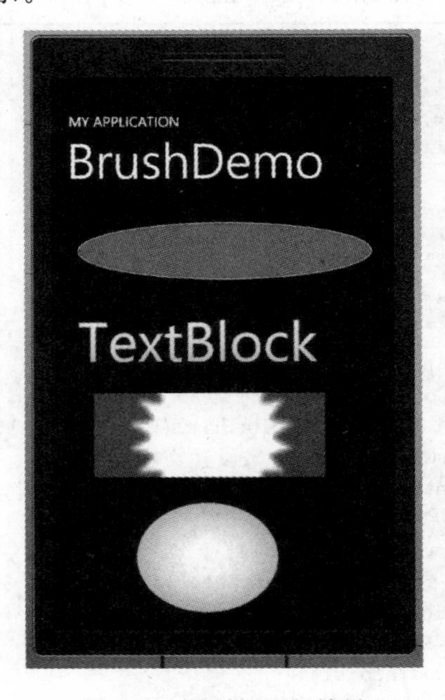

图 7.7　画刷的运行效果

7.2 使用位图编程

Image 控件是用于显示图片的控件，在 Windows Phone 中要在将一张图片显示出来可以添加一个 Image 控件，然后设置图片的路径就行了。使用 Image 控件显示图片的语法如下：

```
<Image Source="myPicture.png" />
```

在 cs 页面中用 c#创建一个图片控件语法如下：

```
Image myImage = new Image();
myImage.Source = new BitmapImage(new Uri("myPicture.jpg", UriKind.RelativeOrAbsolute));
LayoutRoot.Children.Add(myImage);
```

Source 属性用于指定要显示的图像的位置，用户可以设置相对的地址或绝对的地址，还可以在目录结构中向上遍历。用户也可以通过指定一个程序集的方式来引用图片，例如，Source="MySilverlightApp;component/myPicture.png" 引用了应该作为下载程序集提供的程序集 MySilverlightApp。用户指定的程序集可以是 XAP 中的主应用程序程序集，换言之，即 Windows Phone 项目输出路径声明的同一程序集名称。通过引用 component；标记在其概念根位置输入程序集结构，这是最为可靠的 URL 指定方式，因为即便在程序集之外完全重构 XAML(如将此部分 XAML 移入某个附属程序集以进行本地化)，以这种方式仍可解析为图像目标。

7.2.1 拉伸图像

如果没有设置 Image 的 Width 和 Height 值，它将使用 Source 指定图像的自然尺寸显示。设置 Height 和 Width 将创建一个包含矩形区域，图像将显示在该区域中。用户可以通过使用 Stretch 属性指定图像如何填充此包含区域。Stretch 属性接受 Stretch 枚举定义的下列值：None——图像不拉伸以适合输出尺寸；Uniform——对图像进行缩放，以适合输出尺寸，但保留该内容的纵横比，这是默认值；UniformToFill——对图像进行缩放，从而可以完全填充输出区域，但保持其原始纵横比。下面的 XAML 示例显示一个 Image，它填充 300×300 像素的一个输出区域，但保留其原始纵横比，因为 Stretch 属性设置为 UniformToFill。

XAML 示例：

```
<Canvas Width="300" Height="300" Background="Gray">
  <Image Source="myImage.jpg" Stretch="UniformToFill" Width="300" Height="300" />
</Canvas>
```

7.2.2 裁切图像

可以通过使用 Clip 属性设置图像输出的某个区域来裁切图像。将 Clip 属性设置为

Geometry，就意味着可以从图像中裁切掉各种几何形状。下面的示例演示如何将 EllipseGeometry 用作图像的剪辑区域。在此示例中，用 Width 为 200、Height 为 150 定义了一个 Image 对象。一个 RadiusX 值为 100、RadiusY 值为 75、Center 值为“100,75”的 EllipseGeometry 设置为图像的 Clip 属性。只有位于椭圆区域内部的图像部分才会显示。

XAML 示例：

```
<Grid x:Name = "LayoutRoot" Background = "White">
  <Image Source = "Water_lilies.jpg"
    Width = "200" Height = "150">
    <Image.Clip>
      <EllipseGeometry RadiusX = "100" RadiusY = "75" Center = "100,75"/>
        </Image.Clip>
  </Image>
</Grid>
```

7.2.3 动态生成图片

动态生成图片是通过 BitmapImage 位图类来实现的，BitmapImage 类为 Image.Source 和 ImageBrush.ImageSource 属性提供实际的对象源类型。BitmapImage 可用于引用 .JPEG 和.PNG 文件格式的图像。如果将 Image.Source 属性设置为无效格式，或者指定为无法解析的 URI，将引发 ImageFailed 事件。为 BitmapImage 指定图形内容的方法有两种：通过 URI 或通过流。BitmapImage 类的常用事件如表 7.3 所示。

表 7.3 BitmapImage 类的常用事件

名　称	说　明
DownloadProgress	在 BitmapImage 内容的下载进度中已出现显著更改时发生
ImageFailed	当存在与图像检索或格式相关的错误时发生
ImageOpened	在成功下载和解码图像源后发生，可在呈现图像前使用此事件确定图像的大小

下面给出单击屏幕创建矩形位图的示例：单击程序的界面随机地生成一个大小、位置、颜色都不确定的矩形位图。

代码清单 7-8：单击屏幕创建矩形位图(源代码：第 7 章\Examples_7_8)

MainPage.xaml 文件主要代码

```
<Grid x:Name = "ContentPanel" Grid.Row = "1" Margin = "12,0,12,0"
        MouseLeftButtonDown = " ContentPanel _ MouseLeftButtonDown " MouseLeftButtonUp =
"ContentPanel_MouseLeftButtonUp">
    <Image Name = "img" />
    <TextBlock Name = "tb" Text = "你单击屏幕的次数： " />
    <TextBlock Name = "txt" Margin = "182,6,45,563" />
</Grid>
```

MainPage. xaml. cs 文件代码

```
using System;
using Microsoft.Phone.Controls;
using System.Windows.Media.Imaging;
using System.Windows.Media;
using System.Windows.Shapes;
using System.Windows;
namespace WriteableBitmapDemo
{
    public partial class MainPage : PhoneApplicationPage
    {
        Random rand = new Random();
        int count = 0;
        // WriteableBitmap 类提供一个可写入并可更新的 BitmapSource
        WriteableBitmap bitmap;
        public MainPage()
        {
            InitializeComponent();
        }
        //手指按下事件
        private void ContentPanel_MouseLeftButtonDown(object sender, System.Windows.Input.
MouseButtonEventArgs e)
        {
            Random rand = new Random();
            int width = rand.Next(100) % 100;
            //创建一个可写的位图
            bitmap = new WriteableBitmap(width, width);
            //将位图设置为图像控件的资源
            img.Source = bitmap;
        }
        //手指抬起事件
        private void ContentPanel_MouseLeftButtonUp(object sender, System.Windows.Input.
MouseButtonEventArgs e)
        {
            if (bitmap == null)
                return;
            int width = bitmap.PixelWidth;
            int height = bitmap.PixelHeight;
            int x1 = rand.Next(width);
            int x2 = rand.Next(width);
            int y1 = rand.Next(height);
            int y2 = rand.Next(height);
            byte[] bytes = new byte[4];
            rand.NextBytes(bytes);
            //使用随机数来创建一个随机的颜色
            Color clr = Color.FromArgb(bytes[0], bytes[1],
                                       bytes[2], bytes[3]);
```

```
            //使用颜色画刷和矩形几何图形创建一个 Path 图形
            Path path = new Path()
            {
                Fill = new SolidColorBrush(clr),
                Data = new RectangleGeometry()
                {
                    Rect = new Rect(Math.Min(x1, x2),
                                    Math.Min(y1, y2),
                                    Math.Abs(x1 - x2),
                                    Math.Abs(y1 - y2))
                }
            };
            path.Measure(new Size(width, height));
            path.Arrange(new Rect(new Point(), path.DesiredSize));
            bitmap.Render(path, null);
            bitmap.Invalidate();
            //统计你单击的次数
            txt.Text = ( ++ count).ToString();
        }
    }
}
```

程序运行的效果如图 7.8 所示。

图 7.8　随机生成的图片

下面给出屏幕截图的示例：实现了截取当前屏幕界面的图片并且将图片分割成两半进行显示。

代码清单 7-9：单击屏幕创建矩形位图（源代码：第 7 章\Examples_7_9）

MainPage.xaml 文件主要代码

```
<Grid x:Name="ContentPanel" Grid.Row="1" Margin="12,0,12,0">
    <Image Name="imgUL" Margin="2,2,241,339" />
    <Image Name="imgUR" Margin="221,6,6,323" />
    <Image Name="imgLL" Margin="12,274,241,69" />
    <Image Name="imgLR" Margin="212,290,0,69" />
    <TextBlock Text="单击屏幕" HorizontalAlignment="Center" VerticalAlignment="Center"
Height="30" Margin="334,544,26,33" Width="96" />
</Grid>
```

MainPage.xaml.cs 文件代码

```
using System;
using System.Windows;
using System.Windows.Media;
using System.Windows.Shapes;
using Microsoft.Phone.Controls;
using System.Windows.Media.Imaging;
namespace BitmapImageDemo
{
    public partial class MainPage : PhoneApplicationPage
    {
        Image imgBase = new Image();
        public MainPage()
        {
            InitializeComponent();
        }
        protected override void OnManipulationStarted(ManipulationStartedEventArgs args)
        {
            imgBase.Source = new WriteableBitmap(this, null);
            imgBase.Stretch = Stretch.None;
            int width = (new WriteableBitmap(this, null)).PixelWidth;
            int Height = (new WriteableBitmap(this, null)).PixelHeight;
            //左上角图片显示
            WriteableBitmap writeableBitmap = new WriteableBitmap(width / 2, Height/2);
            writeableBitmap.Render(imgBase, null);
            writeableBitmap.Invalidate();
            imgUL.Source = writeableBitmap;
            // 右上角图片显示
            writeableBitmap = new WriteableBitmap(width / 2, Height / 2);
            TranslateTransform translate = new TranslateTransform();
            translate.X = - width / 2;
            writeableBitmap.Render(imgBase, translate);
            writeableBitmap.Invalidate();
            imgUR.Source = writeableBitmap;
            // 左下角图片显示
            writeableBitmap = new WriteableBitmap(width / 2, Height / 2);
            translate.X = 0;
```

```
                translate.Y = - Height / 2;
                writeableBitmap.Render(imgBase, translate);
                writeableBitmap.Invalidate();
                imgLL.Source = writeableBitmap;
                // 右下角图片显示
                writeableBitmap = new WriteableBitmap(width / 2, Height / 2);
                translate.X = - width / 2;
                writeableBitmap.Render(imgBase, translate);
                writeableBitmap.Invalidate();
                imgLR.Source = writeableBitmap;
                args.Complete();
                args.Handled = true;
                base.OnManipulationStarted(args);
            }
        }
    }
```

程序运行的效果如图 7.9 所示。

图 7.9　屏幕截图

7.3　动画

动画是快速播放一系列图像(其中每个图像与下一个图像略微不同)给人造成的一种幻觉。大脑感觉这组图像是一个变化的场景。在电影中,摄像机每秒钟拍摄许多照片(帧),便可使人形成这种幻觉。用投影仪播放这些帧时,观众便可以看电影了。在 Windows Phone 中,通过对对象的个别属性应用动画,可以对对象进行动画处理。例如,若要使 UIElement

增大，需对其 Width 和 Height 属性进行动画处理。若要使 UIElement 逐渐从视野中消失，可以对其 Opacity 属性进行动画处理。可以对 Windows Phone 中许多对象的属性进行动画处理。

在 Windows Phone 中，只能对值类型为 Double、Color 或 Point 的属性执行简单的动画处理。此外，还可以使用 ObjectAnimationUsingKeyFrames 对其他类型的属性进行动画处理，但是这需要使用离散内插（从一个值跳到另一个值），而多数人认为这不是真正的动画。

7.3.1　动画编程中使用的类

动画的类都在空间 System. Windows. Media. Animation 下，在该空间下的类的继承结构如下：

```
System.Object
  System.Windows.DependencyObject
    System.Windows.Media.Animation.Timeline
      System.Windows.Media.Animation.ColorAnimation
      System.Windows.Media.Animation.ColorAnimationUsingKeyFrames
      System.Windows.Media.Animation.DoubleAnimation
      System.Windows.Media.Animation.DoubleAnimationUsingKeyFrames
      System.Windows.Media.Animation.ObjectAnimationUsingKeyFrames
      System.Windows.Media.Animation.PointAnimation
      System.Windows.Media.Animation.PointAnimationUsingKeyFrames
      System.Windows.Media.Animation.Storyboard
```

1. Timeline 类

Timeline 表示时间段，它提供的属性可以让用户指定该时间段的长度、开始时间、重复次数、该时间段内时间进度的快慢等。从 Timeline 派生的类可提供动画功能（如 DoubleAnimation、ColorAnimation 等）。Timeline 类的常用属性和事件分别如表 7.4 和表 7.5 所示。

表 7.4　Timeline 类的常用属性

名　称	说　明
AutoReverse	获取或设置一个值，该值指示时间线在完成向前迭代后是否按相反的顺序播放
BeginTime	获取或设置此 Timeline 将要开始的时间
Duration	获取或设置此时间线播放的时间长度，而不是计数重复
FillBehavior	获取或设置一个值，该值指定动画在其活动期结束后的行为方式
RepeatBehavior	获取或设置此时间线的重复行为
SpeedRatio	获取或设置此 Timeline 的时间相对于其父级的前进速率

表 7.5　Timeline 类的常用事件

名　称	说　明
Completed	当 Storyboard 对象完成播放时发生

2. ColorAnimation 类

ColorAnimation 在指定的 Duration 内使用线性内插对两个目标值之间的 Color 属性值进行动画处理。动画在一段时间内更新属性值。动画的效果可能十分微小，比如将 Shape 左右移动几个像素；也可能十分显著，比如将对象放大到其原始大小的 200 倍，同时对其进行旋转并更改其颜色。若要创建动画，应将动画与对象的属性值关联。

ColorAnimation 类可创建两个目标值之间的过渡。若要设置其目标值，应使用其 From、To 和 By 属性。表 7.6 概括了如何将 From、To 和 By 属性一起使用或单独使用来确定动画的目标值。

表 7.6　ColorAnimation 类的 From 和 To 属性的结果行为

指定的属性	结果行为
From	动画从 From 属性指定的值继续到正在进行动画处理的属性的基值
From 和 To	动画从 From 属性指定的值继续到 To 属性指定的值
From 和 By	动画从 From 属性指定的值继续到 From 与 By 属性之和所指定的值
To	动画从进行动画处理的属性的基值或前一动画的输出值继续到 To 属性指定的值
By	动画从正在进行动画处理的属性的基值或前一动画的输出值继续到该值与 By 属性指定的值之和

3. ColorAnimationUsingKeyFrames 类

ColorAnimationUsingKeyFrames 类的属性 KeyFrames 获取用于定义动画的 ColorKeyFrame 对象的集合。关键帧动画的目标值是通过其 KeyFrames 属性定义的，该属性包含 ColorKeyFrame 对象的集合。每个 ColorKeyFrame 都定义一段动画及其自己的目标 Value 和 KeyTime。动画运行时，它将在指定的关键时间从一个关键值过渡到下一个关键值。ColorKeyFrame 类有 3 种类型，分别对应一个支持的内插方法：LinearColorKeyFrame、DiscreteColorKeyFrame 和 SplineColorKeyFrame。与 ColorAnimation 不同，ColorAnimationUsingKeyFrames 可以有两个以上的目标值。还可以控制单个 ColorKeyFrame 段的内插方法。在 XAML 中声明 ColorAnimationUsingKeyFrames 时，ColorKeyFrame 对象元素的顺序并不重要，这是因为 KeyTime 控制执行时间，从而控制关键帧的执行顺序。不过，保持元素顺序与 KeyTime 序列顺序相同是一种好的标记风格。

4. DoubleAnimation 类

DoubleAnimation 类表示在指定的 Duration 内使用线性内插对两个目标值之间的 Double 属性值进行动画处理。动画在一段时间内更新属性值，动画的效果可能十分微小，比如将 Shape 左右移动几个像素；也可能十分显著，比如将对象放大到其原始大小的 200 倍。DoubleAnimation 类可创建两个目标值之间的过渡。若要设置其目标值，需要使用其 From、To 和 By 属性。表 7.7 概括了如何将 From、To 和 By 属性一起使用或单独使用来确定动画的目标值。

表 7.7 DoubleAnimation 类的 From 和 To 属性的结果行为

指定的属性	结果行为
From	动画从 From 属性指定的值继续到正在进行动画处理的属性的基值或前一动画的输出值,具体取决于前一动画的配置方式
From 和 To	动画从 From 属性指定的值继续到 To 属性指定的值
From 和 By	动画从 From 属性指定的值继续到 From 与 By 属性之和所指定的值
To	动画从进行动画处理的属性的基值或前一动画的输出值继续到 To 属性指定的值
By	动画从正在进行动画处理的属性的基值或前一动画的输出值继续到该值与 By 属性指定的值之和

5. PointAnimation 类

PointAnimation 类与 DoubleAnimation 类相似,PointAnimation 类表示在指定的 Duration 内使用线性内插对两个目标值之间的 Point 属性值进行动画处理。表 7.8 概括了如何将 From、To 和 By 属性一起使用或单独使用来确定动画的目标值。

表 7.8 PointAnimation 类的 From 和 To 属性的结果行为

指定的属性	结果行为
From	动画从 From 属性指定的值继续到正在进行动画处理的属性的基值或前一动画的输出值,具体取决于前一动画的配置方式
From 和 To	动画从 From 属性指定的值继续到 To 属性指定的值
From 和 By	动画从 From 属性指定的值继续到 From 与 By 属性之和所指定的值
To	动画从进行动画处理的属性的基值或前一动画的输出值继续到 To 属性指定的值
By	动画从正在进行动画处理的属性的基值或前一动画的输出值继续到该值与 By 属性指定的值之和

6. Storyboard 类

Storyboard 类表示通过时间线控制动画,并为其子动画提供对象和属性目标信息。可以考虑将 Storyboard 作为其他动画对象(如 DoubleAnimation)以及其他 Storyboard 对象的容器,可以在 Storyboard 对象中彼此嵌套并分别为每个 Storyboard 指定 BeginTime 值。每个子 Storyboard 都会在其父 Storyboard 开始前等待,然后开始倒计时,直到轮到该子对象开始为止,可以使用 Storyboard 对象的交互式方法来启动、暂停、继续和停止动画。

StoryBoard 提供了管理时间线的功能接口,可以用来控制一个或多个 Windows Phone 动画进程,故也称为动画时间线容器。StoryBoard 提供了 6 个常用的动画属性选项,它们分别是:AutoReverse、BeginTime、Duration、FillBehavior、RepeatBehavior 和 SpeedRatio。通过这 6 个属性可以用来控制动画的基本行为动作,比如想要实现动画的缓冲时间,就需要用到 Duration 属性;设置动画的开始时间则用 BeginTime;如果动画执行完后根据执行路线反向执行到原始状态,则需要使用 AutoReverse;如果需要设置动画的运行速度则使用 SpeedRatio 就可以完成。

7.3.2 偏移动画

偏移动画使用的是 TranslateTransform 类来创建的，TranslateTransform 类表示在二维 X-Y 坐标系内平移(移动)对象。可以使用 Canvas. Left 和 Canvas. Top 在 Canvas 上偏移对象的本地(0,0)，但不会将此视为变换；如果是变换，该对象会保留在它自己的本地(0,0)处。通过 TransformGroup 可以应用多个变换。可以使用 MatrixTransform 创建自定义变换。TranslateTransform 定义沿 X 轴和 Y 轴进行的轴对齐平移。

下面给出球的偏移运动的示例：使用 TranslateTransform 偏移动画来实现球的偏移运动。

代码清单 7-10：球的偏移运动(源代码：第 7 章\Examples_7_10)

MainPage. xaml 文件主要代码

```
<!-- 将动画定义为程序的资源 -->
<phone:PhoneApplicationPage.Resources>
    <!-- 创建一个故事画板 -->
        <Storyboard RepeatBehavior="Forever" x:Name="Bounce">
            <!-- 定义球的 X 轴的偏移动画 -->
            <DoubleAnimationUsingKeyFrames BeginTime="00:00:00" Storyboard.TargetName=
"ball" Storyboard.TargetProperty="(UIElement.RenderTransform).(TransformGroup.Children)
[0].(TranslateTransform.X)">
                <SplineDoubleKeyFrame KeyTime="00:00:00" Value="0"/>
                <SplineDoubleKeyFrame KeyTime="00:00:04" Value="297"/>
                <SplineDoubleKeyFrame KeyTime="00:00:06" Value="320"/>
            </DoubleAnimationUsingKeyFrames>
            <!-- 定义球的 Y 轴的偏移动画 -->
            <DoubleAnimationUsingKeyFrames BeginTime="00:00:00" Storyboard.TargetName=
"ball" Storyboard.TargetProperty="(UIElement.RenderTransform).(TransformGroup.Children)
[0].(TranslateTransform.Y)">
                <SplineDoubleKeyFrame KeyTime="00:00:00" Value="0"/>
                <SplineDoubleKeyFrame KeyTime="00:00:02" Value="-206">
                    <SplineDoubleKeyFrame.KeySpline>
                        <KeySpline ControlPoint1="0,1" ControlPoint2="1,1"/>
                    </SplineDoubleKeyFrame.KeySpline>
                </SplineDoubleKeyFrame>
                <SplineDoubleKeyFrame KeyTime="00:00:04" Value="0">
                    <SplineDoubleKeyFrame.KeySpline>
                        <KeySpline ControlPoint1="1,0" ControlPoint2="1,1"/>
                    </SplineDoubleKeyFrame.KeySpline>
                </SplineDoubleKeyFrame>
                <SplineDoubleKeyFrame KeyTime="00:00:05" Value="-20">
                    <SplineDoubleKeyFrame.KeySpline>
                        <KeySpline ControlPoint1="0,1" ControlPoint2="1,1"/>
                    </SplineDoubleKeyFrame.KeySpline>
                </SplineDoubleKeyFrame>
                <SplineDoubleKeyFrame KeyTime="00:00:06" Value="0">
                    <SplineDoubleKeyFrame.KeySpline>
```

```
                    <KeySpline ControlPoint1 = "1,0" ControlPoint2 = "1,1"/>
                </SplineDoubleKeyFrame.KeySpline>
            </SplineDoubleKeyFrame>
        </DoubleAnimationUsingKeyFrames>
    </Storyboard>
</phone:PhoneApplicationPage.Resources>
...
            <Ellipse Height = "85" HorizontalAlignment = "Left" Margin = "71, 0, 0, 151"
VerticalAlignment = "Bottom" Width = "93" Fill = "#FFF40B0B" Stroke = "#FF000000" x:Name =
"ball" RenderTransformOrigin = "0.5,0.5">
            <Ellipse.RenderTransform>
                <TransformGroup>
                    <TranslateTransform/>
                </TransformGroup>
            </Ellipse.RenderTransform>
```

MainPage.xaml.cs 文件主要代码

```
public MainPage()
{
    InitializeComponent();
    //开始运行 Storyboard
   Bounce.Begin();
}
```

程序运行的效果如图 7.10 所示。

图 7.10　运动的小球

7.3.3　旋转动画

旋转动画(RotateTransform)就是一个元素以一个坐标点为旋转中心点旋转，在使用旋转动画(RotateTransform)的时候需要注意两点：旋转中心点(Center)和旋转角度(Angle)。RotateTransform 由以下属性定义：Angle 围绕点 CenterX、CenterY 将对象旋转指定的角度。在使用 RotateTransform 时，变换将围绕某个特定对象的参考框架的原点旋转其坐标系。因此，根据对象的位置，对象可能不会就地(围绕其中心)旋转。例如，如果对象位于 X 轴上距 0 点 200 个单位的位置，则旋转 30°可以让该对象沿着以原点为圆心、半径为 200 的圆摆动 30°。若要就地旋转某个对象，应将 RotateTransform 的 CenterX 和 CenterY 设置为该对象的旋转中心。可以使用 Canvas.Left 和 Canvas.Top 在 Canvas 上偏移对象的本地(0,0)，但不会将此视为变换；如果是变换，该对象会保留在它自己的本地(0,0)处。

下面给出旋转按钮的示例：单击按钮时使用 RotateTransform 旋转动画实现了按钮的旋转运动。

代码清单 7-11：旋转按钮(源代码：第 7 章\Examples_7_11)

MainPage.xaml 文件主要代码

```
<Button Content="会旋转的按钮"
        Grid.Row="0"
        HorizontalAlignment="Center"
        VerticalAlignment="Center"
        RenderTransformOrigin="0.5 0.5"
        Background="Blue"
        Click="OnButtonClick">
    <Button.RenderTransform>
        <RotateTransform />
    </Button.RenderTransform>
</Button>
```

MainPage.xaml.cs 文件主要代码

```
void OnButtonClick(object sender, RoutedEventArgs args)
{
    //获取单击的按钮对象
    Button btn = sender as Button;
    //获取按钮的 RenderTransform 属性
    RotateTransform rotateTransform = btn.RenderTransform as RotateTransform;
    //创建一个 DoubleAnimation 动画
    DoubleAnimation anima = new DoubleAnimation();
    anima.From = 0;//开始的值
    anima.To = 360;//结束的值
    anima.Duration = new Duration(TimeSpan.FromSeconds(0.5));//持续的时间
    // 设置动画的 Target 属性和 TargetProperty 属性
    Storyboard.SetTarget(anima, rotateTransform);
    Storyboard.SetTargetProperty(anima, new PropertyPath(RotateTransform.AngleProperty));
    // 创建 storyboard, 并且添加上 animation, 然后动画开始
    Storyboard storyboard = new Storyboard();
    storyboard.Children.Add(anima);
    storyboard.Begin();
}
```

程序运行的效果如图 7.11 所示。

7.3.4 缩放动画

ScaleTransform 称为缩放变换，它可以对元素沿 X 轴方向和 Y 轴方向按比例进行拉伸或收缩，ScaleX 属性指定使对象沿 X 轴拉伸或收缩的量，ScaleY 属性指定使对象沿 Y 轴拉伸或收缩的量，缩放操作以 CenterX 和 CenterY 属性指定的点为缩放中心，默认值为(0,0)，取值范围不是 0～1，而是相对于控件的偏移像素值。在 Windows Phone 的动画框架中，ScaleTransform 类提供了在二维空间中的坐标内进行缩放操作，通过 ScaleTransform 可以在水平或垂直方向缩放和拉伸对象，以实现一个简单的缩放动画效果，故此将其称为缩放动画(ScaleTransform)。使用 ScaleTransform 需要特别关注两点：中心点坐标和 X、Y 轴方

图 7.11 会旋转的按钮

向的缩放比例，比例值越小则对象元素就越小（即收缩），比例值越大则对象元素就越大（即呈现为放大效果）。

下面给出缩放动画的示例：创建一个矩形的缩放动画。

代码清单 7-12：缩放动画（源代码：第 7 章\Examples_7_12）

MainPage.xaml 文件主要代码

```
<Grid x:Name = "ContentPanel" Grid.Row = "1" Margin = "12,0,12,0">
    <Canvas>
        <Canvas.Resources>
            <Storyboard x:Name = "storyBoard">
                <DoubleAnimation Storyboard.TargetName = "scaleTransform"
                        Storyboard.TargetProperty = "ScaleY"
                        From = "1" To = "2"
                        Duration = "0:0:3"
                        RepeatBehavior = "Forever"
                        AutoReverse = "True">
                </DoubleAnimation>
            </Storyboard>
        </Canvas.Resources>
        <Rectangle x:Name = "rectangle" Height = "50" Width = "50" Canvas.Left = "75" Canvas.
Top = "75" Fill = "Blue">
            <Rectangle.RenderTransform>
                <ScaleTransform x:Name = "scaleTransform"></ScaleTransform>
            </Rectangle.RenderTransform>
        </Rectangle>
```

```
    </Canvas>
</Grid>
```

MainPage.xaml.cs 文件主要代码

```
public MainPage()
{
    InitializeComponent();
    //动画开始
    storyBoard.Begin();
}
```

程序运行的效果如图 7.12 所示。

图 7.12　缩放动画

7.3.5　倾斜动画

SkewTransform 称为倾斜变换或扭曲变换，使用它可以对元素围绕一点进行一定角度的倾斜，从而在二维空间中产生三维的感觉，它是一种以非均匀方式拉伸坐标空间的变换。通过属性 AngleX 和 AngleY 可以分别设置在 X 轴和 Y 轴上的倾斜角度。Windows Phone 中的倾斜变化动画(SkewTransform)能够实现对象元素的水平、垂直方向的倾斜变化动画效果。现实生活中的倾斜变化效果是非常常见的，比如翻书的纸张效果，关门或开门的时候门缝图形倾斜变换。在 Windows Phone 中实现一个倾斜变化的动画效果是非常简单的，如果利用 Blend 这种强大的设计工具来实现那更是锦上添花。倾斜效果的动画应用效果其实非常好看，使用倾斜变换需要注意两点：倾斜方向和倾斜中心点。可以以某点为倾斜中心点进行 X 或 Y 坐标方向进行倾斜。

下面给出倾斜动画的示例：创建一个矩形的倾斜动画。

代码清单 7-13：倾斜动画(源代码：第 7 章\Examples_7_13)

MainPage.xaml 文件主要代码

```
<Canvas >
    <Canvas.Resources>
        <Storyboard x:Name="Storyboard1">
            <PointAnimationUsingKeyFrames BeginTime="00:00:00"
        Storyboard.TargetName="Rectangle1" Storyboard.TargetProperty="(UIElement.
RenderTransformOrigin)">
                <EasingPointKeyFrame KeyTime="00:00:00" Value="1,0.5"/>
                <EasingPointKeyFrame KeyTime="00:00:03" Value="1,0.5"/>
            </PointAnimationUsingKeyFrames>
            <DoubleAnimationUsingKeyFrames BeginTime="00:00:00" Storyboard.TargetName=
"Rectangle1"
        Storyboard.TargetProperty="(UIElement.RenderTransform).(TransformGroup.Children)
[0].(SkewTransform.AngleY)">
                <EasingDoubleKeyFrame KeyTime="00:00:03" Value="-17"/>
            </DoubleAnimationUsingKeyFrames>
        </Storyboard>
```

```
    </Canvas.Resources>
    <Rectangle Width="200" Height="269"
        Canvas.Left="20" Canvas.Top="10"
        Fill="Blue" Stroke="Black"
        StrokeThickness="4"
        x:Name="Rectangle1">
        <Rectangle.RenderTransform>
            <TransformGroup>
                <SkewTransform/>
            </TransformGroup>
        </Rectangle.RenderTransform>
    </Rectangle>
</Canvas>
```

程序运行的效果如图 7.13 所示。

图 7.13 倾斜动画

多 媒 体

Windows Phone 提供了常见媒体的编码、解码机制，因此非常容易地继承音频、视频等多媒体文件到应用程序中。使用 Windows Phone 提供的 API 和 MediaElement 控件，可以实现音乐播放视频播放等应用程序，当然，有些需要硬件的支持。Windows Phone 支持的视频格式有 3GP、3G2、MP4、AVI、ASF（WMV）等格式，支持的音频格式有 m4a、m4b、mp3、wma 等格式。

8.1 MediaElement 元素

MediaElement 可以播放许多不同类型的音频和视频媒体。MediaElement 基本上是一个矩形区域，可以在其图面上显示视频或播放音频（在这种情况下将不显示视频，但 MediaElement 仍然充当具有相应 API 的播放器对象）。因为 MediaElement 是一个 UIElement，所以，它支持输入操作，并可以捕获焦点。使用属性 Height 和 Width 可以指定视频显示图面的高度和宽度。但是，为了获得最佳性能，应避免显式设置 MediaElement 的宽度和高度。而是将这些值保留为未设置。指定源之后，媒体将以其实际大小显示，布局将重新计算该大小。如果需要更改媒体显示的大小，最好使用媒体编码工具将媒体重新编码为所需大小。默认情况下，加载 MediaElement 对象后，将立即播放由 Source 属性定义的媒体。

播放本地视频文件的 XAML 语法如下：

```
< MediaElement Source = "test.wmv" AutoPlay = "True"/> < MediaElement Source = "test.wmv"
AutoPlay = "True"/>
```

播放远程视频文件的 XAML 语法如下：

```
< MediaElement Source = "http://mschannel9.vo.msecnd.net/o9/mix/09/wmv/key01.wmv" AutoPlay =
"True"/>
```

8.1.1 MediaElement 类的属性、事件和方法

MediaElement 类是在命名空间 System.Windows.Controls 下的类，属于 Windows

Phone 的 Silverlight 控件。不过它的属性、方法和事件比较多,下面来看一下它的一些重要的属性、方法和事件,分别如表 8.1、表 8.2 以及表 8.3 所示。

表 8.1 MediaElement 类的一些重要的属性

名 称	说 明
AudioStreamCount	获取当前媒体文件中可用的音频流的数目
AudioStreamIndex	获取或设置与视频组件一起播放的音频流的索引,音频流的集合在运行时组合,并且表示可用于媒体文件内的所有音频流
AutoPlay	获取或设置一个值,该值指示在设置 Source 属性时媒体是否将自动开始播放
Balance	获取或设置立体声扬声器的音量比
BufferingProgress	获取指示当前缓冲进度的值
BufferingTime	获取或设置要缓冲的时间长度
CanPause	获取一个值,该值指示在调用 Pause 方法时媒体是否可暂停
CanSeek	获取一个值,该值指示是否可以通过设置 Position 属性的值来重新定位媒体
CurrentState	获取 MediaElement 的状态
DownloadProgress	获取一个百分比值,该值指示为位于远程服务器上的内容完成的下载量
DownloadProgressOffset	获取下载进度的偏移量
DroppedFramesPerSecond	获取媒体每秒正在丢弃的帧数
IsMuted	获取或设置一个值,该值指示是否已静音
IsUsedForExternalVideoOnly	获取或设置一个值,该值指示是否使用外部视频
NaturalDuration	获取当前打开的媒体文件的持续时间
NaturalVideoHeight	获取与媒体关联的视频的高度
NaturalVideoWidth	获取与媒体关联的视频的宽度
Position	获取或设置媒体播放时间的当前进度位置
RenderedFramesPerSecond	获取媒体每秒正在呈现的帧数
Source	获取或设置 MediaElement 上的媒体来源
Stretch	获取或设置一个 Stretch 值,该值描述 MediaElement 如何填充目标矩形
VideoSessionHandle	获取视频会话的句柄
Volume	获取或设置媒体的音量

表 8.2 MediaElement 类的一些重要的方法

名 称	说 明
Pause	在当前位置暂停媒体
Play	从当前位置播放媒体
RequestLog	发送一个请求,以生成随后将通过 LogReady 事件引发的记录
SetSource(MediaStreamSource)	这会将 MediaElement 的源设置为 MediaStreamSource 的子类

续表

名　　称	说　　明
SetSource(Stream)	使用提供的流设置 Source 属性
Stop	停止媒体并将其重设为从头播放
ToString	返回表示当前 Object 的 String(继承自 Object)

表 8.3　MediaElement 类的一些重要的事件

名　　称	说　　明
BufferingProgressChanged	当 BufferingProgress 属性更改时发生
CurrentStateChanged	当 CurrentState 属性的值更改时发生
DownloadProgressChanged	在 DownloadProgress 属性更改后发生
LogReady	当日志准备就绪时发生
MediaEnded	当 MediaElement 不再播放音频或视频时发生
MediaFailed	在存在与媒体 Source 关联的错误时发生
MediaOpened	当媒体流已被验证和打开且已读取文件头时发生

8.1.2　MediaElement 的状态

MediaElement 的当前状态(Buffering、Closed、Error、Opening、Paused、Playing 或 Stopped)会影响使用媒体的用户，如果某用户正在尝试查看一个大型视频，则 MediaElement 将可能长时间保持在 Buffering 状态。在这种情况下，可能希望用户界面(UI)中提供某种还不能播放媒体的提示。当缓冲完成时，可能希望指示现在可以播放媒体。

表 8.4 概括了 MediaElement 可以处于的不同状态，这些状态与 MediaElementState 枚举的枚举值相对应。

表 8.4　MediaElement 的状态

值	说　　明
AcquiringLicense	仅在播放 DRM 受保护的内容时适用：MediaElement 正在获取播放 DRM 受保护的内容所需的许可证；调用 OnAcquireLicense 后，MediaElement 将保持在此状态下，直到调用了 SetLicenseResponse
Buffering	MediaElement 正在加载要播放的媒体。在此状态中，它的 Position 不前进；如果 MediaElement 已经在播放视频，则它将继续以显示所显示的上一帧
Closed	MediaElement 不包含媒体。MediaElement 显示透明帧
Individualizing	仅在播放 DRM 受保护的内容时适用：MediaElement 正在确保正确的个性化组件(仅在播放 DRM 受保护的内容时适用)安装在用户计算机上；有关更多信息，请参见数字版权管理(DRM)
Opening	MediaElement 正在进行验证，并尝试打开由其 Source 属性指定的统一资源标识符(URI)

续表

值	说明
Paused	MediaElement 不会使它的 Position 前进；如果 MediaElement 正在播放视频，则它将继续以显示当前帧
Playing	MediaElement 正在播放其源属性指定的媒体，它的 Position 向前推进
Stopped	MediaElement 包含媒体，但未播放或已暂停，它的 Position 为 0，并且不前进。如果加载的媒体为视频，则 MediaElement 显示第一帧

表 8.5 概括了这些 MediaElement 状态与在 MediaElement 上采取的操作(如调用 Play 方法、调用 Pause 方法等)。如表 8.5 所示，可供 MediaElement 使用的状态取决于其当前状态。例如，对于当前处于 Playing 状态的 MediaElement，如果更改了 MediaElement 的源，则状态更改为 Opening；如果调用了 Play 方法，则没有任何情况发生(无选项)；如果调用了 Pause 方法，则状态更改为 Paused，等。避免“未指定”状态。例如，当媒体处于 Opening 状态时不应调用 Play 方法。为避免发生这种情况，可以在允许调用 Play 之前检查 MediaElement 的 CurrentState。可以通过注册 CurrentStateChanged 事件跟踪到媒体的状态变化。CurrentStateChanged 事件可能会没有按预期工作。当状态迅速更改时，事件可能合并到一个事件引发中。例如，CurrentState 属性可能会从 Playing 切换为 Buffering 并很快又切换回 Playing，以致只引发了单个 CurrentStateChanged 事件，在这种情况下，该属性将并不表现为具有已更改的值。此外，应用程序不应采用事件发生的顺序，尤其是针对 Buffering 之类的瞬态。在事件报告中可能会跳过某一瞬态，因为该瞬态发生得太快。

表 8.5 MediaElement 状态与在 MediaElement 上采取的操作

状态	源设置	Play()	Pause()	Stop()	Seek()	默认退出条件
Closed (default)	Opening	未指定	未指定	未指定	未指定	
Opening	Opening (新源)	未指定	未指定	未指定	未指定	如果源有效：Buffering (如果 AutoPlay == true) 或 Stopped (如果 AutoPlay == false) (MediaOpened) 如果源无效：Opening (MediaFailed)
Buffering	Opening	Playing	Paused	Stopped	Buffering (新位置)	BufferingTime 到达：Playing
Playing	Opening	无选项	Paused	Stopped	Buffering (新位置)	流结尾：Paused 缓冲区结尾：Buffering
Paused	Opening	Buffering	无选项	Stopped	Paused (新位置)	
Stopped	Opening	Buffering	Paused	无选项	Paused (新位置)	

8.2 本地音频播放

在 Windows Phone 应用程序中，我们可以将音频文件直接放到程序包里面，然后在程序中就可以通过 MediaElement 元素，用相对的路径进行加载音频文件来播放。

下面给出播放程序音乐的示例：播放应用程序本地加载的音乐。

代码清单 8-1：播放程序音乐（源代码：第 8 章\Examples_8_1）

MainPage. xaml 文件主要代码

```
<Grid x:Name = "ContentPanel" Grid.Row = "1" Margin = "12,0,12,0">
    <MediaElement x:Name = "sound" />
    <TextBlock Text = "请选择你要播放的歌曲"
FontSize = " 30" Height = " 59" HorizontalAlignment = " Left" Margin = " 6, 6, 0, 0" Name =
"textBlock1" VerticalAlignment = "Top" Width = "409" />
    <RadioButton Content = "罗志祥 - Touch My Heart"
Height = " 72" HorizontalAlignment = " Left" Margin = " 0, 71, 0, 0" Name = " radioButton1"
VerticalAlignment = "Top" Width = "394" IsChecked = "True" />
    <RadioButton Content = "陈小春 - 独家记忆"
Height = " 72" HorizontalAlignment = " Left" Margin = " 0, 149, 0, 0" Name = " radioButton2"
VerticalAlignment = "Top" />
            <RadioButton Content = "大灿 - 贝多芬的悲伤"
Height = " 72" HorizontalAlignment = " Left" Margin = " 0, 227, 0, 0" Name = " radioButton3"
VerticalAlignment = "Top" />
            <RadioButton Content = "筷子兄弟 - 老男孩"
Height = " 72" HorizontalAlignment = " Left" Margin = " 0, 305, 0, 0" Name = " radioButton4"
VerticalAlignment = "Top" />
            <RadioButton Content = "梁静茹 - 比较爱"
Height = " 72" HorizontalAlignment = " Left" Margin = " 0, 383, 0, 0" Name = " radioButton5"
VerticalAlignment = "Top" />
    <!-- 播放按钮触发播放的事件 -->
    <Button Content = "播放"
Height = "72" HorizontalAlignment = "Left" Margin = "0,476,0,0" Name = "play" VerticalAlignment =
"Top" Width = "160" Click = "play_Click" />
    <!-- 暂停按钮触发暂停的事件 -->
    <Button Content = "暂停"
Height = " 72" HorizontalAlignment = " Left" Margin = " 135, 476, 0, 0" Name = " pause"
VerticalAlignment = "Top" Width = "160" Click = "pause_Click" />
 <!-- 停止按钮触发停止的事件 -->
    <Button Content = "停止"
Height = " 72" HorizontalAlignment = " Left" Margin = " 272, 476, 0, 0" Name = " stop"
VerticalAlignment = "Top" Width = "160" Click = "stop_Click" />
</Grid>
```

MainPage.xaml.cs 文件主要代码

```
//播放音乐
private void play_Click(object sender, RoutedEventArgs e)
{
    //设置 MediaElement 的源
    if (radioButton1.IsChecked == true)
    {
        sound.Source = new Uri("TouchMyHeart.mp3", UriKind.Relative);
    }
    else if (radioButton2.IsChecked == true)
    {
        sound.Source = new Uri("2.mp3", UriKind.Relative);
    }
    else if (radioButton3.IsChecked == true)
    {
        sound.Source = new Uri("3.mp3", UriKind.Relative);
    }
    else if (radioButton4.IsChecked == true)
    {
        sound.Source = new Uri("4.mp3", UriKind.Relative);
    }
    else if (radioButton5.IsChecked == true)
    {
        sound.Source = new Uri("5.mp3", UriKind.Relative);
    }
    else
    {
        sound.Source = new Uri("TouchMyHeart.mp3", UriKind.Relative);
    }
    //播放
    sound.Play();
}
//暂停播放音乐
private void pause_Click(object sender, RoutedEventArgs e)
{
    sound.Pause();
}
//停止播放音乐
private void stop_Click(object sender, RoutedEventArgs e)
{
    sound.Stop();
}
```

程序运行的效果如图 8.1 所示。

图 8.1　播放本地 MP3

8.3　网络音频播放

在 Windows Phone 应用程序中，播放网络音频和播放本地音频的处理方式差不多，只不过播放本地音频的时候使用的是音频本地相对路径，而播放网络音频的时候使用的是网络音频地址。需要注意的是，播放网络音频需要在手机联网的状态下才能够成功地播放，否则会抛出异常，所以在播放网络音频的时候需要检查手机的联网状态或者捕获该异常进行处理。

下面给出播放网络音乐的示例：播放网络上的 mp3 音乐，并将播放的记录保存到程序的独立存储里面。

代码清单 8-2：播放网络音乐(源代码：第 8 章\Examples_8_2)

MainPage.xaml 文件主要代码

```
<Grid x:Name = "ContentPanel" Grid.Row = "1" Margin = "12,0,12,0">
    <ListBox Height = "200" HorizontalAlignment = "Left" Margin = "6,79,0,0" Name = "listBox1"
VerticalAlignment = "Top" Width = "444" />
    <TextBlock FontSize = "30" Height = "43" HorizontalAlignment = "Left" Margin = "12,285,0,
0" Name = "textBlock2" Text = "请输入 mp3 的网络地址：" VerticalAlignment = "Top" Width = "374" />
    <TextBox Height = "72" HorizontalAlignment = "Left" Margin = "0,334,0,0" Name = "mp3Uri"
VerticalAlignment = "Top" Width = "460" Text = "http://localhost/2.mp3" />
    <Button Content = "播放" Height = "72" HorizontalAlignment = "Left" Margin = "0,428,0,0"
Name = "play" VerticalAlignment = "Top" Width = "220" Click = "play_Click" />
    <Button Content = "停止" Height = "72" HorizontalAlignment = "Left" Margin = "224,428,0,0"
Name = "stop" VerticalAlignment = "Top" Width = "220" Click = "stop_Click" />
    <MediaElement Height = "43" HorizontalAlignment = "Left" Margin = "6,555,0,0" Name =
"media" VerticalAlignment = "Top" Width = "438" />
```

```
    < TextBlock Height = " 54 " HorizontalAlignment = " Left " Margin = " 12, 23, 0, 0 " Name =
"textBlock1" FontSize = "30" Text = "播放的历史记录: " VerticalAlignment = "Top" Width = "317" />
</Grid>
```

MainPage. xaml. cs 文件代码

```
using System;
using System.Windows;
using Microsoft.Phone.Controls;
using System.IO.IsolatedStorage;
using System.Xml.Linq;
namespace OnlineMp3
{
    public partial class MainPage : PhoneApplicationPage
    {
        public MainPage()
        {
            InitializeComponent();
            RefreshIsoFiles();
        }
        //刷新播放的历史记录
        private void RefreshIsoFiles()
        {
            string[] fileList;
            //读取独立存储中的文件名
            using (var store = IsolatedStorageFile.GetUserStoreForApplication())
            {
                fileList = store.GetFileNames();
            }
            //将文件名绑定到 ListBox 控件上
            listBox1.ItemsSource = fileList;
        }
        //保存播放的历史记录
        private void savehistory()
        {
            //使用音频的 url 地址作为文件名
            string fileName = System.IO.Path.GetFileName(mp3Uri.Text);
            using (var store = IsolatedStorageFile.GetUserStoreForApplication())
            {
                if (!store.FileExists(fileName))
                {
                  //创建独立存储文件
                  IsolatedStorageFileStream file = store.CreateFile(fileName);
                }
            }
            //刷新播放的历史记录
            RefreshIsoFiles();
        }
        //播放音乐
        private void play_Click(object sender, RoutedEventArgs e)
        {
```

```
            try
            {
                if (!string.IsNullOrEmpty(mp3Uri.Text))
                {
                    media.Source = new Uri(mp3Uri.Text, UriKind.Absolute);
                    media.Play();
                    savehistory();
                }
                else
                {
                    MessageBox.Show("请输入 mp3 的网络地址!");
                }

            }
            catch (Exception)
            {
                //捕获播放失败的异常
                MessageBox.Show("无法播放!");
            }
        }
        //停止播放
        private void stop_Click(object sender, RoutedEventArgs e)
        {
            media.Stop();
        }
    }
}
```

程序运行的效果如图 8.2 所示。

图 8.2 播放网络 MP3

8.4　本地视频播放

在 Windows Phone 应用程序中播放视频依然还是使用 MediaElement 元素去实现，与音频播放不同的是视频的播放还产生一个显示效果，也就是说 MediaElement 元素就必须要有一定的宽度和高度才能够把视频展现出来。

下面给出播放本地视频的示例：播放程序本地的视频，并将播放的进度用进度条显示出来。

代码清单 8-3：播放本地视频（源代码：第 8 章\Examples_8_3）

MainPage.xaml 文件主要代码

```
<Grid x:Name = "LayoutRoot" Background = "Transparent">
    …
    <Grid x:Name = "ContentPanel" Grid.Row = "1" Margin = "12,0,12,0">
        …
        <!-- 添加 MediaElement 多媒体播放控件 -->
        <MediaElement Name = "myMediaElement" AutoPlay = "True" Grid.Row = "0" />
        <ProgressBar Name = "pbVideo" Grid.Row = "1" />
    </Grid>
</Grid>
<!-- 3 个菜单栏：播放、暂停和停止 -->
<phone:PhoneApplicationPage.ApplicationBar>
    <shell:ApplicationBar IsVisible = "True" IsMenuEnabled = "True" >
        <shell:ApplicationBarIconButton IconUri = "/icons/play.png" Click = "Play_Click"
Text = "播放"/>
        <shell:ApplicationBarIconButton IconUri = "/icons/pause.png" Click = "Pause_Click"
Text = "暂停"/>
        <shell:ApplicationBarIconButton IconUri = "/icons/stop.png" Click = "Stop_Click"
Text = "停止"/>
    </shell:ApplicationBar>
</phone:PhoneApplicationPage.ApplicationBar>
```

MainPage.xaml.cs 文件代码

```
using System;
using System.Windows;
using System.Windows.Media;
using Microsoft.Phone.Controls;
using System.Windows.Threading;
using Microsoft.Phone.Shell;
namespace MediaPlayer
{
    public partial class MainPage : PhoneApplicationPage
    {
        // 使用定时器来处理视频播放的进度条
        DispatcherTimer currentPosition = new DispatcherTimer();
        // 页面的初始化
```

```
        public MainPage()
        {
            InitializeComponent();
            //定义多媒体流可用并被打开时触发的事件
            myMediaElement. MediaOpened  +=  new  RoutedEventHandler ( myMediaElement _
MediaOpened);
            //定义多媒体停止时触发的事件
            myMediaElement. MediaEnded  +=  new  RoutedEventHandler  ( myMediaElement _
MediaEnded);
            //定义多媒体播放状态改变时触发的事件
            myMediaElement.CurrentStateChanged += new RoutedEventHandler(myMediaElement_
CurrentStateChanged);
            //定义定时器触发的事件
            currentPosition.Tick += new EventHandler(currentPosition_Tick);
            //设置多媒体控件的网络视频资源
            myMediaElement.Source = new Uri("123.wmv", UriKind.Relative);
        }
        //视频状态改变时的处理事件
        void myMediaElement_CurrentStateChanged(object sender, RoutedEventArgs e)
        {
            if (myMediaElement.CurrentState == MediaElementState.Playing)
            {//播放视频时各菜单的状态
                currentPosition.Start();
                ((ApplicationBarIconButton)ApplicationBar.Buttons[0]).IsEnabled = false;
                                                        //播放
                ((ApplicationBarIconButton)ApplicationBar.Buttons[1]).IsEnabled = true;
                                                        //暂停
                ((ApplicationBarIconButton)ApplicationBar.Buttons[2]).IsEnabled = true;
                                                        //停止
            }
            else if (myMediaElement.CurrentState == MediaElementState.Paused)
            { //暂停视频时各菜单的状态
                currentPosition.Stop();
                ((ApplicationBarIconButton)ApplicationBar.Buttons[0]).IsEnabled = true;
                ((ApplicationBarIconButton)ApplicationBar.Buttons[1]).IsEnabled = false;
                ((ApplicationBarIconButton)ApplicationBar.Buttons[2]).IsEnabled = true;
            }
            else
            {//停止视频时各菜单的状态
                currentPosition.Stop();
                ((ApplicationBarIconButton)ApplicationBar.Buttons[0]).IsEnabled = true;
                ((ApplicationBarIconButton)ApplicationBar.Buttons[1]).IsEnabled = false;
                ((ApplicationBarIconButton)ApplicationBar.Buttons[2]).IsEnabled = false;
            }
        }
        //多媒体停止时触发的事件
        void myMediaElement_MediaEnded(object sender, RoutedEventArgs e)
        {
            //停止播放
            myMediaElement.Stop();
```

```
        }
        //多媒体流可用并被打开时触发的事件
        void myMediaElement_MediaOpened(object sender, RoutedEventArgs e)
        {
            //获取多媒体视频的总时长来设置进度条的最大值
            pbVideo.Maximum = (int)myMediaElement.NaturalDuration.TimeSpan.TotalMilliseconds;
            //播放视频
            myMediaElement.Play();
        }
        //定时器触发的事件
        void currentPosition_Tick(object sender, EventArgs e)
        {
            //获取当前视频播放了的时长来设置进度条的值
            pbVideo.Value = (int)myMediaElement.Position.TotalMilliseconds;
        }
        //播放视频菜单事件
        private void Play_Click(object sender, EventArgs e)
        {
            myMediaElement.Play();
        }
        //暂停视频菜单事件
        private void Pause_Click(object sender, EventArgs e)
        {
            myMediaElement.Pause();
        }
        //停止视频菜单事件
        private void Stop_Click(object sender, EventArgs e)
        {
            myMediaElement.Stop();
        }
    }
}
```

程序运行的效果如图 8.3 所示。

图 8.3　播放本地视频

8.5　网络视频播放

网络视频的播放处理方式是和本地视频播放相类似的，只不过是 MediaElement 的源换成了网络的视频文件。因为播放网络视频会受到网络的好坏影响，不过 Windows Phone 8 系统内部已经做了缓冲的处理，可以通过 MediaElement 类的 BufferingProgress 属性来获取视频缓冲的信息。

下面给出播放网络视频的示例：播放网络上的视频文件，并监控视频播放缓冲的情况和播放的进度。

代码清单 8-4：播放网络视频(源代码：第 8 章\Examples_8_4)

MainPage.xaml 文件主要代码

```
<Grid x:Name = "ContentGrid" Grid.Row = "1">
        <MediaElement Height = "289" HorizontalAlignment = "Left"
                  Margin = "26,148,0,0" x:Name = "mediaPlayer"
                  VerticalAlignment = "Top" Width = "417"
                  AutoPlay = "False"/>
    <Button Content = "&gt;" Height = "72"
            HorizontalAlignment = "Left" Margin = "13,527,0,0"
            x:Name = "btnPlay" VerticalAlignment = "Top" Width = "87"
            Click = "btnPlay_Click" />
    <Button Content = "O" Height = "72"
            HorizontalAlignment = "Right" Margin = "0,527,243,0"
            x:Name = "btnStop" VerticalAlignment = "Top" Width = "87"
            Click = "btnStop_Click" />
    <Button Content = "||" Height = "72" Margin = "0,527,313,0"
            x:Name = "btnPause" VerticalAlignment = "Top"
            Click = "btnPause_Click" HorizontalAlignment = "Right" Width = "87" />
    <!-- 播放的进度条 -->
    <Slider Height = "84" HorizontalAlignment = "Left"
            Margin = "13,423,0,0" Name = "mediaTimeline"
            VerticalAlignment = "Top" Width = "443"
            ValueChanged = "mediaTimeline_ValueChanged"
            Maximum = "1" LargeChange = "0.1" />
    <TextBlock Height = "30" HorizontalAlignment = "Left"
               Margin = "26,472,0,0" Name = "lblStatus"
               Text = "00:00" VerticalAlignment = "Top" Width = "88" FontSize = "16" />
    <TextBlock Height = "30"
           Margin = "118,472,222,0" x:Name = "lblBuffering"
           Text = "缓冲" VerticalAlignment = "Top" FontSize = "16" />
    <TextBlock Height = "30"
           Margin = "0,472,82,0" x:Name = "lblDownload"
           Text = "下载" VerticalAlignment = "Top" FontSize = "16" HorizontalAlignment = "Right"
Width = "140" />
    <Button Content = "声音开关" Height = "72"
            HorizontalAlignment = "Left" Margin = "217,527,0,0"
            Name = "btnMute" VerticalAlignment = "Top" Width = "115"
            FontSize = "16" Click = "btnMute_Click" />
    <TextBlock Height = "30" HorizontalAlignment = "Left"
               Margin = "324,551,0,0" Name = "lblSoundStatus"
               Text = "声音开" VerticalAlignment = "Top" Width = "128" />
    <TextBox x:Name = "txtUrl" Height = "57" Margin = "91,33,8,0"
             TextWrapping = "Wrap" VerticalAlignment = "Top" FontSize = "16"
             Text = "http://localhost/123.wmv"/>
    <TextBlock x:Name = "lblUrl" HorizontalAlignment = "Left" Height = "25"
               Margin = "8,48,0,0" TextWrapping = "Wrap" Text = "视频地址:"
               VerticalAlignment = "Top" Width = "83" FontSize = "16"/>
    <Button Content = "加载视频" Height = "72" HorizontalAlignment = "Left" Margin = "91,78,0,
0" Name = "load" VerticalAlignment = "Top" Width = "339" Click = "load_Click" />
</Grid>
```

MainPage. xaml. cs 文件代码

```
using System;
using System.Windows;
using System.Windows.Media;
using Microsoft.Phone.Controls;
using Microsoft.Phone.Tasks;
namespace MediaPlayerDemo
{
    public partial class MainPage : PhoneApplicationPage
    {
        //是否更新播放的进度显示
        private bool _updatingMediaTimeline;
        public MainPage()
        {
            InitializeComponent();
        }
        //暂停播放
        private void btnPause_Click(object sender, RoutedEventArgs e)
        {
            if (mediaPlayer.CanPause)
            {
                mediaPlayer.Pause();
                lblStatus.Text = "暂停";
            }
            else
            {
                lblStatus.Text = "不能暂停,请重试!";
            }
        }
        //停止播放
        private void btnStop_Click(object sender, RoutedEventArgs e)
        {
            mediaPlayer.Stop();
            mediaPlayer.Position = System.TimeSpan.FromSeconds(0);
            lblStatus.Text = "停止";
        }
        //开始播放
        private void btnPlay_Click(object sender, RoutedEventArgs e)
        {
            mediaPlayer.Play();
        }
        //声音开关
        private void btnMute_Click(object sender, RoutedEventArgs e)
        {
            if (lblSoundStatus.Text.Equals("声音开", StringComparison.CurrentCultureIgnoreCase))
            {
                lblSoundStatus.Text = "声音关";
                mediaPlayer.IsMuted = true;
            }
```

```
                else
                {
                    lblSoundStatus.Text = "声音开";
                    mediaPlayer.IsMuted = false;
                }
            }
            //播放进度
            private void mediaTimeline_ValueChanged(object sender, RoutedPropertyChangedEventArgs
<double> e)
            {
                if (!_updatingMediaTimeline && mediaPlayer.CanSeek)
                {
                    TimeSpan duration = mediaPlayer.NaturalDuration.TimeSpan;
                    int newPosition = (int)(duration.TotalSeconds * mediaTimeline.Value);
                    mediaPlayer.Position = new TimeSpan(0, 0, newPosition);
                }
            }
            //加载视频
            private void load_Click(object sender, RoutedEventArgs e)
            {
                _updatingMediaTimeline = false;
                mediaPlayer.Source = new Uri(txtUrl.Text);
                // 视频播放的位置从第 0 秒开始
                mediaPlayer.Position = System.TimeSpan.FromSeconds(0);
                //更新播放的进度百分比
                mediaPlayer.DownloadProgressChanged += (s, ee) =>
                {
                    lblDownload.Text = string.Format("下载 {0:0.0%}", mediaPlayer.DownloadProgress);
                };
                // 预设一分钟缓冲视频的时间
                mediaPlayer.BufferingTime = TimeSpan.FromSeconds(60);
                //更新缓冲的进度百分比
                mediaPlayer.BufferingProgressChanged += (s, ee) =>
                {
                    lblBuffering.Text = string.Format("缓冲 {0:0.0%}", mediaPlayer.BufferingProgress);
                };
                // 更新视频播放的进度条
                CompositionTarget.Rendering += (s, ee) =>
                {
                    _updatingMediaTimeline = true;
                    TimeSpan duration = mediaPlayer.NaturalDuration.TimeSpan;
                    if (duration.TotalSeconds != 0)
                    {
                        double percentComplete = mediaPlayer.Position.TotalSeconds / duration.
TotalSeconds;
                        mediaTimeline.Value = percentComplete;
                        TimeSpan mediaTime = mediaPlayer.Position;
                        string text = string.Format("{0:00}:{1:00}",
                            (mediaTime.Hours * 60) + mediaTime.Minutes, mediaTime.Seconds);
                        if (lblStatus.Text != text)
```

```
                        lblStatus.Text = text;
                    _updatingMediaTimeline = false;
                }
            };
            mediaPlayer.Play();
        }
    }
}
```

程序运行的效果如图 8.4 所示。

图 8.4　播放网络视频

启动器与选择器

手机中会有很多系统自身的功能，比如拍照、发短信和打电话等，这类功能都是手机系统里面的一些系统级别的功能，那么在应用程序里面调用这些功能应该怎么处理呢？这就需要用到 Windows Phone 系统中的启动器和选择器，其实就是这些手机系统的功能提供的一些接口，供用户在自己的应用程序里面可以通过这些接口来访问系统的一些功能。

启动器和选择器框架使得 Windows Phone 应用程序能够向用户提供一套通用的任务，如打电话，发送电子邮件和拍照片。Windows Phone 应用程序模型将每个应用分离成各自独立的沙箱，包括运行时(包括内存的隔离)和文件存储。应用程序不能直接访问通用存储区的信息(如联系人列表)来直接调用电话或短信等其他应用。为了适应需要这些通用任务的场景，Windows 手机公布了一套启动器和选择器的 API，允许应用程序间接访问这些常用的手机功能。启动器和选择器的 API 调用独立的内置应用程序，取代当前运行的应用程序。只要实施正确，启动器和选择器框架可以为最终用户提供一个完全无缝的体验，使其完全感觉不到应用程序之间的切换。

9.1 使用启动器

启动器是一个“点火后不再理会”的动作，你可以使用它启动一个指定的 Windows Phone 功能，如发送短信，打开一个网页，或是打电话。启动器只是负责把相应的应用程序启动起来就可以了。

在 Windows Phone 8 中支持的启动器有以下的一些：

(1) EmailComposeTask ：允许应用程序启动电子邮件应用程序并创建一条新消息，以此来让用户从应用程序发送电子邮件。

(2) PhoneCallTask ：允许应用程序启动电话应用程序，使得用户能够在应用程序中开始打电话。

(3) SearchTask：允许应用程序启动 Web 搜索应用程序。

(4) SmsComposeTask：允许应用程序启动 SMS 应用程序。

(5) WebBrowserTask：允许应用程序启动 Web 浏览器应用程序。

(6) MediaPlayerLauncher：允许应用程序启动媒体播放器。

(7) MarketplaceDetailTask：允许应用程序启动 Windows Phone Market 客户端应用程序并显示指定产品的详细信息页面。

(8) MarketplaceHubTask：允许应用程序启动 Windows Phone Market 的客户端应用程序。

(9) MarketplaceReviewTask：允许应用程序启动 Windows Phone Market 客户端应用程序并显示指定产品的评论信息页面。

(10) MarketplaceSearchTask：允许应用程序启动 Windows Phone Market 客户端应用程序并显示指定搜索条件的检索结果。

(11) BingMapsTask：允许应用程序启动 Bing 地图。

(12) BingMapsDirectionsTask：允许应用程序启动 Bing 地图应用程序，以此指定起始位置或结束位置，或两者都指定，用于显示驾驶的方向。

(13) ConnectionSettingsTask：允许应用程序启动一个"设置"对话框，该对话框允许用户更改设备的网络连接设置。

(14) SaveAppointmentTask：允许启动日程安排的保存任务。

(15) MapDownloaderTask：允许在应用程序里面启动地图设置界面。

(16) MapsTask：允许应用程序启动诺基亚地图。

(17) MapsDirectionsTask：允许应用程序启动诺基亚地图应用程序，以此指定起始位置或结束位置，或两者都指定，用于显示驾驶的方向。

(18) ShareMediaTask：允许在应用程序里面共享一个多媒体文件到社交网络。

(19) ShareLinkTask：允许在应用程序里面共享一个链接到社交网络。

(20) ShareStatusTask：允许在应用程序里面共享一个状态信息到社交网络。

9.1.1　发邮件(EmailComposeTask)

EmailComposeTask 让你可以呼叫出系统预设的发送 Email 功能，并且在呼叫之前，能够设定收件人、邮件内容等信息，但是在模拟器中执行的时候会报错，这是因为内建的开发用模拟器没有设定 email 相关的账号，因此无法做发送 email 的动作，只能将应用程式部属到机器上做实际的测试的才能真正地把邮件发送出去。EmailComposeTask 类的相关属性如表 9.1 所示。

表 9.1　EmailComposeTask 类的相关属性

名　称	说　明
Bcc	获取或设置新的电子邮件线的收件人
Body	获取或设置新的电子邮件消息正文
Cc	获取或设置新的电子邮件抄送行中的收件人
Subject	获取或设置新的电子邮件的主题
To	获取或设置新的电子邮件消息的收件人地址

下面给出发邮件的示例：使用 EmailComposeTask 启动器发送邮件。

代码清单 9-1：播放程序音乐（源代码：第 9 章\Examples_9_1）

MainPage. xaml 文件主要代码

```
<Grid x:Name = "ContentPanel" Grid.Row = "1" Margin = "12,0,0,0">
    <TextBlock FontSize = "30" Height = "51" HorizontalAlignment = "Left" Margin = "12,41,0,0"
Name = "to" Text = "收件人：" VerticalAlignment = "Top" Width = "136" />
    <TextBox Height = "72" HorizontalAlignment = "Left" Margin = "154, 19, 0, 0" Name =
"textBox1" Text = "" VerticalAlignment = "Top" Width = "308" />
    <TextBlock FontSize = "30" Height = "50" HorizontalAlignment = "Left" Margin = "9,148,0,0"
Name = "title" Text = "邮件标题：" VerticalAlignment = "Top" />
    <TextBox Height = "72" HorizontalAlignment = "Left" Margin = "154, 126, 0, 0" Name =
"textBox2" Text = "" VerticalAlignment = "Top" Width = "308" />
    <TextBlock FontSize = "30" Height = "44" HorizontalAlignment = "Left" Margin = "12,217,0,
0" Name = "textBlock3" Text = "邮件内容" VerticalAlignment = "Top" Width = "147" />
    <TextBox Height = "248" HorizontalAlignment = "Left" Margin = "6,267,0,0" Name = "body"
Text = "" VerticalAlignment = "Top" Width = "460" />
    <Button Content = "发送" Height = "72" HorizontalAlignment = "Left" Margin = "154,521,0,0"
Name = "send" VerticalAlignment = "Top" Width = "160" Click = "send_Click" />
</Grid>
```

MainPage. xaml. cs 文件主要代码

```
private void send_Click(object sender, RoutedEventArgs e)
{
    //创建一个 EmailComposeTask 任务
    EmailComposeTask emailComposeTask = new EmailComposeTask();
    //设置收件人
    emailComposeTask.To = to.Text;
    //设置邮件的标题
    emailComposeTask.Subject = title.Text;
    //设置邮件的内容
    emailComposeTask.Body = body.Text;
    //发送邮件
    emailComposeTask.Show();
}
```

程序运行的效果如图 9.1 所示。

9.1.2 打电话（PhoneCallTask）

PhoneCallTask 是能够让你在应用程式中去执行拨打电话的功能，执行 PhoneCallTask 时，需要先指定电话号码以及显示在画面上的名称（DisplayName），之后呼叫 Show 的方法；呼叫 Show 方法之后，首先会请使用者确认是否要拨打电话，之后便会进行拨打电话的动作了。PhoneCallTask 类的相关属性如表 9.2 所示。

图 9.1　发送邮件

表 9.2 PhoneCallTask 类的相关属性

名 称	说 明
DisplayName	获取或设置时显示的电话应用程序启动的名称
PhoneNumber	获取或设置的待拨打的电话号码

下面给出打电话的示例：使用 PhoneCallTask 启动器拨打电话。

代码清单 9-2：打电话(源代码：第 9 章\Examples_9_2)

MainPage. xaml 文件主要代码

```
<StackPanel x:Name="ContentPanel" Margin="12,0,12,0" Grid.Row="1" >
    <TextBlock FontSize="30" Height="35" Margin="8,0" TextWrapping="Wrap" Text="请输入你的电话号码: "/>
    <TextBox x:Name="txtPhoneNo" Height="72" Margin="8,0" TextWrapping="Wrap" Text=""/>
    <Button Content="拨号" Height="80" Margin="196,0,27,0" Name="call" Click="call_Click" />
</StackPanel>
```

MainPage. xaml. cs 文件主要代码

```
private void call_Click(object sender, RoutedEventArgs e)
{
    //创建拨打电话任务
    PhoneCallTask pct = new PhoneCallTask();
    pct.DisplayName = "拨打电话";
    pct.PhoneNumber = txtPhoneNo.Text;
    pct.Show();
}
```

程序运行的效果如图 9.2～图 9.4 所示。

图 9.2 打电话界面

图 9.3 确定打电话提示

图 9.4 拨打电话中

9.1.3 搜索(SearchTask)

SearchTask 能够让应用程序指定查询的关键字,并且去启动系统预设的查询功能。当第一次去启动 SearchTask 时,会询问使用者,是不是允许装置利用 GPS/AGPS 等方式去取得目前所在位置的一些相关资讯,确认之后就出现收寻结果的画面。SearchTask 类的相关属性如表 9.3 所示。

表 9.3 SearchTask 类的属性

名　称	说　明
SearchQuery	获取或设置的搜索的内容
PhoneNumber	获取或设置的你要拨打的电话号码

下面给出搜索内容的示例：使用 SearchTask 启动器搜索内容。

代码清单 9-3：搜索内容(源代码：第 9 章\Examples_9_3)

MainPage.xaml 文件主要代码

```
<StackPanel x:Name = "ContentPanel" Margin = "12,0,12,0" Grid.Row = "1" >
    <TextBlock FontSize = "30" Height = "35" Margin = "8,0" TextWrapping = "Wrap" Text = "请输入你的搜索内容: "/>
    <TextBox x:Name = "txtSearch" Height = "72" Margin = "8,0" TextWrapping = "Wrap" Text = ""/>
    <Button Content = "搜索" Height = "80" Margin = "196,0,27,0" Name = "search" Click = "search_Click" />
</StackPanel>
```

MainPage.xaml.cs 文件主要代码

```
private void search_Click(object sender, RoutedEventArgs e)
{
    //新建一个搜索任务
    SearchTask st = new SearchTask();
    //设置搜索的条件
    st.SearchQuery = txtSearch.Text;
    //开始搜索
    st.Show();
}
```

程序运行的效果如下图 9.5～图 9.7 所示。

9.1.4 发送短信(SmscomposeTask)

SmscomposeTask 让应用程序可以呼叫系统的简讯发送功能,使用的方式跟 EmailComposeTask 相当的相似,只要设定接收端的号码以及 SMS 内容之后,就可以启动系统预设的简讯发送界面了。SmscomposeTask 类的相关属性如表 9.4 所示。

图 9.5　搜索界面

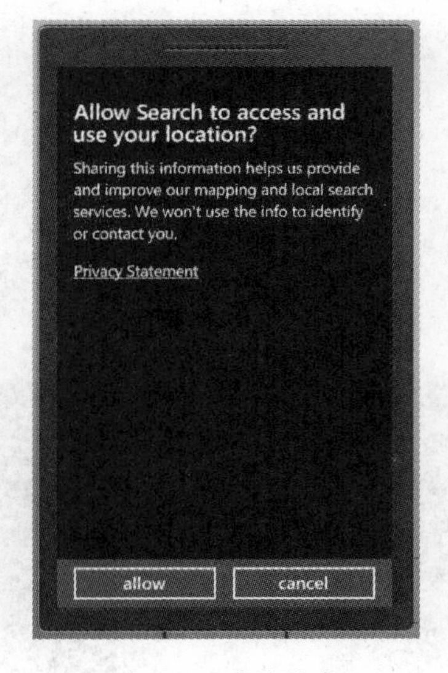

图 9.6　确认搜索

图 9.7　搜索的结果

表 9.4　SmscomposeTask 类的相关属性

名　称	说　明
Body	获取或设置短信发送的内容
To	获取或设置发送人列表

下面给出发送短信的示例：使用 SmscomposeTask 启动器发送短信。

代码清单 9-4：搜索内容(源代码：第 9 章\Examples_9_4)

MainPage.xaml 文件主要代码

```
<Grid x:Name = "ContentPanel" Grid.Row = "1" Margin = "12,0,12,0">
    <TextBlock FontSize = "30" Height = "51" HorizontalAlignment = "Left" Margin = "12,41,0,0"
Name = "to" Text = "收信人：" VerticalAlignment = "Top" Width = "136" />
    <TextBox Height = "72" HorizontalAlignment = "Left" Margin = "154,19,0,0" Name =
"textBox1" Text = "" VerticalAlignment = "Top" Width = "308" />
    <TextBlock FontSize = "30" Height = "44" HorizontalAlignment = "Left" Margin = "12,127,0,
0" Name = "textBlock3" Text = "短信内容:" VerticalAlignment = "Top" Width = "147" />
    <TextBox Height = "248" HorizontalAlignment = "Left" Margin = " - 4,191,0,0" Name = "body"
Text = "" VerticalAlignment = "Top" Width = "460" />
    <Button Content = "发送" Height = "72" HorizontalAlignment = "Left" Margin = "125,474,0,0"
Name = "send" VerticalAlignment = "Top" Width = "160" Click = "send_Click" />
</Grid>
```

MainPage.xaml.cs 文件主要代码

```
private void send_Click(object sender, RoutedEventArgs e)
{
    //创建一个发送短信的任务
```

```
        SmsComposeTask sct = new SmsComposeTask();
        //发送人
        sct.To = to.Text;
        //发送的短信内容
        sct.Body = body.Text;
        //开始启动发送任务
        sct.Show();
    }
```

程序运行的效果如图 9.8 和图 9.9 所示。

图 9.8　发送短信界面

图 9.9　发送短信成功界面

9.1.5　启动浏览器(WebBrowserTask)

WebBrowserTask 是启动系统内置浏览器的功能,并且前往你指定的 URL 位置。调用 WebBrowserTask 启动器将会自动使用系统内置的 IE 浏览器打开你的网址。WebBrowserTask 类的相关属性如表 9.5 所示。

表 9.5　WebBrowserTask 类的相关属性

名　　称	说　　明
URI	获取或设置的 Web 浏览器浏览的 URI
URL	获取或设置的 Web 浏览器浏览的 URL,绝对地址

下面给出启动浏览器的示例：使用 WebBrowserTask 启动器打开浏览器浏览指定的网页。

代码清单 9-5：启动浏览器(源代码：第 9 章\Examples_9_5)

MainPage.xaml 文件主要代码

```
<StackPanel x:Name = "ContentPanel" Margin = "12,0,12,0" Grid.Row = "1" >
    <TextBlock FontSize = "30" Height = "35" Margin = "8,0" TextWrapping = "Wrap" Text = "请输入你的网址: "/
    <TextBox x:Name = "txtUrl" Height = "72" Margin = "8,0" TextWrapping = "Wrap" Text = ""/>
    <Button Content = "前进" Height = "80" Margin = "196,0,27,0" Name = "go" Click = "go_Click" />
</StackPanel>
```

MainPage.xaml.cs 文件主要代码

```
private void go_Click(object sender, RoutedEventArgs e)
{
    //创建一个浏览器任务
    WebBrowserTask wbt = new WebBrowserTask();
    //设置浏览器的网址
    wbt.URL = txtUrl.Text;
    //启动浏览器任务
    wbt.Show();
}
```

程序运行的效果如图 9.10 和 9.11 所示。

图 9.10　启动浏览器

图 9.11　打开的网页浏览效果

9.1.6　播放多媒体(MediaPlayerLanucher)

MediaPlayerLanucher 的功能是启动和播放多媒体文件。第 8 章介绍过使用 MediaElement 元素来播放多媒体文件,那么使用 MediaPlayerLanucher 启动器是另外的一种播放多媒体文件的方式,这是利用了系统内部的多媒体播放器直接全屏显示播放多媒体文件。下面来看一下 MediaPlayerLanucher 类的一些重要的属性。

1) Location 属性

Location 是描述文件是放置在什么样的位置，有下面三种类型。

(1) MediaLocationType. Install：指的就是跟着你的 xap 文件一起部署过去的相关文件，也就是位于程序安装的目录中。

(2) MediaLocationType. Data：指的是位于隔离储存区当中的文件，也就是说如果你的文件是执行之后才会取得或是产生的(例如说从网络下载)，而会将档案写入到隔离储存区当中，这个时候就要设定为这个属性。

(3) MediaLocationType. None：这个属性目前来说是没有作用的，如果设定为 None，那么呼叫 Show 的方法之后，直接就会丢出异常 FileNotFroundException。

2) Meida 属性

Media 是文件的位置以及文件名称，是以 URI 的方式来表示。

3) Controls 属性

Controls 是设定 MediaPlayer 出现之后，在画面上会出现哪一些控制按钮，而各个项目也可以利用 OR 的方式去设定。

下面给出播放视频的示例：使用 MediaPlayerLanucher 启动器播放程序中的视频文件。

代码清单 9-6：播放视频(源代码：第 9 章\Examples_9_6)

MainPage. xaml 文件主要代码

```
<Grid x:Name = "ContentPanel" Grid.Row = "1" Margin = "12,0,12,0">
    <Button Content = "播放视频" Height = "116" HorizontalAlignment = "Left" Margin = "100,81,
0,0" Name = "Start" VerticalAlignment = "Top" Width = "273" Click = "Start_Click" />
</Grid>
```

MainPage. xaml. cs 文件主要代码

```
private void Start_Click(object sender, RoutedEventArgs e)
{
    //创建一个多媒体的启动器
    MediaPlayerLauncher mpl = new MediaPlayerLauncher();
    //设置播放文件放置的位置属性
    mpl.Location = MediaLocationType.Install;
    //设置所有控制钮都出现
    mpl.Controls = MediaPlaybackControls.All;
    //设置出现停止按钮以及暂停按钮
    mpl.Controls = MediaPlaybackControls.Pause | MediaPlaybackControls.Stop;
    //设置播放的文件
    mpl.Media = new Uri(@"Media\123.wmv", UriKind.Relative);
    //启动播放
    mpl.Show();
}
```

程序运行的效果如图 9.12 和图 9.13 所示。

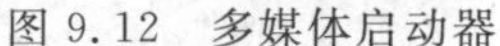
图 9.12　多媒体启动器

图 9.13　多媒体视频的播放效果

9.1.7　应用的详细情况（MarketPlaceDetailTask）

MarketPlaceDetailTask 主要的动作会启动系统内建的 MarketPlace 应用程序，并且可以指定要浏览的应用程序 ID。应特别注意两个属性：一是 ContentType 的属性，目前来说 ContentType 一定要指定为 Applications，如果指定为 Music 的话，在呼叫 Show 的方法时就会直接丢出错误了；二是 ContentIdentifier，这个属性是要指定应用程式 ID（是一个 GUID 值），如果没有指定（也就是 null）的话，便会以目前执行的应用程式为目标。

下面给出查看应用信息的示例：使用 MarketPlaceDetailTask 启动器去查看应用在应用市场中的详细信息。

代码清单 9-7：查看应用信息（源代码：第 9 章\Examples_9_7）

MainPage.xaml 文件主要代码

```
<StackPanel x:Name = "ContentPanel" Margin = "12,0,12,0" Grid.Row = "1" >
    <TextBlock FontSize = "30" Height = "35" Margin = "8,0" TextWrapping = "Wrap" Text = "请输
入应用程序的 ID: "/>
    <TextBox x:Name = "txtID" Height = "72" Margin = "8,0" TextWrapping = "Wrap" Text = ""/>
    <Button Content = "前进" Height = "80" Margin = "196,0,27,0" Name = "go" Click = "go_Click" />
</StackPanel>
```

MainPage.xaml.cs 文件主要代码

```
private void go_Click(object sender, RoutedEventArgs e)
{
    try
    {
        //创建一个查看应用程序详细的任务
```

```
            MarketplaceDetailTask mdt = new MarketplaceDetailTask();
            //只能设定为 Applcations
            mdt.ContentType = MarketplaceContentType.Applications;
            //当没有指定值则会以目前的应用程式为目标
            //如果格式检查不符合 GUID 的格式则会丢出 exception
            mdt.ContentIdentifier = txtID.Text;
            //启动任务
            mdt.Show();
        }
        catch
        {
            MessageBox.Show("找不到应用程序!");
        }
}
```

图 9.14　应用详细查看

程序运行的效果如图 9.14 所示。

9.1.8　应用市场(MarketplaceHubTask)

MarketlaceHubTask 的主要功用是启动后便会带领使用者直接连线到 Marketplace，操作的方法与其他 Lanucher 类似，也是相当简单，要注意的是 ContentType 属性，可以设定为 Application 与 Music，设置为 Application 可以带领用户进入应用程序的市场，设置为 Music 将会带领用户进入音乐的市场。

下面给出进入应用市场的示例：使用 MarketlaceHubTask 启动器进入应用市场。

代码清单 9-8：进入应用市场(源代码：第 9 章\Examples_9_8)

MainPage.xaml 文件主要代码

```
<Grid x:Name = "ContentPanel" Grid.Row = "1" Margin = "12,0,12,0">
    <Button Content = "应用程序类别" Height = "101" HorizontalAlignment = "Left" Margin = "64,
69,0,0" Name = "app" VerticalAlignment = "Top" Width = "313" Click = "app_Click" />
    <Button Content = "音乐类别" Height = "89" HorizontalAlignment = "Left" Margin = "64,214,
0,0" Name = "music" VerticalAlignment = "Top" Width = "313" Click = "music_Click" />
</Grid>
```

MainPage.xaml.cs 文件主要代码

```
private void music_Click(object sender, RoutedEventArgs e)
{
    //创建一个进入应用市场的任务
    MarketplaceHubTask mht = new MarketplaceHubTask();
    //设置为音乐类别
    mht.ContentType = MarketplaceContentType.Music;
    //启动任务
    mht.Show();
}
private void app_Click(object sender, RoutedEventArgs e)
{
```

```
    //创建一个进入应用市场的任务
    MarketplaceHubTask mht = new MarketplaceHubTask();
    //设置为应用程序类别
    mht.ContentType = MarketplaceContentType.Applications;
    //启动任务
    mht.Show();
}
```

程序运行的效果如图 9.15～图 9.17 所示。

图 9.15　进入 Hub

图 9.16　应用程序 Hub 的界面

图 9.17　音乐 Hub 的界面

9.1.9　当前应用在应用市场的信息(MarketplaceReviewTask)

MarketplcaeReviewTask 的用途是在启动之后连到 Marketplace 的页面，并直接地为应用程序做评分、建议等动作。

下面给出应用评价的示例：使用 MarketplcaeReviewTask 启动器进入当前应用的信息评价页面。

代码清单 9-9：应用评价(源代码：第 9 章\Examples_9_9)

MainPage.xaml 文件主要代码

```
<StackPanel x:Name = "ContentPanel" Margin = "12,0,12,0" Grid.Row = "1" >
    <Button Content = "查看本应用的评论和简介" Height = "80" Margin = "19,0,27,0" Name = "go"
Click = "go_Click" Width = "357" />
</StackPanel >
```

MainPage.xaml.cs 文件主要代码

```
private void go_Click(object sender, RoutedEventArgs e)
{
```

```
    try
    {
        //创建一个查看本应用程序简介和评论的任务
        MarketplaceReviewTask mdr = new MarketplaceReviewTask();
        //启动任务
        mdr.Show();
    }
    catch
    {
        MessageBox.Show("找不到应用程序!");
    }
}
```

程序运行的效果如图 9.18 所示。

图 9.18　查看应用程序的详细信息

9.1.10　应用市场搜索（MarketPlaceSearchTask）

MarketplaceSearchTask 可以让你搜寻 Marketplace 上的应用程式或是音乐(一样是透过 ContentType 的属性来设定),另外 SearchTerms 可以指定搜索的关键字。

下面给出应用搜索的示例：使用 MarketplaceSearchTask 启动器来搜索应用。

代码清单 9-10：应用搜索(源代码：第 9 章\Examples_9_10)

MainPage. xaml 文件主要代码

```
<StackPanel x:Name = "ContentPanel" Margin = "12,0,12,0" Grid.Row = "1" >
    <TextBlock FontSize = "30" Height = "35" Margin = "8,0" TextWrapping = "Wrap" Text = "请输入搜索关键字: "/>
    <TextBox x:Name = "txt" Height = "72" Margin = "8,0" TextWrapping = "Wrap" Text = ""/>
    <RadioButton IsChecked = "True"  Content = "搜索音乐" Height = "71" Name = "music" />
    <RadioButton Content = "搜索应用程序" Height = "71" Name = "app" />
    <Button Content = "前进" Height = "80" Margin = "196,0,27,0" Name = "go" Click = "go_Click" />
</StackPanel>
```

MainPage. xaml. cs 文件主要代码

```
private void go_Click(object sender, RoutedEventArgs e)
{
    //创建一个市场搜索任务
    MarketplaceSearchTask mst = new MarketplaceSearchTask();
    if (music.IsChecked == true)
    {//搜索音乐类别
        mst.ContentType = MarketplaceContentType.Music;
    }
    else
    {//搜索应用程序类表
        mst.ContentType = MarketplaceContentType.Applications;
    }
```

```
    //搜索的关键字
    mst.SearchTerms = txt.Text;
    //启动搜索
    mst.Show();
}
```

程序的运行效果如图 9.19～图 9.21 所示。

图 9.19 应用搜索

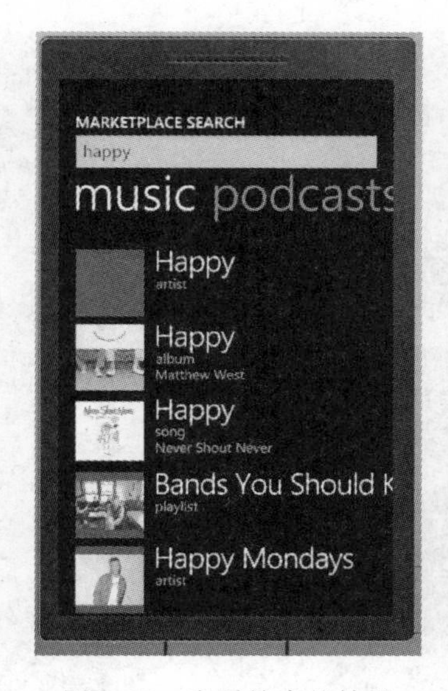

图 9.20 音乐搜索的结果

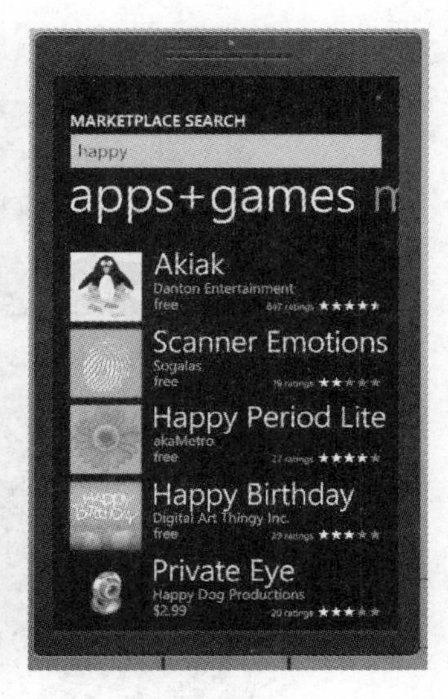

图 9.21 程序的搜索结果

9.1.11 地图(BingMapsTask)

BingMapsTask 启动 Bing 地图来搜索你的地点，通过 SearchTerm 属性来设置搜索的地点关键字。

下面给出启动地图的示例：使用 BingMapsTask 启动器来启动 Bing 地图搜索地点。

代码清单 9-11：启动地图（源代码：第 9 章\Examples_9_11）

MainPage.xaml 文件主要代码

```
<StackPanel x:Name = "ContentPanel" Margin = "12,0,12,0" Grid.Row = "1" >
    <TextBlock FontSize = "30" Height = "35" Margin = "8,0" TextWrapping = "Wrap" Text = "请输入地点关键字: "/>
    <TextBox x:Name = "txt" Height = "72" Margin = "8,0" TextWrapping = "Wrap" Text = ""/>
    <Button Content = "启动地图搜索" Height = "80" Margin = "196,0,27,0" Name = "go" Click = "go_Click" />
</StackPanel>
```

MainPage.xaml.cs 文件主要代码

```
private void go_Click(object sender, RoutedEventArgs e)
{
```

```
    //创建一个 Bing 地图搜索任务
    BingMapsTask _tskBingmap = new BingMapsTask();
    //设置搜索的地址关键字
    _tskBingmap.SearchTerm = txt.Text;
    //启动 bing 地图搜索任务
    _tskBingmap.Show();
}
```

程序的运行效果如图 9.22 和图 9.23 所示。

图 9.22　启动 Bing 地图

图 9.23　Bing 地图的效果

9.1.12　地图方向(BingMapsDirectionsTask)

BingMapsDirectionsTask 启动 Bing 地图应用程序，通过传入出发点位置和终点位置来查看你现在的驾驶方向。

下面给出地图导航的示例：设置起点和终点，使用 BingMapsDirectionsTask 启动器来查看驾驶方向。

代码清单 9-12：地图导航(源代码：第 9 章\Examples_9_12)

MainPage.xaml 文件主要代码

```
<Grid x:Name = "ContentPanel" Grid.Row = "1" Margin = "12,0,12,0">
    <TextBlock FontSize = "30" Height = "47" HorizontalAlignment = "Left" Margin = "23,19,0,0"
Name = "textBlock1" Text = "起点" VerticalAlignment = "Top" Width = "164" />
    <TextBlock FontSize = "30" Height = "46" HorizontalAlignment = "Left" Margin = "15,78,0,0"
Name = "textBlock2" Text = "纬度: " VerticalAlignment = "Top" Width = "116" />
    <TextBox Height = "72" HorizontalAlignment = "Left" Margin = "111,52,0,0" Name = "lat1"
Text = "30.24" VerticalAlignment = "Top" Width = "327" />
    <TextBlock FontSize = "30" Height = "47" HorizontalAlignment = "Left" Margin = "15,169,0,
```

```
0" Name = "textBlock3" Text = "经度: " VerticalAlignment = "Top" />
    <TextBox Height = "72" HorizontalAlignment = "Left" Margin = "111,169,0,0" Name = "lon1"
Text = "120.123" VerticalAlignment = "Top" Width = "327" />
    <TextBlock FontSize = "30" Height = "47" HorizontalAlignment = "Left" Margin = "17,298,0,
0" Name = "textBlock4" Text = "终点" VerticalAlignment = "Top" Width = "82" />
    <TextBlock FontSize = "30" Height = "55" HorizontalAlignment = "Left" Margin = "12,361,0,
0" Name = "textBlock5" Text = "纬度: " VerticalAlignment = "Top" Width = "93" />
    <TextBox Height = "72" HorizontalAlignment = "Left" Margin = "111,344,0,0" Name = "lat2"
Text = "30.25" VerticalAlignment = "Top" Width = "327" />
    <TextBlock FontSize = "30" Height = "47" HorizontalAlignment = "Left" Margin = "9,457,0,0"
Name = "textBlock6" Text = "经度: " VerticalAlignment = "Top" />
    <TextBox Height = "72" HorizontalAlignment = "Left" Margin = "111,439,0,0" Name = "lon2"
Text = "120.223" VerticalAlignment = "Top" Width = "327" />
    <Button Content = "前进" Height = "72" HorizontalAlignment = "Left" Margin = "137,526,0,0"
Name = "go" VerticalAlignment = "Top" Width = "160" Click = "go_Click" />
</Grid>
```

MainPage.xaml.cs 文件主要代码

```
private void go_Click(object sender, RoutedEventArgs e)
{
    try
    {
        //创建一个 Bing 地图方向的任务
        BingMapsDirectionsTask bmd = new BingMapsDirectionsTask();
        //设置起点位置
        bmd.Start = new LabeledMapLocation("start", new GeoCoordinate(Double.Parse(lat1.
Text), Double.Parse(lon1.Text)));
        //设置终点位置
        bmd.End = new LabeledMapLocation("end", new GeoCoordinate(Double.Parse(lat2.Text),
Double.Parse(lon2.Text)));
        //启动任务
        bmd.Show();
    }
    catch
    {
        MessageBox.Show("输入有误!");
    }
}
```

程序的运行效果如图 9.24 和图 9.25 所示。

9.1.13 连接设置(ConnectionSettingsTask)

ConnectionSettingsTask 表示使用连接设置任务使用户能够调整设备的网络连接设置,允许应用程序启动一个“设置”对话框,该对话框允许用户更改设备的网络连接设置。通过 ConnectionSettingsTask 类的 ConnectionSettingsType 属性来设置将显示的网络连接设置的类型,包含 Wi-Fi 设置、蓝牙设置、手机网络设置和飞行模式设置 4 种类型。

图 9.24 查看地图的方向

图 9.25 地图方向的查看结果

下面给出连接设置的示例：可以通过单击按钮直接进入 Wi-Fi 设置、蓝牙设置、手机网络设置和飞行模式设置的系统页面。

代码清单 9-13：连接设置(源代码：第 9 章\Examples_9_13)

MainPage.xaml 文件主要代码

```
<Grid x:Name="ContentPanel" Grid.Row="1" Margin="12,0,12,0">
    <StackPanel>
        <Button x:Name="btWiFi"  Content="WiFi 设置" Click="btWiFi_Click"/>
        <Button x:Name="btBluetooth"  Content="蓝牙设置" Click="btBluetooth_Click"/>
        <Button x:Name="btCellular"  Content="手机网络设置" Click="btCellular_Click"/>
        <Button x:Name="btAirplaneMode"  Content="飞行模式设置 " Click="btAirplaneMode_
Click"/>
    </StackPanel>
</Grid>
```

MainPage.xaml.cs 文件代码

```
using System.Windows;
using Microsoft.Phone.Controls;
using Microsoft.Phone.Tasks;
namespace ConnectionSettingsTaskDemo
{
    public partial class MainPage : PhoneApplicationPage
    {
        ConnectionSettingsTask connectionSettingsTask;
        public MainPage()
        {
            InitializeComponent();
```

```
            //创建一个连接设置任务
            connectionSettingsTask = new ConnectionSettingsTask();
        }
        // Wi-Fi 设置
        private void btWiFi_Click(object sender, RoutedEventArgs e)
        {
            connectionSettingsTask.ConnectionSettingsType = ConnectionSettingsType.WiFi;
            connectionSettingsTask.Show();
        }
        // 蓝牙设置
        private void btBluetooth_Click(object sender, RoutedEventArgs e)
        {
            connectionSettingsTask.ConnectionSettingsType = ConnectionSettingsType.Bluetooth;
            connectionSettingsTask.Show();
        }
        // 手机网络设置
        private void btCellular_Click(object sender, RoutedEventArgs e)
        {
            connectionSettingsTask.ConnectionSettingsType = ConnectionSettingsType.Cellular;
            connectionSettingsTask.Show();
        }
        // 飞行模式设置
        private void btAirplaneMode_Click(object sender, RoutedEventArgs e)
        {
            connectionSettingsTask.ConnectionSettingsType = ConnectionSettingsType.AirplaneMode;
            connectionSettingsTask.Show();

        }
    }
}
```

程序的运行效果如图 9.26 所示。

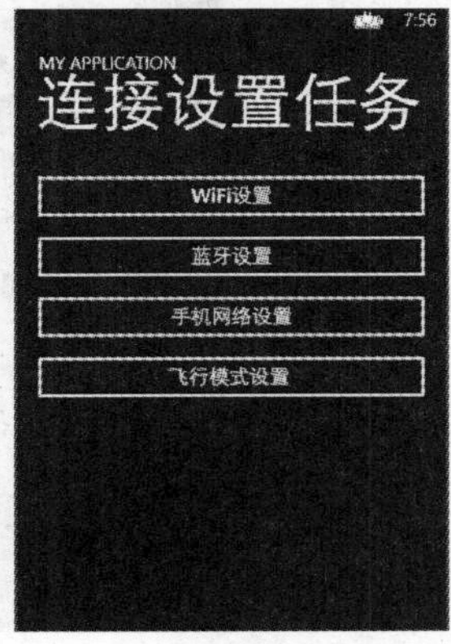

图 9.26　连接设置任务

9.1.14　保存日程安排(SaveAppointmentTask)

SaveAppointmentTask 启动日程安排的保存任务,可以在程序中给日程安排的相关参数赋值,然后通过 SaveAppointmentTask 任务把这些数值带到日程安排的保存页面。

下面给出保存日程安排的示例：设置一个小时后进行提醒的任务。

代码清单 9-14：保存日程安排(源代码：第 9 章\Examples_9_14)

MainPage.xaml 文件主要代码

```
<Grid x:Name = "ContentPanel" Grid.Row = "1" Margin = "12,0,12,0">
    <StackPanel>
        <TextBlock Text = "任务内容: "/>
        <TextBox x:Name = "tbTask"/>
        <Button x:Name = "btAdd"  Content = "添加任务" Click = "btAdd_Click" />
```

```
    </StackPanel>
</Grid>
```

MainPage.xaml.cs 文件主要代码

```
//添加日程安排
private void btAdd_Click(object sender, RoutedEventArgs e)
{
    SaveAppointmentTask saveAppointmentTask = new SaveAppointmentTask();
    saveAppointmentTask.Subject = tbTask.Text;
    saveAppointmentTask.Details = tbTask.Text;
    saveAppointmentTask.StartTime = DateTime.Now;
    saveAppointmentTask.Reminder = Reminder.OneHour;
    saveAppointmentTask.Show();
}
```

程序的运行效果如图 9.27 和图 9.28 所示。

图 9.27　一小时的任务

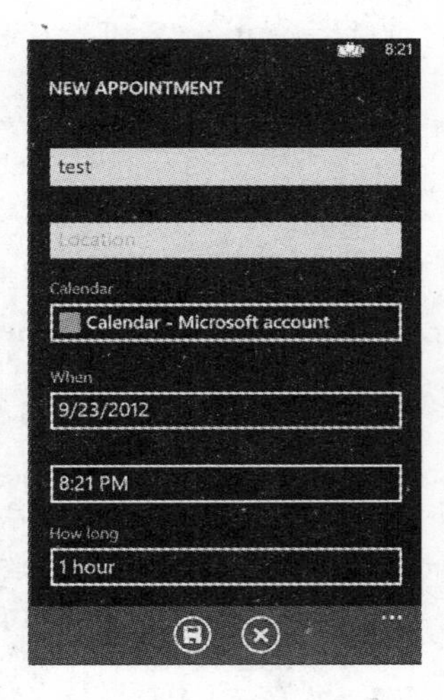

图 9.28　保存界面

9.1.15　诺基亚地图加载（MapDownloaderTask）

MapDownloaderTask 允许在应用程序里面启动地图设置界面，主要是方便用户进行离线地图的下载，直接调用 MapDownloaderTask 类的 Show 方法便可以跳转到地图的设置界面。使用的语法如下：

```
MapDownloaderTask mapDownloaderTask = new MapDownloaderTask();
mapDownloaderTask.Show();
```

9.1.16　诺基亚地图（MapsTask）

MapsTask 是指允许应用程序启动诺基亚地图，与 BingMapsTask 启动 Bing 地图的用法类似。可以通过 SearchTerm 属性来设置搜索的地点关键字和通过 Center 属性来指定当前用的中心位置。使用的语法如下：

```
MapsTask mapsTask = new MapsTask();
mapsTask.SearchTerm = "超市";
mapsTask.Show();
```

9.1.17　地图方向(MapsDirectionsTask)

MapsDirectionsTask 启动诺基亚地图应用程序，通过传入出发点位置和终点位置来查看现在的驾驶方向，BingMapsDirectionsTask 启动 Bing 地图应用程序的用法类似，区别是 BingMapsDirectionsTask 启动的是 Bing 地图而 MapsDirectionsTask 启动的是诺基亚地图。使用的语法如下：

```
MapsDirectionsTask mapsDirectionsTask = new MapsDirectionsTask();
mapsDirectionsTask.Start = new LabeledMapLocation("start", new GeoCoordinate(Double.Parse
(lat1.Text), Double.Parse(lon1.Text)));
mapsDirectionsTask.End = new LabeledMapLocation("end", new GeoCoordinate(Double.Parse(lat2.
Text), Double.Parse(lon2.Text)));
mapsDirectionsTask.Show();
```

9.1.18　共享多媒体(ShareMediaTask)

ShareMediaTask 是指在应用程序里面共享一个多媒体文件到社交网络，FilePath 属性表示文件的位置。使用的语法如下：

```
ShareMediaTask shareMediaTask = new ShareMediaTask();
shareMediaTask.FilePath = "Images/Test.png";
shareMediaTask.Show();
```

9.1.19　共享链接(ShareLinkTask)

ShareLinkTask 是指在应用程序里面共享一个链接到社交网络，LinkUri 属性表示链接的地址，Message 属性表示链接的消息内容，Title 属性表示分享的标题。使用的语法如下：

```
ShareLinkTask shareLinkTask = new ShareLinkTask();
shareLinkTask.LinkUri = new Uri("http://wwww.cnblogs.com/linzheng");
shareLinkTask.Message = "欢迎来到我的博客";
shareLinkTask.Title = "博客";
shareLinkTask.Show();
```

9.1.20　共享状态(ShareStatusTask)

ShareStatusTask 是指在应用程序里面共享一个状态信息到社交网络，Status 属性表示状态的信息。使用的语法如下：

```
ShareStatusTask shareStatusTask = new ShareStatusTask();
```

```
shareStatusTask.Status = "busy";
shareStatusTask.Show();
```

9.2 使用选择器

选择器是一个“打开文件对话框”动作，可以使用它从手机选择一些信息，并带回到用户的应用程序，例如，选取一个电子邮件地址或联系人，或选择一张照片。选择器比启动器略复杂一些，因为它们要将数据带回到程序中，而启动器只是让用户去完成一个任务。

在 Windows Phone 8 中支持的选择器有以下的一些：

(1) CaptureCameraTask：允许应用程序启动照相机应用程序，使用户能够从自己的应用中拍照片。

(2) EmailAddressChooserTask：允许应用程序启动联系人应用程序，使用它来获取用户选定的联系人的电子邮件地址。

(3) PhoneNumberChooserTask：允许应用程序启动联系人应用程序，使用它来获取用户选定的联系人的电话号码。

(4) PhotoChooserTask：允许应用程序启动照片选择应用程序，使用它来让用户选择照片。

(5) SaveEmailAddressTask：允许应用程序启动联系人应用程序，以此允许用户从应用程序中保存电子邮件地址到一个新的或现有的联系人。

(6) SavePhoneNumberTask：允许应用程序启动联系人应用程序，以此允许用户从应用程序中保存电话号码到一个新的或现有的联系人。

(7) GameInviteTask：允许应用程序来显示游戏邀请屏幕，使用户可以邀请玩家多人游戏。

(8) SaveRingtoneTask：使应用程序能够启动铃声，并可以将铃声保存到系统铃声列表。

(9) AddWalletItemTask：允许应用程序启动电子钱包应用程序，并且添加电子钱包的项目。

(10) AddressChooserTask：允许应用程序启动“联系人”应用程序来选择联系人地址。

(11) SaveContactTask：允许启动联系人保存页面保存联系人。

9.2.1 照相机(CameraCaptureTask)

CameraCaptureTask 是照相机拍照的选择器，通过 CameraCaptureTask 可以调用相机拍照并且可以获取拍完后的照片。获取拍照之后的照片是通过 Completed 的事件来处理的，在这个事件中，必须要先判断 TaskResult 属性，在这个属性当中，可以取得拍照动作的结果，例如当使用者按下确定(Accept) 的按钮时，会回应 OK，而如果使用者按下返回键呢？这时候回传的就会是 Cancel 的状态了，由这个状态，可以去判断接下来应用程式当中要处理的动作。而怎么取得拍摄的照片呢？主要便是利用 ChoosenPhoto 的属性，ChoosenPhoto 是一个 Stream，是指向实体照片位置的资料流，拍照后图片是不会直接的储存到应用程式所属的隔离储存区中的，因为 Chooser 所叫出的是另外的应用程式，不同应用程式之间是不能去交叉存取隔离储存区中的文件的，因此 Chooser 会由这种方式来让我们的应用程式取得结果。

下面给出使用照相机的示例：使用 CameraCaptureTask 选择器拍照并将照片展现在应用上。

代码清单 9-15：使用照相机(源代码：第 9 章\Examples_9_15)

MainPage.xaml 文件主要代码

```
<Grid x:Name="ContentPanel" Grid.Row="1" Margin="12,0,12,0">
    <Image Height="489" HorizontalAlignment="Left" Margin="21,19,0,0" Name="image1"
Stretch="Fill" VerticalAlignment="Top" Width="447" />
    <Button Content="拍照" Height="82" HorizontalAlignment="Left" Margin="191,535,0,0"
Name="btnShot" VerticalAlignment="Top" Width="259" Click="btnShot_Click" />
</Grid>
```

MainPage.xaml.cs 文件代码

```
CameraCaptureTask cct;
public MainPage()
{
    InitializeComponent();
    //创建一个捕获相机拍照的选择器
    cct = new CameraCaptureTask();
    //注册选择器完成的事件
    cct.Completed += new EventHandler<PhotoResult>(cct_Completed);
}
private void btnShot_Click(object sender, RoutedEventArgs e)
{
    //启动拍照选择器
    cct.Show();
}
private void cct_Completed(object sender, PhotoResult e)
{
    //判断结果是否成功
    if (e.TaskResult == TaskResult.OK)
    {
        BitmapImage bmpSource = new BitmapImage();
        bmpSource.SetSource(e.ChosenPhoto);
        image1.Source = bmpSource;
    }
    else
    {
        image1.Source = null;
    }
}
```

程序的运行效果如图 9.29～图 9.31 所示。

图 9.29　捕获相机拍照

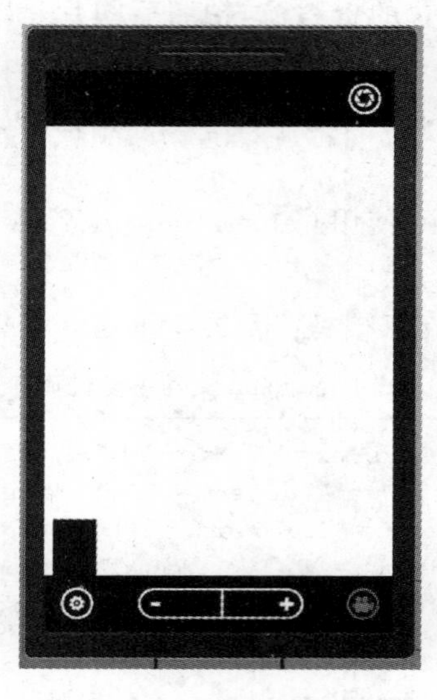

图 9.30　拍照的模拟器效果

图 9.31　捕获后的照片的效果

9.2.2　邮箱地址(EmailAddressChooserTask)

EmailAddressChooserTask 主要是用来取得手机联系人的电子邮件资料，然后就可以利用获取电子邮件来做一些其他的事情。

下面给出选择联系人邮箱的示例：使用 EmailAddressChooserTask 选择器来选择通讯录的联系人邮箱。

代码清单 9-16：选择联系人邮箱(源代码：第 9 章\Examples_9_16)

MainPage. xaml 文件主要代码

```
<Grid x:Name = "ContentPanel" Grid.Row = "1" Margin = "12,0,12,0">
    <Button Content = "选择邮箱" Height = "87" HorizontalAlignment = "Left" Margin = "12,6,0,
0" Name = "choose" VerticalAlignment = "Top" Width = "438" Click = "choose_Click" />
    <TextBox Height = "75" HorizontalAlignment = "Left" Margin = "12,153,0,0" Name = "address"
Text = "" VerticalAlignment = "Top" Width = "438" />
    <TextBlock FontSize = "30" Height = "39" HorizontalAlignment = "Left" Margin = "37,108,0,
0" Name = "textBlock1" Text = "你选择的邮箱地址如下:" VerticalAlignment = "Top" Width = "334" />
</Grid>
```

MainPage. xaml. cs 文件主要代码

```
public partial class MainPage : PhoneApplicationPage
{
    EmailAddressChooserTask eac;
    public MainPage()
    {
        InitializeComponent();
```

```
        //创建一个获取邮箱地址的选择器
        eac = new EmailAddressChooserTask();
        //注册选择器完成的事件
        eac.Completed += new EventHandler<EmailResult>(eac_Completed);
    }
    private void choose_Click(object sender, RoutedEventArgs e)
    {
        //启动邮箱地址选择器
        eac.Show();
    }
    void eac_Completed(object sender, EmailResult e)
    {
        if (e.TaskResult == TaskResult.OK)
        {
            address.Text = e.Email;
        }
    }
}
```

程序的运行效果如图 9.32 所示。

图 9.32　选择联系人的邮箱

9.2.3　电话号码(PhoneNumberChooserTask)

PhoneNumberChooserTask 主要是用来选择手机联系人的电话号码，与联系人邮件选择器相似，只不过 PhoneNumberChooserTask 获取的是联系人的电话号码。

下面给出选择联系人电话号码的示例：使用 PhoneNumberChooserTask 选择器来选取联系人的电话号码。

代码清单 9-17：选择联系人电话号码(源代码：第 9 章\Examples_9_17)

MainPage.xaml 文件主要代码

```
<Grid x:Name="ContentPanel" Grid.Row="1" Margin="12,0,12,0">
```

```
    <Button Content="选择电话号码" Height="87" HorizontalAlignment="Left" Margin="12,
6,0,0" Name="choose" VerticalAlignment="Top" Width="438" Click="choose_Click" />
    <TextBox Height="75" HorizontalAlignment="Left" Margin="12,153,0,0" Name="number"
Text="" VerticalAlignment="Top" Width="438" />
    <TextBlock FontSize="30" Height="39" HorizontalAlignment="Left" Margin="37,108,0,
0" Name="textBlock1" Text="你选择的电话号码如下:" VerticalAlignment="Top" Width="334" />
</Grid>
```

MainPage.xaml.cs 文件主要代码

```
PhoneNumberChooserTask pnc;
public MainPage()
{
    InitializeComponent();
    //创建一个选择电话号码的选择器
    pnc = new PhoneNumberChooserTask();
    //注册选择器完成的事件
    pnc.Completed += new EventHandler<PhoneNumberResult>(pnc_Completed);
}
private void choose_Click(object sender, RoutedEventArgs e)
{
    //启动电话号码选择器
    pnc.Show();
}
void pnc_Completed(object sender, PhoneNumberResult e)
{
    if (e.TaskResult == TaskResult.OK)
    {
        number.Text = e.PhoneNumber;
    }
}
```

程序的运行效果如图 9.33～图 9.35 所示。

图 9.33 选择电话号码

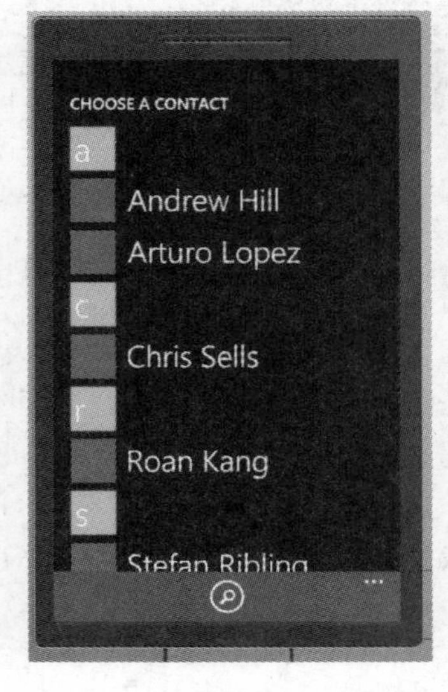

图 9.34 联系人电话号码界面

图 9.35 选择电话号码后的效果

9.2.4　选取图片(PhotoChooserTask)

PhotoChooserTask 是用来选择图片用的，这部分使用上跟 CameraCaptureTask 是极其相似的，主要的区别是 CameraCaptureTask 只是用相机拍照返回照片，而 PhotoChooserTask 还可以直接选取手机里面的照片然后返回给应用程序。

PhotoChooserTask 的动作都与 CameraCaptureTask 相同，下面来看看在 PhotoChooser 中特殊的属性，首先是 ShowCamera 的属性，ShowCamera 的属性是一个布尔型态，当设定为真时，在选择图片的画面中，会出现拍照的按钮，让使用者也可以透过照相机来做为图片的来源。接下来是 PixelHeight、PixelWidth 的属性，这两个属性是让使用者可以裁切原始的图形，比如说，现在应用程式要让使用者设定头像，头像的尺寸只需要 100×100，这时候过大的图形并没有用处，就可以通过设置两个属性来裁剪选择的图片。

下面给出选取照片的示例：使用 PhotoChooserTask 选择器来选取照片。

代码清单 9-18：选取照片(源代码：第 9 章\Examples_9_18)

MainPage. xaml 文件主要代码

```
<Grid x:Name="ContentPanel" Grid.Row="1" Margin="12,0,12,0">
    <Button Content="选择图片" Height="85" HorizontalAlignment="Left" Margin="24,23,0,
0" Name="button1" VerticalAlignment="Top" Width="417" Click="button1_Click" />
    <Image Height="336" HorizontalAlignment="Left" Margin="38,139,0,0" Name="image1"
Stretch="Uniform" VerticalAlignment="Top" Width="394" />
    <TextBlock Height="90" HorizontalAlignment="Left" Margin="38,490,0,0" Name=
"textBlock1" Text="TextBlock" VerticalAlignment="Top" Width="394" TextWrapping="Wrap"
/>
</Grid>
```

MainPage. xaml. cs 文件主要代码

```
public partial class MainPage : PhoneApplicationPage
{
    PhotoChooserTask pc;
    public MainPage()
    {
        InitializeComponent();
        pc = new PhotoChooserTask();
        //注册选择器完成的事件
        pc.Completed += new EventHandler<PhotoResult>(pc_Completed);
    }
    void pc_Completed(object sender, PhotoResult e)
    {
        if (e.TaskResult == TaskResult.OK)
        {
            //新建一个位图类
            BitmapImage  bmpSource = new BitmapImage();
            //设置位图类的源为选择器选择的照片
            bmpSource.SetSource(e.ChosenPhoto);
```

```
            //设置图片空间的源为该位图
            image1.Source = bmpSource;
            //在文本框中显示照片的文件路径
            textBlock1.Text = e.OriginalFileName;
        }
        else
        {
            image1.Source = null;
        }
    }
    private void button1_Click(object sender, RoutedEventArgs e)
    {
        //是否裁切相片，设置裁剪图片的最大高度和宽度
        pc.PixelHeight = 30;
        pc.PixelWidth = 80;
        pc.Show();
    }
}
```

图 9.36　选择手机的图片

程序的运行效果如图 9.36 所示。

9.2.5　保存邮箱地址(SaveEmailAddressTask)

SaveEmailAddressTask 是用来储存联络人中电子邮件的相关资料。启动了 SaveEmailAddressTask 之后，你可以选择要将这个电子邮件储存到哪一个联络人，或是说要建立新的联络人都可以。

下面给出保存邮箱地址的示例：使用 SaveEmailAddressTask 选择器来保存邮箱地址到联系人中。

代码清单 9-19：保存邮箱地址(源代码：第 9 章\Examples_9_18)

MainPage.xaml 文件主要代码

```
        <Grid x:Name = "ContentPanel" Grid.Row = "1" Margin = "12,0,12,0">
    <TextBox Height = "76" HorizontalAlignment = "Left" Margin = "0,72,0,0" Name = "txtEmail"
Text = "" VerticalAlignment = "Top" Width = "444" />
    <TextBlock HorizontalAlignment = "Left" Margin = "12,33,0,0" Name = "textBlock1" Text =
"邮箱地址如下：" Width = "350" Height = "33" VerticalAlignment = "Top" />
    <Button Content = "保存" Height = "84" HorizontalAlignment = "Left" Margin = "249,167,0,0"
Name = "button1" VerticalAlignment = "Top" Width = "183" Click = "button1_Click" />
</Grid>
```

MainPage.xaml.cs 文件主要代码

```
SaveEmailAddressTask sea;
public MainPage()
{
    InitializeComponent();
    //设置文本输入框为邮件输入格式
```

```
    txtEmail.InputScope = new InputScope()
    {
        Names = { new InputScopeName() { NameValue = InputScopeNameValue.EmailNameOrAddress
} }
    };
    sea = new SaveEmailAddressTask();
    sea.Completed += new EventHandler<TaskEventArgs>(sea_Completed);
}
void sea_Completed(object sender, TaskEventArgs e)
{
    if (e.TaskResult == TaskResult.OK)
    {
        MessageBox.Show("保存成功!");
    }
    else
    {
        MessageBox.Show("保存失败!");
    }
}
private void button1_Click(object sender, RoutedEventArgs e)
{
    sea.Email = txtEmail.Text;
    sea.Show();
}
```

图 9.37　保存邮箱的地址

程序的运行效果如图 9.37 所示。

9.2.6　保存电话号码(SavePhoneNumberTask)

SavePhoneNumberTask 则是用来储存联络人的电话号码，与 SaveEmailAddressTask 选择器使用的方式是相同的，都是将要保存的信息保存到手机的通讯录里面。

下面给出保存电话号码的示例：使用 SavePhoneNumberTask 选择器来保存电话号码。

代码清单 9-20：保存电话号码(源代码：第 9 章\Examples_9_20)

MainPage.xaml 文件主要代码

```
<Grid x:Name="ContentPanel" Grid.Row="1" Margin="12,0,12,0">
    <TextBox Height="76" HorizontalAlignment="Left" Margin="10,65,0,0" Name=
"txtPhoneNo" Text="" VerticalAlignment="Top" Width="444" />
    <TextBlock Height="33" HorizontalAlignment="Left" Margin="22,10,0,0" Name=
"textBlock1" Text="电话号码" FontSize="30" VerticalAlignment="Top" Width="350" />
    <Button Content="保存" Height="84" HorizontalAlignment="Left" Margin="259,144,0,0"
Name="button1" VerticalAlignment="Top" Width="183" Click="button1_Click" />
</Grid>
```

MainPage.xaml.cs 文件主要代码

```
SavePhoneNumberTask spn;
public MainPage()
```

```
{
    InitializeComponent();
    txtPhoneNo.InputScope = new InputScope()
    {
        //设置文本框为数字输入模式
        Names = { new InputScopeName() { NameValue = InputScopeNameValue.TelephoneNumber } }
    };
    spn = new SavePhoneNumberTask();
    spn.Completed += new EventHandler<TaskEventArgs>(spn_Completed);
}
void spn_Completed(object sender, TaskEventArgs e)
{
    if (e.TaskResult == TaskResult.OK)
    {
        MessageBox.Show("保存成功!");
    }
    else
    {
        MessageBox.Show("保存失败!");
    }
}
private void button1_Click(object sender, RoutedEventArgs e)
{
    spn.PhoneNumber = txtPhoneNo.Text;
    spn.Show();
}
```

程序的运行效果如图 9.38 所示。

图 9.38 保存电话号码

9.2.7 游戏邀请(GameInviteTask)

GameInviteTask 可以让用户去邀请其他的玩家来和你一起来玩游戏，需要传递用户的游戏的 SessionId 过去。

下面给出发送游戏邀请的示例：演示使用 GameInviteTask 选择器。

代码清单 9-21：发送游戏邀请(源代码：第 9 章\Examples_9_21)

MainPage.xaml 文件主要代码

```
<Grid x:Name = "ContentPanel" Grid.Row = "1" Margin = "12,0,12,0">
    <Button Content = "发起邀请" Height = "73" HorizontalAlignment = "Left" Margin = "40,55,0,
0" Name = "button1" VerticalAlignment = "Top" Width = "302" Click = "button1_Click" />
    <TextBlock Height = "83" HorizontalAlignment = "Left" Margin = "55,173,0,0" Name =
"textBlock1" Text = "" VerticalAlignment = "Top" Width = "328" />
</Grid>
```

MainPage.xaml.cs 文件主要代码

```
GameInviteTask _tskGame;
public MainPage()
{
```

```
        InitializeComponent();
        //新建一个游戏邀请选择器
        _tskGame = new GameInviteTask();
        //注册邀请完成事件
        _tskGame.Completed += new EventHandler<TaskEventArgs>(_tskGame_Completed);
        //设置游戏的 SessionId
        _tskGame.SessionId = "<my session id>";
    }
    //游戏邀请处理事件
    private void button1_Click(object sender, RoutedEventArgs e)
    {
        //启动选择器
        _tskGame.Show();
    }
    //邀请完成处理事件
    void _tskGame_Completed(object sender, TaskEventArgs e)
    {
       //显示选择器的结果
       textBlock1.Text = e.TaskResult.ToString();
    }
```

程序的运行效果如图 9.39 和图 9.40 所示。

图 9.39 游戏邀请

图 9.40 启动游戏邀请后的界面

9.2.8 保存铃声(SaveRingtoneTask)

SaveRingtoneTask 可以将音乐保存到系统的铃声列表,并可以设置为系统的铃声。下面给出保存铃声的示例:演示使用 SaveRingtoneTask 选择器。

代码清单 9-22：保存铃声(源代码：第 9 章\Examples_9_22)

MainPage. xaml 文件主要代码

```
<Grid x:Name = "ContentPanel" Grid.Row = "1" Margin = "12,0,12,0">
    <Button Content = "选择铃声" Height = "73" HorizontalAlignment = "Left" Margin = "40,55,0,
0" Name = "button1" VerticalAlignment = "Top" Width = "302" Click = "button1_Click" />
    < TextBlock Height = "83" HorizontalAlignment = "Left" Margin = "55,173,0,0" Name =
"textBlock1" Text = "" VerticalAlignment = "Top" Width = "328" />
</Grid>
```

MainPage. xaml. cs 文件主要代码

```
SaveRingtoneTask _tskSaveRingTone;
public MainPage()
{
    InitializeComponent();
    //新建一个铃声选择器
    _tskSaveRingTone = new SaveRingtoneTask();
    //注册完成事件
    _tskSaveRingTone.Completed += new EventHandler<TaskEventArgs>(_tskSaveRingTone_Completed);
}
private void button1_Click(object sender, RoutedEventArgs e)
{
    //设置铃声显示的名字
    _tskSaveRingTone.DisplayName = "mp3 - 2";
    //铃声存储位置
    _tskSaveRingTone. Source = new Uri("appdata:/mp3/2.mp3", UriKind.Absolute);
    //是否振动
    _tskSaveRingTone.IsShareable = true;
    //启动选择器
    _tskSaveRingTone.Show();
}
//保存铃声完成事件
void _tskSaveRingTone_Completed(object sender, TaskEventArgs e)
{
    textBlock1.Text = e.TaskResult.ToString();
}
```

程序的运行效果如图 9.41 和图 9.42 所示。

9.2.9 添加钱包项目(AddWalletItemTask)

AddWalletItemTask 允许应用程序启动电子钱包应用程序，并且添加电子钱包的项目。Item 表示的是一个电子钱包的项目，使用电子钱包功能需要在配置文件里面添加 ID_CAP_WALLET 功能。

下面给出添加电子钱包的示例：演示使用 AddWalletItemTask 选择器添加电子钱包的项目。

图 9.41　选择铃声

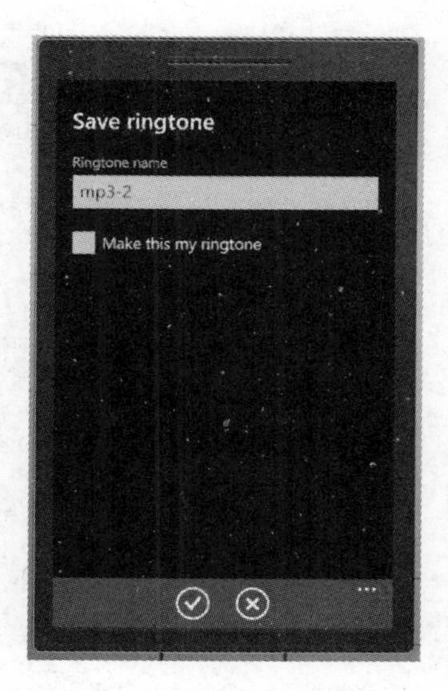

图 9.42　保存铃声

代码清单 9-23：添加电子钱包(源代码：第 9 章\Examples_9_23)

MainPage.xaml 文件主要代码

```
<Grid x:Name = "ContentPanel" Grid.Row = "1" Margin = "12,0,12,0">
    <StackPanel>
        <TextBlock Text = "客户名称：" />
        <TextBox x:Name = "tbCustomerName"/>
        <TextBlock Text = "发行人名称：" />
        <TextBox x:Name = "tbIssuerName"/>
        <TextBlock Text = "备注：" />
        <TextBox x:Name = "tbNotes"/>
        <TextBlock Text = "发行网站：" />
        <TextBox x:Name = "tbIssuerWebsite"/>
        <Button x:Name = "btAdd" Content = "添加" Click = "btAdd_Click" />
    </StackPanel>
</Grid>
```

MainPage.xaml.cs 文件代码

```
//添加电子钱包项目
private void btAdd_Click(object sender, RoutedEventArgs e)
{
    try
    {
        AddWalletItemTask addWalletItemTask = new AddWalletItemTask();
        //赋值电子钱包的项目
        WalletTransactionItem membershipItem;
        membershipItem = new WalletTransactionItem("membership");
        membershipItem.IssuerName = tbIssuerName.Text;
        membershipItem.DisplayName = tbIssuerName.Text + " Membership Card";
```

```
            membershipItem.CustomerName = tbCustomerName.Text;
            membershipItem.IssuerWebsite = new Uri(tbIssuerWebsite.Text);
            membershipItem.DisplayAvailableBalance = "1000 points";
            BitmapImage bmp = new BitmapImage();
            //获取项目中的文件
            using (System.IO.Stream stream = Assembly.GetExecutingAssembly().
    GetManifestResourceStream("AddWalletItemTaskDemo.Assets.adventure.jpg"))
                bmp.SetSource(stream);
            //添加项目的 logo
            membershipItem.Logo99x99 = bmp;
            membershipItem.Logo159x159 = bmp;
            membershipItem.Logo336x336 = bmp;
            addWalletItemTask.Item = membershipItem;
            //订阅完成事件
            addWalletItemTask.Completed += new EventHandler<AddWalletItemResult>(awic_Completed);
            addWalletItemTask.Show();
        }
        catch (Exception ee)
        {
            MessageBox.Show(ee.Message);
        }
    }
    //添加电子钱包项目完成事件
    void awic_Completed(object sender, Microsoft.Phone.Tasks.AddWalletItemResult e)
    {
        if (e.TaskResult == Microsoft.Phone.Tasks.TaskResult.OK)
        {
            MessageBox.Show(e.Item.DisplayName + " 已经被添加到你的电子钱包里面!");
        }
        else if (e.TaskResult == Microsoft.Phone.Tasks.TaskResult.Cancel)
        {
            MessageBox.Show("Cancelled");
        }
        else if (e.TaskResult == Microsoft.Phone.Tasks.TaskResult.None)
        {
            MessageBox.Show("None");
        }
    }
```

程序的运行效果如图 9.43 所示。

图 9.43　添加钱包

9.2.10　选择地址(AddressChooserTask)

AddressChooserTask 允许应用程序启动“联系人”应用程序。使用此方法获取用户选择的联系人的物理地址。示例代码如下：

```
void AddressChooserTask()
{
    AddressChoserTask addressChooserTask = new AddressChooserTask();
    //订阅选择地址事件
```

```
    addressChooserTask.Completed += new EventHandler<AddressResult>(addressChooserTask_
Completed);
    try
    {
        addressChooserTask.Show();
    }
    catch (System.InvalidOperationException ex)
    {
        MessageBox.Show("An error occurred.");
    }
}
// 选择地址完成事件
void addressChooserTask_Completed(object sender, AddressResult e)
{
    if (e.TaskResult == TaskResult.OK)
    {
        MessageBox.Show(e.DisplayName + " : " + e.Address);
    }
}
```

9.2.11 保存手机联系人(SaveContactTask)

使用保存联系人任务 SaveContactTask 可使用户能够通过应用程序保存联系人。此任务启动"联系人"应用程序。示例代码如下：

```
void SaveContactTask()
{
    SaveContactTask saveContactTask = new SaveContactTask();
    saveContactTask.Completed += new EventHandler<SaveContactResult>(saveContactTask_
Completed);
    try
    {
        saveContactTask.FirstName = "李";
        saveContactTask.LastName = "明";
        saveContactTask.MobilePhone = "12345678";
        saveContactTask.Show();
    }
    catch (System.InvalidOperationException ex)
    {
        MessageBox.Show("An error occurred.");
    }

}
// 保存联系人完成事件
void saveContactTask_Completed(object sender, SaveContactResult e)
{
    switch (e.TaskResult)
    {
        case TaskResult.OK:
```

```
            MessageBox.Show("保存成功");
            break;
        case TaskResult.Cancel:
            MessageBox.Show("取消保存");
            break;
        case TaskResult.None:
            MessageBox.Show("保存失败");
            break;
    }
}
```

第10章

手机感应编程

随着智能手机的发展，像重力感应、手势触摸感、指南针、陀螺仪这些功能渐渐地成为了智能手机必备的基本功能。在 Windows Phone 8 里面还有新增的语音控制功能，让 Windows Phone 8 手机的功能更加的强大。在应用程序开发里面我们可以充分地利用手机提供的这些感应的功能来开发出功能强大和体验更加好的应用程序，比如通过语音命令来进行一些操作等。

10.1 加速器

加速器是 Windows Phone 8 中一种标准的系统自带功能，通过加速器可以在手机中模拟出来现实中的重力感应，可以感应到手机的各种方向的变化，可以获取到手机在各个方向的模拟加速度。加速器通常会在 Windows Phone 8 的游戏上进行使用，例如赛车游戏就是一个典型的例子，利用了加速器的原理来设计，玩过 Windows Phone 8 中的赛车游戏的人肯定会知道，车的方向是通过左右摆动手机来控制的，有部分的赛车游戏还可以通过上下摆动来控制车的速度，这些都是运用了加速器来实现的。下面将详细地介绍 Windows Phone 8 的加速器的原理以及如何在应用程序中使用加速器。

10.1.1 加速器原理

加速器在 API 里面是用 Accelerometer 类来表示，Accelerometer 类在空间 Microsoft. Devices. Sensors 里面，需要使用系统加速器的功能，必须创建一个 Accelerometer 类的对象，然后通过这个对象来捕获手机当前的加速模拟状态。Accelerometer 类的 CurrentValueChanged 事件就是用来监控手机加速模拟状态的，在 CurrentValueChanged 事件中传递的 AccelerometerReadingEventArgs 参数会传递当前手机的加速模拟状态的数据变化，手机的加速模拟状态是通过一个三维的空间来表示的，这个三维空间一直都是以手机为中心点，通过 X 轴、Y 轴和 Z 轴的值来反映手机当前的状态。在使用加速度传感器时，可以把手机想象成一个三维的坐标系统。无论电话放置的方向是什么，Y 坐标轴是电话的底端（包含按钮的那端）到顶端的方向，而且这个走向是 Y 轴正方向。X 坐标轴则是从左至右的走向，这个走向亦是X轴正方向，Z坐标轴正走向则是面对用户的方向。这是一个在实际生活和数学中都经常使用的经典三维坐标系统，XNA中的3D编程也采用了这种坐标方法。这种坐标系统有一个专业术语，被称为笛卡儿右手坐标系统。笛卡儿右手坐标系统的意思就是将右手背对着手机屏幕放置，拇指即指向X轴的正方向。伸出食指和中指，食指指向Y轴的正方向，中指所指示的方向即是Z轴的正方向。有点类似面对自己的兰花指造型。这种坐标朝向永远是固定的，无论是将手机横拿还是竖放，又或者游戏是在Landscape和Portrait模式下运行，均是如此。

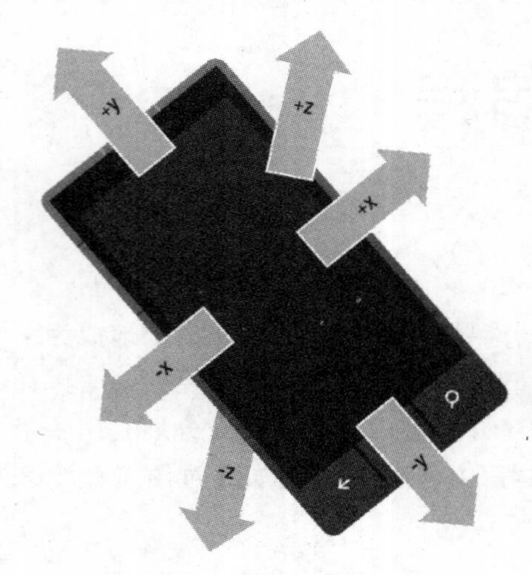

图10.1 手机的坐标方向

下面来介绍一下手机的加速模拟状态是怎么反映到X轴、Y轴和Z轴上的。当用户拿着手机面对自己的时候，那么Z轴的维度就是指向自己的方向，Y轴维度就是手机的上下方向，X轴就是手机的左右方向，如图10.1所示显示了这些三个轴相对位置的设备。

如果将手机放在一个理想状态的水平桌子上，那么加速器的X轴、Y轴和Z轴的值在以下6种特殊情况下的标准值如下：

(1) 手机竖着立起来放在桌子上，与桌子处于垂直的状态，如图10.2所示，手机的开始按键在下方，那么这时候对应的X轴、Y轴和Z轴的值是(x, y, z)=(0, −1, 0)。

(2) 手机竖着立起来放在桌子上，与桌子处于垂直的状态，手机的开始按键在上方，如图10.3所示，那么这时候对应的X轴、Y轴和Z轴的值是(x, y, z)=(0, 1, 0)。

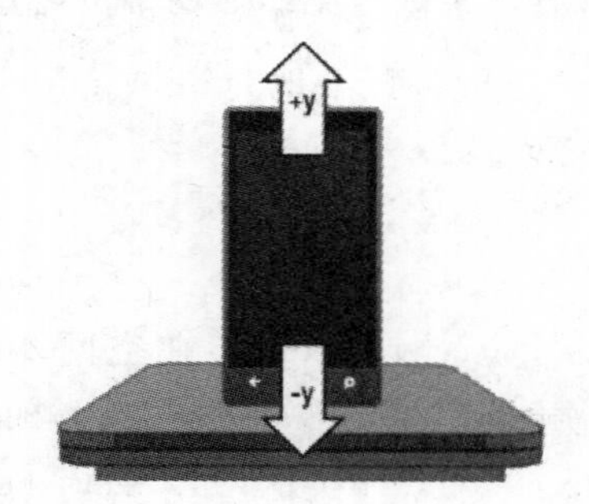

图10.2 手机竖着立起来放在桌子上，开始按键在下方

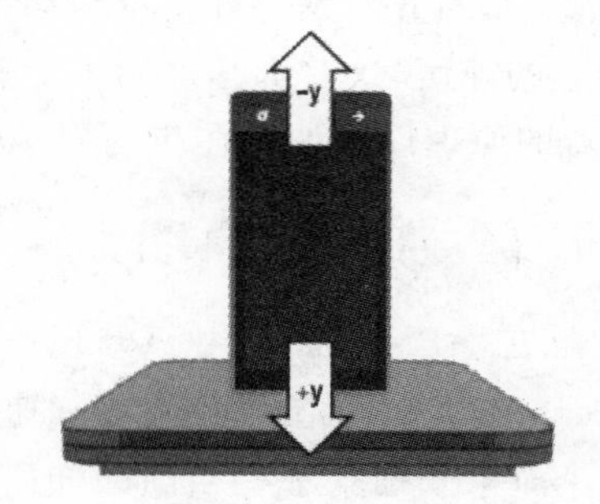

图10.3 手机竖着立起来放在桌子上，手机的开始按键在上方

(3) 手机横着立起来放在桌子上，与桌子处于垂直的状态，手机的返回按键在下方，如图10.4所示，那么这时候对应的X轴、Y轴和Z轴的值是(x, y, z)=(−1, 0, 0)。

(4) 手机横着立起来放在桌子上，与桌子处于垂直的状态，手机的返回按键在上方，如图10.5所示，那么这时候对应的X轴、Y轴和Z轴的值是(x, y, z)=(1, 0, 0)。

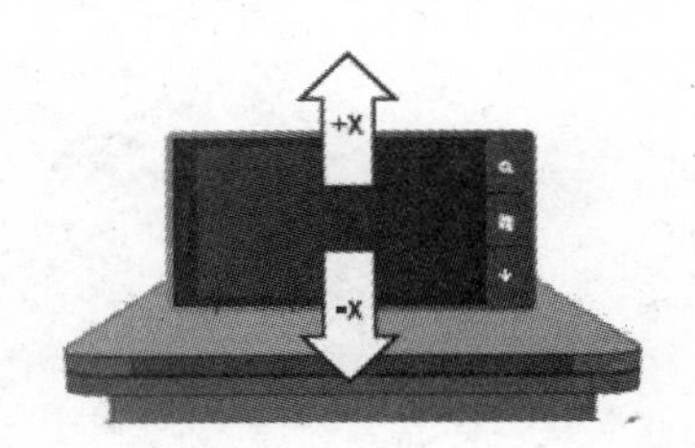

图 10.4 手机横着立起来放在桌子上,手机的返回按键在下方

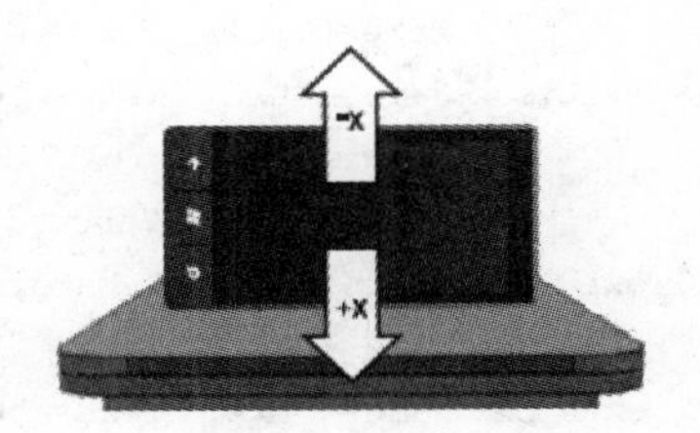

图 10.5 手机横着立起来放在桌子上,手机的返回按键在上方

(5) 手机平放在桌子上,手机的屏幕朝上,如图 10.6 所示,那么这时候对应的 X 轴、Y 轴和 Z 轴的值是(x, y, z)=(0, 0, −1)。

(6) 手机平放在桌子上,手机的屏幕朝下,如图 10.7 所示,那么这时候对应的 X 轴、Y 轴和 Z 轴的值是(x, y, z)=(0, 0, 1)。

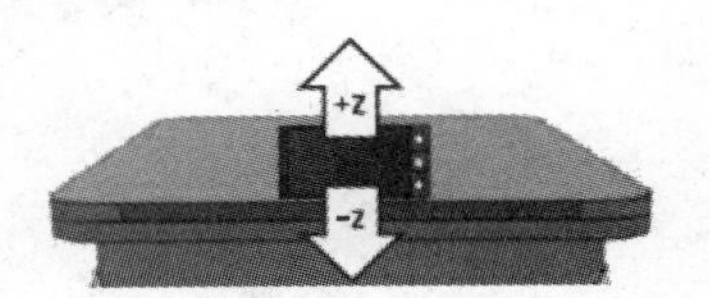

图 10.6 手机平放在桌子上,手机的屏幕朝上

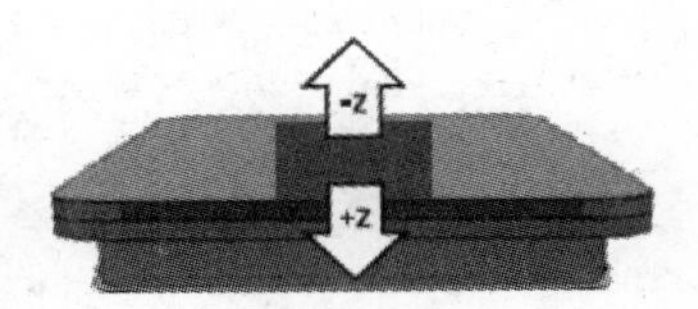

图 10.7 手机平放在桌子上,手机的屏幕朝下

这 6 种情况都是在完全标准情况下的值,在现实中这些值是有误差的,并且和手机的硬件质量和所处环境有关。

当一个三维的点坐标(x,y,z)表示空间一个特定的位置时,矢量(x,y,z)代表的意义则更加丰富,它包含了方向和长度的概念。很明显,点坐标和矢量是有关联的。矢量(x,y,z)的方向就是点(0,0,0)到点(x,y,z)的方向。但是矢量(x,y,z)并不是由点(0,0,0)到点(x,y,z)构成的那条直线,而只是代表这条直线的方向。那么矢量(x,y,z)的长度是 $\sqrt{x^2+y^2+z^2}$,如果当前的坐标是(x_1,y_1,z_1),移动手机后产生了一个新的坐标(x_2,y_2,z_2),那么可以使用计算空间两点的距离方法来计算加速器三维空间的这两点的距离,即距离是 $\sqrt{(x_1-x_2)^2+(y_1-y_2)^2+(z_1-z_2)^2}$,注意这个不是现实的三维空间的距离,而是手机加速器的三维空间的两点之间的距离。当手机静止的时候其加速器的矢量长度的值是等于 1 的,即 $x_2+y_2+z_2$ 等于 1;当手机正在做自由落体运动的时候其加速器的矢量长度的值是等于 0 的,即 $x_2+y_2+z_2$ 等于 0。

下面给出测试加速器的示例:用一个实例来测试加速器的三维空间坐标值的变化。

代码清单 10-1:测试加速器(源代码:第 10 章\Examples_10_1)

MainPage.xaml 文件主要代码

```
<Grid x:Name = "ContentPanel" Grid.Row = "1">
    <TextBlock FontSize = "40" Margin = "36,34,331,520"
```

```
                Name = "lblX" Text = "X 轴值:" />
        <TextBlock Name = "txtX" Text = "0"
                Margin = "155,51,17,0" FontSize = "20"
                Height = "36" VerticalAlignment = "Top" />
        <TextBlock FontSize = "40" Margin = "36,97,331,448"
                Name = "lblY" Text = "Y 轴值:" />
        <TextBlock Name = "txtY" Text = "0"
                Margin = "160,119,12,459" FontSize = "20" />
        <TextBlock FontSize = "40" Margin = "36,159,331,391"
                Name = "lblZ" Text = "Z 轴值:" />
        <TextBlock Name = "txtZ" Text = "0"
                Margin = "155,181,12,391" FontSize = "20" />
        <TextBlock FontSize = "40" Margin = "36,222,331,330"
                Name = "lblTime" Text = "Time" />
        <TextBlock Name = "txtTime" Text = "0"
                Margin = "155,244,12,322" FontSize = "20" />
        <TextBlock FontSize = "40" Margin = "0,283,350,257"
                Name = "textBlock3"
                Text = "ρ 角度:" TextAlignment = "Right" />
        <TextBlock FontSize = "20" Margin = "155,305,12,259"
                Name = "txtPitch" Text = "0" />
        <TextBlock FontSize = "40" Margin = "0,343,350,207"
                Name = "textBlock4" Text = "ϕ 角度:" TextAlignment = "Right" />
        <TextBlock FontSize = "20" Margin = "155,357,12,207"
                Name = "txtRoll" Text = "0" />
        <TextBlock FontSize = "40" Margin = "0,406,350,145"
                Name = "textBlock6" Text = "θ 角度:" TextAlignment = "Right" />
        <TextBlock FontSize = "20" Margin = "155,419,12,145"
                Name = "txtTheta" Text = "0" />
        <Button Content = "开始" Name = "btnStart" Width = "160"
            Margin = "36,514,284,6" Click = "btnStart_Click" />
        <Button Content = "停止" Name = "btnStop" Width = "160"
            Margin = "207,514,113,6" Click = "btnStop_Click" />
    </Grid>
```

MainPage.xaml.cs 文件主要代码

```
    public partial class MainPage : PhoneApplicationPage
    {
        Accelerometer _ac;
        public MainPage()
        {
            InitializeComponent();
            //新建一个加速器类
            _ac = new Accelerometer();
           //注册加速度变化的处理事件
```

```
        _ac.ReadingChanged += new EventHandler<AccelerometerReadingEventArgs>(_ac_
ReadingChanged);
    }
    //加速器变化的事件处理
    private void _ac_ReadingChanged(object sender, AccelerometerReadingEventArgs e)
    {
        //通过 Dispatcher.BeginInvoke 方法来更新 UI,传入事件变量 AccelerometerReadingEventArgs
        Deployment.Current.Dispatcher.BeginInvoke(() => ProcessAccelerometerReading(e));
    }
    //将各个方向的加速度变化值显示出来
    private void ProcessAccelerometerReading(AccelerometerReadingEventArgs e)
    {
        txtTime.Text = e.Timestamp.ToString();
        txtX.Text = e.X.ToString();
        txtY.Text = e.Y.ToString();
        txtZ.Text = e.Z.ToString();
        txtPitch.Text = ((Math.Atan(e.X / Math.Sqrt(Math.Pow(e.Y, 2) + Math.Pow(e.Z,
2))))).ToString();
        txtRoll.Text = ((Math.Atan(e.Y / Math.Sqrt(Math.Pow(e.X, 2) + Math.Pow(e.Z, 2))))).
ToString();
        txtTheta.Text = ((Math.Atan(Math.Sqrt(Math.Pow(e.X, 2) + Math.Pow(e.Y, 2)) / e.
Z))).ToString();
    }
    //开始运行加速器
    private void btnStart_Click(object sender, RoutedEventArgs e)
    {
        try
        {
            _ac.Start();
        }
        catch (AccelerometerFailedException)
        {
            MessageBox.Show("加速器开始失败!");
        }
    }
    //停止运行加速器
    private void btnStop_Click(object sender, RoutedEventArgs e)
    {
        try
        {
            _ac.Stop();
        }
        catch (AccelerometerFailedException)
        {
            MessageBox.Show("加速器停止失败!");
        }
    }
}
```

程序运行的效果如图 10.8 所示。

图 10.8　加速器测试

10.1.2　使用加速器实例编程

下面给出重力球的示例：通过重力感应来控制球的运动，用 X 轴和 Y 轴的加速感应来控制球的运动方向和运动的速度，用 Z 轴加速感应来改变球的大小。

代码清单 10-2：重力球(源代码：第 10 章\Examples_10_2)

MainPage.xaml 文件主要代码

```
<Canvas x:Name = "ContentGrid" Margin = "0,8,8,0" HorizontalAlignment = "Right"
        Width = "448" Height = "593" VerticalAlignment = "Top">
    <Ellipse x:Name = "ball" Canvas.Left = "126"
             Fill = "#FF963C3C" HorizontalAlignment = "Left"
             Height = "47" Stroke = "Black" StrokeThickness = "1"
             VerticalAlignment = "Top" Width = "46"
             Canvas.Top = "222"/>
</Canvas>
```

MainPage.xaml.cs 文件主要代码

```
public partial class MainPage : PhoneApplicationPage
{
    private Accelerometer _ac;
    public MainPage()
    {
        InitializeComponent();
```

```
        //设置应用的方向为水平方向
        SupportedOrientations = SupportedPageOrientation.Portrait;
        //把球设置在画布的中间
        ball.SetValue(Canvas.LeftProperty, ContentGrid.Width / 2);
        ball.SetValue(Canvas.TopProperty, ContentGrid.Height / 2);
        //创建一个加速器
        _ac = new Accelerometer();
        //绑定加速器的值的变化
        _ac.ReadingChanged += new EventHandler<AccelerometerReadingEventArgs>(ac_
ReadingChanged);
        _ac.Start();
    }
    //加速器变化的事件处理
    private void ac_ReadingChanged(object sender, AccelerometerReadingEventArgs e)
    {
        //通过 Dispatcher.BeginInvoke 方法来更新 UI,传入事件变量 AccelerometerReadingEventArgs
        Deployment.Current.Dispatcher.BeginInvoke(() => MyReadingChanged(e));
    }
    private void MyReadingChanged(AccelerometerReadingEventArgs e)
    {
        //取 z 轴的绝对值
        double accelerationFactor = Math.Abs(e.Z) == 0 ? 0.1 : Math.Abs(e.Z);
        double ballX = (double)ball.GetValue(Canvas.LeftProperty) + e.X / accelerationFactor;
        double ballY = (double)ball.GetValue(Canvas.TopProperty) - e.Y / accelerationFactor;
        double ballZ = (double)ball.GetValue(Canvas.HeightProperty) - e.Z/10;
        if (ballX < 0)
        {
            ballX = 0;
        }
        else if (ballX > ContentGrid.Width)
        {
            ballX = ContentGrid.Width;
        }
        if (ballY < 0)
        {
            ballY = 0;
        }
        else if (ballY > ContentGrid.Height)
        {
            ballY = ContentGrid.Height;
        }
        if (ballZ < 0)
        {
            ballZ = 10;
        }
        else if (ballZ > ContentGrid.Width)
```

```
            {
                ballZ = ContentGrid.Width;
            }
            //设置球在画布中的 Left 属性,即距离左边的距离
            ball.SetValue(Canvas.LeftProperty, ballX);
            //设置球在画布中的 Top 属性,即距离上边的距离
            ball.SetValue(Canvas.TopProperty, ballY);
            //设置球的宽度
            ball.SetValue(Canvas.WidthProperty, ballZ);
            //设置球的长度
            ball.SetValue(Canvas.HeightProperty, ballZ);
        }
    }
```

程序运行的效果如图 10.9 所示。

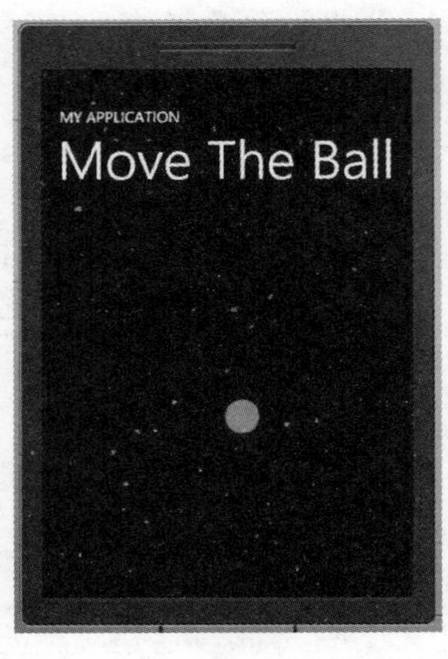

图 10.9 有重力感应的小球

下面给出拿起手机试试的示例：通过 Z 轴的值变化来判断手机的状态,从而判断是否把手机拿起来了。

代码清单 10-3：拿起手机试试(源代码：第 10 章\Examples_10_3)

MainPage.xaml 文件主要代码

```
<Grid x:Name = "ContentPanel" Grid.Row = "1" Margin = "12,0,12,0">
    <Ellipse x:Name = "ellipse" Fill = "White" Stroke = "Black"/>
</Grid>
```

MainPage.xaml.cs 文件主要代码

```
public partial class MainPage : PhoneApplicationPage
{
    private Accelerometer _ac;
    public MainPage()
    {
        InitializeComponent();
        //创建一个加速器
        _ac = new Accelerometer();
        //绑定加速器的值的变化
         _ac.ReadingChanged += new EventHandler < AccelerometerReadingEventArgs >(ac_
ReadingChanged);
        _ac.Start();
    }
    //加速器变化的事件处理
    private void ac_ReadingChanged(object sender, AccelerometerReadingEventArgs e)
    {
        //通过 Dispatcher.BeginInvoke 方法来更新 UI,传入事件变量 AccelerometerReadingEventArgs
        Deployment.Current.Dispatcher.BeginInvoke(() => MyReadingChanged(e));
    }
    private void MyReadingChanged(AccelerometerReadingEventArgs e)
```

```
    {
        if (Math.Abs(e.Z) < 0.9)
        {
            Storyboard shakeAnimation = Resources["ShakeAnimation"] as Storyboard;
            shakeAnimation.Begin();
        }
    }
}
```

程序运行的效果如图10.10所示。

图10.10　拿起手机试试

10.2 触摸感应

触控感应是智能手机系统的基本功能，Windows Phone系统里面提供了两种事件来感应手指的触控，一种是UIElement中的Manipulation事件，这种是基于控件元素级别的触控感应，另外一种是Touch.FrameReported事件，这是基于应用程序级别的触控感应事件。下面将介绍则两种不同类型的触控事件，并结合实例来演示如何进行触控感应编程。

10.2.1 Manipulation事件

在Windows Phone里面的控件都会继承UIElement基类，在UIElement类里面提供了ManipulationStarted、ManipulationDelta和ManipulationCompleted 3个触控事件，分别

表示触控开始、触控过程和触控结束触发的事件，这只是基于具体的元素级别的，也就是说这些事件是定义在某个元素上的，并且只对定义元素的触控才会触发这些事件。它们被触发的顺序是：先是一个 ManipulationStarted 事件，然后是 0 个或多个 ManipulationDelta 事件，最后是一个 ManipulationCompleted 事件。事件方法中 ManipulationStartedEventArgs 类、ManipulationCompletedEventArgs 类和 ManipulationCompletedEventArgs 类的参数包含了手指触控感应的一些信息。它们的一些共同的参数如下：

（1）OriginalSource：类型为 object，它是定义在 RoutedEventArgs 类中的，通过它可以获取到触发这个事件的原始对象。

（2）ManipulationContainer：类型为 UIElement，可以获取到定义当前这个触控操作坐标的对象（通常与 OriginalSource 是相同的）。

（3）ManipulationOrigin：类型为 Point。获取操作的起源坐标，即手指触控到的那一点的坐标，此坐标的数值就是相对于 ManipulationContainer 对象左上角的。如果有两个或多个手指在操作一个元素，那么 ManipulationOrigin 属性会给出多个手指的平均坐标。

（4）Handled：类型为 bool，是用来指示路由事件在路由过程中的事件处理状态。如果不想让当前事件沿着 Visual Tree 继续传播可以将其设置为 true。

1. ManipulationStartedEventArgs 类

除了上面说的 4 种共有的属性外，它还有一个 Complete 方法，它是用来告诉系统将 ManipulationStarted 事件结束掉，这样的话即使手指在屏幕上移动 ManipulationDelta 也不会被触发。

2. ManipulationDeltaEventArgs 类

除了共有属性外，这个类还包含两个 ManipulationDelta 类型的属性——CumulativeManipulation 和 DeltaManipulation。ManipulationDelta 类包含两个 Point 类型的属性——Scale 和 Translation。Scale 和 Translation 属性帮用户将一个或多个手指在某个元素上的复合动作解析成了元素自身的移动和尺寸变化。Scale 表示的是缩放因子，Translation 表示的是平移距离。用一个手指操作时就可以改变 Translation 的值，但如果要改变 Scale 需要用两个手指操作。当手指在一个元素上移动时，手指所在的新位置和原来位置的差值就会反映在 Translation 中。如果是用两个手指进行缩放，原来手指之间的距离与缩放后手指间距离差值会反映在 Scale 中。

CumulativeManipulation 属性和 DeltaManipulation 属性的区别：虽然这两个属性都包含 Scale 和 Translation。但 CumulativeManipulation 中的值是从 ManipulationStarted 事件开始到当前事件为止累加得到的（通过属性的名字就可以看出来），而 DeltaManipulation 中的值是本次 ManipulationDelta 事件相对于上一次 ManipulationDelta 或 ManipulationStarted 事件而言的，是单次的改变。除了 CumulativeManipulation 和 DeltaManipulation 属性，ManipulationDeltaEventArgs 类中还有 Complete 方法，它的作用和 ManipulationStarted 中 Complete 方法一样，都是通知系统当前的操作结束，调用此方法后即便手指在元素上移动也只会触发一次 ManipulationDelta 事件，本系列 Manipulation 操作中不会再有后续的 ManipulationDelta 事件被

触发，但可以触发 ManipulationCompleted 事件。

3. ManipulationCompletedEventArgs 类

除了共有属性外，ManipulationCompletdEventArgs 还包含下面 3 个属性：

(1) FinalVelocities：类型和 ManipulationDeltaEventArgs 类中的 Velocities 一样，都是 ManipulationVelocities，通过它可以获取手指离开屏幕时的速度。

(2) IsInertial：类型为 bool，和 ManipulationDeltaEventArgs 类中的 IsInertial 一样(IsInertial 属性在后面会详细说明)。

(3) TotalManipulation：类似于 ManipulationDeltaEventArgs 类的 CumulativeManipulation，类型也是 ManipulationDelta，它是从 ManipulationStarted 事件到 ManipulationCompleted 事件全过程的累加值。

当多个手指触控一个元素时它们会被转化为一系列的 Manipulation 事件，当不同的手指在不同的元素上时则会产生两个系列的 Manipulation 事件，这两个系列是独立的。当然它们可以通过 ManipulationContainer 属性来区分。例如，将一个手指放在一个元素上，首先一个 ManipulatedStarted 事件会被触发，如果手指移动那么就会触发 ManipulationDelta 事件。保持这个手指不动，将另一个手指放在相同的元素上不会再触发一个新的 ManipulatonStarted 事件。但如果此时我将另一个手指放在其他的元素上，则会触发那个元素的 ManipulationStarted 事件。

下面给出触控信息测试的示例：演示 Manipulation 事件的触发过程以及传递的信息。

代码清单 10-4：触摸信息测试(源代码：第 10 章\Examples_10_4)

MainPage.xaml 文件主要代码

```
<StackPanel x:Name = "TitlePanel" Grid.Row = "0" Margin = "12,17,0,28">
    <TextBlock x:Name = "ApplicationTitle" Text = "MY APPLICATION" Style = "{StaticResource
PhoneTextNormalStyle}"/>
    <TextBlock x:Name = "PageTitle" Text = "Manipulation"
               ManipulationStarted = "PageTitle_ManipulationStarted"
               ManipulationDelta = "PageTitle_ManipulationDelta"
               ManipulationCompleted = "PageTitle_ManipulationCompleted"
               Margin = "9, -7,0,0" Style = "{StaticResource PhoneTextTitle1Style}"/>
</StackPanel>
<Grid x:Name = "ContentPanel" Grid.Row = "1" Margin = "12,0,12,0">
    <ListBox Name = "list"></ListBox>
</Grid>
```

MainPage.xaml.cs 文件主要代码

```
//触摸开始事件处理
private void PageTitle_ManipulationStarted(object sender, ManipulationStartedEventArgs e)
{
    //将触摸的信息添加到 ListBox 控件中
    list.Items.Add("ManipulationStarted 你的手指刚接触到 PageTitle 控件");
    list.Items.Add("接触点   X:" + e.ManipulationOrigin.X + "   Y:" + e.ManipulationOrigin.Y);
    list.Items.Add("-------------------------------");
```

```
}
//触摸过程事件处理
private void PageTitle_ManipulationDelta(object sender, ManipulationDeltaEventArgs e)
{
    //将触摸的信息添加到 ListBox 控件中
    list.Items.Add("ManipulationDelta 的手指在滑动的过程中");
    list.Items.Add("变化 Translation X:" + e.DeltaManipulation.Translation.X + "  Y:" +
e.DeltaManipulation.Translation.Y);
    list.Items.Add("累增 Cumulative X:" + e.CumulativeManipulation.Translation.X + "  Y:" +
e.CumulativeManipulation.Translation.Y);
    list.Items.Add("线速度 LinearVelocity X:" + e.Velocities.LinearVelocity.X + "  Y:" +
e.Velocities.LinearVelocity.Y + " IsInertial:" + e.IsInertial);
    list.Items.Add("------------------------------");
}
//触摸结束事件处理
private void PageTitle_ManipulationCompleted(object sender, ManipulationCompletedEventArgs e)
{
    //将触摸的信息添加到 ListBox 控件中
    list.Items.Add("ManipulationCompleted 手指离开了屏幕");
    list.Items.Add("总的变化 Total Translation X:" + e.TotalManipulation.Translation.X +
" Y:" + e.TotalManipulation.Translation.Y);
    list.Items.Add("最后的线速度 FinalVelocities X:" + e.FinalVelocities.LinearVelocity.X
+ "  Y:" + e.FinalVelocities.LinearVelocity.Y + " IsInertial: " + e.IsInertial);
    list.Items.Add("------------------------------");
}
```

程序运行的效果如图 10.11 所示。

图 10.11　Manipulation 测试

10.2.2　应用示例：画图形

下面给出画图形的示例：可以选择三种正方形、圆形和直线三种图形，点击屏幕开始画画，当手指触摸到的是之前画出来的图画元素时，则是移动状态，手指的移动就会把之前的动画给移动位置。

代码清单 10-5：画图形(源代码：第 10 章\Examples_10_5)

MainPage.xaml 文件主要代码

```
<Grid x:Name = "ContentPanel" Grid.Row = "1" Margin = "12,0,12,0" Background = "Transparent">
</Grid>
</Grid>
<phone:PhoneApplicationPage.ApplicationBar>
    <shell:ApplicationBar>
        <!-- 菜单栏 -->
        <shell:ApplicationBar.MenuItems>
            <shell:ApplicationBarMenuItem Text = "画圆形"
                                          Click = "OnAppbarSelectGraphClick" />
            <shell:ApplicationBarMenuItem Text = "画正方形"
                                          Click = "OnAppbarSelectGraphClick" />
            <shell:ApplicationBarMenuItem Text = "画直线"
                                          Click = "OnAppbarSelectGraphClick" />
        </shell:ApplicationBar.MenuItems>
    </shell:ApplicationBar>
</phone:PhoneApplicationPage.ApplicationBar>
```

MainPage.xaml.cs 文件代码

```
using System;
using System.Windows;
using System.Windows.Input;
using System.Windows.Media;
using System.Windows.Shapes;
using Microsoft.Phone.Controls;
using Microsoft.Phone.Shell;
namespace TouchAndDraw
{
    public partial class MainPage : PhoneApplicationPage
    {
        //创建一个随机数产生类的对象,用于随机产生一种颜色
        Random rand = new Random();
        //一个画图标识符、一个拖动图片标识符
        bool isDrawing, isDragging;
        //Path 用于封装图形
        Path path;
        //graph 用于标识你选择要画的图形,初始化为画圆形
        string graph = "画圆形";
        //draggingGraph 用于标识你正在拖动的图形的形状
        string draggingGraph = "";
```

```
//椭圆图形
EllipseGeometry ellipseGeo;
//矩形图形
RectangleGeometry rectangleGeo;
//线形图形
LineGeometry lineGeo;
public MainPage()
{
    InitializeComponent();
}
///< summary >
///点击应用程序将会触发该事件,即手指接触到手机屏幕
///</summary >
protected override void OnManipulationStarted(ManipulationStartedEventArgs args)
{
    if (args.OriginalSource is Path)
    {//如果是点击在 Path 元素上则表示是拖动图形
        if ((args.OriginalSource as Path).Data is EllipseGeometry)
        {
            ellipseGeo = (args.OriginalSource as Path).Data as EllipseGeometry;
            draggingGraph = "圆形";
        }
        else if ((args.OriginalSource as Path).Data is RectangleGeometry)
        {
            rectangleGeo = (args.OriginalSource as Path).Data as RectangleGeometry;
            draggingGraph = "正方形";
        }
        else if ((args.OriginalSource as Path).Data is LineGeometry)
        {
            lineGeo = (args.OriginalSource as Path).Data as LineGeometry;
            draggingGraph = "线形";
        }
        //设置拖动图片状态为 true
        isDragging = true;
        args.ManipulationContainer = ContentPanel;
        args.Handled = true;
    }
    else if (args.OriginalSource == ContentPanel)
    {//如果是点击在 ContentPanel 控件上,则表示开始画图
        if (graph == "画圆形")
        {
            ellipseGeo = new EllipseGeometry();
            ellipseGeo.Center = args.ManipulationOrigin;
            path = new Path();
            path.Stroke = this.Resources["PhoneForegroundBrush"] as Brush;
            path.Data = ellipseGeo;
        }
        else if (graph == "画正方形")
        {
            rectangleGeo = new RectangleGeometry();
```

```
            Rect re = new Rect(args.ManipulationOrigin,args.ManipulationOrigin);
            rectangleGeo.Rect = re;
            path = new Path();
            path.Stroke = this.Resources["PhoneForegroundBrush"] as Brush;
            path.Data = rectangleGeo;
        }
        else if (graph == "画直线")
        {
            Color clr = Color.FromArgb(255, (byte)rand.Next(256),
                                            (byte)rand.Next(256),
                                            (byte)rand.Next(256));
            lineGeo = new LineGeometry();
            lineGeo.StartPoint = args.ManipulationOrigin;
            lineGeo.EndPoint = args.ManipulationOrigin;
            path = new Path();
            path.Stroke = new SolidColorBrush(clr);
            path.Data = lineGeo;
        }
        ContentPanel.Children.Add(path);
        isDrawing = true;
        args.Handled = true;
    }
    base.OnManipulationStarted(args);
}
///<summary>
///在画图或者拖动图片的过程中触发该事件,即手指还是屏幕上移动
///</summary>
protected override void OnManipulationDelta(ManipulationDeltaEventArgs args)
{
    if (isDragging)//如果是拖动图片
    {
        if (draggingGraph == "圆形")
        {
            Point center = ellipseGeo.Center;
            center.X += args.DeltaManipulation.Translation.X;
            center.Y += args.DeltaManipulation.Translation.Y;
            ellipseGeo.Center = center;
        }
        else if (draggingGraph == "正方形")
        {
            Rect re = rectangleGeo.Rect;
            re.X += args.DeltaManipulation.Translation.X;
            re.Y += args.DeltaManipulation.Translation.Y;
            rectangleGeo.Rect = re;
        }
        else if (draggingGraph == "线形")
        {
            Point start = lineGeo.StartPoint;
            start.X += args.DeltaManipulation.Translation.X;
            start.Y += args.DeltaManipulation.Translation.Y;
```

```
                lineGeo.StartPoint = start;
            }
            args.Handled = true;
        }
        else if (isDrawing)//如果是画图
        {
            Point translation = args.CumulativeManipulation.Translation;
            double radius = Math.Max(Math.Abs(translation.X),
                                     Math.Abs(translation.Y));
            if (graph == "画圆形")
            {
                ellipseGeo.RadiusX = radius;
                ellipseGeo.RadiusY = radius;
            }
            else if (graph == "画正方形")
            {
                Rect re = rectangleGeo.Rect;
                Rect re2 = new Rect(re.X, re.Y, radius, radius);
                rectangleGeo.Rect = re2;
            }
            else if (graph == "画直线")
            {
                Point end = lineGeo.StartPoint;
                end.X += translation.X;
                end.Y += translation.Y;
                lineGeo.EndPoint = end;
            }
            args.Handled = true;
        }
        base.OnManipulationDelta(args);
    }
    ///<summary>
    ///画图结束,即手指离开手机屏幕
    ///</summary>
    protected override void OnManipulationCompleted(ManipulationCompletedEventArgs args)
    {
        if (isDragging)
        {
            isDragging = false;
            args.Handled = true;
        }
        else if (isDrawing)
        {
            Color clr = Color.FromArgb(255, (byte)rand.Next(256),
                                            (byte)rand.Next(256),
                                            (byte)rand.Next(256));
            path.Fill = new SolidColorBrush(clr);
            isDrawing = false;
            args.Handled = true;
        }
```

```
            base.OnManipulationCompleted(args);
        }
        ///<summary>
        ///设置图形的类型
        ///</summary>
        void OnAppbarSelectGraphClick(object sender, EventArgs args)
        {
            ApplicationBarMenuItem item = sender as ApplicationBarMenuItem;
            graph = item.Text;
        }
    }
}
```

程序的运行效果如图 10.12 和图 10.13 所示。

图 10.12　画图形应用界面

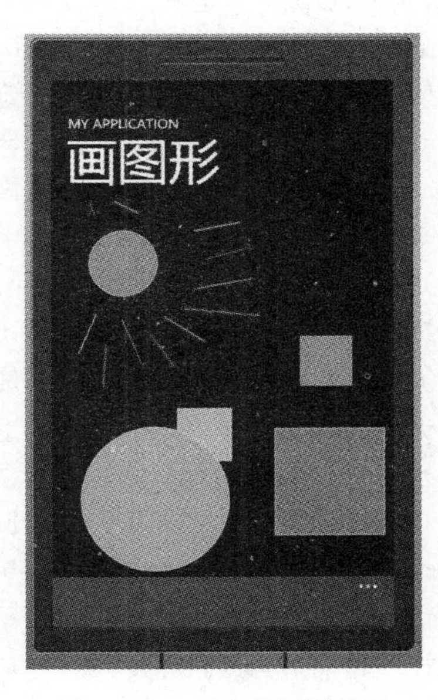

图 10.13　画出的图形效果

10.2.3　Touch.FrameReported 事件实现多点触摸

多点触控输入是一种输入类型，它依赖于触控屏输入概念，它需要有对触控敏感的硬件设备，以及支持将触控事件向各应用程序公开所需基础结构的环境。多点触控输入与 Windows Phone 支持的其他输入方法之间的一个主要差异在于，注册多点触控事件是基于应用程序范围的，而不是向特定输入元素(UIElement 对象)添加处理程序。

在 Windows Phone 中可以使用 System.Windows.Input.Touch 类的 FrameReported 事件来实现多点触控的编程，Touch 是一个出于此目的而存在的静态服务类，Touch.FrameReported 是其仅有的 API。下面介绍一下多点触控的一些关键的类和事件。

1. Touch 类

Touch 类是在 System.Windows.Input 空间下的，提供应用程序级服务，用以处理来自操作系统的多点触控输入并引发 FrameReported 事件。多点触控事件与其他 Windows Phone 输入事件(如 MouseLeftButtonDown)使用的是不同的事件模型。多点触控输入事件是在应用程序级别处理的单一事件，而不是公开为可能通过 UI 的对象树路由的特定于元素的事件。然后，使用 TouchFrameEventArgs 并调用 GetPrimaryTouchPoint 来获取第一个接触点。

2. TouchPoint 类

TouchPoint 类表示来自多点触控消息源的单个触控点。可以从这两种方法之一获取 TouchPoint 值，这两种方法都要求处理 FrameReported 事件，然后再使用 TouchFrame-EventArgs 事件数据。调用 GetPrimaryTouchPoint 以获取主触控点，如果用户不处理自己的手势，主接触点通常是唯一需要的数据点。调用 GetTouchPoints，将返回的集合同与应用程序处理触控输入的方式相关的 TouchPoint 项一起使用。TouchPoint 的 Position 始终是其接触区域的中心。若要获取指定 TouchPoint 的边界框，可以将边界框居中放置在 Position 点上，该边界框的尺寸基于 Size。

TouchPoint 是在屏幕上触控的手指的一个抽象。它有 4 个只读属性：

(1) Action：类型为 TouchAction 枚举，包含 3 个值，即 Down、Move 和 Up。

(2) Position：类型为 Point，它是相对于引用元素的左上角而言，这里说的引用元素就是前面提到的 GetPrimaryTouchPoint 和 GetTouchPoints 的 UIElement 参数，如果传入的参数是 null，那么得到的 Position 就是相对于屏幕左上角的(在传入非 null 值时，Position 中的数据很有可能会是负的)。

(3) Size：类型是 Size，它是要计算出屏幕中被触控的一个矩形区域。

(4) TouchDevice：类型为 TouchDevice，它包含两个只读属性：一个用于区分不同手指的 int 型 Id 属性；另一个是 UIElement 类型的 DirectlyOver 属性，它是紧贴手指的最上层 UI 元素。注意：如果需要在多个手指之间进行区分，Id 属性是至关重要的。在特定的手指触控屏幕时，与该手指关联的一系列特定事件总是以 Down 操作开始，接着是 Move 事件，最后是 Up 事件。所有这些事件都将与相同的 Id 关联。

3. Touch. FrameReported 事件

Touch. FrameReported 事件提供了应用程序级的服务，是 Silverlight for Windows Phone 中提供的底层触控编程接口。可以通过它来获取操作系统或整个应用程序中每个触控点的信息。Touch 是一个静态类，它只包含一个静态成员就是 FrameReported 事件。在程序中对此事件进行订阅时可以通过事件处理程序中的 TouchFrameEventArgs 参数获得想要的数据。

在订阅 Touch. FrameReported 后，通过调用事件处理程序中 TouchFrameEventArgs 类型参数的 GetPrimaryTouchPoint 和 GetTouchPoints 方法来获得相应的触控点数据。GetPrimaryTouchPoint 返回的是 TouchPoint，GetTouchPoints 返回的是 TouchPointCollection (此集合包含 0 个或多个 TouchPoint)。

在 GetPrimaryTouchPoint 返回的 TouchPoint 或 GetTouchPoints 方法返回的

是说，当按下手指然后抬起的时候分别会激发一次 FrameReported 事件，动作信息分别是 Down 和 up，如果你的手指在屏幕上移动了，那么 FrameReported 就会被触发至少 3 次，动作信息为 Down、Move 和 up(其中有可能移动了多次，而我们区分不出来，因为带 Move 信息的 FrameReported 事件可能会被触发多次)。另外，如果你用两个手指触控屏幕，FrameReported 会分别被触发，但此时如果调用 GetPrimaryTouchPoint 方法它只会返回第一个碰到屏幕的手指的触控信息，对于第二个手指 GetPrimaryTouchPoint 会返回 null。

TouchFrameEventArgs 类的 SuspendMousePromotionUntilTouchUp 方法。这个方法是用来禁止鼠标事件提升的。鼠标事件提升是源于桌面版的 Silverlight，目的是使多点触控用户可以使用触控和手势来代替鼠标移动或鼠标单击。比如当用户使用多点触控设备点击一个按钮时，按钮的预期行为与鼠标被单击时的相同，所以桌面版 Silverlight 提供了将触控输入自动提升为鼠标事件的机制，它被延续到了 Silverlight for Windows Phone 中。但是鼠标事件提升只是针对主触控点(第一个触控屏幕的手指同时没有其他手指触控屏幕时)，如果不想让这个手指的动作被提升，就可以使用 SuspendMousePromotionUntilTouchUp 方法来挂起鼠标事件提升。

下面给出 Touch. FrameReported 事件测试的示例：测试用 Touch. FrameReported 事件实现多点触控，并获取触控传递的详细信息。

代码清单 10-6：Touch. FrameReported 事件测试(源代码：第 10 章\Examples_10_6)

MainPage. xaml 文件主要代码

```
<Grid x:Name = "ContentPanel" Grid.Row = "1" Margin = "12,0,12,0">
    <ListBox Name = "list"></ListBox>
</Grid>
```

MainPage. xaml. cs 文件代码

```
using System.Windows;
using System.Windows.Input;
using Microsoft.Phone.Controls;
namespace FrameReportedTest
{
    public partial class MainPage : PhoneApplicationPage
    {
        public MainPage()
        {
            InitializeComponent();
            //注册 FrameReported 事件
            Touch.FrameReported += new TouchFrameEventHandler(Touch_FrameReported);
        }
        void Touch_FrameReported(object sender, TouchFrameEventArgs e)
        {
            //传入 null 表明获取到的触控点信息是以屏幕左上角为原点的
            var primaryPoint = e.GetPrimaryTouchPoint(null);
            if (primaryPoint != null)
            {
```

```
                list.Items.Add("第一个接触点的信息如下：");
                list.Items.Add("触控点的位置 X:" + primaryPoint.Position.X.ToString() +
" Y:" + primaryPoint.Position.Y.ToString());
                list.Items.Add("触控点的动作：" + primaryPoint.Action.ToString());
                list.Items.Add("触控点的设备：" + primaryPoint.TouchDevice.Id.ToString());
                list.Items.Add("触控点的元素：" + (primaryPoint.TouchDevice.DirectlyOver
as FrameworkElement).Name);
                list.Items.Add("--------------------------------");
            }
            //获取所有接触点
            TouchPointCollection touchPoints = e.GetTouchPoints(null);
            foreach (TouchPoint touchPoint in touchPoints)
            {
                list.Items.Add("接触点的信息如下：");
                list.Items.Add("触控点的位置 X:" + touchPoint.Position.X.ToString() +
" Y:" + touchPoint.Position.Y.ToString());
                list.Items.Add("触控点的动作：" + touchPoint.Action.ToString());
                list.Items.Add("触控点的设备：" + touchPoint.TouchDevice.Id.ToString());
                list.Items.Add("触控点的元素：" + (touchPoint.TouchDevice.DirectlyOver as
FrameworkElement).Name);
                list.Items.Add("--------------------------------");
            }
        }
    }
}
```

程序运行的效果如图 10.14 所示。

图 10.14　FrameReported 事件测试

10.2.4　应用实例：涂鸦板

下面给出涂鸦板的示例：涂鸦板应用使用 Touch.FrameReported 事件实现了多点触摸绘画的功能，可以多只手指在画板上绘画，最后可以将绘画的内容保存到手机独立存储空间里面。

代码清单 10-7：涂鸦板（源代码：第 10 章\Examples_10_7）

MainPage.xaml 文件主要代码：使用了 InkPresenter 墨迹控件作为绘画的画板

```
    <Grid x:Name = "ContentPanel" Grid.Row = "1" Margin = "12,0,12,0">
        <InkPresenter Name = "inkPresenter" />
    </Grid>
    …
<phone:PhoneApplicationPage.ApplicationBar>
    <shell:ApplicationBar>
        <!-- 工具条 -->
        <shell:ApplicationBarIconButton x:Name = "appbarAddButton"
                                        IconUri = "/Images/appbar.add.rest.png"
                                        Text = "新增" Click = "OnAppbarAddClick" />
        <shell:ApplicationBarIconButton x:Name = "appbarLastButton"
                                        IconUri = "/Images/appbar.back.rest.png"
                                        Text = "上一页" Click = "OnAppbarLastClick" />
        <shell:ApplicationBarIconButton x:Name = "appbarNextButton"
                                        IconUri = "/Images/appbar.next.rest.png"
                                        Text = "下一页" Click = "OnAppbarNextClick" />
        <shell:ApplicationBarIconButton x:Name = "appbarDeleteButton"
                                        IconUri = "/Images/appbar.delete.rest.png"
                                        Text = "删除" Click = "OnAppbarDeleteClick" />
        <!-- 菜单栏 -->
        <shell:ApplicationBar.MenuItems>
            <shell:ApplicationBarMenuItem Text = "红色画笔"
                                          Click = "OnAppbarSwapColorsClick" />
            <shell:ApplicationBarMenuItem Text = "蓝色画笔"
                                          Click = "OnAppbarSwapColorsClick" />
            <shell:ApplicationBarMenuItem Text = "黑色画笔"
                                          Click = "OnAppbarSwapColorsClick" />
            <shell:ApplicationBarMenuItem Text = "细画笔"
                                          Click = "OnAppbarSetStrokeWidthClick" />
            <shell:ApplicationBarMenuItem Text = "粗画笔"
                                          Click = "OnAppbarSetStrokeWidthClick" />
            <shell:ApplicationBarMenuItem Text = "橡皮擦"
                                          Click = "OnAppbarEraserClick" />
            <shell:ApplicationBarMenuItem Text = "保存涂鸦板"
                                          Click = "OnAppbarSaveClick" />
        </shell:ApplicationBar.MenuItems>
    </shell:ApplicationBar>
</phone:PhoneApplicationPage.ApplicationBar>
```

MainPage. xaml. cs 文件代码：在这里处理多点触摸绘画的事件

```
using System;
using System.Collections.Generic;
using System.Windows;
using System.Windows.Input;
using System.Windows.Media;
using System.Windows.Ink;
using System.IO.IsolatedStorage;
using System.IO;
using System.Xml.Serialization;
using Microsoft.Phone.Shell;
namespace Palette
{
    public partial class MainPage : PhoneApplicationPage
    {
        //涂鸦板对象
        PaletteAppSettings appSettings;
        //用一个字典类型来存储当前图画的每次画图的图形
        Dictionary<int, Stroke> activeStrokes = new Dictionary<int, Stroke>();
        // 是否是擦除操作
        private bool _isEraser = false;
        public MainPage()
        {
            InitializeComponent();
            //加载涂鸦板对象
            appSettings = PaletteAppSettings.Load();
            //获取当前涂鸦板的图画
            inkPresenter.Strokes = appSettings.StrokeCollections[appSettings.PageNumber];
            //设置涂鸦板的背景颜色
            inkPresenter.Background = new SolidColorBrush(appSettings.Background);
            //获取上一页按钮的对象
            appbarLastButton = this.ApplicationBar.Buttons[1] as ApplicationBarIconButton;
            //获取下一页按钮的对象
            appbarNextButton = this.ApplicationBar.Buttons[2] as ApplicationBarIconButton;
            //获取删除按钮的对象
            appbarDeleteButton = this.ApplicationBar.Buttons[3] as ApplicationBarIconButton;
            //更新应用的标题显示
            TitleAndAppbarUpdate();
            //注册 Touch.FrameReported 事件来响应多点触控
            Touch.FrameReported += OnTouchFrameReported;
        }
        ///<summary>
        ///处理多点触摸事件
        ///</summary>
        void OnTouchFrameReported(object sender, TouchFrameEventArgs args)
        {
            //获取涂鸦板的所有接触点
            TouchPointCollection touchPoints = args.GetTouchPoints(inkPresenter);
            foreach (TouchPoint touchPoint in touchPoints)
```

```
        {
            //获取接触点的位置
            Point pt = touchPoint.Position;
            //获取这一次触摸操作的唯一设备 id
            int id = touchPoint.TouchDevice.Id;
            //根据手指的动作来处理图画
            switch (touchPoint.Action)
            {
                case TouchAction.Down:
                    Stroke stroke = new Stroke();
                    if (_isEraser)
                    {
                        stroke.DrawingAttributes.Color = appSettings.Background;
                        stroke.DrawingAttributes.Height = 20;
                        stroke.DrawingAttributes.Width = 20;
                    }
                    else
                    {
                        stroke.DrawingAttributes.Color = appSettings.Foreground;
                        stroke.DrawingAttributes.Height = appSettings.StrokeWidth;
                        stroke.DrawingAttributes.Width = appSettings.StrokeWidth;
                    }
                    stroke.StylusPoints.Add(new StylusPoint(pt.X, pt.Y));
                    inkPresenter.Strokes.Add(stroke);
                    activeStrokes.Add(id, stroke);
                    break;
                case TouchAction.Move:
                     activeStrokes[id].StylusPoints.Add(new StylusPoint(pt.X, pt.Y));
                    break;
                case TouchAction.Up:
                     activeStrokes[id].StylusPoints.Add(new StylusPoint(pt.X, pt.Y));
                     activeStrokes.Remove(id);
                     TitleAndAppbarUpdate();
                    break;
            }
        }
    }
    ///<summary>
    ///新增一个页面
    ///</summary>
    void OnAppbarAddClick(object sender, EventArgs args)
    {
        StrokeCollection strokes = new StrokeCollection();
        appSettings.PageNumber += 1;
        appSettings.StrokeCollections.Insert(appSettings.PageNumber, strokes);
        inkPresenter.Strokes = strokes;
        TitleAndAppbarUpdate();
    }
    ///<summary>
    ///上一个页面
```

```
///</summary>
void OnAppbarLastClick(object sender, EventArgs args)
{
    appSettings.PageNumber -= 1;
    inkPresenter.Strokes = appSettings.StrokeCollections[appSettings.PageNumber];
    TitleAndAppbarUpdate();
}
///<summary>
///下一个页面
///</summary>
void OnAppbarNextClick(object sender, EventArgs args)
{
    appSettings.PageNumber += 1;
    inkPresenter.Strokes = appSettings.StrokeCollections[appSettings.PageNumber];
    TitleAndAppbarUpdate();
}
///<summary>
///删除当前的页面
///</summary>
void OnAppbarDeleteClick(object sender, EventArgs args)
{
    MessageBoxResult result = MessageBox.Show("是否删除当前的页面?", "涂鸦板",
                                              MessageBoxButton.OKCancel);
    if (result == MessageBoxResult.OK)
    {
        if (appSettings.StrokeCollections.Count == 1)
        {
            appSettings.StrokeCollections[0].Clear();
        }
        else
        {
            appSettings.StrokeCollections.RemoveAt(appSettings.PageNumber);
            if (appSettings.PageNumber == appSettings.StrokeCollections.Count)
                appSettings.PageNumber -= 1;
            inkPresenter.Strokes = appSettings.StrokeCollections[appSettings.PageNumber];
        }
        TitleAndAppbarUpdate();
    }
}
///<summary>
///设置画笔的颜色
///</summary>
void OnAppbarSwapColorsClick(object sender, EventArgs args)
{
    ApplicationBarMenuItem item = sender as ApplicationBarMenuItem;
    if (item.Text.StartsWith("红色画笔"))
        appSettings.Foreground = Colors.Red;
    else if (item.Text.StartsWith("蓝色画笔"))
        appSettings.Foreground = Colors.Blue;
    else if (item.Text.StartsWith("黑色画笔"))
```

```
            appSettings.Foreground = Colors.Black;
        _isEraser = false;
    }
    ///< summary>
    ///设置画笔的粗细
    ///</summary>
    void OnAppbarSetStrokeWidthClick(object sender, EventArgs args)
    {
        ApplicationBarMenuItem item = sender as ApplicationBarMenuItem;
        if (item.Text.StartsWith("细画笔"))
            appSettings.StrokeWidth = 1;
        else if (item.Text.StartsWith("粗画笔"))
            appSettings.StrokeWidth = 5;
        _isEraser = false;
    }
    /// < summary>
    /// 使用橡皮擦
    /// </summary>
    void OnAppbarEraserClick(object sender, EventArgs args)
    {
        ApplicationBarMenuItem item = sender as ApplicationBarMenuItem;
        // 单击了橡皮擦按钮
        inkPresenter.Cursor = Cursors.Eraser;
        _isEraser = true;
    }
    /// < summary>
    /// 保存涂鸦板
    /// </summary>
    void OnAppbarSaveClick(object sender, EventArgs args)
    {
        appSettings.Save();
    }
    /// < summary>
    /// 更新程序标题显示页码
    /// </summary>
    void TitleAndAppbarUpdate()
    {
        ApplicationTitle.Text = String.Format(" 当前页码 {0} --- 总页数 {1}",
                                        appSettings.PageNumber + 1,
                                        appSettings.StrokeCollections.Count);
        //设置上一页、下一页以及删除按钮是否可用
        appbarLastButton.IsEnabled = appSettings.PageNumber > 0;
        appbarNextButton.IsEnabled =
                appSettings.PageNumber < appSettings.StrokeCollections.Count - 1;
        appbarDeleteButton.IsEnabled = (appSettings.StrokeCollections.Count > 1) ||
                                    (appSettings.StrokeCollections[0].Count > 0);
    }
}
```

PaletteAppSettings.cs 文件代码：PaletteAppSettings 类用于处理将图画保存到独立存储空间中和从独立存储空间中获取以前的图画

```
public class PaletteAppSettings
{
    ///< summary >
    ///初始化一个涂鸦板
    ///</summary >
    public PaletteAppSettings()
    {
        this.PageNumber = 0;
        this.Foreground = Colors.Black;
        this.Background = Colors.White;
        this.StrokeWidth = 1;
    }
    //涂鸦板的图画数组
    public List < StrokeCollection > StrokeCollections { get; set; }
    //涂鸦板的当前页码数
    public int PageNumber { set; get; }
    //涂鸦板的画笔颜色
    public Color Foreground { set; get; }
    //涂鸦板的背景颜色
    public Color Background { set; get; }
    //涂鸦板的画笔宽度
    public int StrokeWidth { set; get; }
    ///< summary >
    ///加载之前的涂鸦板信息
    ///</summary >
    public static PaletteAppSettings Load()
    {
        //创建一个涂鸦板对象
        PaletteAppSettings settings;
        //获取当前应用程序的独立存储
        IsolatedStorageFile iso = IsolatedStorageFile.GetUserStoreForApplication();
        //判断独立存储中是否存在 settings.xml 文件,在第一次保存操作之前,这个文件
        //是不存在的
        if (iso.FileExists("settings.xml"))
        {
            //用独立存储的文件流打开 settings.xml 文件
            IsolatedStorageFileStream stream = iso.OpenFile("settings.xml", FileMode.Open);
            //转化为可读流
            StreamReader reader = new StreamReader(stream);
            //创建一个涂鸦板类的 xml 可序列化对象
            XmlSerializer ser = new XmlSerializer(typeof(PaletteAppSettings));
            //反序列化 xml 文件,转为 PaletteAppSettings 对象
            settings = ser.Deserialize(reader) as PaletteAppSettings;
            //关闭可读流
            reader.Close();
        }
        else
        {
            //第一次使用涂鸦板创建一个新的涂鸦板对象
```

```
                settings = new PaletteAppSettings();
                //添加一个空的 StrokeCollection 数组到涂鸦板对象中
                settings.StrokeCollections = new List<StrokeCollection>();
                settings.StrokeCollections.Add(new StrokeCollection());
            }
            //释放独立存储资源
            iso.Dispose();
            return settings;
        }
        /// <summary>
        /// 保存涂鸦板的图画
        /// </summary>
        public void Save()
        {
            //获取当前应用程序的独立存储
            IsolatedStorageFile iso = IsolatedStorageFile.GetUserStoreForApplication();
            //创建一个 XML 文件来保存涂鸦板的信息
            IsolatedStorageFileStream stream = iso.CreateFile("settings.xml");
            //转化为可写流
            StreamWriter writer = new StreamWriter(stream);
            //创建一个涂鸦板类的 xml 可序列化对象
            XmlSerializer ser = new XmlSerializer(typeof(PaletteAppSettings));
            //用 xml 格式序列化当前的涂鸦板类对象
            ser.Serialize(writer, this);
            //关闭可写流
            writer.Close();
            //释放独立存储资源
            iso.Dispose();
        }
    }
}
```

程序运行的效果如图 10.15 和图 10.16 所示。

图 10.15　涂鸦板的画图效果

图 10.16　涂鸦板的画图选项

10.3 电子罗盘

电子罗盘，也叫数字指南针，是利用地磁场来定北极的一种方法。指南针是用来指示方向的一种工具。常见的机械式指南针，是一种根据地球磁场的有极性制作的地磁指南针，但这种指南针指示的南北方向与真正的南北方向不同，存在一个磁偏角。电子器件的飞速发展，为我们带来了电子指南针，也就是所谓的电子罗盘，它采用了磁场传感器的磁阻技术，可很好地修正磁偏角的问题。Windows Phone 手机设备里面可以利用罗盘传感器的 API 来实现电子罗盘的功能。

10.3.1 罗盘传感器原理

应用程序可以使用罗盘或磁力计传感器来确定设备相对于地球磁场北极旋转的角度，也可以使用原始磁力计读数来检测设备周围的磁力。罗盘传感器对于所有 Windows Phone 设备来说都不是必需的，设计和实现应用程序时应该考虑此内容，这一点非常重要。应用程序应该始终检查传感器是否可用，如果不可用，将会引发相关的异常。判断罗盘传感器是否可以用可以使用 Compass 类的 IsSupported 属性来判断，true 表示支持，false 表示不支持。

罗盘 API 根据设备的方向使用单个轴来计算航向。如果要创建一个使用所有轴上的设备方向的应用程序，则应该使用 Motion 类的 RotationMatrix 属性。设备中的罗盘传感器可能随时间变得不精确，尤其是暴露在磁场中时更是如此。有一个重新校准罗盘的简单用户操作，只要系统检测到航向精度大于± 20°，就引发 Calibrate 事件。

Compass 类为 Windows Phone 应用程序提供对设备罗盘传感器的访问。可以使用罗盘或磁力计传感器来确定设备相对于地球磁场北极的角度。Compass 类的主要成员如表 10.1 所示。

表 10.1 Compass 类的主要成员

名　称	说　明
属性 CurrentValue	获取一个对象，该对象实现包含传感器当前值的 ISensorReading；此对象将为以下类型之一（取决于引用的传感器）：AccelerometerReading、CompassReading、GyroscopeReading、MotionReading
属性 IsDataValid	获取传感器数据的有效性
属性 IsSupported	获取其上运行应用程序的设备是否支持罗盘传感器
属性 TimeBetweenUpdates	获取或设置 CurrentValueChanged 事件之间的首选时间
方法 Dispose	释放由传感器使用的托管资源和非托管资源
方法 Finalize	允许 Object 在垃圾回收器回收该对象之前尝试释放资源并执行其他清理操作
方法 Start	开始从传感器获取数据

续表

名　称	说　明
方法 Stop	停止从传感器获取数据
事件 Calibrate	当操作系统检测到罗盘需要校准时发生
事件 CurrentValueChanged	在从传感器获得新数据时发生

10.3.2　创建一个指南针应用

下面给出指南针应用的示例：使用电子罗盘传感器来创建一个指南针应用，可以及时调整方向的指向和进行精度的校准。

代码清单 10-8：指南针(源代码：第 10 章\Examples_10_8)

MainPage.xaml 文件主要代码

```
<Grid x:Name = "ContentPanel" Grid.Row = "1" Margin = "12,0,12,0">
    <!-- 精度和角度显示 -->
    <TextBlock Height = "70" HorizontalAlignment = "Left" Margin = "198,100,0,0" Name =
"RecipLabelTextBlock" Text = "角度: " VerticalAlignment = "Top" Width = "85" FontSize = "32"
Foreground = "#EE8F908D" />
    <TextBlock Height = "71" HorizontalAlignment = "Right" Margin = "0,100,111,0" Name =
"RecipTextBlock" Text = "000" VerticalAlignment = "Top" Width = "56" FontSize = "32" Foreground
= "#FFC8C7CC" />
    <TextBlock Height = "76" HorizontalAlignment = "Left" Margin = "165,36,0,0" Name =
"HeadingLabelTextBlock" Text = "精度: " VerticalAlignment = "Top" Width = "105" FontSize = "48"
Foreground = "#EE8F908D" />
    <TextBlock HorizontalAlignment = "Right" Margin = "0,36,73,655" Name = "HeadingTextBlock"
Text = "000" Width = "94" FontSize = "48" Foreground = "#FFC8C7CC" />
    <!-- 指南针罗盘 UI -->
    <Ellipse Height = "385" HorizontalAlignment = "Left" Margin = "31,0,0,176" Name =
"EllipseBorder" Stroke = "#FFF80D0D" StrokeThickness = "2" VerticalAlignment = "Bottom" Width =
"385" Fill = "White" />
    <Image Height = "263" HorizontalAlignment = "Left" Margin = "91,266,0,0" Name =
"CompassFace" VerticalAlignment = "Top" Width = "263" Source = "compass.png" Stretch = "None" />
    <Image Height = "24" HorizontalAlignment = "Left" Margin = "209,163,0,0" Name =
"PointerImage" Stretch = "Fill" VerticalAlignment = "Top" Width = "29" Source = "pointer.png" />
    <Ellipse Height = "263"  Width = "263" x:Name = "EllipseGlass" Margin = "91,266,102,239"
Stroke = "Black" StrokeThickness = "1">
        <Ellipse.Fill>
            <LinearGradientBrush EndPoint = "1,0.5" StartPoint = "0,0.5">
                <GradientStop Color = "#A5000000" Offset = "0" />
                <GradientStop Color = "#BFFFFFFF" Offset = "1" />
            </LinearGradientBrush>
        </Ellipse.Fill>
    </Ellipse>
</Grid>
<!-- 校准 UI -->
<StackPanel Name = "calibrationStackPanel" Background = "White" Opacity = "1" Visibility =
```

```
"Collapsed">
    < Image Source = "calibrate_compass3.png" Opacity = ".95" HorizontalAlignment = "Center"/>
    < TextBlock TextWrapping = "Wrap" TextAlignment = "Center" Foreground = "Black">设备上的指
南针需要校准,校准完成后,请单击“完成”按钮。</TextBlock >
    < StackPanel Orientation = "Horizontal" Margin = "0,10" HorizontalAlignment = "Center">
        < TextBlock Foreground = "Black">精度调整:</TextBlock >
        < TextBlock Name = "CalibrationTextBlock" Foreground = "Red"> 0.0°</TextBlock >
    </StackPanel >
    < Button Name = "calibrationButton" Content = "完成" Click = "calibrationButton_Click"
Foreground = "Black" HorizontalAlignment = "Center" BorderBrush = "Black"></Button >
</StackPanel >
```

MainPage.xaml.cs 文件代码

```
using System;
using System.Windows;
using System.Windows.Media;
using Microsoft.Phone.Controls;
using Microsoft.Devices.Sensors;
namespace CompassDemo
{
    public partial class MainPage : PhoneApplicationPage
    {
        Compass compass;                    //电子罗盘对象
        RotateTransform transform = new RotateTransform();   //旋转偏移动画
        double TrueHeading;                 //与北极偏角
        double ReciprocalHeading;           //旋转的倒角度
        double HeadingAccuracy;             //精度
        bool Calibrating = false;           //是否在校准
        public MainPage()
        {
            InitializeComponent();
            if (Compass.IsSupported)
            {
                compass = new Compass();
                //订阅更新数据事件
                 compass.CurrentValueChanged += new EventHandler < SensorReadingEventArgs
< CompassReading >>(compass_CurrentValueChanged);
                //订阅校准事件
                compass.Calibrate += new EventHandler < CalibrationEventArgs >(compass_Calibrate);
                //设置更新的时间间隔,必须是 20 的整数倍
                compass.TimeBetweenUpdates = TimeSpan.FromMilliseconds(400);
                //启动电子罗盘
                compass.Start();
            }
            else
                MessageBox.Show("设备不支持电子罗盘");
        }
        //校准数据处理
        void compass_Calibrate(object sender, CalibrationEventArgs e)
        {
```

```
            Dispatcher.BeginInvoke(() => { calibrationStackPanel.Visibility =
Visibility.Visible; });
            Calibrating = true;
        }
        //校准完成单击事件处理
        private void calibrationButton_Click(object sender, RoutedEventArgs e)
        {
            //隐藏校准 UI
            calibrationStackPanel.Visibility = Visibility.Collapsed;
            Calibrating = false;
        }
        //电子罗盘感应器数据更新事件处理
        void compass_CurrentValueChanged(object sender, SensorReadingEventArgs
<CompassReading> e)
        {
            if (compass.IsDataValid)
            {
                //启动 UI 线程进行更新界面数据
                Deployment.Current.Dispatcher.BeginInvoke(() =>
                {
                    //获取精度
                    HeadingAccuracy = e.SensorReading.HeadingAccuracy;
                    //不需要校准则开始更新数据
                    if (!Calibrating)
                    {
                        //获取与地理北极的顺时针方向的偏角
                        TrueHeading = e.SensorReading.TrueHeading;
                        if ((180 <= TrueHeading) && (TrueHeading <= 360))
                            ReciprocalHeading = TrueHeading - 180;
                        else
                            ReciprocalHeading = TrueHeading + 180;
                        //旋转动画的中心点
                        CompassFace.RenderTransformOrigin = new Point(0.5, 0.5);
                        EllipseGlass.RenderTransformOrigin = new Point(0.5, 0.5);
                        //计算旋转的角度
                        transform.Angle = 360 - TrueHeading;
                        CompassFace.RenderTransform = transform;
                        EllipseGlass.RenderTransform = transform;
                    }
                    else
                    {
                        //当精度小于 10 则需要校准
                        if (HeadingAccuracy <= 10)
                        {
                                CalibrationTextBlock.Foreground = new SolidColorBrush(Colors.
Green);
                            CalibrationTextBlock.Text = "完成";
                        }
                        else
                        {
```

```
                            CalibrationTextBlock.Foreground = new SolidColorBrush(Colors.Red);
                            CalibrationTextBlock.Text = HeadingAccuracy.ToString("0.0");
                        }
                    }
                });
            }
        }
    }
}
```

程序运行的效果如图 10.17 所示。

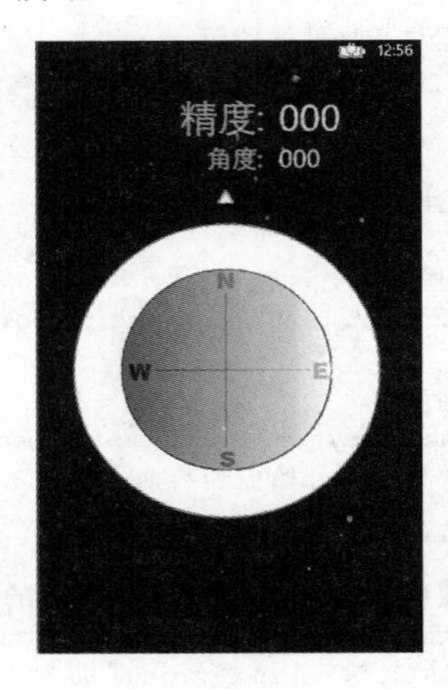

图 10.17 指南针应用

10.4 陀螺仪

陀螺仪又叫角速度传感器，测量物理量是偏转、倾斜时的转动角速度。陀螺仪可以对转动、偏转的动作做很好的测量，这样可以精确分析判断出使用者的实际动作，而后根据动作，可以对手机做相应的操作。

10.4.1 陀螺仪原理

陀螺仪传感器测量设备沿着其三个主轴的旋转速度。当设备静止时，所有轴的陀螺仪读数都为零。如果设备面向使用者并围绕其中心点旋转，就像飞机螺旋桨一样，那么 Z 轴上的旋转速度值将大于零，设备旋转的速度越快，该值越大。旋转速度的测量以弧度/秒为单位，其中 2 * Pi 弧度就是全程旋转。手机坐标系可以用相对于手机位置的右手坐标系来理解：以手机位置为参照，假设手机垂直水平面放(竖着放)，屏幕对着使用者，那么左右是 X 轴，右侧为正方向，左侧为负方向，上下是 Y 轴，上侧为正方向，下侧为负方向，里外是 Z 轴，靠近使用者为正方向，远离使用者为负方向。

Gyroscope 类为 Windows Phone 应用程序提供对设备陀螺仪传感器的访问。Gyroscope 类的主要成员如表 10.2 所示。

表 10.2 Gyroscope 类的主要成员

名 称	说 明
Gyroscope	用于访问设备中的陀螺仪
IsSupported	设备是否支持陀螺仪
IsDataValid	是否可从陀螺仪中获取到有效数据
CurrentValue	陀螺仪当前的数据，GyroscopeReading 类型
TimeBetweenUpdates	触发 CurrentValueChanged 事件的时间间隔，如果设置的值小于 Gyroscope 允许的最小值，则此属性的值将被设置为 Gyroscope 允许的最小值
Start()	打开陀螺仪
Stop()	关闭陀螺仪
CurrentValueChanged	陀螺仪传感器获取到的数据发生改变时所触发的事件，属性 TimeBetweenUpdates 的值决定触发此事件的时间间隔
GyroscopeReading	陀螺仪传感器数据
RotationRate	获取围绕设备各轴旋转的旋转速率(单位：弧度/秒)
DateTimeOffset	从陀螺仪传感器中获取到数据的时间点

10.4.2 创建一个陀螺仪应用

下面给出陀螺仪应用的示例：使用陀螺仪传感器来观察陀螺仪的 X、Y、Z 轴的数值变化。

代码清单 10-9：陀螺仪应用(源代码：第 10 章\Examples_10_9)

MainPage.xaml 文件主要代码

```
<Grid x:Name = "ContentPanel" Grid.Row = "1" Margin = "12,0,12,0">
    <TextBlock Height = "30" HorizontalAlignment = "Left" Margin = "20,100,0,0" Name =
"xTextBlock" Text = "X: 1.0" VerticalAlignment = "Top" Foreground = "Red" FontSize = "28"
FontWeight = "Bold"/>
    <TextBlock Height = "30" HorizontalAlignment = "Center" Margin = "0,100,0,0" Name =
"yTextBlock" Text = "Y: 1.0" VerticalAlignment = "Top" Foreground = "Yellow" FontSize = "28"
FontWeight = "Bold"/>
    <TextBlock Height = "30" HorizontalAlignment = "Right" Margin = "0,100,20,0" Name =
"zTextBlock" Text = "Z: 1.0" VerticalAlignment = "Top" Foreground = "Blue" FontSize = "28"
FontWeight = "Bold"/>
    <Line x:Name = "xLine" X1 = "240" Y1 = "350" X2 = "340" Y2 = "350" Stroke = "Red"
StrokeThickness = "4"></Line>
    <Line x:Name = "yLine" X1 = "240" Y1 = "350" X2 = "240" Y2 = "270" Stroke = "Yellow"
StrokeThickness = "4"></Line>
```

```
    <Line x:Name = "zLine" X1 = "240" Y1 = "350" X2 = "190" Y2 = "400" Stroke = "Blue"
StrokeThickness = "4"></Line>
    <TextBlock Height = "30" HorizontalAlignment = "Center" Margin = "6,571,6,0" Name =
"statusTextBlock" Text = "" VerticalAlignment = "Top" Width = "444" />
</Grid>
```

MainPage.xaml.cs 文件代码

```
using System;
using Microsoft.Phone.Controls;
using Microsoft.Devices.Sensors;
using Microsoft.Xna.Framework;
namespace GyroscopeDemo
{
    public partial class MainPage : PhoneApplicationPage
    {
        Gyroscope g;
        public MainPage()
        {
            InitializeComponent();
            //判断设备是否支持陀螺仪
            if (Gyroscope.IsSupported)
            {
                g = new Gyroscope();
                //设置陀螺仪计算的时间间隔
                g.TimeBetweenUpdates = TimeSpan.FromMilliseconds(20);
                //订阅陀螺仪的数值变化
                    g.CurrentValueChanged += new EventHandler < SensorReadingEventArgs
<GyroscopeReading>>(g_CurrentValueChanged);
                //开始启动陀螺仪
                g.Start();
            }
            else statusTextBlock.Text = "不支持陀螺仪";
        }
        void g_CurrentValueChanged(object sender, SensorReadingEventArgs<GyroscopeReading> e)
        {
            //启动UI线程来更新数据
            Dispatcher.BeginInvoke(() => UpdateUI(e.SensorReading));
        }
        private void UpdateUI(GyroscopeReading gyroscopeReading)
        {
            statusTextBlock.Text = "获取数据";
            Vector3 rotationReading = gyroscopeReading.RotationRate;
            // X,Y,Z轴的数据
            xTextBlock.Text = "X " + rotationReading.X.ToString("0.00");
            yTextBlock.Text = "Y " + rotationReading.Y.ToString("0.00");
            zTextBlock.Text = "Z " + rotationReading.Z.ToString("0.00");
            xLine.X2 = xLine.X1 + rotationReading.X * 200;
            yLine.Y2 = yLine.Y1 - rotationReading.Y * 200;
            zLine.X2 = zLine.X1 - rotationReading.Z * 100;
```

```
            zLine.Y2 = zLine.Y1 + rotationReading.Z * 100;
        }
    }
}
```

程序运行的效果如图 10.18 所示。

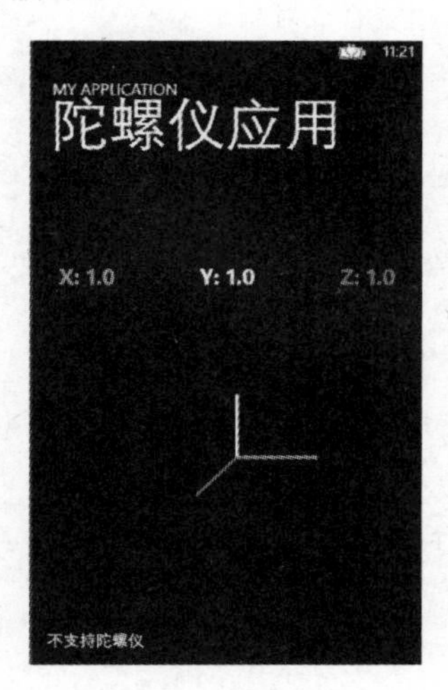

图 10.18　陀螺仪应用

10.5　语音控制

Windows Phone 从一开始就具有了强大的语音功能，长按开始键就可以调用手机的语音识别界面，然后可以通过语音来进行启动一些任务。那么在 Windows Phone 8 里面，语音控制的编程开放了相关的 API 给应用程序调用，所以在应用程序里面也一样可以实现语音的控制。

10.5.1　发音合成

发音的合成是指把文本转化为语音由手机系统进行发音，从而实现了把文本自动转化为了更加自然化的声音。在 Windows Phone 8 里面可以使用 SpeechSynthesizer 类来实现发音合成的功能，通过 SpeakTextAsync 方法可以直接文本转化为声音并且播放。

下面给出发音合成的示例：使用发音合成对文本的内容进行发音。

代码清单 10-10：发音合成(源代码：第 10 章\Examples_10_10)

MainPage.xaml 文件主要代码

```
<Grid x:Name = "ContentPanel" Grid.Row = "1" Margin = "12,0,12,0">
    <StackPanel >
        <TextBox Name = "textBox1" Text = "Hello World! "/>
        <Button Content = "发音" Name = "button1"  Click = "button1_Click" />
```

```
        <TextBlock x:Name="erro"/>
    </StackPanel>
</Grid>
```

MainPage.xaml.cs 文件主要代码

```
SpeechSynthesizer voice;    //语音合成对象
public MainPage()
{
    InitializeComponent();
    this.voice = new SpeechSynthesizer();
}
private async void button1_Click(object sender, RoutedEventArgs e)
{
    try
    {
        if (textBox1.Text! = "")
        {
            button1.IsEnabled = false;
            await voice.SpeakTextAsync(textBox1.Text); //文本语音合成
            button1.IsEnabled = true;
        }
        else
        {
            MessageBox.Show("请输入要读取的内容");
        }
    }
    catch (Exception ex)
    {
        erro.Text = ex.ToString();
    }
}
```

程序运行的效果如图 10.19 所示。

图 10.19 发音合成

10.5.2 语音识别

语音识别是指让手机通过识别和理解过程把语音信号转变为相应的文本或命令。在 Windows Phone 8 里面语音识别分为两种类型：一种是使用用户自定义的 UI 页面，另一种是使用系统默认的语音识别界面也就是我们长按开始键的语音识别界面。使用语音识别的功能需要在 WMAppManifest.xml 文件中添加两种功能要求 ID_CAP_SPEECH_RECOGNITION 和 ID_CAP_MICROPHONE。下面分别来介绍一下这两种语音识别的编程。

自定义语音识别界面可以通过 SpeechRecognizer 类来实现，首先需要先添加监听的语法，然后通过使用 SpeechRecognizer 类 RecognizeAsync 方法来监听语音的识别。

下面给出数字语音识别的示例：对 1～10 的英文数字发音进行监控，如果监听到数字的发音则把英文数字单词显示出来。

代码清单 10-11：数字语音识别(源代码：第 10 章\Examples_10_11)

MainPage. xaml 文件主要代码

```
<Grid x:Name = "ContentPanel" Grid.Row = "1" Margin = "12,0,12,0">
    <StackPanel>
        <TextBlock  Text = "语音识别的内容: "/>
        <TextBlock x:Name = "tbOutPut" Text = ""/>
           < Button  Content  =  " 开 始 识 别 "    Name  =  " continuousRecoButton "  Click  =
"continuousRecoButton_Click" />
    </StackPanel>
</Grid>
```

MainPage. xaml. cs 文件主要代码

```
using System;
using System.Collections.Generic;
using System.Windows;
using Microsoft.Phone.Controls;
using Windows.Foundation;
using Windows.Phone.Speech.Recognition;
namespace SpeechRecognizerDemo
{
    public partial class MainPage : PhoneApplicationPage
    {
        SpeechRecognizer recognizer;      //语音识别对象
        IAsyncOperation<SpeechRecognitionResult> recoOperation; //语音识别操作任务
        bool recoEnabled = false;     //判断是否停止监听
        public MainPage()
        {
            InitializeComponent();
            try
            {
                //创建一个语音识别类
                this.recognizer = new SpeechRecognizer();
                // 添加监听的单词列表
                this.recognizer.Grammars.AddGrammarFromList("Number", new List<string>()
{ "one", "two", "three", "four", "five", "six", "seven", "eight", "nine", "ten" });
            }
            catch (Exception err)
            {
                tbOutPut.Text = "Error: " + err.Message;
            }
        }
        //按钮单击事件处理
        private async void continuousRecoButton_Click(object sender, RoutedEventArgs e)
        {
            if (this.recoEnabled)
```

```
{
    this.recoEnabled = false;
    this.continuousRecoButton.Content = "开始识别";
    this.recoOperation.Cancel();
    return;
}
else
{
    this.recoEnabled = true;
    this.continuousRecoButton.Content = "取消识别";
}
do
{
    try
    {
        //捕获语音的结果
        this.recoOperation = recognizer.RecognizeAsync();
        var recoResult = await this.recoOperation;
        //音量过低无法识别
        if ((int)recoResult.TextConfidence < (int)SpeechRecognitionConfidence.Medium)
        {
            tbOutPut.Text = "说话声音太小";
        }
        else
        {
            tbOutPut.Text = recoResult.Text;
        }
    }
    catch (System.Threading.Tasks.TaskCanceledException)
    {
        //忽略语音识别的取消异常
    }
    catch (Exception err)
    {
        const int privacyPolicyHResult = unchecked((int)0x80045509);
        if (err.HResult == privacyPolicyHResult)
        {
            MessageBox.Show("尚未接受语音隐私协议。");
            this.recoEnabled = false;
            this.continuousRecoButton.Content = "开始识别";
        }
        else
        {
            tbOutPut.Text = "Error: " + err.Message;
        }
    }
} while (this.recoEnabled);  //循环进行监听语音
```

```
            }
        }
    }
```

程序运行的效果如图 10.20 所示。

图 10.20 语音识别数字

系统语音识别界面可以通过 SpeechRecognizerUI 类来实现，使用的语法与 SpeechRecognizer 类相类似，系统的语音识别界面通过 SpeechRecognizerUI 类 Settings 属性来设置，Settings. ListenText 表示界面的标题，Settings. ExampleText 表示界面的示例内容。

下面给出数字语音识别系统界面的示例：使用系统的语音识别界面，对 1～10 的英文数字发音进行监控，如果监听到数字的发音则把英文数字单词显示出来。

代码清单 10-12：数字语音识别系统界面（源代码：第 10 章\Examples_10_12）

MainPage. xaml 文件主要代码

```
<Grid x:Name="ContentPanel" Grid.Row="1" Margin="12,0,12,0">
    <StackPanel>
        <TextBlock  Text="语音识别的内容："/>
        <TextBlock x:Name="tbOutPut" Text=""/>
            <Button Content="开始识别"  Name="continuousRecoButton" Click=
"continuousRecoButton_Click" />
    </StackPanel>
</Grid>
```

MainPage. xaml. cs 文件主要代码

```
using System;
using System.Collections.Generic;
using System.Windows;
```

```
using Microsoft.Phone.Controls;
using Windows.Phone.Speech.Recognition;
namespace SpeechRecognizerUIDemo
{
    public partial class MainPage : PhoneApplicationPage
    {
        SpeechRecognizerUI recognizer;      //语音识别对象
        public MainPage()
        {
            InitializeComponent();
            try
            {
                //创建一个语音识别类
                this.recognizer = new SpeechRecognizerUI();
                // 语音弹出框的标题
                this.recognizer.Settings.ListenText = "说出一个 1 到 10 的英文单词";
                // 语音弹出框的示例内容
                this.recognizer.Settings.ExampleText = "例如, 'one' or 'two'";
                // 添加监听的单词列表
                this.recognizer.Recognizer.Grammars.AddGrammarFromList("Number", new List
<string>() { "one", "two", "three", "four", "five", "six", "seven", "eight", "nine",
"ten" });
            }
            catch (Exception err)
            {
                tbOutPut.Text = "Error: " + err.Message;
            }
        }
        //按钮单击事件处理
        private async void continuousRecoButton_Click(object sender, RoutedEventArgs e)
        {
            // 开始启动系统的语音识别 UI 并且等待用户的回答
            SpeechRecognitionUIResult recognizerResult = await this.recognizer.
RecognizeWithUIAsync();
            // 确认识别是否成功和音量是否达到要求
            if (recognizerResult.ResultStatus == SpeechRecognitionUIStatus.Succeeded&&
recognizerResult.RecognitionResult.TextConfidence == SpeechRecognitionConfidence.High)
            {
                tbOutPut.Text = recognizerResult.RecognitionResult.Text;
            }
            else
            {
                tbOutPut.Text = "音量太小";
            }
        }
    }
}
```

程序运行的效果如图 10.21 所示。

图 10.21　系统语音识别界面

MVVM模式

MVVM模式(Mode View View Model)是由MVC模式和MVP模式发展而来，这种模式的诞生是为了解决Silverlight和WPF这两方面的应用程序的架构而产生的，其核心思想与MVC模式类似，都是为了使应用程序行为与用户界面的分离和打造可维护性、可扩展性及可测试性更高架构体系。应用MVVM模式，把应用程序的状态和行为进行封装，隔离了用户界面和用户体验，从而使得设计者与开发者可独立工作，并易于协作。开发者可迅速投入代码开发中，只关注应用逻辑。如果实现一些相对较小、后期变化不大的应用程序，可能并不关注设计模式，使用设计模式反而更加复杂。

MVVM模式同样适用于Windows Phone的手机应用开发，使用MVVM模式来构建一个Windows Phone 8手机应用可以让程序代码带来与用MVVM模式构建Silverlight和WPF程序时一样的好处，对于大型的Windows Phone的开发团队，应用这些好处更加明显。

11.1 MVVM模式简介

MVVM模式分为Model模型、View视图和ViewModel视图模型。三者的关系如图11.1所示。

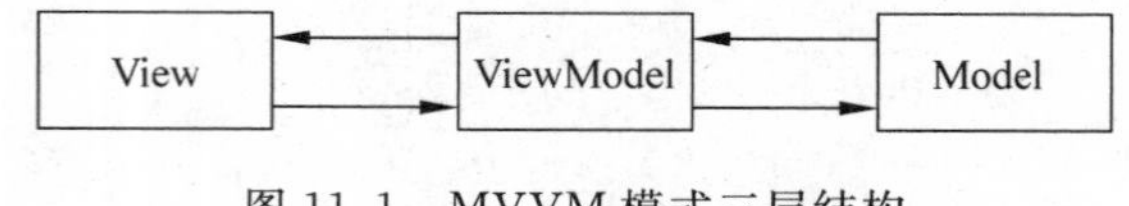

图11.1 MVVM模式三层结构

下面分别介绍MVVM模式的3个组成部分的含义和作用。

1. View

View是指UI界面层。View层所做的事情就是展现在用户面前的应用程序的界面设计，所以它在Windows Phone 8中是由XAML文件来表示的。

2. ViewModel

ViewModel是View的抽象，负责View与Model之间的信息转换，将View的Command

传送到 Model。View 与 ViewModule 连接可以通过 Binding Data 实现数据的传递、通过 Command 实现操作的调用和通过 AttachBehavior 实现控件加载过程中的操作。ViewModel 是为这个 View 量身订制的，它包含了 Binding 所需的相关信息，比如 Converter 以及为 View 的 Binding 提供 DataContext；它包含了 Command 的定义以便 View 层可以直接使用；它得负责业务流程的调度。ViewModel 是 MVVM 架构中最重要的部分，ViewModel 中包含属性、命令、方法、事件、属性验证等逻辑。ViewModel 的命令是指 ViewModel 中的命令用于接受 View 的用户输入，并作相应的处理；也可以通过方法实现相同的功能。ViewModel 的事件是指 ViewModel 中的事件主要用来通知 View 做相应的 UI 变换。ViewModel 的方法是指有些事件是没有直接提供命令调用的，如自定义的事件。这时候可以通过 CallMethodAction 来调用 ViewModel 中的方法来完成相应的操作。

3. Model

Model 是指数据访问层。为了解决现实世界中的问题，需要将现实世界中的事物加以抽象，即"由现实世界抽象出来的模型"，那么这时候就是用 Model 来表示，也就是面向对象的方法，Model 的意义就是将程序用面向对象的方式来将数据表示出来。

在 Windows Phone 8 中采用 MVVM 的架构来开发应用程序可以获得以下好处：

(1) 项目可测试更高，从而可以执行单元测试。View 绑定到 ViewModel，然后执行一些命令再向它请求一个动作。而反过来，ViewModel 跟 Model 通信，告诉它更新来响应 UI。这样便使得为应用构建 UI 非常容易。往一个应用程序上贴一个界面越容易，外观设计师就越容易使用 Blend 来创建一个漂亮的界面。同时，当 UI 和功能越来越松耦合的时候，功能的可测试性就越来越强。

(2) 将 UI 和业务的设计完全分开，View 和 UnitTest 只是 ViewModel 的两个不同形式的消费者。

(3) 有助于区别哪些是 UI 操作，哪些是业务操作，而不是将它们全混在 CodeBehind 中。

11.2 数据绑定

数据绑定提供了一种从数据源对象中获取信息，然后显示在应用程序的界面上，同时不需要写冗长的代码就可以完成所有工作的方式。通常情况下，数据绑定会提供两种方向的绑定方式，使数据绑定不仅可以从数据源兑现获取数据，显示到界面上，也可以将界面的数据传回给数据源对象，从界面上自动更新了数据源的数据。数据绑定(Binding Data)是用户界面 UI 和业务对象或者其他数据提供者的链接。数据绑定是 MVVM 模式中 View 层数据展现的主要方式，在 Windows Phone 8 中数据绑定是一种很常用的操作，不只是 MVVM 模式才有的特性。

下面介绍数据绑定的一些关键的概念：

1) 数据源(Source)

数据源充当了一个数据中心的角色，是数据绑定的数据提供者，可以理解为最底下数据层。数据源保有数据的实体，是数据的来源和源头。它可以是一个 UI 元素或者某个类的实例，也可以是一个集合。通过其他扩展标记制定任何类型的对象实例为数据源，可以使用 StaticResource 标记设置。例如，“{Binding Source={StaticResource xKeyElement}, XPath=item}”，xKeyElement 是 XMLDataProvider 的对象，意思是获取以 xKeyElement 中所有以 item 为标签节点的数据为数据源。

2）路径(Path)

数据源作为一个实体可能保存着很多数据，具体关注它的哪个数值呢？这个数值就是 Path。例如，要用一个 Slider 控件作为一个数据源，那么这个 Slider 控件会有很多属性，这些属性都是作为数据源来提供的，它拥有很多数据，除了 Value 之外，还有 Width、Height 等，这时候数据绑定就要选择一个最关心的属性来作为绑定的路径。例如，使用的数据绑定是为了监测 Slider 控件的值的变化，那么就需要把 Path 设为 Value 了。使用集合作为数据源的道理也是一样，Path 的值就是集合里面的某个字段。

3）元素名称(ElementName)

如果数据绑定的数据源是某一个 UI 元素的控件，那么就需要设置 ElementName 属性，设置为要绑定 UI 元素控件的 Name 值。例如，“{Binding ElementName=ListBox1, Path=SelectedItem}”的意思是把 ListBox1 的 SelectedItem 属性作为当前控件的数据源。

4）目标(Target)

数据将传送到哪里去？这就是数据的目标，也就是数据源对应的绑定对象。Target 一定是数据的接收者、被驱动者，但它不一定是数据的显示者。例如，<TextBox Name="theTextBox" /> <TextBlock Text="{Binding ElementName=theTextBox, Path=Text}"/>，在这个绑定当中<TextBox Name="theTextBox" />就是绑定的目标了。

5）关联(Binding)

数据源与目标之间的通道。正是这个通道，使 Source 数据源和数据绑定目标 Target 之间关联了起来、使数据源的数据能够在界面上进行显示和通信。

6）设定关联(Set Binding)

为 Target 指定 Binding，并将 Binding 指向 Target 的一个属性，完成数据的“端对端”传输。

7）关联的模式(Model)

设置数据绑定的模式，即数据源和绑定目标的交互形式。有 3 种形式，OneWay 仅当源属性发生更改时更新目标属性。OneTime 仅当应用程序启动时或 DataContext 进行更改时更新目标属性。TwoWay 表示数据源和更新目标同步进行更新，互相影响，在这种模式下，在目标属性更改时更新源属性，可以理解为数据绑定的目标也成为一个数据源，这是双向的绑定方式。默认的绑定方式是 OneWay。

11.2.1 用元素值绑定

用元素值进行绑定就是将某一个控件元素作为绑定的数据源，绑定的对象是控件元素，而绑定的数据源同时也是控件元素，这种绑定的方式，可以轻松地实现两个控件之间的值的交互影响。用元素值进行绑定时通过设置 Binding 的 ElementName 属性和 Path 属性来实现的，ElementName 属性赋值为数据源控件的 Name，Path 属性则赋值为数据源控件的某个属性，这个属性就是数据源控件的一个数据变化的反映。

下面给出控制圆的半径的示例：圆形的半径绑定到 Slider 控件的值，从而实现通过即时改变 Slider 控件的值来改变圆的大小。

代码清单 11-1：控制圆的半径(源代码：第 11 章\Examples_11_1)

MainPage. xaml 文件主要代码

```
<Grid x:Name = "ContentPanel" Grid.Row = "1" Margin = "12,0,12,0">
    <TextBlock FontSize = "25" Height = "44" HorizontalAlignment = "Left" Margin = "12,27,0,0"
Name = "textBlock1" Text = "圆形的半径会根据 slider 控件的值而改变" VerticalAlignment = "Top"
Width = "438" />
    <!-- Slider 控件作为 Ellipse 的宽度和高度的绑定源 -->
    <Slider Name = "slider" Value = "50" Maximum = "400" Margin = "12,93,36,423" />
    <TextBlock FontSize = "25"  Height = "30" HorizontalAlignment = "Left" Margin = "74,237,
0,0" Name = "textBlock2" Text = "半径为:" VerticalAlignment = "Top" Width = "95" />
    <TextBlock Name = "txtblk" Text = "{Binding ElementName = slider, Path = Value}" FontSize = "48"
HorizontalAlignment = "Center" VerticalAlignment = "Center" Margin = "189,222,171,321" Width =
"96" />
    <!-- Ellipse 的宽度和高度绑定了 slider 的值 -->
    <Ellipse Height = "{Binding ElementName = slider, Path = Value}"
        Width = "{Binding ElementName = slider, Path = Value}"
        Fill = "Red" HorizontalAlignment = "Left" Margin = "171,406,0,0" Name = "ellipse1"
Stroke = "Black" StrokeThickness = "1" VerticalAlignment = "Top" />
</Grid>
```

程序的运行效果如图 11.2 所示。

11.2.2 三种绑定模式

Binding 对象的 Mode 属性是表示绑定的模式，有 3 种不同的赋值，也就是 3 种不同的绑定模式：第一种是赋值为 OneTime 表示一次绑定，在绑定创建时使用源数据更新目标，适用于只显示数据而不进行数据的更新；第二种是赋值为 OneWay 表示单向绑定，在绑定创建时或者源数据发生变化时更新到目标，适用于显示变化的数据；第三种是赋值为 TwoWay 表示双向绑定，在任何时候都可以同时更新源数据和目标。

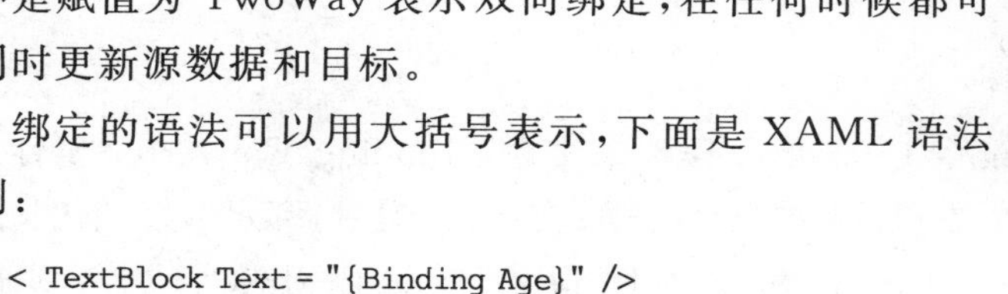

绑定的语法可以用大括号表示，下面是 XAML 语法示例：

```
< TextBlock Text = "{Binding Age}" />
```

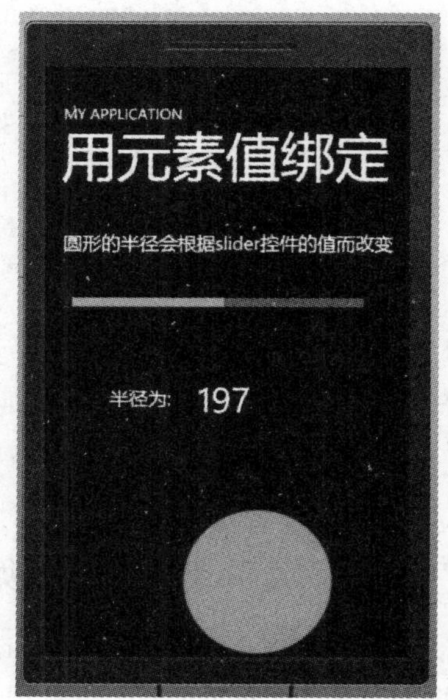

图 11.2 用元素值绑定

等同于

```
< TextBlock Text = "{Binding Path = Age}" />
```

或者显式写出绑定方向

```
< TextBlock Text = "{Binding Path = Age, Mode = OneWay}" />
```

按照数据绑定的语义，默认是 OneWay 的，也就是说如果后台的数据发生变化，前台建立了绑定关系的相关控件也应该发生更新。

下面给出三种绑定模式的示例：演示了 OneTime、OneWay 和 TwoWay 三种绑定模式的区别。

代码清单 11-2：三种绑定模式(源代码：第 11 章\Examples_11_2)

MainPage. xaml 文件主要代码

```
    < Grid x:Name = "ContentPanel" Grid.Row = "1" Margin = "12,0,12,0">
        < Slider Name = "slider" Value = "50" Maximum = "400" Margin = "9,20,39,496" />
        < TextBlock FontSize = "25" Height = "41" HorizontalAlignment = "Left" Margin = "34,
238,0,0" Name = "textBlock1" Text = "OneTime" VerticalAlignment = "Top" Width = "112" />
<! -- OneTime 绑定模式,第一次绑定 Slider 控件的值会影响文本的值 -->
        < TextBox  Height = "72" HorizontalAlignment = "Left" Margin = "152,220,0,0" Name =
"textBox1" Text = " {Binding ElementName = slider, Path = Value, Mode = OneTime }"
VerticalAlignment = "Top" Width = "269" />
        < TextBlock FontSize = "25" Height = "46" HorizontalAlignment = "Left" Margin = "34,
338,0,0" Name = "textBlock2" Text = "OneWay" VerticalAlignment = "Top" Width = "99" />
 <! -- OneWay 绑定模式,Slider 控件的值会影响文本的值 -->
        < TextBox Height = "72" HorizontalAlignment = "Left" Margin = "151,323,0,0" Name =
"textBox2" Text = " {Binding ElementName = slider, Path = Value, Mode = OneWay }"
VerticalAlignment = "Top" Width = "269" />
        < TextBlock FontSize = "25" Height = "40" HorizontalAlignment = "Left" Margin = "39,
436,0,0" Name = "textBlock3" Text = "TwoWay" VerticalAlignment = "Top" Width = "94" />
<! -- TwoWay 绑定模式,Slider 控件的值和文本的值互相影响 -->
        < TextBox Height = "72" HorizontalAlignment = "Left" Margin = "152,420,0,0" Name =
"textBox3" Text = " {Binding ElementName = slider, Path = Value, Mode = TwoWay }"
VerticalAlignment = "Top" Width = "268" />
        < TextBlock FontSize = "25" Height = "43" HorizontalAlignment = "Left" Margin = "12,
142,0,0" Name = "textBlock4" Text = "slider 控件的值: " VerticalAlignment = "Top" />
        < TextBlock FontSize = "25" Height = "43" HorizontalAlignment = "Left" Margin = "203,
142, 0, 0" Name = " textBlock5" Text = " {Binding ElementName = slider, Path = Value}"
VerticalAlignment = "Top" Width = "185" />
    </Grid>
```

程序运行的效果如图 11.3 所示。

11.2.3 绑定值转换

绑定值转换需要通过 Binding 的 Converter 属性来实现。Binding. Converter 属性表示获取或设置转换器对象，当数据在源和目标之间(或相反方向)传递时，绑定引擎调用该对象来修改数据。对实现 IValueConverter 并作为转换器的现有对象引用。通常在

ResourceDictionary 中创建该对象，并为它指定一个键，然后使用 StaticResource 标记扩展来引用它。例如，<Binding Converter＝"{StaticResource myConverter}" .../> 对数据进行修改的 IValueConverter 对象。通过实现 IValueConverter 接口以及实现 Convert 和 ConvertBack 方法创建转换器。

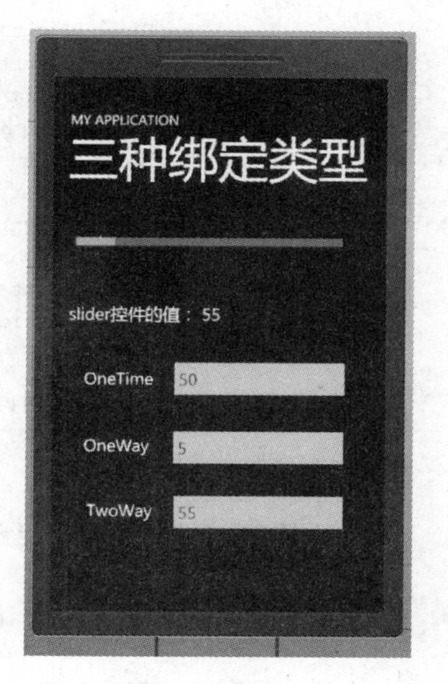

图 11.3　三种绑定类型

若要在绑定中使用转换器，首先要创建转换器类的实例。将转换器类的实例设置为程序中资源的 XAML 语法如下：

```
<phone:PhoneApplicationPage.Resources>
  <local:DateToStringConverter x:Key="Converter1"/>
</phone:PhoneApplicationPage.Resources>
```

然后设置该实例的绑定的 Converter 属性，例如：

```
<TextBlock Grid.Column="0"  Text="{Binding Month,
Converter={StaticResource Converter1}}"/>
```

下面给出绑定转换的示例：将绑定的时间转化为中文的时间问候语。

代码清单 11-3：绑定转换(源代码：第 11 章\Examples_11_3)

MainPage.xaml 文件主要代码

```
<phone:PhoneApplicationPage
    ...
    xmlns:my="clr-namespace:ConverterDemo;assembly=ConverterDemo">
    <!--添加值转换类所在的空间引用-->
    ...
    <!--将引用的时间类和值转换类定义为应用程序的资源-->
        <phone:PhoneApplicationPage.Resources>
            <my:Clock x:Key="clock" />
            <my:HoursToDayStringConverter x:Key="booleanToDayString" />
        </phone:PhoneApplicationPage.Resources>
        ...
        <!--设置Grid控件的上下文数据源为上面定义的时间类-->
        <Grid x:Name="ContentPanel" Grid.Row="1" Margin="12,0,12,0"  DataContext=
"{StaticResource clock}">
            <!--将绑定的小时时间转化为了值转换类定义的转换字符串-->
            <TextBlock FontSize="30" Height="61" HorizontalAlignment="Left" Margin=
"12,35,0,0" Name=" textBlock1" Text="{Binding Hour,  Converter={StaticResource
booleanToDayString}}" VerticalAlignment="Top" Width="392" />
            <TextBlock FontSize="30" Height="56" HorizontalAlignment="Left" Margin="9,
133,0,0" Name="textBlock2" Text="现在的时间是:" VerticalAlignment="Top" Width="211" />
            <!--显示绑定的时间-->
            <TextBlock Height="30" HorizontalAlignment="Left" Margin="29,205,0,0" Name
="textBlock3" Text="{Binding Hour}" VerticalAlignment="Top" />
```

```
            <TextBlock Height = "30" HorizontalAlignment = "Left" Margin = "135,205,0,0" Name =
"textBlock4" Text = "小时" VerticalAlignment = "Top" />
            <TextBlock Height = "30" HorizontalAlignment = "Left" Margin = "29,241,0,0" Name =
"textBlock5" Text = "{Binding Minute}" VerticalAlignment = "Top" />
            <TextBlock Height = "30" HorizontalAlignment = "Left" Margin = "135,241,0,0" Name =
"textBlock6" Text = "分钟" VerticalAlignment = "Top" />
            <TextBlock Height = "30" HorizontalAlignment = "Left" Margin = "29,282,0,0"
Name = "textBlock7" Text = "{Binding Second}" VerticalAlignment = "Top" />
            <TextBlock Height = "30" HorizontalAlignment = "Left" Margin = "135,277,0,0" Name =
"textBlock8" Text = "秒" VerticalAlignment = "Top" />
        </Grid>
```

Clock.cs 文件代码：时间的信息类，即时更新类的属性以匹配当前最新的时间

```
using System;
using System.ComponentModel;
using System.Windows.Threading;
namespace ConverterDemo
{
    public class Clock : INotifyPropertyChanged
    {
        int hour, min, sec;
        //属性值改变事件
        public event PropertyChangedEventHandler PropertyChanged;
        //类初始化方法
        public Clock()
        {
            //获取当前的时间
            OnTimerTick(null, null);
            //使用定时器来触发时间来改变类的时分秒属性
            //每 0.1 秒获取一次当前的时间
            DispatcherTimer tmr = new DispatcherTimer();
            tmr.Interval = TimeSpan.FromSeconds(0.1);
            tmr.Tick += OnTimerTick;
            tmr.Start();
        }
        //小时属性
        public int Hour
        {
            protected set
            {
                if (value != hour)
                {
                    hour = value;
                    OnPropertyChanged(new PropertyChangedEventArgs("Hour"));
                }
            }
            get
            {
                return hour;
            }
```

```
        }
        //分钟属性
        public int Minute
        {
            protected set
            {
                if (value != min)
                {
                    min = value;
                    OnPropertyChanged(new PropertyChangedEventArgs("Minute"));
                }
            }
            get
            {
                return min;
            }
        }
        //秒属性
        public int Second
        {
            protected set
            {
                if (value != sec)
                {
                    sec = value;
                    OnPropertyChanged(new PropertyChangedEventArgs("Second"));
                }
            }
            get
            {
                return sec;
            }
        }
        //属性改变事件
        protected virtual void OnPropertyChanged(PropertyChangedEventArgs args)
        {
            if (PropertyChanged != null)
                PropertyChanged(this, args);
        }
        //时间触发器
        void OnTimerTick(object sender, EventArgs args)
        {

            DateTime dt = DateTime.Now;
            Hour = dt.Hour;
            Minute = dt.Minute;
            Second = dt.Second;
        }
    }
}
```

ConverterDemo.cs 文件代码：将时间转化为中文问候语的值转换类

```
using System;
using System.Windows.Data;
using System.Globalization;
namespace ConverterDemo
{
    public class HoursToDayStringConverter : IValueConverter
{
    //定义转换方法
        public object Convert(object value, Type targetType,
                                object parameter, CultureInfo culture)
        {
            if (Int16.Parse(value.ToString()) < 12)
            {
                return "尊敬的用户,上午好。";
            }
            else if (Int16.Parse(value.ToString()) > 12)
            {
                return "尊敬的用户,下午好。";
            }
            else
            {
                return "尊敬的用户,中午好。";
            }
        }
        //定义反向转换的方法
        public object ConvertBack(object value, Type targetType,
                                    object parameter, CultureInfo culture)
        {
            return DateTime.Now.Hour;
        }
    }
}
```

程序的运行效果如图 11.4 所示。

11.2.4 绑定集合

绑定集合是指将数据的集合绑定到控件中的值，在 ListBox 这类型的列表控件中会经常使用到绑定集合的方法。绑定集合有两种方式：一种是通过 XAML 进行，在 XAML 中创建绑定；另外一种是使用代码创建绑定。下面介绍这两种绑定方式的实现步骤。

第一种方式通过 XAML 进行，在 XAML 中创建绑定的步骤：

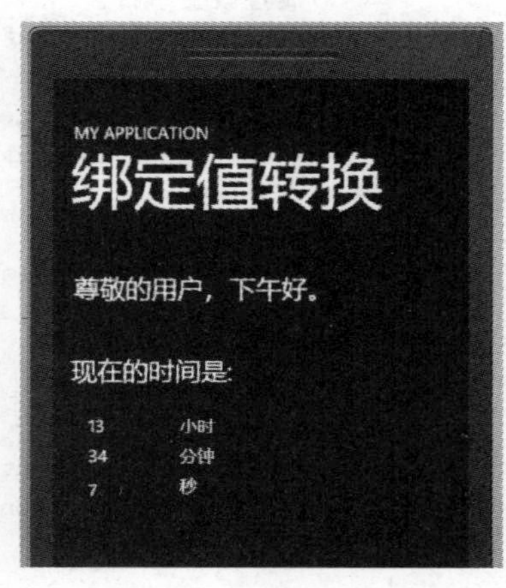

图 11.4 绑定值转换

(1) 定义源对象。

代码示例：

```
public class Model
{
    public string ModelName { get; set; }
}
```

(2) 在 XAML 中创建对源对象的命名空间的引用。

代码示例：

```
xmlns:my = "clr - namespace:BindingXAML"
```

(3) 在 Resources 创建源对象的实例。

代码示例：

```
< Grid.Resources >
    < my:Dog x:Name = "myModel" ModelName  = "myName"/>
</Grid.Resources >
```

(4) 通过设置 Source 属性或 DataContext 属性绑定到源对象，该元素的所有子级都继承 DataContext。

代码示例：

```
< TextBlock Text = "{Binding ModelName, Source = {StaticResource myModel }, Mode = OneTime}"/>
```

或者

```
< TextBlock Text = "{Binding ModelName, Mode = OneTime}" DataContext = "{StaticResource myModel }"/>
```

第二种方式使用代码创建绑定的步骤：

(1) 添加 System.Windows.Data 命名空间。

代码示例：

```
using System.Windows.Data; (Binding 类的命名空间)
```

(2) 定义源对象。

代码示例：

```
public class Model
{
    public string ModelName { get; set; }
}
```

(3) 创建要绑定到的 FrameworkElement。

代码示例：

```
< Grid x:Name = "LayoutRoot" Background = "White">
    < TextBlock x:Name = "MyTextBlock" Text = "Test"/>
</Grid >
```

（4）创建源对象的实例。

代码示例：

```
Model MyModel = new Model ();
MyModel. ModelName = "myName";
```

（5）创建绑定对象。

代码示例：

```
Binding MyBinding = new Binding();
```

（6）对绑定对象设置绑定属性。

代码示例：

```
MyBinding.Path = new PropertyPath("ModelName ");
MyBinding.Mode = BindingMode.OneTime;
```

（7）通过设置 Source 属性或 DataContext 属性来设置绑定源。该元素的所有子级都继承 DataContext。

代码示例：

```
MyBinding.Source = MyModel;
```

或者

```
MyTextBlock.DataContext = MyModel;
```

（8）将此绑定附加到 FrameworkElement 的属性。

代码示例：

```
MyTextBlock.SetBinding(TextBlock.TextProperty, MyBinding);
```

在两个示例中都可以看到，指定数据源有两种方法，以 TextBlock 为例，可以指定 Binding 类的 Source 属性到数据源，也可以将 TextBlock 的 DataContext 指定到数据源，效果一样。如果 TextBlock 没有指定数据源，会在其绑定的 Binding 中寻找是否有数据源。在实际开发中，为了让程序更易读更有条理，可以混合使用两种方法，并在后台创建数据源，在 XAML 中绑定。这样前台就免去了创建数据源的 XAML 元素，后台省去了创建 Binding 类的代码，并且通过查看前台的 XAML 代码就可以很容易地判断出各个控件的绑定数据。

在数据绑定的过程中，经常需要写一个类，并且拿它的实例当作数据源。怎样才能让一个类成为“合格的”数据源呢？要诀就是：为这个类定义一些 Property，相当于为 Binding 提供 Path，然后让这个类实现 INotifyPropertyChanged 接口。实现这个接口的目的是当 Source 的属性值改变后通知 Binding，这样就可以让 Binding 把数据传输给 Target，本质上还是使用事件机制来做，只是掩盖在底层，不用程序员去写 event handler 了。INotifyPropertyChanged 接口在 System. ComponentModel 名称空间中，在数据源类中继承 INotifyPropertyChanged 接口就需要引入 System. ComponentModel 名称空间。当然一些实现了 INotifyPropertyChanged 接口的集合类就可以直接使用来作为数据绑定的数据源，比如 ObservableCollection<T>

类就可以直接用来作为数据绑定的数据源。下面是这两者的介绍和区别。

1）INotifyPropertyChanged

它的作用是向客户端发出某一属性值已更改的通知。当属性改变时，它可以通知客户端，并进行界面数据更新。而不用写很多复杂的代码来更新界面数据，这样可以做到方法简洁而清晰，松耦合让方法变得更通用。例如，考虑一个带有名为 FirstName 属性的 Person 对象。若要提供一般性属性更改通知，则 Person 类型实现 INotifyPropertyChanged 接口并在 FirstName 更改时引发 PropertyChanged 事件。若要在将客户端与数据源进行绑定时发出更改通知，则绑定类型应具有下列任一功能：实现 INotifyPropertyChanged 接口（首选）或者为绑定类型的每个属性提供更改事件。

2）ObservableCollection<T>

它是从 Collection<T>（泛型集合的基类）继承而来，可实现 INotifyCollectionChanged 和 INotifyPropertyChanged 两种接口。INotifyCollectionChanged 接口增加了集合的趣味性，同时也是允许绑定对象（和代码）确定集合是否已发生更改的接口。值得注意的是，虽然 ObservableCollection 类会广播有关对其元素所做的更改的信息，但它并不了解也不关心对其元素的属性所做的更改。也就是说，它并不关注有关其集合中项目的属性更改通知。如果需要了解是否有人更改了集合中某个项目的属性，则需要确保集合中的项目可以实现 INotifyPropertyChanged 接口，并需要手动附加这些对象的属性更改事件处理程序。无论如何更改此集合中的对象属性，都不会触发该集合的 PropertyChanged 事件。事实上，ObservableCollection 的 PropertyChanged 事件处理程序已受到保护，除非从此类中继承并亲自将其公开，否则甚至无法对其作出反应。

下面给出食物分类的示例：演示 MVVM 模式的数据绑定。

代码清单 11-4：食物分类（源代码：第 11 章\Examples_11_4）

Food.cs 文件代码：创建 Model 层

```
namespace BindingDataDemo.Model
{
    public class Food
    {
        //食物名字
        public string Name
        {
            get;
            set;
        }
        //食物描述
        public string Description
        {
            get;
            set;
        }
        //食物图片地址
```

```
        public string IconUri
        {
            get;
            set;
        }
        //食物类型
        public string Type
        {
            get;
            set;
        }
    }
}
```

FoodViewModel.cs 文件代码：创建 ViewModel 层

```
using System;
using System.ComponentModel;
using BindingDataDemo.Model;
using System.Collections.ObjectModel;
namespace BindingDataDemo.ViewModel
{
    public class FoodViewModel : INotifyPropertyChanged
    {
        //定义食物类的集合
        private  ObservableCollection<Food> _allFood;
        //将集合作为 ViewModel 层的属性
        public ObservableCollection<Food> AllFood
        {
            get
            {
                if (_allFood == null)
                    _allFood = new ObservableCollection<Food>();
                return _allFood;
            }
            set
            {
                if (_allFood != value)
                {
                    _allFood = value;
                    NotifyPropertyChanged("AllFood");
                }
            }
        }
        //初始化 ViewModel 层的数据
        public FoodViewModel()
        {
            try
            {
                Food item0 = new Food() { Name = "西红柿", IconUri = "Images/Tomato.png",
Type = "Healthy",Description = "西红柿的味道不错。" };
```

```
                Food item1 = new Food() { Name = "茄子", IconUri = "Images/Beer.png", Type =
"NotDetermined", Description = "不知道这个是否好吃." };
                Food item2 = new Food() { Name = "火腿", IconUri = "Images/fries.png",
Type = "Unhealthy", Description = "这是不健康的食品." };
                Food item3 = new Food() { Name = "三明治", IconUri = "Images/Hamburger.
png", Type = "Unhealthy",Description = "肯德基的好吃?" };
                Food item4 = new Food() { Name = "冰激凌", IconUri = "Images/icecream.
png", Type = "Healthy", Description = "给小朋友吃的." };
                Food item5 = new Food() { Name = "Pizza", IconUri = "Images/Pizza.png",
Type = "Unhealthy",Description = "这个非常不错." };
                Food item6 = new Food() { Name = "辣椒", IconUri = "Images/Pepper.png",
Type = "Healthy", Description = "我不喜欢吃这东西." };
                AllFood.Add(item0);
                AllFood.Add(item1);
                AllFood.Add(item2);
                AllFood.Add(item3);
                AllFood.Add(item4);
                AllFood.Add(item5);
                AllFood.Add(item6);
            }
            catch ( Exception e )
            {
                System.Windows.MessageBox.Show( "Exception: " + e.Message );
            }
        }
        //定义属性改变事件
        public event PropertyChangedEventHandler PropertyChanged;
        //实现属性改变事件
        public void NotifyPropertyChanged(string propertyName)
        {
            if (PropertyChanged != null)
            {
                PropertyChanged(this, new PropertyChangedEventArgs(propertyName));
            }
        }
    }
}
```

MainPage.xaml 文件主要代码：创建 View 层

```
<phone:PhoneApplicationPage
    ...
    xmlns:my = "clr-namespace:BindingDataDemo.ViewModel" >
    <!-- 添加 ViewModel 层的 FoodViewModel 类为资源 -->
    <phone:PhoneApplicationPage.Resources>
        <my:FoodViewModel x:Key = "food" />
    </phone:PhoneApplicationPage.Resources>
    ...
        <!-- 在 Grid 控件中将上面定义好的 FoodViewModel 类资源赋值给 DataContent 属性,表示
Grid 控件内使用 FoodViewModel 类作为上下文数据 -->
        <Grid x:Name = "ContentPanel" Grid.Row = "1" Margin = "12,0,12,0" DataContext =
```

```
"{StaticResource food }">
            <!-- 在 ListBox 控件中绑定 FoodViewModel 类的 AllFood 属性，AllFood 是
ObservableCollection<Food>类型 -->
            <ListBox x:Name = "listBox" HorizontalContentAlignment = "Stretch" ItemsSource
= "{Binding AllFood}">
                <ListBox.ItemTemplate>
                    <DataTemplate>
                        <StackPanel Orientation = "Horizontal" Background = "Gray" Width =
"450" Margin = "10">
                            <!-- 绑定 Food 类的 IconUri 属性 -->
                            <Image Source = "{Binding IconUri}" Stretch = "None"/>
                            <!-- 绑定 Food 类的 Name 属性 -->
                            <TextBlock Text = "{Binding Name}" FontSize = "40" Width = "150"/>
                            <!-- 绑定 Food 类的 Description 属性 -->
                            <TextBlock Text = "{Binding Description}" FontSize = "20" Width =
"280"/>
                        </StackPanel>
                    </DataTemplate>
                </ListBox.ItemTemplate>
            </ListBox>
        </Grid>
    ...
</phone:PhoneApplicationPage>
```

程序运行的效果如图 11.5 所示。

图 11.5 数据绑定

11.3 Command 的实现

Command 是 MVVM 模式中的操作命令的处理，在 code-behind 模式中操作命令的处理方式很简单，例如，单击按钮直接在控件元素上添加 click 事件，然后在控件元素对应的 xaml. cs 文件中进行处理就行了，但是在 MVVM 模式中使用的是另外一种完全不同的方式，也就是通过 Command 命令来处理这些事件，事件的处理代码不是写在 xaml. cs 文件中的，而是写在 ViewModel 层的类中。通过 Command 的方式将程序操作的处理放到了 ViewModel 层中，使得 View 层和 ViewModel 层的互相独立性更加强了，在小项目中使用也许会增加程序的复杂度，但是在大项目中使用，会让项目的逻辑更加清晰，可维护性更好。

自定义的 Command 需要使用附件属性的方法来实现，下面介绍附件属性类 DependencyProperty 的含义和用法。

DependencyProperty 表示向 Silverlight 依赖项属性系统注册的依赖项属性。依赖项属性为值表达式、数据绑定、动画和属性更改通知提供支持。其功能是充当一种引用已注册到特定 DependencyObject 所有者类型的依赖项属性的方式。当所有者类型注册属性时，所有者类型会将 DependencyProperty 实例公开为标识符。当依赖项属性参与 Silverlight 属性系统时，所有者 DependencyObject 会为该依赖项属性提供属性存储。在代码中使用依赖项属性时，可以使用 DependencyProperty 标识符作为对属性系统方法(如 SetValue)的调用的输入。

如果要为某一元素添加一个自定义的属性，那么先通过声明一个 DependencyProperty 类型的 public static readonly 字段可以将附加属性定义为依赖项属性。通过使用 RegisterAttached 方法的返回值来定义此字段。此字段名称必须与附加属性名称一致，并且后面追加字符串 Property，以便遵循就其所表示的属性命名标识字段时所建立的约定。附加属性提供程序还必须以访问器的形式为这些附加属性提供静态 GetPropertyName 和 SetPropertyName 方法。不这样做将导致 XAML 处理器无法在某些情况下使用该附加属性，如在使用样式 Setter 设置属性时。

下面是附加属性的一些重要的方法。

1. DependencyProperty. RegisterAttached 方法

该方法使用属性的指定属性名称、属性类型、所有者类型和属性元数据注册附加依赖项属性。

方法的语法如下：

```
public static DependencyProperty RegisterAttached(
 string name,
 Type propertyType,
 Type ownerType,
 PropertyMetadata defaultMetadata
)
```

其中，参数含义如下：

（1）name 类型：System. String，要注册的依赖项对象的名称。

（2）propertyType 类型：System. Type，属性的类型。

（3）ownerType 类型：System. Type，正注册依赖项对象的所有者类型。

（4）defaultMetadata 类型：System. Windows. PropertyMetadata，属性元数据实例。实例中可以包含一个 PropertyChangedCallback 实现引用。

返回值类型：System. Windows. DependencyProperty。

一个依赖项对象标识符，应使用它在你的类中设置 public static readonly 字段的值。随后可以使用该标识符在将来引用附加属性，用于某些操作，比如以编程方式设置该属性的值，或者附加 Binding。

2. Get 访问器

GetPropertyName 访问器的签名必须如下：

```
public static valueType GetPropertyName(DependencyObject target)
```

在实现中，可以将 target 对象指定为更具体的类型，但必须从 DependencyObject 派生。可以在实现中将 valueType 返回值指定为更具体的类型。基本的 Object 类型是可接受的，但常常希望附加属性比该基本类型更为类型安全，因此在 getter 和 setter 签名中强制该类型非常适合于强制类型安全。

3. Set 访问器

Set PropertyName 访问器的签名必须如下：

```
public static void SetPropertyName(DependencyObject target, valueType value)
```

在实现中，可以将 target 对象指定为更具体的类型，但必须从 DependencyObject 派生。可以在实现中将 value 对象及其 valueType 指定为更具体的类型。请记住，此方法的值是 XAML 处理器在标记中遇到附加属性时来自 XAML 处理器的输入。因此，必须存在支持你所使用类型的类型转换或现有标记扩展，以便可以从属性值创建相应的类型（最终只是一个字符串）。

下面给出 Command 控制圆的大小的示例：实现 MVVM 模式的 Command 操作。

代码清单 11-5：Command 控制圆的大小（源代码：第 11 章\Examples_11_5）

MainPage. xaml 文件主要代码

```
<phone:PhoneApplicationPage
    …
    xmlns:my = "clr - namespace:CommandDemo.ViewModel"
    xmlns:my_Interactivity = "clr - namespace:CommandDemo.Command"
xmlns:Custom = " clr - namespace: System. Windows. Interactivity; assembly = System. Windows.
Interactivity"
xmlns:ic = " clr - namespace: Microsoft. Expression. Interactivity. Core; assembly = Microsoft.
Expression.Interactions" >
    <! -- 设置整个页面的上下文数据 DataContext 为 RadiusViewModel -- >
```

```
    <phone:PhoneApplicationPage.DataContext>
        <my:RadiusViewModel/>
    </phone:PhoneApplicationPage.DataContext>
    ...
            <!--绑定圆的宽度和高度-->
            <Ellipse Fill="Red" Height="{Binding Radius}" Width="{Binding Radius}"
HorizontalAlignment="Left" Margin="119,84,0,0" Name="ellipse1" Stroke="Black"
StrokeThickness="1" VerticalAlignment="Top" />
            <Button Content="小" Height="72" HorizontalAlignment="Left" Margin="0,385,
0,0" Name="button1" VerticalAlignment="Top" Width="160">
                <!-添加Command的单击事件-->
                <Custom:Interaction.Triggers>
                    <Custom:EventTrigger EventName="Click">
                        <my_Interactivity:ExecuteCommandAction CommandName="MinRadius"/>
                    </Custom:EventTrigger>
                </Custom:Interaction.Triggers>
            </Button>
            <Button Content="中" Height="72" HorizontalAlignment="Left" Margin="149,
384,0,0" Name="button2" VerticalAlignment="Top" Width="160" >
                <Custom:Interaction.Triggers>
                    <Custom:EventTrigger EventName="Click">
                        <my_Interactivity:ExecuteCommandAction CommandName="MedRadius"/>
                    </Custom:EventTrigger>
                </Custom:Interaction.Triggers>
            </Button>
            <Button Content="大" Height="72" HorizontalAlignment="Left" Margin="299,
382,0,0" Name="button3" VerticalAlignment="Top" Width="160" >
                <Custom:Interaction.Triggers>
                    <Custom:EventTrigger EventName="Click">
                        <my_Interactivity:ExecuteCommandAction  CommandName="MaxRadius"/>
                    </Custom:EventTrigger>
                </Custom:Interaction.Triggers>
            </Button>
...
</phone:PhoneApplicationPage>
```

RadiusViewModel.cs文件代码：实现ViewModel层

```
using System;
using System.Windows.Input;
using System.ComponentModel;
using Microsoft.Expression.Interactivity.Core;
namespace CommandDemo.ViewModel
{
    public class RadiusViewModel : INotifyPropertyChanged
{
        //圆的半径
        private Double radius;
        //初始化ViewModel
        public RadiusViewModel()
```

```
        {
            Radius = 0;
            //注册小、中、大的单击 Command 事件
            MinRadius = new ActionCommand(p => Radius = 100);
            MedRadius = new ActionCommand(p => Radius = 200);
            MaxRadius = new ActionCommand(p => Radius = 300);
        }
        //定义属性改变事件
        public event PropertyChangedEventHandler PropertyChanged;
        //定义小按钮的 Command 接口
        public ICommand MinRadius
        {
            get; private set;
        }
        //定义中按钮的 Command 接口
        public ICommand MedRadius
        {
            get;
            private set;
        }
        //定义大按钮的 Command 接口
        public ICommand MaxRadius
        {
            get;
            private set;
        }
        //半径属性
        public Double Radius
        {
            get
            {
                return radius;
            }
            set
            {
                radius = value;
                OnPropertyChanged("Radius");
            }
        }
        //实现属性改变事件
        protected virtual void OnPropertyChanged(string propertyName)
        {
            var propertyChanged = PropertyChanged;
            if(propertyChanged != null)
                propertyChanged(this, new PropertyChangedEventArgs(propertyName));
        }
    }
}
```

ExecuteCommandAction.cs 文件代码：定义 Command 操作

```
using System;
using System.Windows;
using System.Windows.Input;
using System.Windows.Interactivity;
using System.Reflection;
namespace CommandDemo.Command
{
    public class ExecuteCommandAction : TriggerAction<FrameworkElement>
{
    //注册一个附加 Command
        public static readonly DependencyProperty CommandNameProperty =
            DependencyProperty. Register ( " CommandName ",  typeof ( string ),  typeof
(ExecuteCommandAction), null);
        //注册一个附加 Command 参数
        public static readonly DependencyProperty CommandParameterProperty =
            DependencyProperty. Register ( " CommandParameter ",  typeof ( object ),  typeof
(ExecuteCommandAction), null);
        //重载 Invoke 方法
        protected override void Invoke(object parameter)
        {
            if (AssociatedObject == null)

                return;
            ICommand command = null;
            var dataContext = AssociatedObject.DataContext;
            foreach (var info in dataContext.GetType().GetProperties(BindingFlags.Public |
BindingFlags.Instance))
            {
                if ( IsCommandProperty ( info )  &&  String. Equals ( info. Name,  CommandName,
StringComparison.Ordinal))
                {
                    command = (ICommand)info.GetValue(dataContext, null);
                    break;
                }
            }
            if ((command != null) && command.CanExecute(CommandParameter))
            {
                command.Execute(CommandParameter);
            }
        }
        //判断 Command 属性
        private static bool IsCommandProperty(PropertyInfo property)
        {
            return typeof(ICommand).IsAssignableFrom(property.PropertyType);
        }
        //CommandName 属性
        public string CommandName
        {
```

```
                get
                {
                    return (string)GetValue(CommandNameProperty);
                }
                set
                {
                    SetValue(CommandNameProperty, value);
                }
            }
            // CommandParameter 属性
            public object CommandParameter
            {
                get
                {
                    return GetValue(CommandParameterProperty);
                }
                set
                {
                    SetValue(CommandParameterProperty, value);
                }
            }
        }
    }
```

图 11.6　Command 实现效果

程序的运行效果如图 11.6 所示。

11.4　Attached Behaviors 的实现

前面介绍了 MVVM 模式的 Binding 和 Command，利用 Binding 可以完成 View 和 ViewModel 之间数据的通信，利用 Command 可以实现 View 和 ViewModel 之间操作命令的通信，一般情况下有这两项的通信可以完成大部分的需求，可以使 View 和 ViewModel 之间能配合工作得很好。但是还有一种情况，使用 Binding 和 Command 是无法实现的，那就是控件元素在加载时候的一些初始化工作。Binding 只是在数据上的绑定和初始化，Command 的命令只是在控件元素加载完之后才绑定好的，需要输入操作的命令才能触发 Command 的操作，所以 Command 也无法完成控件在初始化过程中的操作。假设有一个 Button，当该 Button 被点击的时候要完成一些操作，那么只要将该操作封装成一个 Command 并绑定到该 Button 上就可以了，但如果要在 Button 被 Load 的时候执行另外一些操作呢？由于 Button 没有直接被 Load 事件所触发的 Command，所以不能使用 Command。如果使用 code-behind 模式的话，就可以在 Button 所在的 Xaml 所对应的 xaml. cs 文件下面使用 Load 事件处理 Button 控件加载过程的操作，那么在 MVVM 模式里面的处理方法就是通过 Attached Behaviors 来实现的。

下面给出 AttachedBehavior 实现单击颜色属性的示例：演示在控件加载的过程中初始化控件的事件和附件属性。

代码清单 11-6：AttachedBehavior 实现单击颜色属性（源代码：第 11 章\Examples_11_6）

MainPage. xaml 文件主要代码

```
<phone:PhoneApplicationPage
    …
    xmlns:local = "clr-namespace:AttachedBehaviorDemo" >
    …
        <Grid x:Name = "ContentPanel" Grid.Row = "1" Margin = "12,0,12,0">
            <TextBlock FontSize = "50" Text = "金色 Gold" local:Behavior.Brush = "Gold"
Margin = "0,6,0,504" />
            <TextBlock FontSize = "50"  Text = "绿色 Green" local:Behavior.Brush = "Green"
Margin = "0,94,0,416" />
            <TextBlock FontSize = "50"  Text = "蓝色 Blue" local:Behavior.Brush = "Blue"
Margin = "0,184,0,339" />
            <TextBlock FontSize = "50"  Text = "橙色 Orange" local:Behavior.Brush = "Orange"
Margin = "-3,274,3,243" />
            <TextBlock FontSize = "50"  Text = "紫色 Purple" local:Behavior.Brush = "Purple"
Margin = "0,386,0,133" />
            <TextBlock FontSize = "50"  Text = "橄榄色 Olive" local:Behavior.Brush = "Olive"
Margin = "0,497,0,0" Height = "110" VerticalAlignment = "Top" />
        </Grid>
    …
</phone:PhoneApplicationPage>
```

Behavior. cs 文件代码：处理附加的属性和事件

```
using System.Windows;
using System.Windows.Controls;
using System.Windows.Input;
using System.Windows.Media;
namespace AttachedBehaviorDemo
{
    static public class Behavior
    {
        //注册一个附加属性 BrushProperty,在 XAML 中名字为 Brush,是 Brush 类型,在 Hover 类中,
//PropertyMetadata 初始化元数据
        public static readonly DependencyProperty BrushProperty = DependencyProperty.
RegisterAttached(
            "Brush",
            typeof(Brush),
            typeof(Behavior),
            new PropertyMetadata (null, new PropertyChangedCallback(OnHoverBrushChanged)));
        //获取 Brush 的属性值
        public static Brush GetBrush(DependencyObject obj)
        {
            return (Brush)obj.GetValue(BrushProperty);
        }
        //设置属性的值
        public static void SetBrush(DependencyObject obj, Brush value)
        {
            obj.SetValue(BrushProperty, value);
        }
```

```
        //属性初始化
        private static void OnHoverBrushChanged(DependencyObject obj,
DependencyPropertyChangedEventArgs args)
        {
            //获取属性所在的 TextBlock 控件
            TextBlock control = obj as TextBlock;
            //注册控件的事件
            if (control != null)
            {
                //注册鼠标进入事件
                control.MouseEnter += new MouseEventHandler(OnControlEnter);
                //注册鼠标离开事件
                control.MouseLeave += new MouseEventHandler(OnControlLeave);
            }
        }
        //鼠标进入事件
        static void OnControlEnter(object sender, MouseEventArgs e)
        {
            //获取当前的 TextBlock 控件
            TextBlock control = (TextBlock)e.OriginalSource;
            //设置控件的前景颜色为红色
            control.Foreground = new SolidColorBrush(Colors.Red);
        }
        //鼠标离开事件
        static void OnControlLeave(object sender, MouseEventArgs e)
        {
            //获取当前的 TextBlock 控件
            TextBlock control = (TextBlock)e.OriginalSource;
            //设置控件的前景颜色为当前控件的 Brush 属性的值
            control.Foreground = GetBrush(control);
        }
    }
}
```

程序运行的效果如图 11.7 所示。

图 11.7　单击会变色的字体

11.5 MVVM Light Toolkit 组件的使用

MVVM Light Toolkit 组件是一个 MVVM 模式的框架，它目前有支持 WPF、Silverlight 和 Windows Phone 的版本，并且是开源的，可以在项目中使用该组件进行编程，官方的网址：http://www.galasoft.ch/mvvm/。使用 MVVM Light Toolkit 组件来编写 MVVM 模式的 Windows Phone 的应用程序可以更加方便和快速地完成想要的功能。MVVM Light Toolkit 组件中封装了很多 MVVM 模式中一些基本的操作和基类，直接继承或者调用就可以使用其功能，而不需要再手工地去编写很多类。MVVM Light Toolkit 组件有专门针对微软 visual studio 开发工具的开发包，可以快速地运用该组件打造出 MVVM 模式的应用程序框架，该工具包可以到其官方网站上进行下载，不过要注意其对应的版本要求。当然，也可以直接在工程项目中引用 MVVM Light Toolkit 组件的 DLL 来进行编程，同样也可以使用其所有的功能。

下面给出 MVVM 模式客户列表的示例：演示如果使用 MVVM Light Toolkit 组件来创建一个 MVVM 模式的 Windows Phone 的应用程序，该示例是一个客户列表绑定了客户的名字、QQ、地址信息，单击的时候通过 Command 传递参数来显示客户的全部详细信息。该示例的数据绑定使用了 ObservableCollection<T> 类来实现，ViewModel 通过继承 GalaSoft.MvvmLight 命名空间下的 ViewModelBase 类来实现，Command 使用 GalaSoft.MvvmLight.Command 命名空间下的 RelayCommand<T>来实现。

代码清单 11-7：MVVM 模式客户列表（源代码：第 11 章\Examples_11_7）

Customers.cs 文件代码：Model 层，创建一个客户列表类和客户实体类

```
using System;
using System.Collections.Generic;
using System.Collections.ObjectModel;
namespace MvvmLight5.Model
{
    public class CustomerCollection
    {
        //在这里绑定数据使用了 ObservableCollection<T> 类
        private readonly ObservableCollection < Customer > _customers = new ObservableCollection
< Customer >();
        public ObservableCollection < Customer > Customers
        {
            get { return _customers; }
        }
        public Customer GetCustomerByID( int id )
        {
            return _customers[ id ];
        }
        public CustomerCollection()
        {
```

```
            try
            {
                GenerateCustomers();
            }
            catch ( Exception e )
            {
                System.Windows.MessageBox.Show( "Exception: " + e.Message );
            }
        }
        //初始化数据
        public void GenerateCustomers()
        {
                _customers.Add( new Customer(1,"黄小琥","台湾高雄市十六街 8 号","高雄",
"13457789907","3232","huangxiaohu@qq.com") );
                _customers.Add(new Customer(2, "李开复", "北京市东城区十六街 8 号", "北京",
"136589907", "787222894", "huasdsdu@qq.com"));
                _customers.Add(new Customer(3, "周杰伦", "台湾台北市十六街 8 号", "台北",
"145455779907", "2323266", "232@qq.com"));
                _customers.Add(new Customer(4, "郎咸平", "香港十六街 8 号", "香港",
"145489907", "787222894", "6ggg@qq.com"));
                _customers.Add(new Customer(5, "加菲猫", "高雄市十六街 8 号", "高雄市",
"15777789907", "333434", "2323@qq.com"));
                _customers.Add(new Customer(6, "灰太狼", "台湾第三代市十六街 8 号", "高雄市",
"134357789907", "23232", "3232@qq.com"));
                _customers.Add(new Customer(7, "喜洋洋", "台湾高雄市十六街 8 号", "高雄市",
"134544589907", "23232777", "88sds@qq.com"));
                _customers.Add(new Customer(8, "春哥", "台湾所得税十六街 8 号", "高雄市",
"13453445907", "888888", "sdsgg@qq.com"));
        }
    }
    public class Customer
    {
        public int ID { get; set; }
        public string Name { get ; set; }
        public string Address { get; set; }
        public string City { get; set; }
        public string Phone { get; set; }
        public string QQ { get; set; }
        public string Email { get; set; }
        public Customer()
        { }
        public Customer(
            int id,
              string name,
              string address,
              string city,
              string phone,
              string qq,
              string email )
        {
```

```
            ID = id;
            Name = name;
            Address = address;
            City = city;
            Phone = Phone;
            QQ = qq;
            Email = email;
        }
    }
}
```

MainViewModel.cs 文件代码：ViewModel 层，创建一个 MainViewModel 类作为客户列表的 ViewModel 层，该类继承了 MvvmLight5.Model. ViewModelBase 类，并实现了 Command

```
using System.Collections.ObjectModel;
using GalaSoft.MvvmLight;
using GalaSoft.MvvmLight.Command;
using MvvmLight5.Model;
namespace MvvmLight5.ViewModel
{
    public class MainViewModel : ViewModelBase
    {
        //数据绑定的客户列表
        public ObservableCollection<Customer> Customers
        {
            get
            {
                var customerCollection = new CustomerCollection();
                return customerCollection.Customers;
            }
        }
        //定义 Command
        public RelayCommand<Customer> DetailsPageCommand
        {
            get;
            private set;
        }
        public string ApplicationTitle
        {
            get
            {
                return "MVVM LIGHT";
            }
        }
        public string PageName
        {
            get
            {
                return "客户列表如下:";
```

```
            }
        }
        public MainViewModel()
        {
            //初始化 Command
            DetailsPageCommand = new RelayCommand<Customer>( ( msg ) => GoToDetailsPage( msg ) );
        }
        private object GoToDetailsPage( Customer msg )
        {
            System.Windows.MessageBox.Show("客户的详细资料如下名字:" + msg.Name + "城市:" + msg.City + "地址:" + msg.Address + "电话:" + msg.Phone + "QQ:" + msg.QQ);
            return null;
        }
    }
}
```

ViewModelLocator.cs 文件代码：ViewModelLocator 类对 ViewModel 进行初始化和清理的集中处理的类，添加资源的时候只需要添加这一个类就可以了

```
namespace MvvmLight5.ViewModel
{
    public class ViewModelLocator
    {
        private static MainViewModel _main;
        //初始化,在这里创建 ViewModel,可以将多个 ViewModel 在这里一起创建
        public ViewModelLocator()
        {
            CreateMain();
        }
        //获取 MainViewModel 的静态的实例对象
        public static MainViewModel MainStatic
        {
            get
            {
                if (_main == null)
                {
                    CreateMain();
                }
                return _main;
            }
        }
        //获取 MainViewModel 的实例对象
        public MainViewModel Main
        {
            get
            {
```

```
                return MainStatic;
            }
        }
        //清理 MainViewModel,退出程序的时候进行清理,在 App.xmal.cs 中调用
        public static void ClearMain()
        {
            _main.Cleanup();
            _main = null;
        }
        //创建 MainViewModel
        public static void CreateMain()
        {
            if ( _main == null )
            {
                _main = new MainViewModel();
            }
        }
    }
}
```

MainPage.xaml 文件主要代码：View 层

```
<phone:PhoneApplicationPage
    ...
    xmlns:Custom = "clr - namespace:System.Windows.Interactivity; assembly = System.Windows.
Interactivity"
    xmlns:GalaSoft_MvvmL ight_Command = "clr - namespace:GalaSoft.MvvmLight.Command; assembly
= GalaSoft.MvvmLight.Extras.WP8"
    DataContext = "{Binding Main, Source = {StaticResource Locator}}">
    <!-- 绑定 ViewModel 中的 Main 实例对象,资源在 App.xaml 中进行了加载 -->
    ...
        <StackPanel x:Name = "TitlePanel"  Grid.Row = "0" Margin = "24,24,0,12">
            <TextBlock x:Name = "ApplicationTitle" Text = "{Binding ApplicationTitle}"
                Style = "{StaticResource PhoneTextNormalStyle}" />
            <TextBlock x:Name = "PageTitle" Text = "{Binding PageName}"
                Margin = " - 3, - 8,0,0" Style = "{StaticResource PhoneTextTitle1Style}" />
        </StackPanel>
        <Grid  x:Name = "ContentGrid" Grid.Row = "1">
            <ListBox x:Name = "PersonListBox" Margin = "10" ItemsSource = "{Binding Customers}">
                    <!-- 绑定 MainViewModel 的 Customers 数据 -->
                <ListBox.ItemTemplate>
                    <DataTemplate>
                        <StackPanel>
                            <StackPanel x:Name = "DataTemplateStackPanel"
Orientation = "Horizontal">
                                <TextBlock x:Name = "Name"
                                    Text = "{Binding Name}"
                                    Margin = "0,0,5,0"
```

```
                    Style = "{StaticResource PhoneTextExtraLargeStyle}" />
                <TextBlock
                    x:Name = "QQ"
                    Text = "{Binding QQ}"
                    Margin = "0"
                    Style = "{StaticResource PhoneTextExtraLargeStyle}" />
            </StackPanel>
            <TextBlock
                x:Name = "Email"
                Text = "{Binding Address}"
                Margin = "0"
                Style = "{StaticResource PhoneTextSubtleStyle}" />
        </StackPanel>
    </DataTemplate>
</ListBox.ItemTemplate>
    <!-- 添加 System.Windows.Interactivity 触发器处理事件,放在 ListBox 里面 -->
  <Custom:Interaction.Triggers>
      <!-- 当选中客户的时候触发该事件 -->
    <Custom:EventTrigger EventName = "SelectionChanged">
                    < GalaSoft _ MvvmLight _ Command: EventToCommand x: Name =
"SelectionCommand" Command = "{Binding DetailsPageCommand, Mode = OneWay}" CommandParameter =
"{Binding SelectedItem, ElementName = PersonListBox}"/>
                <!-- 传递的参数为 ListBox 选中的 Customer 对象 -->
      </Custom:EventTrigger>
  </Custom:Interaction.Triggers>
    </ListBox>
  </Grid>
...
</phone:PhoneApplicationPage>
```

App.xaml 文件主要代码：程序初始化处理

```
<Application
            ...
            xmlns:vm = "clr-namespace:MvvmLight5.ViewModel">
    <!-- 应用程序资源 -->
    <Application.Resources>
        <vm:ViewModelLocator x:Key = "Locator"
                             d:IsDataSource = "True" />
    </Application.Resources>
...
</Application>
```

App.xaml.cs 文件主要代码：关闭程序事件中处理清理 ViewModel 资源

```
private void Application_Closing( object sender, ClosingEventArgs e )
{
    ViewModelLocator.ClearMain();
}
```

程序运行的效果如图 11.8 和图 11.9 所示。

图 11.8　客户列表

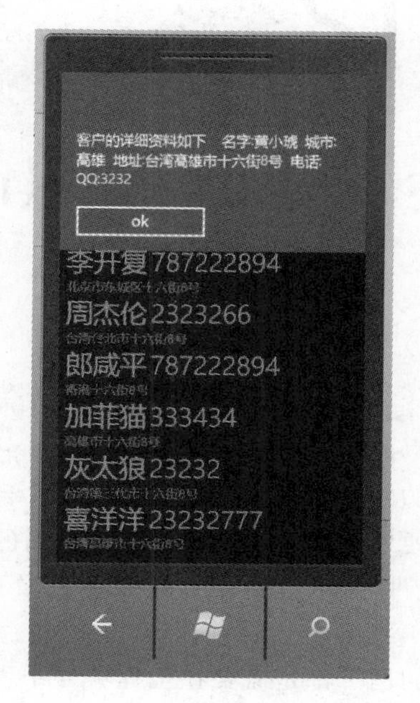

图 11.9　客户列表传递参数的效果

第12章

Silverlight Toolkit组件

Silverlight Toolkit 组件是由微软官方提供的针对于 Silverlight 程序开发的一系列控件，Windows Phone 系统发布之后，微软同时也发布了一套针对于 Windows Phone 的 Silverlight Toolkit 组件，该组件提供了一系列各种各样的精彩的控件，利用这些控件你就可以很方便地给应用程序实现一些炫丽的效果和更好的用户体验。本章将介绍 Silverlight Toolkit 组件里面一些常用的控件以及如何在 Windows Phone 的应用程序上使用这些控件。

本章的例子都是需要在工程项目中引用 Microsoft. Phone. Controls. Toolkit. dll 组件，并在调用的页面添加 Toolkit 组件类库的引用 xmlns：toolkit＝"clr-namespace：Microsoft. Phone. Controls；assembly＝Microsoft. Phone. Controls. Toolkit"，该组件的 Windows Phone 的最新版本可以到 http://silverlight. codeplex. com/中去下载。

12.1 自动完成文本框（AutoCompleteBox）

自动完成文本框控件（AutoCompleteBox）是一种很常用的控件，它实现了文本框的输入的自动搜索的功能，可以加快用户的输入效率。AutoCompleteBox 控件常用属性如表 12.1 所示。

表 12.1 AutoCompleteBox 控件常用属性

名 称	说 明
FilterMode	获取或设置文本框中的文本怎样被用来过滤具体由为下拉内容准备的 ItemsSource 属性的项
IsDropDownOpen	获取或设置一个值用以确定该组件的下拉部分是否已打开
IsTextCompletionEnabled	该属性的作用是获取或设置一个值用以确定在过滤过程中第一个可能的匹配结果是否自动地被填充至 AutoCompleteBox 组件中；有时我们需要第一个匹配的预选对象自动填充至 AutoCompleteBox 的文本框中，这时就需要通过设置该属性以达到所需效果
ItemFilter	获取或设置自定义方法用来使用用户输入的文本来过滤在下拉框中显示的具体由 ItemsSource 属性所决定的 items

续表

名　称	说　明
MaxDropDownHeight	获取或设置该组件的下拉部分高度的最大值
MinimumPopulateDelay	获取或设置第一个匹配结果出现的最小延迟时间
MinimumPrefixLength	获取或设置需要被输入该组件文本框中的最小字符数量,在该组件显示可能的匹配之前
SearchText	获取在 itemsSource 项目集合中被用作过滤项目的文本
SelectionAdapter	获取或设置用以生成下拉选项部分的选择适配器,通过一个可选择的项目列表
Text	获取或设置该组件文本框中的值
TextBoxStyle	获取或设置该组件的文本框的样式
TextFilter	获取或设置自定义方法用来使用用户输入的文本以基于文本的方式过滤在下拉框中显示的具体由 ItemsSource 属性所决定的 items
ValueMemberBinding	获取或设置用于获取对在 AutoCompleteBox 控制文本框部分显示的值的绑定,为显示在下拉过滤项目
ValueMemberPath	获取或设置用来获取为在 AutoCompleteBox 控制文本框部分显示的值的属性的路径,为显示在下拉筛选项目

在 Windows Phone 应用程序中使用 AutoCompleteBox 控件的 XAML 语法如下:

```
< toolkit:AutoCompleteBox x:Name = "myBox" Height = "80"/>
```

在 XAML.CS 使用 List<T> 绑定 AutoCompleteBox 的数据源的语法如下:

```
List < string >  names  =  new List < string >();
names.Add("names1");
names.Add("names 2");
this. myBox.ItemsSource  =  names;
```

这样简单的两步,一个简单的 AutoCompleteBox 自动完成文本框就完成了。

下面给出自动完成文本框的示例:通过两种方式来在 Windows Phone 应用程序上实现 AutoCompleteBox 控件,第一种方式是直接的 code-behind 用 List<T>绑定控件,第二种方式是使用 MVVM 模式用面向对象的方式绑定控件。

代码清单 12-1:自动完成文本框(源代码:第 12 章\Examples_12_1)

MainPage.xaml 文件主要代码

```
< phone:PhoneApplicationPage
    …
    xmlns:toolkit = "clr - namespace:Microsoft.Phone.Controls; assembly = Microsoft.Phone.
Controls.Toolkit"
    xmlns:My = "clr - namespace:TestingAutoComplete" >
    <! -- 引用 Toolkit 组件空间 -- >
    <! -- 引用 MVVM 定义的 ViewModel 空间,第二种处理方式使用 -- >
    …
    <! -- 定义数据绑定资源,第二种处理方式使用 -- >
```

```
    <phone:PhoneApplicationPage.Resources>
        <My:Names x:Key="Names"></My:Names>
    </phone:PhoneApplicationPage.Resources>
    …
        <!--界面设计,使用 Pivot 控件布局-->
        <controls:Pivot Title="AutoCompleteBox 控件">
            <controls:PivotItem Header="模糊" Margin="0">
            <!--第一种处理方式: code-behind 模式-->
                <Grid>
                    <toolkit:AutoCompleteBox Grid.Row="0" x:Name="people1" Height="70"/>
                </Grid>
            </controls:PivotItem>
            <controls:PivotItem Header="前缀" Margin="0">
            <!--第二种处理方式: MVVM 模式-->
                <Grid DataContext="{StaticResource Names}">
                    <toolkit:AutoCompleteBox  Name="people2"
                                            ValueMemberBinding="{Binding MyName}"
                                            ItemsSource="{ Binding ListOfNames}" >
                        <toolkit:AutoCompleteBox.ItemTemplate>
                            <DataTemplate>
                                <TextBlock Text="{Binding MyName}"></TextBlock>
                            </DataTemplate>
                        </toolkit:AutoCompleteBox.ItemTemplate>
                    </toolkit:AutoCompleteBox>
                </Grid>
            </controls:PivotItem>
        </controls:Pivot>
    …
</phone:PhoneApplicationPage>
```

MainPage.xaml.cs 文件代码：第一种方式 code-behind 模式实现自动完成文本框控件

```
namespace TestingAutoComplete
{
    public partial class MainPage : PhoneApplicationPage
    {
        public MainPage()
        {
            InitializeComponent();
            List<string> names = new List<string>();
            string namesString = " Fernando Sucre, Scofield, Alexander Mahone, Theodore
Bagwell,Sara Tancredi,Lincoln Burrows,John Abruzzi,Fluorine";
            foreach (var name in namesString.Split(','))
                names.Add(name);
            //给自动完成文本框控件的源赋值
            this.people1.ItemsSource = names;
            this.people1.ItemFilter += SearchCountry;
        }
```

```
        //全局模糊搜索
        bool SearchCountry(string search, object value)
        {
            if (value != null)
            {
                //如果包含了搜索的字符串则返回 true
                if (value.ToString().ToLower().IndexOf(search) >= 0)
                    return true;
            }
            //如果不匹配 返回 false
            return false;
        }
    }
}
```

Name.cs 文件代码：第二种方式 MVVM 模式的 Model 的实现

```
namespace TestingAutoComplete
{
    public class Name
    {
        //名字属性
        public string MyName { get; set; }
        public override string ToString()
        {
            return MyName;
        }
    }
}
```

Names.cs 文件代码：第二种方式 MVVM 模式的 ViewModel 的实现

```
using System.ComponentModel;
using System.Collections.ObjectModel;
namespace TestingAutoComplete
{
    public class Names : INotifyPropertyChanged
    {
    //定义属性改变事件
        public event PropertyChangedEventHandler PropertyChanged;
    //名字集合属性
        private ObservableCollection<Name> _listOfnames;
        public ObservableCollection<Name> ListOfNames
        {
            get { return _listOfnames; }
            set
            {
                _listOfnames = value;
```

```
                if (PropertyChanged != null)
                    PropertyChanged(this, new PropertyChangedEventArgs("ListOfNames"));
            }
        }
        public Names()
        {
            //初始化名字集合属性
            ListOfNames = new ObservableCollection<Name>();
            string namesString = "Fernando Sucre,Scofield,Alexander Mahone,Theodore
            Bagwell,Sara Tancredi,Lincoln Burrows,John Abruzzi,Fluorine";
            foreach (var name in namesString.Split(','))
                ListOfNames.Add(new Name() { MyName = name });
        }
    }
}
```

程序运行的效果如图 12.1 所示。

图 12.1 模糊查询

12.2 上下文菜单(ContextMenu)

上下文菜单(ContextMenu)在 Windows Phone 里面是指手指长触摸带出来的菜单，原理相当于鼠标的右键的菜单。ContextMenu 控件附加到特定的控件里面，使用 ContextMenu 元素可以向用户呈现一个项列表，这些项指定与特定控件(例如 Button)相关联的命令或选项。ContextMenu 控件的常用属性和事件分别如表 12.2 和表 12.3 所示。

表 12.2 ContextMenu 控件常用属性

名　　称	说　　明
IsOpen	获取或设置一个值，该值指示 ContextMenu 是否可见
ItemContainerStyle	获取或设置 Style，它应用于为每个项生成的容器元素，这个是给 ContextMenu 添加附加样式的属性
DataContext	获取或设置元素参与数据绑定时的数据上下文，绑定数据使用
ActualHeight	获取此元素的呈现高度
ActualWidth	获取此元素的呈现宽度
ItemTemplate	获取或设置用于显示每个项的 DataTemplate，数据模板

表 12.3 ContextMenu 控件常用事件

名　　称	说　　明
KeyDown	在焦点位于此元素上并且用户按下键时发生
KeyUp	在焦点位于此元素上并且用户释放键时发生
Closed	在 ContextMenu 的特定实例关闭时发生
Opened	在上下文菜单的特定实例打开时发生

实现 ContextMenu 的语法，如在 Button 控件中加入上下文菜单的 XAML 语法如下：

```
<Button Content = "OpenContextMenu" Height = "100" Width = "270">
    <toolkit:ContextMenuService.ContextMenu>
        <toolkit:ContextMenu VerticalOffset = "100.0" IsZoomEnabled = "True"  x:Name = "menu">
            <toolkit:MenuItem Header = "新增"  Click = "MenuItem_Click"/>
            <toolkit:MenuItem Header = "删除"  Click = "MenuItem_Click"/>
        </toolkit:ContextMenu>
    </toolkit:ContextMenuService.ContextMenu>
</Button>
```

XAML.CS 语法如下：

```
private void MenuItem_Click(object sender, RoutedEventArgs e)
{
    MenuItem menuItem = (MenuItem)sender;
    if (menuItem.Header.ToString() == "新增")
     {
         //处理新增
     }
     else if (menuItem.Header.ToString() == "删除")
     {
         //处理删除
     }
}
```

下面给出上下文菜单的示例：在 ListBox 控件中使用 ContextMenu 来操作 ListBox 的选项和删除数据。

代码清单 12-2：上下文菜单(源代码：第 12 章\Examples_12_2)

MainPage.xaml 文件主要代码

```
<ListBox  Name = "lbNames" Height = "437"
          HorizontalAlignment = "Left"
          Margin = "10,119,0,0"
          VerticalAlignment = "Top"
          Width = "460" ItemsSource = "{Binding}" Grid.RowSpan = "2">
    <ListBox.ItemTemplate>
        <DataTemplate>
            <StackPanel>
                <TextBlock  FontSize = "40"  Text = "{Binding MyName}" />
                <TextBlock  FontSize = "40"  Text = "{Binding Company}" />
                <toolkit:ContextMenuService.ContextMenu>
                    <toolkit:ContextMenu>
                        <toolkit:MenuItem Header = "变色"
                                          Click = "MenuItem_Click"/>
                        <toolkit:MenuItem Header = "删除"
                                          Click = "MenuItem_Click"/>
                        <toolkit:MenuItem Header = "全部删除"
                                          Click = "MenuItem_Click"/>
                    </toolkit:ContextMenu>
                </toolkit:ContextMenuService.ContextMenu>
            </StackPanel>
        </DataTemplate>
    </ListBox.ItemTemplate>
</ListBox>
```

MainPage.xaml.cs 文件主要代码

```
namespace ContextMenuDemo
{
    public partial class MainPage : PhoneApplicationPage
    {
        //联系人的列表
        ObservableCollection<Person> personList = new ObservableCollection<Person>();
        //选中的联系人
        Person selectedPerson = null;
        public MainPage()
        {
            InitializeComponent();
            //页面加载事件在这里初始化 ListBox 控件
            this.Loaded += new RoutedEventHandler(MainPage_Loaded);
        }
        //这个是 MenuItem 的单击事件
        private void MenuItem_Click(object sender, RoutedEventArgs e)
        {
```

```
            //获取选中的 menuItem 对象
            MenuItem menuItem = (MenuItem)sender;
            //获取对象的标题头的内容
            string header = (sender as MenuItem).Header.ToString();
            //获取选中的 ListBoxItem
            ListBoxItem selectedListBoxItem = this.lbNames.ItemContainerGenerator.
ContainerFromItem((sender as MenuItem).DataContext) as ListBoxItem;
            //如果没有选中则返回
            if (selectedListBoxItem == null)
            {
                return;
            }
            if (menuItem.Header.ToString() == "变色")
            {
                //改变选中的 ListBox 选项的颜色
                selectedListBoxItem.Background = new SolidColorBrush(Colors.Red);
            }
            else if (menuItem.Header.ToString() == "删除")
            {
                //删除短触摸选中的 Person
                selectedPerson = lbNames.SelectedItem as Person;
                if (selectedPerson == null)
                {
                    MessageBox.Show("请先短触摸选中一个选项!再进行删除!");
                }
                else
                {
                    personList.Remove(selectedPerson);
                    lbNames.ItemsSource = personList;
                }
            }
            else if (menuItem.Header.ToString() == "全部删除")
            {
                //删除所有的数据
                personList.Clear();
                lbNames.ItemsSource = personList;

            }
        }
        //初始数据
        void MainPage_Loaded(object sender, RoutedEventArgs e)
        {
            personList.Add(new Person() { MyName = "张三", Company = "大大有限公司" });
            personList.Add(new Person() { MyName = "李四", Company = "测试有限公司" });
            personList.Add(new Person() { MyName = "小明", Company = "逛逛有限公司" });
            personList.Add(new Person() { MyName = "中宗", Company = "大所是有限公司" });
            personList.Add(new Person() { MyName = "离开", Company = "测试88有限公司" });
            personList.Add(new Person() { MyName = "灭里", Company = "逛逛9有限公司" });
```

```
                lbNames.ItemsSource = personList;
            }
        }
        //定义联系人的类
        public class Person
        {
            public string MyName { get; set; }
            public string Company { get; set; }
        }
    }
```

程序的运行效果如图 12.2 所示。

图 12.2　上下文菜单的效果

12.3　日期采集器(DatePicker)

日期采集器(DatePicker)表示一个允许用户选择日期的控件。DatePicker 控件在界面上现实效果如图 12.3 所示，单击后会弹出一个全屏的日期选择界面，效果如图 12.4 所示。DatePicker 控件的常用属性如表 12.4 所示。

图 12.3　日期控件的显示效果

图 12.4 日历控件的打开效果

表 12.4 DatePicker 控件的常用属性

名　称	说　明
Header	控件的标题显示
PickerPageUri	这个属性的值是一个 URL 的地址，指向于扩展 DatePicker 的控件，当要扩展 DatePicker 控件的时候，需要继承 IDateTimePickerPage 接口
Value	设置和获取时间的值，DateTime 类型
ValueString	获取时间的字符串值
ValueStringFormat	日期格式化，默认的格式化值是"{}{0:d}"，2011 年 4 月 16 日则显示 16/4/2011，可以通过改变 ValueStringFormat 的值来改变时间的显示格式

下面给出日期采集器的示例：测试 DatePicker 控件的效果以及扩展 DatePicker 控件实现自定义的日期选择页面。

代码清单 12-3：日期采集器(源代码：第 12 章\Examples_12_3)

MainPage.xaml 文件主要代码

```
<StackPanel x:Name = "ContentPanel" Grid.Row = "1" Margin = "12,0,12,0">
    <toolkit:DatePicker x:Name = "datePicker" Header = "DatePicker 日期选择控件" Value = "4/
16/2010" />
    <toolkit:DatePicker x:Name = "datePicker1" Header = "扩展的 DatePicker 控件" Value = "4/
16/2010" PickerPageUri = "/DatePickerDemo;component/CustomPage.xaml" />
</StackPanel>
```

MainPage. xaml. cs 文件代码

```
namespace DatePickerDemo
{
    public partial class MainPage : PhoneApplicationPage
    {
        public MainPage()
        {
            InitializeComponent();
            //控件值改变触发的事件
            this.datePicker.ValueChanged += new EventHandler<DateTimeValueChangedEventArgs>
(picker_ValueChanged);
        }
        void picker_ValueChanged(object sender, DateTimeValueChangedEventArgs e)
        {
            DateTime date = (DateTime)e.NewDateTime;
            MessageBox.Show(date.ToString("d"));
        }
    }
}
```

CustomPage. xaml 文件主要代码：扩展的日期控件的页面，单击按钮，返回当前的日期

```
<Grid x:Name="ContentPanel" Grid.Row="1" Margin="12,0,12,0">
    <Border BorderBrush="Red" Background="Orange" Height="200" Width="300">
        <Button Content="DateTime Now" x:Name="btn" Click="btn_Click"/>
    </Border>
</Grid>
```

CustomPage. xaml. cs 文件代码

```
namespace DatePickerDemo
{
    public partial class CustomPage : PhoneApplicationPage, IDateTimePickerPage
    {
        public CustomPage()
        {
            InitializeComponent();
        }
        public DateTime? Value
        {
            get;
            set;
        }
        private void btn_Click(object sender, RoutedEventArgs e)
        {
            //选择时间并关闭页面
            Value = DateTime.Now;
            NavigationService.GoBack();//返回
        }
    }
}
```

程序运行的效果如图 12.5 所示。

图 12.5 日期控件的测试效果

12.4 手势服务/监听(GestureService/GestureListener)

手势服务/监听(GestureService/GestureListener)是 Silverlight Tookit 组件提供的一个监听手指在控件或者屏幕上的手势和动作事件的服务。使用 GestureService/GestureListener 就可以捕获到手指在程序中的各种操作动作，比如拖动、单击等。

下面将详细介绍 GestureService/GestureListener 能够监听到的一些手指操作动作。

1）单击

单击对应 GestureListener. Tap 事件，动作的模拟图如图 12.6 所示。

2）长时间单击触摸

长时间单击触摸对应 GestureListener. Hold 事件，动作的模拟图如图 12.7 所示。

3）双击

双击对应 GestureListener. DoubleTap 事件，动作的模拟图如图 12.8 所示。

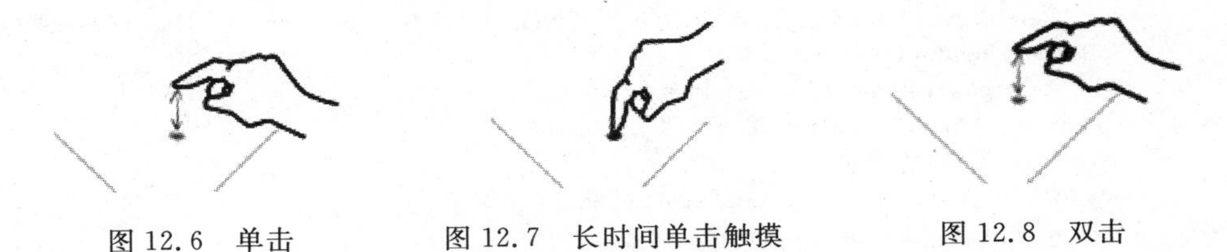

图 12.6 单击　　图 12.7 长时间单击触摸　　图 12.8 双击

4）拖动

拖动开始对应 GestureListener. DragStarted 事件，拖动过程对应 GestureListener. OnDragDelta 事件，拖动结束对应 GestureListener. OnDragCompleted 事件，动作的模拟图如图 12.9 所示。

5）轻击

轻击对应 GestureListener. Flick 事件，动作的模拟图如图 12.10 所示。

6）捏动作(放大收缩)

捏动作开始对应 GestureListener. OnPinchStarted 事件，捏动作过程对应 GestureListener.

OnPinchDelta 事件，捏动作结束对应 GestureListener. OnPinchCompleted 事件，动作的模拟图如图 12.11 所示。

图 12.9 拖动

图 12.10 轻击

图 12.11 捏动作

GestureService/GestureListener 使用的 XAML 语法如下所示：

```
<Rectangle Fill = "Orange" x:Name = "rect">
   <toolkit:GestureService.GestureListener>
       <toolkit:GestureListener  Tap = "GestureListener_Tap" Hold = "GestureListener_Hold"  />
   </toolkit:GestureService.GestureListener>
</Rectangle>
```

下面给出手势演示的示例：在一个矩形里使用了这 6 个手势动作并且将这些手势动作反应到矩形的颜色大小等属性上面。

代码清单 12-4：手势演示(源代码：第 12 章\Examples_12_4)

MainPage.xaml 文件主要代码

```
<Grid x:Name = "ContentPanel" Grid.Row = "1" Margin = "12,0,12,0">
    <TextBlock x: Name = " flickData " Text = " Flick:" Style = " { StaticResource
PhoneTextNormalStyle}" VerticalAlignment = "Bottom"/>
    <Border x:Name = "border" Width = "300" Height = "200"
            BorderBrush = "{StaticResource PhoneForegroundBrush}"
            BorderThickness = "4,16,4,4"
            Background = "{StaticResource PhoneAccentBrush}"
            RenderTransformOrigin = "0.5,0.5" Opacity = ".5" CacheMode = "BitmapCache">
        <Border.RenderTransform>
            <CompositeTransform x:Name = "transform"/>
        </Border.RenderTransform>
        <!-- 加入手势的监听 -->
        <toolkit:GestureService.GestureListener>
            <toolkit:GestureListener
                Tap = "OnTap" Hold = "OnHold"
                DoubleTap = "OnDoubleTap"
                DragStarted = " OnDragStarted" DragDelta = " OnDragDelta" DragCompleted =
"OnDragCompleted"
                Flick = "OnFlick"
                PinchStarted = " OnPinchStarted" PinchDelta = " OnPinchDelta" PinchCompleted =
"OnPinchCompleted"
                />
        </toolkit:GestureService.GestureListener>
    </Border>
    <TextBlock Name = "Gestures" Text = "事件" Margin = "0,489,0,50" />
</Grid>
```

MainPage.xaml.cs 文件主要代码

```
public partial class MainPage : PhoneApplicationPage
{
    SolidColorBrush greenBrush = new SolidColorBrush(Colors.Green);
    SolidColorBrush redBrush = new SolidColorBrush(Colors.Red);
    SolidColorBrush normalBrush;
    double initialAngle;
    double initialScale;
    public MainPage()
    {
        InitializeComponent();
        normalBrush = (SolidColorBrush) Resources["PhoneAccentBrush"];
    }
    //Tap 单击事件
    private void OnTap(object sender, GestureEventArgs e)
    {
        transform.TranslateX = transform.TranslateY = 0;
        Gestures.Text = "Tap 单击事件";
    }
    //DoubleTap 双击事件
    private void OnDoubleTap(object sender, GestureEventArgs e)
    {
        transform.ScaleX = transform.ScaleY = 1;
        Gestures.Text = "DoubleTap 双击事件";
    }
    //Hold 长单击事件
    private void OnHold(object sender, GestureEventArgs e)
    {
        transform.TranslateX = transform.TranslateY = 0;
        transform.ScaleX = transform.ScaleY = 1;
        transform.Rotation = 0;
        Gestures.Text = "Hold 长单击事件";
    }
    //DragStarted 拖动开始
    private void OnDragStarted(object sender, DragStartedGestureEventArgs e)
    {
        border.Background = greenBrush;
        Gestures.Text = "DragStarted 拖动开始";
    }
    //DragDelta 拖动过程
    private void OnDragDelta(object sender, DragDeltaGestureEventArgs e)
    {
        transform.TranslateX += e.HorizontalChange;
        transform.TranslateY += e.VerticalChange;
        Gestures.Text = "DragDelta 拖动过程";
    }
    //DragCompleted 拖动结束
    private void OnDragCompleted(object sender, DragCompletedGestureEventArgs e)
    {
        border.Background = normalBrush;
        Gestures.Text = "DragCompleted 拖动结束";
```

```
        }
        //PinchStarted 捏动作开始
        private void OnPinchStarted(object sender, PinchStartedGestureEventArgs e)
        {
            border.Background = redBrush;
            initialAngle = transform.Rotation;
            initialScale = transform.ScaleX;
            Gestures.Text = "PinchStarted 捏动作开始";
        }
        //PinchDelta 捏动作过程
        private void OnPinchDelta(object sender, PinchGestureEventArgs e)
        {
            transform.Rotation = initialAngle + e.TotalAngleDelta;
            transform.ScaleX = transform.ScaleY = initialScale * e.DistanceRatio;
            Gestures.Text = "PinchDelta 捏动作过程";
        }
        //PinchCompleted 捏动作结束
        private void OnPinchCompleted(object sender, PinchGestureEventArgs e)
        {
            border.Background = normalBrush;
            Gestures.Text = "PinchCompleted 捏动作结束";
        }
        //Flick 轻击事件
        private void OnFlick(object sender, FlickGestureEventArgs e)
        {
            flickData.Text = string.Format("{0} Flick: Angle {1} Velocity {2},{3}",
                e.Direction, Math.Round(e.Angle), e.HorizontalVelocity, e.VerticalVelocity);
            Gestures.Text = "Flick 轻击事件";
        }
    }
```

程序运行的效果如图 12.12 所示。

图 12.12　Gestures 测试

12.5 列表采集器(ListPicker)

列表采集器(ListPicker)相当于一个下拉菜单的效果,列表的数据可以静态或者动态地添加,然后可以通过 ListPicker 采用下拉的形式或者弹出一个页面的形式来选择列表的选项。ListPicker 在界面实现的效果如图 12.13 所示,单击后出现的下拉效果如图 12.14 所示。

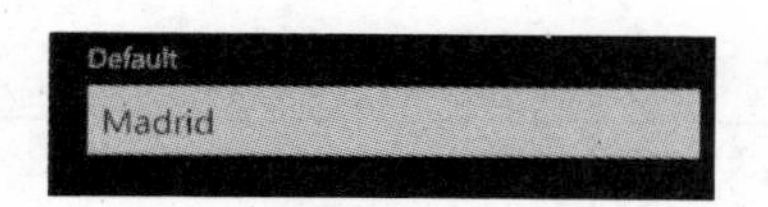

图 12.13　列表采集器的效果

图 12.14　列表采集器打开的效果

ListPicker 使用的 XAML 语法如下所示:

```
<toolkit:ListPicker Header = "Default" Grid.Row = "1" x:Name = "defaultPicker"/>
```

CS 绑定下拉的数据示例如下:

```
this.defaultPicker.ItemsSource = new List<string>() { "Madrid", "London","Mexico" };
```

下面给出列表采集器的示例:利用 ListPicker 的 ItemTemplate 属性(文本框显示的单个选项模板)和 FullModeItemTemplate 属性(所有选项列表的模板)来实现一个弹出全屏的选项列表。

代码清单 12-5:列表采集器(源代码:第 12 章\Examples_12_5)

MainPage.xaml 文件主要代码

```
<!-- 定义模板资源 -->
<Grid.Resources>
    <DataTemplate x:Name = "PickerItemTemplate">
        <StackPanel Orientation = "Horizontal">
            <Border Background = "LightGreen" Width = "34" Height = "34">
                <TextBlock Text = "{Binding Country}" FontSize = "16" HorizontalAlignment =
"Center" VerticalAlignment = "Center"/>
                </Border>
                <TextBlock Text = "{Binding Name}" Margin = "12 0 0 0"/>
            </StackPanel>
        </DataTemplate>
        <DataTemplate x:Name = "PickerFullModeItemTemplate">
            <StackPanel Orientation = "Horizontal" Margin = "16 21 0 20">
                <TextBlock Text = "{Binding Name}" Margin = "16 0 0 0" FontSize = "43"
FontFamily = "{StaticResource PhoneFontFamilyLight}"/>
                <TextBlock Text = "language: "/>
                <TextBlock Text = "{Binding Language}" Foreground = "Green"/>
            </StackPanel>
```

```
        </DataTemplate>
    </Grid.Resources>
    …
    <Grid x:Name="ContentPanel" Grid.Row="1" Margin="12,0,12,0">
        <!-- 使用 ListPicker 控件 -->
            <toolkit:ListPicker Grid.Row="0" x:Name="listPicker" ItemTemplate=
"{StaticResource PickerItemTemplate}" ItemCountThreshold="3"
                                FullModeItemTemplate="{StaticResource PickerFullModeItemTemplate}"
FullModeHeader="Cities" SelectedIndex="2"CacheMode="BitmapCache"
                                Header="Cities"/>
        <toolkit:ListPicker Header="Default" Grid.Row="1" x:Name="defaultPicker"/>
    </Grid>
```

MainPage.xaml.cs 文件主要代码

```
public MainPage()
{
    InitializeComponent();
    //绑定有模板的 ListPicker
    List<Cities> source = new List<Cities>();
    source.Add(new Cities(){Name = "Madrid",Country = "ES",Language = "Spanish"});
    source.Add(new Cities() { Name = "Las Vegas", Country = "US", Language = "English" });
    source.Add(new Cities() { Name = "London", Country = "UK", Language = "English" });
    source.Add(new Cities() { Name = "Mexico", Country = "MX", Language = "Spanish" });
    this.listPicker.ItemsSource = source;
    //绑定没有模板的 ListPicker
    this.defaultPicker.ItemsSource = new List<string>() { "Madrid", "London","Mexico" };
}
```

Cities.cs 文件代码：ListPicker 控件的源集合子对象

```
public class Cities
{
    //城市名称
    public string Name
    {
        get;
        set;
    }
    //城市所属国家
    public string Country
    {
        get;
        set;
    }
    //城市语言
    public string Language
    {
        get;
        set;
    }
}
```

程序运行的效果如图 12.15 和图 12.16 所示。

图 12.15　列表采集器界面

图 12.16　列表采集器打开的效果

12.6　列表选择框(LongListSelector)

LongListSelector(注意：8.0 的 SDK 已经内置了，不需要再使用 toolkit 里面的。)是一种比 ListBox 更加强大的列表控件，可以根据列表的信息来分类排列，根据类别快速定位到选中的类别的列表下，在数据量很大的情况下这种分类的优势很明显。LongListSelector 可以自定义列表头、列表尾、类表头、类表尾等的样式和数据，可以实现各种个性化的列表样式和不同的数据的展现方式。Windows Phone 手机的联系人列表就是基于 LongListSelector 控件设计的。LongListSelector 控件的常用属性和常用事件分别如表 12.5 和表 12.6 所示。

表 12.5　LongListSelector 控件常用属性

名　称	说　明
DisplayAllGroups	bool 类型的属性，当值为 true 时，它显示所有的分组无论该组中是否有选项或者数据，默认值为 false
GroupFooterTemplate	DataTemplate 类型的属性，它是负责绑定每个组的底部的数据和样式的模板
GroupHeaderTemplate	DataTemplate 类型的属性，它是负责绑定每个组的顶部的数据和样式的模板
GroupItemsPanel	ItemsPanelTemplate 类型的属性，设置组的内部的 Panel 面板的内容
GroupItemTemplate	DataTemplate 类型的属性，它是负责绑定每个组里面的元素的数据和样式的模板

续表

名　称	说　明
ItemTemplate	DataTemplate类型的属性，它是负责绑定所有选项或者元素的数据和样式的模板
ListFooterTemplate	DataTemplate类型的属性，它是负责绑定整个List底部的数据和样式的模板
ListHeaderTemplate	DataTemplate类型的属性，它是负责绑定整个List顶部的数据和样式的模板
SelectedItem	获取或者设置选中的选项
ShowListFooter	bool类型的属性，是否显示列脚，默认值为true
ShowListHeader	bool类型的属性，是否显示列头，默认值为true

表12.6　LongListSelector控件常用事件

名　称	说　明

Link	当查找的内容被找到时，触发的事件；用法示例：
	`selector.Link += new EventHandler<LinkUnlinkEventArgs>(selector_Link);`
	`void selector_Link(object sender, LinkUnlinkEventArgs e) {...}`

Unlink	查找的内容没有被找到时，触发的事件；用法示例：
	`selector.Unlink += new EventHandler<LinkUnlinkEventArgs>(selector_Unlink);`
	`void selector_Unlink(object sender, LinkUnlinkEventArgs e) {... }`

SelectionChanged	选择的选项改变时触发的事件；用法示例：
	`selector.SelectionChanged += new SelectionChangedEventHandler(selector_SelectionChanged);`
	`void selector_SelectionChanged(object sender, SelectionChangedEventArgs e) {... }`

ScrollingCompleted	当列表滚动结束的时候触发的事件；用法示例：
	`selector.ScrollingCompleted += new EventHandler(selector_ScrollingCompleted);`
	`void selector_ScrollingCompleted(object sender, EventArgs e) {...}`

ScrollingStarted	当列表滚动开始的时候触发的事件；用法示例：
	`selector.ScrollingStarted += new EventHandler(selector_ScrollingStarted);`
	`void selector_ScrollingStarted(object sender, EventArgs e){...}`

下面给出列表选择框的示例：演示如何使用LongListSelector控件进行列表信息分类（注意：源代码里面给出7.1和8.0两种实现方式）。

代码清单12-6：列表选择框(源代码：第12章\Examples_12_6)

MainPage.xaml文件主要代码

```
<phone:PhoneApplicationPage.Resources>
    <!--定义组头绑定模板-->
    <DataTemplate x:Key="GroupHeader">
        <Border Background="{StaticResource PhoneAccentBrush}" Margin=
```

```
"{StaticResource PhoneTouchTargetOverhang}" Padding = "{StaticResource PhoneTouchTargetOverhang}">
                <TextBlock Text = "{Binding Key}"/>
            </Border>
        </DataTemplate>
        <!-- 定义组选项绑定模板 -->
        <DataTemplate x:Key = "GroupItem">
                < Border Background = " {StaticResource PhoneAccentBrush}" Margin =
"{StaticResource PhoneTouchTargetOverhang}" Padding = "{StaticResource PhoneTouchTargetOverhang}">
                <TextBlock Text = "{Binding Key}" Style = "{StaticResource PhoneTextLargeStyle}"/>
            </Border>
        </DataTemplate>
        <!-- 定义列头绑定模板 -->
        <DataTemplate x:Key = "ListHeader">
            <TextBlock Text = "Header" Style = "{StaticResource PhoneTextTitle1Style}"/>
        </DataTemplate>
         <!-- 定义列表选项绑定模板 -->
        <DataTemplate x:Key = "ItemTmpl">
            <Grid>
                <TextBlock Text = "{Binding Title}"></TextBlock>
            </Grid>
        </DataTemplate>
    </phone:PhoneApplicationPage.Resources>
    …
        <!-- 添加 LongListSelector 控件 -->
        <Grid x:Name = "ContentPanel" Grid.Row = "1" Margin = "12,0,12,0">
            <toolkit:LongListSelector x:Name = "LongList" Background = "Transparent"
                    ItemTemplate = "{StaticResource ItemTmpl}"
                    ListHeaderTemplate = "{StaticResource ListHeader}"
                    GroupHeaderTemplate = "{StaticResource GroupHeader}"
                    GroupItemTemplate = "{StaticResource GroupItem}" >
            </toolkit:LongListSelector>
        </Grid>
    </Grid>
</phone:PhoneApplicationPage>
```

MainPage.xaml.cs 文件代码

```
using System;
using System.Collections.Generic;
using System.Linq;
using System.Windows;
using System.Windows.Controls;
using Microsoft.Phone.Controls;
namespace LongListSelectorDemo
{
    public partial class MainPage : PhoneApplicationPage
    {
        public MainPage()
        {
            InitializeComponent();
```

```
            //使用 List<T>来初始化数据
            List<Item> mainItem = new List<Item>();
            for (int i = 0; i < 10; i++)
            {
                mainItem.Add(new Item() { Content = "A类别", Title = "测试A" + i.ToString() });
                mainItem.Add(new Item() { Content = "B类别", Title = "测试B" + i.ToString() });
                mainItem.Add(new Item() { Content = "C类别", Title = "测试C" + i.ToString() });
            }
            //使用 Linq 来查询 List<Item>数据按照 Content 来进行分组
            var selected = from c in mainItem group c by c.Content into n select new
GroupingLayer<string, Item>(n);
            this.LongList.ItemsSource = selected;

        }
        //继承 Linq 的 IGrouping 接口来存储分组的数据
        public class GroupingLayer<TKey, TElement> : IGrouping<TKey, TElement>
        {
            //分组数据
            private readonly IGrouping<TKey, TElement> grouping;
            //初始化
            public GroupingLayer(IGrouping<TKey, TElement> unit)
            {
                grouping = unit;
            }
            //唯一的键值
            public TKey Key
            {
                get { return grouping.Key; }
            }
            //重载判断相等方法
            public override bool Equals(object obj)
            {
                GroupingLayer<TKey, TElement> that = obj as GroupingLayer<TKey, TElement>;
                return (that != null) && (this.Key.Equals(that.Key));
            }
            public IEnumerator<TElement> GetEnumerator()
            {
                return grouping.GetEnumerator();
            }
            System.Collections.IEnumerator System.Collections.IEnumerable.GetEnumerator()
            {
                return grouping.GetEnumerator();
            }
        }
        //List 选项的类 Content 表示类别 Title 表示选项的标题
        public class Item
        {
            public string Title { get; set; }
            public string Content { get; set; }
```

```
            }
        }
    }
```

程序的运行效果如图 12.17 和图 12.18 所示。

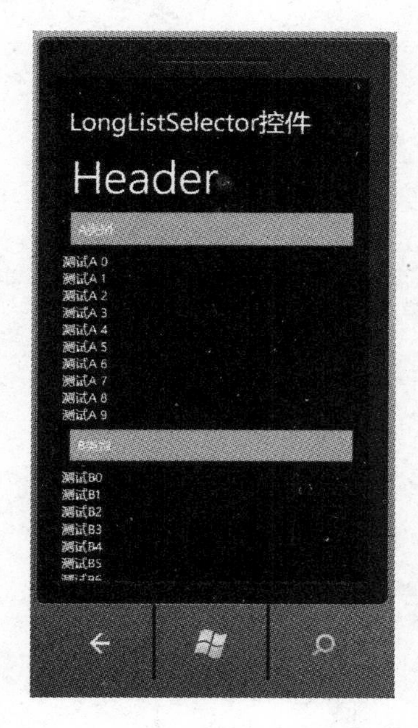

图 12.17 LongListSelector 控件效果

图 12.18 LongListSelector 控件类别选择

12.7 页面转换(Page Transitions)

页面转换(Page Transitions)提供了页面之间转换的一些动画的效果,丰富了 Windows Phone 应用程序的用户体验。Windows Phone 手机有种动画的变动方向 forward in(向前进),forward out(向前出),backward in(向后进),backward out(向后出)。当设置了进入页面的动画的时候,单击后退按钮页面会按照进入页面的动画逆向返回。Toolkit 里面定义动画的类型有 turnstile(轴旋转效果)、turnstile feather (羽毛式轴旋转效果)、continuum(继承动画效果)、slide(滑动效果)、rotate(旋转效果)等。

在应用程序中使用这些页面转换效果很简单,只需要在 XAML 页面上定义上这些页面转换的效果就行了。

使用 Turnstile 轴旋转的效果的 XAML 语法:

```
<toolkit:TransitionService.NavigationInTransition>
    <toolkit:NavigationInTransition>
        <toolkit:NavigationInTransition.Backward>
            <toolkit:TurnstileTransition Mode="BackwardIn"/>
        </toolkit:NavigationInTransition.Backward>
```

```
            <toolkit:NavigationInTransition.Forward>
                <toolkit:TurnstileTransition Mode = "ForwardIn"/>
            </toolkit:NavigationInTransition.Forward>
        </toolkit:NavigationInTransition>
    </toolkit:TransitionService.NavigationInTransition>
    <toolkit:TransitionService.NavigationOutTransition>
        <toolkit:NavigationOutTransition>
            <toolkit:NavigationOutTransition.Backward>
                <toolkit:TurnstileTransition Mode = "BackwardOut"/>
            </toolkit:NavigationOutTransition.Backward>
            <toolkit:NavigationOutTransition.Forward>
                <toolkit:TurnstileTransition Mode = "ForwardOut"/>
            </toolkit:NavigationOutTransition.Forward>
        </toolkit:NavigationOutTransition>
    </toolkit:TransitionService.NavigationOutTransition>
```

使用 Slide 滑动的效果的 XAML 语法如下：

```
<toolkit:TransitionService.NavigationInTransition>
    <toolkit:NavigationInTransition>
        <toolkit:NavigationInTransition.Backward>
            <toolkit:SlideTransition Mode = "SlideRightFadeIn"/>
        </toolkit:NavigationInTransition.Backward>
        <toolkit:NavigationInTransition.Forward>
            <toolkit:SlideTransition Mode = "SlideLeftFadeIn"/>
        </toolkit:NavigationInTransition.Forward>
    </toolkit:NavigationInTransition>
</toolkit:TransitionService.NavigationInTransition>
<toolkit:TransitionService.NavigationOutTransition>
    <toolkit:NavigationOutTransition>
        <toolkit:NavigationOutTransition.Backward>
            <toolkit:SlideTransition Mode = "SlideRightFadeOut"/>
        </toolkit:NavigationOutTransition.Backward>
        <toolkit:NavigationOutTransition.Forward>
            <toolkit:SlideTransition Mode = "SlideLeftFadeOut"/>
        </toolkit:NavigationOutTransition.Forward>
    </toolkit:NavigationOutTransition>
</toolkit:TransitionService.NavigationOutTransition>
```

除了使用组件里面预定义好的页面转换效果，还可以选择自己自定义页面转换的效果，下面介绍自定义页面转换的效果的步骤。

1. 定义一个继承 ITransition 的转换类

ITransition 接口是 transition framework 一个很重要的接口，所有控制着转换的效果的类都需要继承这个接口，如果你要定义一个转换效果的类，那么你就需要继承这个接口，然后就可以利重载 ITransition 接口里面的方法来控制动画的过程和效果。ITransition 接口里面的方法如表 12.7 所示。

表 12.7 ITransition 接口的方法说明

方　法	说　明
Begin()	初始化 ITransition
GetCurrentState()	获取当前转换动画的 ClockState
GetCurrentTime()	获取当前转换动画的时间
Pause()	暂定转换动画
Resume()	恢复转换动画
Seek()	查找下一个动画
SeekAlignedToLastTick()	快速开始
SkipToFill()	快速跳转到动画结束
Stop()	停止转换动画

2. 定义一个继承 TransitionElement 类的具体的效果转换类

在效果转换类中重写 public override ITransition GetTransition(UIElement element)方法，并在这个方法里面你要实现你自己自定义的动画效果。

3. 调用你自定义的转换效果

在 XAML 页面上引入你在第二步里面定义好的效果转换类的空间 xmlns:local="clr-namespace:空间名"，并在 XAML 页面中调用自定义的转换效果。

XAML 的语法如下：

```
<toolkit:TransitionService.NavigationOutTransition>
    <toolkit:NavigationOutTransition>
      <toolkit:NavigationOutTransition.Forward>
          <local:第二步定义的效果类>
          </local:第二步定义的效果类>
      </toolkit:NavigationOutTransition.Forward>
  </toolkit:NavigationOutTransition>
</toolkit:TransitionService.NavigationOutTransition>
```

下面给出自定义页面转换的示例：实现按钮控件的渐渐消失以及页面的渐渐消失然后到达第二页面。

代码清单 12-7：自定义页面转换(源代码：第 12 章\Examples_12_7)

MainPage.xaml 文件主要代码

```
<phone:PhoneApplicationPage
    ...
    xmlns:toolkit = "clr-namespace:Microsoft.Phone.Controls;assembly = Microsoft.Phone.
Controls.Toolkit"
    xmlns:local = "clr-namespace:WP7TransitionsDemo">
    <!--引入页面离开的时候转换效果处理类-->
    <toolkit:TransitionService.NavigationOutTransition>
        <toolkit:NavigationOutTransition>
            <toolkit:NavigationOutTransition.Forward>
                <local:TransitionPage>
```

```
                </local:TransitionPage>
            </toolkit:NavigationOutTransition.Forward>
        </toolkit:NavigationOutTransition>
</toolkit:TransitionService.NavigationOutTransition>
…
        <!--通过按钮的单击事件处理来实现页面转换的效果-->
        <StackPanel x:Name="ContentPanel" Grid.Row="1" Margin="12,100,12,0">
            <Button x:Name="btnTransition2" Content="TransitionButton" Height="100"
Click="btnTransition2_Click"/>
            <Button x:Name="btnTransition3" Content="TransitionPage" Height="100" Click
="btnTransition3_Click"/>
        </StackPanel>
    …
</phone:PhoneApplicationPage>
```

MainPage.xaml.cs 文件代码

```
using System;
using System.Windows;
using Microsoft.Phone.Controls;
namespace WP7TransitionsDemo
{
    public partial class MainPage : PhoneApplicationPage
    {
        public MainPage()
        {
            InitializeComponent();
        }
        //按钮的转换效果
        private void btnTransition2_Click(object sender, RoutedEventArgs e)
        {
            TransitionButton transition2 = new TransitionButton();
            ITransition transition = transition2.GetTransition(this.btnTransition2);
            transition.Begin();
        }
        //页面跳转的转换效果
        private void btnTransition3_Click(object sender, RoutedEventArgs e)
        {
            //页面离开的时候将会调用XAML页面中引入的转换效果
            NavigationService.Navigate(new Uri("/Page1.xaml", UriKind.Relative));
        }
    }
}
```

CustomTransition.cs 文件代码：定义继承 ITransition 接口的转换类

```
using System;
using System.Windows.Media.Animation;
using Microsoft.Phone.Controls;
namespace WP7TransitionsDemo
{
```

```
public class CustomTransition : ITransition
{
    //利用 Storyboard 来实现动画的效果
     Storyboard storyboard;
     public CustomTransition(Storyboard sb)
     {
       storyboard = sb;
     }

    #region ITransition 接口的成员,需要重载这部分的方法
    public void Begin()
    {
        storyboard.Begin();
    }
    public event EventHandler Completed
    {
        add
        {
            storyboard.Completed += value;
        }
        remove
        {
            storyboard.Completed -= value;
        }
    }
    public ClockState GetCurrentState()
    {
        return storyboard.GetCurrentState();
    }
    public TimeSpan GetCurrentTime()
    {
        throw new NotImplementedException();
    }
    public void Pause()
    {
        storyboard.Pause();
    }
    public void Resume()
    {
        storyboard.Resume();
    }
    public void Seek(TimeSpan offset)
    {
    }
    public void SeekAlignedToLastTick(TimeSpan offset)
    {
        throw new NotImplementedException();
```

```
        }
        public void SkipToFill()
        {
            storyboard.SkipToFill();
        }
        public void Stop()
        {
            storyboard.Stop();
        }
        #endregion
    }
}
```

TransitionButton.cs 文件代码：定义第一个按钮单击的时候实现按钮转换效果的类

```
using System.Windows;
using System.Windows.Media.Animation;
using Microsoft.Phone.Controls;
namespace WP7TransitionsDemo
{
    public class TransitionButton : TransitionElement
    {
        public override ITransition GetTransition(UIElement element)
        {
            //用一个故事板来实现动画的效果
            Storyboard myStoryboard = CreateStoryboard(1.0,0.0);
            //将动画效果附加到控件上
            Storyboard.SetTarget(myStoryboard, element);
            //返回一个新的 Transition 类
            return new CustomTransition(myStoryboard);
        }
        //创建一个 Storyboard 面板
        private Storyboard CreateStoryboard(double from, double to)
        {
            //新建一个故事板
            Storyboard result = new Storyboard();
            //定义一个 DoubleAnimation 动画效果
            DoubleAnimation animation = new DoubleAnimation();
            animation.From = from;
            animation.To = to;
            //将元素的可视性从 1 变成 0 就是慢慢隐藏掉
            Storyboard.SetTargetProperty(animation, new PropertyPath(UIElement.OpacityProperty));
            //添加动画到故事板
            result.Children.Add(animation);
            //返回故事板
            return result;
        }
    }
}
```

TransitionPage.cs 文件主要代码：定义第二个按钮单击的时候实现页面转换效果的类

```
public class TransitionPage : TransitionElement
{
    public override ITransition GetTransition(UIElement element)
    {
        Storyboard myStoryboard = CreateStoryboard(1.0, 0.0);
        Storyboard.SetTarget(myStoryboard, element);
        return new CustomTransition(myStoryboard);
    }
    //创建一个 Storyboard 面板
    private Storyboard CreateStoryboard(double from, double to)
    {
        Storyboard result = new Storyboard();
        //定义一个 DoubleAnimation 动画效果
        DoubleAnimation animation = new DoubleAnimation();
        animation.From = from;
        animation.To = to;
        //将元素的可视性从 1 变成 0 就是慢慢隐藏掉
        Storyboard.SetTargetProperty(animation, new PropertyPath(UIElement.OpacityProperty));
        result.Children.Add(animation);
        return result;
    }
}
```

程序运行的效果如图 12.19 所示。

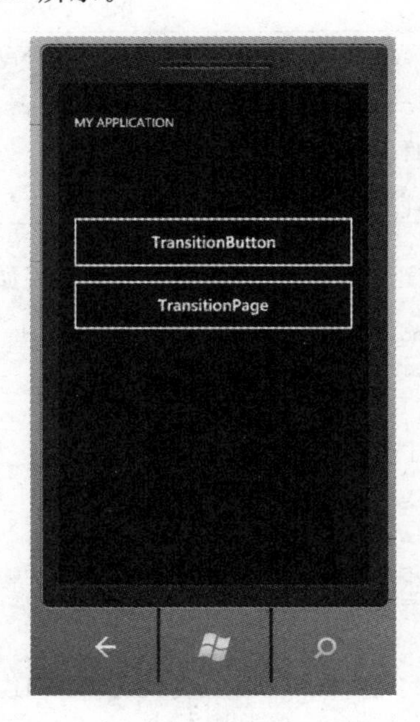

图 12.19　页面转换的效果测试界面

12.8 性能进度条(PerformanceProgressBar)

性能进度条(PerformanceProgressBar)是一种高性能的进度条，是 ProgressBar 进度条的升级版。PerformanceProgressBar 通过属性 IsIndeterminate 来控制进度条是否运行，true 表示运行，false 表示关闭。实现的效果如图 12.20 所示。

图 12.20 性能进度条的显示效果

PerformanceProgressBar 的 XAML 语法如下：

```
<toolkit:PerformanceProgressBar x:Name = "performanceProgressBar"/>
```

12.9 倾斜效果(TiltEffect)

倾斜效果(TiltEffect)是一种触摸反映出来的用户体验的效果，使用这种效果的时候，可以使单击控件时产生使控件动一下的效果，表示已经单击了该控件。在 Windows Phone 控件里面使用 TiltEffect 效果很简单，只需要添加 toolkit：TiltEffect. IsTiltEnabled = "True"到控件里面即可。TiltEffect 的常用属性如表 12.8 所示。

表 12.8 TiltEffect 控件的常用属性

名　称	说　明
IsTiltEnabled	true 表示包含 TiltEffect 的控件里面以及其下级的所有的控件都能产生效果
SuppressTilt	true 表示在一个控件里面的子控件除了它本身之外其他的都生效
TiltStrength	表示倾斜的程度
PressStrength	表示压下去的深度

在 Button 控件里面添加 TiltEffect 效果的 XAML 语法如下：

```
<Button Content = "Button" toolkit:TiltEffect.IsTiltEnabled = "True"/>
            <CheckBox Content = "CheckBox"/>
            <TextBlock Text = "ListBox:"/>
```

在 ListBox 控件里面添加 TiltEffect 效果的 XAML 语法如下：

```
<ListBox toolkit:TiltEffect.IsTiltEnabled = "True">
       <ListBoxItem Content = "Item1"/>
       <ListBoxItem Content = "Item2" toolkit:TiltEffect.SuppressTilt = "True"/>
       <ListBoxItem Content = "Item3"/>
</ListBox>
```

在 ContextMenu 控件里面添加 TiltEffect 效果的 XAML 语法如下：

```
<Button Content = "ContextMenu">
       <toolkit:ContextMenuService.ContextMenu>
          <toolkit:ContextMenu >
```

```
                <toolkit:MenuItem toolkit:TiltEffect.IsTiltEnabled = "true" Name = "item1"
Header = "Item1" />
                <toolkit:MenuItem toolkit:TiltEffect.IsTiltEnabled = "true" Name = "item2"
Header = "Item2"  />
            </toolkit:ContextMenu>
        </toolkit:ContextMenuService.ContextMenu>
    </Button>
```

12.10 时间采集器(TimePicker)

时间采集器表示一个允许用户选择时间的控件。时间采集器和日期采集器(DatePicker)相类似,使用的方法和实现的效果也差不多,只是一个显示的是时间,一个显示的是日期,它们的格式表示如表 12.9 所示。TimePicker 在界面的显示效果如图 12.21 所示,单击后会弹出一个全屏的时间选择界面效果如图 12.22 所示。TimePicker 控件的属性如表 12.10 所示。

图 12.21 TimePicker 的显示效果

图 12.22 TimePicker 的打开效果

表 12.9 日期时间格式表示

描 述	XAML	例 子
简洁日期格式 (DatePicker 默认的)	"{}{0:d}"	9/20/2010
简洁时间格式(TimePicker 默认的)	"{}{0:t}"	4:00 PM
长日期格式	"{}{0:D}"	Monday, September 20, 2010
长时间格式	"{}{0:T}"	4:00:00 PM

续表

描　　述	XAML	例　　子
自定义的日期格式	"{}{0:MM-dd-yyyy}"	09-20-2010
自定义的时间格式	"{}时间是{0:HH:mm}."	时间是 16:00

表 12.10　TimePicker 控件的常用属性

名　　称	说　　明
Header	控件的标题显示
PickerPageUri	这个属性的值是一个 URL 的地址，指向扩展 DatePicker 的控件，当要扩展 TimePicker 控件的时候，需要继承 IDateTimePickerPage 接口
Value	设置和获取时间的值，DateTime 类型
ValueString	获取时间的字符串值
ValueStringFormat	日期格式化

TimePicker 控件的 XAML 语法：

```
<toolkit:TimePicker x:Name = "timePicker" Header = " TimePicker 时间选择控件" />
```

下面给出时间采集器的示例：实现一个原生态的 TimePicker 控件以及一个扩展了时间选择页面的 TimePicker 控件。

代码清单 12-8：时间采集器（源代码：第 12 章\Examples_12_8）

MainPage.xaml 文件主要代码

```
<TextBlock Grid.Row = "1" x:Name = "CurrentTimeTextBlock" Text = "[CURRENT TIME]"
VerticalAlignment = "Center" Margin = "{StaticResource PhoneHorizontalMargin}"/>
<toolkit:TimePicker x:Name = "timepicker" Grid.Row = "3" Header = "当前时间 Time Picker"
VerticalAlignment = "Center"/>
<toolkit:TimePicker Grid.Row = "5" Value = "{x:Null}" Header = "自定义 Time Picker 选择页面"
PickerPageUri = "/CustomDateTimePickerPage; component/CustomTimePickerPage. xaml "
VerticalAlignment = "Center"/>
```

MainPage.xaml.cs 文件主要代码

```
public partial class MainPage : PhoneApplicationPage
{
    public MainPage()
    {
        InitializeComponent();
        //使用一个定时器每 0.5 秒更新一下当前的时间
        var timer = new DispatcherTimer { Interval = TimeSpan.FromSeconds(0.5) };
        timer.Tick += new EventHandler(HandleTimerTick);
        timer.Start();
        HandleTimerTick(null, null);
    }
    private void HandleTimerTick(object sender, EventArgs e)
    {
```

```
            var now = DateTime.Now;
            CurrentTimeTextBlock.Text = now.ToString("d") + " " + now.ToString("T");
            timepicker.Value = now;
        }
    }
```

CustomTimePickerPage.xaml 文件主要代码：自定义时间选择页面

```
<TextBlock Text="请选择时间" Margin="12" FontWeight="Bold"/>
<StackPanel Grid.Row="1" VerticalAlignment="Center">
    <!-- 选择时间按钮,选择后时间的值将返回到控件里面 -->
    <Button Content="30 分钟后" Click="PlusThirtyMinutesButtonClick"/>
    <Button Content="1 小时后" Click="PlusOneHourButtonClick"/>
    <Button Content="2 小时后" Click="PlusTwoHoursButtonClick"/>
    <!-- 单击取消将退出时间选择页面 -->
    <Button Content="[取消]" Click="CancelButtonClick"/>
</StackPanel>
```

CustomTimePickerPage.xaml.cs 文件主要代码

```
using System;
using System.Windows;
using Microsoft.Phone.Controls;
using Microsoft.Phone.Controls.Primitives;
namespace CustomDateTimePickerPage
{
    //需要继承 IDateTimePickerPage 接口
    public partial class CustomTimePickerPage : PhoneApplicationPage, IDateTimePickerPage
    {
        //重定义了 IDateTimePickerPage 接口的返回值
        public DateTime? Value { get; set; }
        public CustomTimePickerPage()
        {
            InitializeComponent();
        }
        //30 分钟后
        private void PlusThirtyMinutesButtonClick(object sender, RoutedEventArgs e)
        {
            ChooseTime(DateTime.Now.AddMinutes(30));
        }
        //1 小时后
        private void PlusOneHourButtonClick(object sender, RoutedEventArgs e)
        {
            ChooseTime(DateTime.Now.AddHours(1));
        }
        //2 小时后
        private void PlusTwoHoursButtonClick(object sender, RoutedEventArgs e)
        {
            ChooseTime(DateTime.Now.AddHours(2));
        }
        //提交选择的时间并且关闭当前的时间选择页面
```

```
            private void ChooseTime(DateTime time)
            {
                Value = time;
                NavigationService.GoBack();
            }
            //取消选择并且关闭当前的时间选择页面
            private void CancelButtonClick(object sender, RoutedEventArgs e)
            {
                Value = null;
                NavigationService.GoBack();
            }
        }
    }
```

程序运行的效果如图 12.23 和图 12.24 所示。

图 12.23　TimePicker 控件

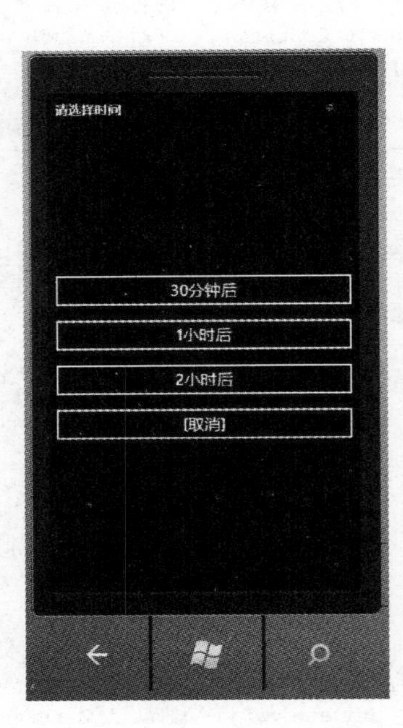

图 12.24　自定义时间选择页面

12.11　棒形开关(ToggleSwitch)

棒形开关(ToggleSwitch)是一个开关控件，ToggleSwitch 控件在 Windows Phone 手机里面很常见，比如闹钟开关，系统设置里面的开关等。ToggleSwitch 控件的界面效果如图 12.25 所示。

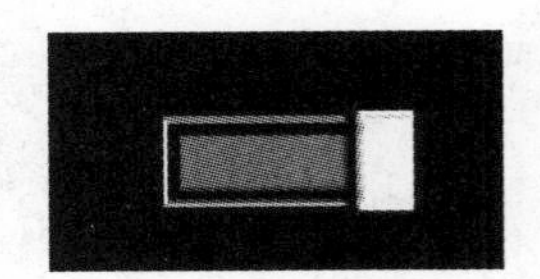

图 12.25　开关的效果

ToggleSwitch 控件的 XAML 语法如下：

```
<toolkit:ToggleSwitch x:Name = "toggle" Content = "ToggleSwitch is on" IsChecked = "True"  Header = "ToggleSwitch"/>
```

12.12 折叠容器(WrapPanel)

折叠容器(WrapPanel)是一个自动布局的面板,该控件的作用是从左至右或从上至下依次安排位于其中的元素的位置,当元素超过该组件边缘时,它们将会被自动安排至下一行或列。WrapPanel一般用于文本布局、拾色器、图片选择等。WrapPanel是和StackPanel最相近的一个控制面板,StackPanel把其中的UI元素按行或列排列,而WrapPanel则可根据其中UI元素的尺寸和其自身可能的大小自动地把其中的UI元素排列到下一行或下一列。

WrapPanel控件的XAML语法如下:

```
<toolkit:WrapPanel ItemHeight = "100" ItemWidth = "100"  Height = "250">
</toolkit:WrapPanel>
```

WrapPanel的三个重要的属性说明如下:

(1) ItemHeight:获取或设置包含在WrapPanel组件中的每一个项目布局区域的高度。每个子元素在其中显示的宽度由子元素自己的Width及HorizontalAlignment属性确定,若子元素的宽度大于ItemWidth,WrapPanel就会自动剪掉子元素超过ItemWidth的部分。ItemWidth的默认值为NaN,在这种情况下,WrapPanel使用其中最大子元素的宽度来作为列的宽度。

(2) ItemWidth:获取或设置包含在WrapPanel组件中的每一个项目布局区域的宽度。每个子元素在其中显示的高度由子元素自己的Height及VerticalAlignment确定,若子元素的宽度大于ItemHeight,WrapPanel就会自动剪掉子元素超过ItemHeight的部分。ItemHeight的默认值为NaN,在这种情况下,WrapPanel使用其中最大子元素的高度来作为行的高度。

(3) Orientation:获取或设置子元素被安排布局的方向。Orientation的默认值为Horizontal(水平放置)。

下面给出折叠容器的示例:实现WrapPanel控件的水平排版和垂直排版以及在ListBox控件中用WrapPanel控件进行排版。

代码清单12-9:折叠容器(源代码:第12章\Examples_12_9)

MainPage.xaml文件主要代码

```
<StackPanel x:Name = "ContentPanel" Grid.Row = "1" Margin = "12,0,12,0">
        <TextBlock Text = "水平排列面板"/>
        <toolkit:WrapPanel ItemHeight = "100" ItemWidth = "100"  Height = "250">
            <Rectangle Fill = "Aqua" Height = "80" Width = "80"/>
            <Rectangle Fill = "Pink" Height = "80" Width = "80"/>
            <Rectangle Fill = "Green" Height = "80" Width = "80"/>
            <Rectangle Fill = "YellowGreen" Height = "80" Width = "80"/>
            <Rectangle Fill = "Red" Height = "80" Width = "80"/>
        </toolkit:WrapPanel>
        <TextBlock Text = "垂直排列面板"/>
```

```
        <toolkit:WrapPanel ItemHeight = "100" ItemWidth = "100" Orientation = "Vertical" Height = "250">
            <Button Background = "BlueViolet"></Button>
            <Button Background = "BlueViolet"></Button>
            <Button Background = "BlueViolet"></Button>
            <Button Background = "BlueViolet"></Button>
            <Button Background = "BlueViolet"></Button>
        </toolkit:WrapPanel>
        <TextBlock Text = "ListBox 控件选项水平排列"/>
        <ListBox>
            <ListBox.ItemsPanel>
                <ItemsPanelTemplate>
                    <toolkit:WrapPanel ItemWidth = "100" ItemHeight = "50"/>
                </ItemsPanelTemplate>
            </ListBox.ItemsPanel>
            <ListBoxItem Content = "选项 1"/>
            <ListBoxItem Content = "选项 2"/>
            <ListBoxItem Content = "选项 3"/>
            <ListBoxItem Content = "选项 4"/>
            <ListBoxItem Content = "选项 5"/>
            <ListBoxItem Content = "选项 6"/>
        </ListBox>
</StackPanel>
```

程序运行的效果如图 12.26 所示。

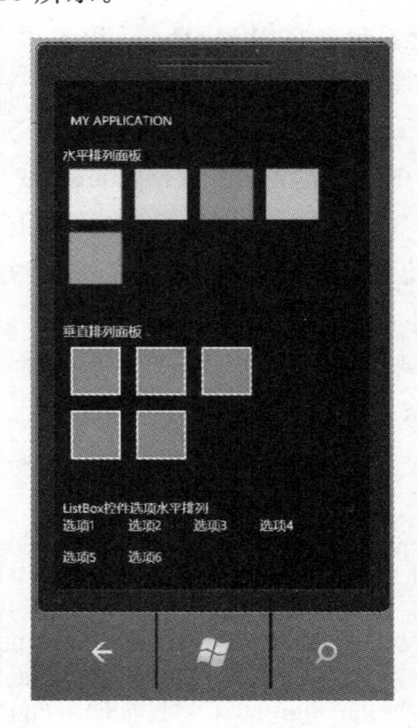

图 12.26　WrapPanel 控件运行的各种效果

第13章

网络编程

Windows Phone 操作系统有出色的网络编程的功能，它本身也融入了很多互联网的元素，例如内置的 msn、Xbox Live 等服务都是与互联网紧密联系在一起的，在 Windows Phone 以后的发展方向中会更加偏重于往互联网手机的方向发展，所以 Windows Phone 的网络编程的功能随着操作系统的发展将会越来越强大。

13.1 HTTP 协议网络编程

HTTP 协议是 HyperText Transfer Protocol（超文本传送协议）的缩写，它是万维网（World Wide Web，简称为 WWW 或 Web）的基础，也是手机联网常用的协议之一，HTTP 协议是建立在 TCP 协议之上的一种应用。

在浏览器的地址栏里输入的网站地址叫做 URL（UniformResourceLocator，统一资源定位符）。就像每家每户都有一个门牌地址一样，每个网页也都有一个 Internet 地址。当在浏览器的地址框中输入一个 URL 或是单击一个超级链接时，URL 就确定了要浏览的地址。浏览器通过超文本传输协议（HTTP），将 Web 服务器上站点的网页代码提取出来，并呈现出客户端需要的网页。

HTTP 连接最显著的特点是客户端发送的每次请求都需要服务器回送响应，在请求结束后，会主动释放连接。从建立连接到关闭连接的过程称为“一次连接”。

由于 HTTP 在每次请求结束后都会主动释放连接，因此 HTTP 连接是一种“短连接”，要保持客户端程序的在线状态，需要不断地向服务器发起连接请求。通常的做法是即使不需要获得任何数据，客户端也保持每隔一段固定的时间向服务器发送一次“保持连接”的请求，服务器在收到该请求后对客户端进行回复，表明知道客户端“在线”。若服务器长时间无法收到客户端的请求，则认为客户端“下线”，若客户端长时间无法收到服务器的回复，则认为网络已经断开。

13.1.1 WebClient 类和 HttpWebRequest 类

1. WebClient 类

WebClient 类在 System. Net 空间下，提供向 URI 标识的资源发送数据和从 URI 标识的资源接收数据的公共方法。WebClient 类提供向 URI（支持以“http:”、“https:”、“ftp:”和“file:”方案标识符开头的 URI）标识的任何本地、Intranet 或 Internet 资源发送数据以及从这些资源接收数据的公共方法。WebClient 类使用 WebRequest 类提供对资源的访问，WebClient 实例可以通过任何已向 WebRequest.RegisterPrefix 方法注册的 WebRequest 子代码访问数据。WebClient 类的一些常用的方法、属性和事件分别如表 13.1～表 13.3 所示。

表 13.1 WebClient 类常用的方法

名称	说明
CancelAsync	取消一个挂起的异步操作
DownloadStringAsync(Uri)	以字符串形式下载位于指定 Uri 的资源
GetWebRequest	为指定资源返回一个 WebRequest 对象
GetWebResponse	使用指定的 IAsyncResult 获取对指定 WebRequest 的 WebResponse
OpenReadAsync(Uri)	打开流向指定资源的可读流
OpenReadAsync(Uri, Object)	打开流向指定资源的可读流
OpenWriteAsync(Uri)	打开一个流以将数据写入指定的资源，此方法不会阻止调用线程
OpenWriteAsync(Uri, String)	打开一个流以将数据写入指定的资源，此方法不会阻止调用线程
OpenWriteAsync(Uri, String, Object)	打开一个流以使用指定的方法向指定的资源写入数据，此方法不会阻止调用线程
UploadStringAsync(Uri, String)	将指定的字符串上载到指定的资源，此方法不会阻止调用线程
UploadStringAsync(Uri, String, String)	将指定的字符串上载到指定的资源，此方法不会阻止调用线程
UploadStringAsync(Uri, String, String, Object)	将指定的字符串上载到指定的资源，此方法不会阻止调用线程

表 13.2 WebClient 类常用的常用属性

名称	说明
AllowReadStreamBuffering	获取或设置一个值，该值指示是否对从某一 WebClient 实例的 Internet 资源读取的数据进行缓冲处理
AllowWriteStreamBuffering	获取或设置一个值，该值指示是否对写入到 WebClient 实例的 Internet 资源的数据进行缓冲处理
BaseAddress	获取或设置 WebClient 发出请求的基 URI
Credentials	获取或设置发送到主机并用于对请求进行身份验证的网络凭据
Encoding	获取和设置用于上载和下载字符串的 Encoding
Headers	获取或设置与请求关联的标头名称/值对集合
IsBusy	获取一个值，该值指示某一 Web 请求是否处于进行中

续表

名称	说明
ResponseHeaders	获取与响应关联的标头名称/值对集合
UseDefaultCredentials	获取或设置一个 Boolean 值，该值控制默认凭据是否随请求一起发送

表 13.3 WebClient 类常用的常用事件

名称	说明
DownloadProgressChanged	在异步下载操作成功传输部分或全部数据后发生
DownloadStringCompleted	在异步资源下载操作完成时发生
OpenReadCompleted	在异步资源读取操作完成时发生
OpenWriteCompleted	在打开流以将数据写入资源的异步操作完成时发生
UploadProgressChanged	在异步上载操作成功转换部分或全部数据后发生
UploadStringCompleted	在异步字符串上载操作完成时发生
WriteStreamClosed	在异步写入流操作完成时发生

2. HttpWebRequest 类

HttpWebRequest 类在 System.Net 空间下，提供 WebRequest 类的 HTTP 特定的实现。HttpWebRequest 类对 WebRequest 中定义的属性和方法提供支持，也给用户能够直接使用 HTTP 服务器交互的附加属性和方法提供支持。不要使用 HttpWebRequest 构造函数，而是使用 WebRequest.Create 方法初始化新的 HttpWebRequest 对象。如果 URI 的方案是“http:”或“https:”，则 WebRequest.Create 返回 HttpWebRequest 对象。可以使用 BeginGetResponse 和 EndGetResponse 方法对资源发出异步请求。BeginGetRequestStream 和 EndGetRequestStream 方法提供对发送数据流的异步访问。如果在访问资源时发生错误，则 HttpWebRequest 类将引发 WebException。WebException.Status 属性包含指示错误源的 WebExceptionStatus 值。HttpWebRequest 对 HTTP 协议进行了完整的封装，程序使用 HTTP 协议和服务器交互主要是进行数据的提交，通常数据的提交是通过 GET 和 POST 两种方式来完成。HttpWebRequest 类的一些常用的方法和属性分别如表 13.4 和表 13.5 所示。

表 13.4 HttpWebRequest 类常用方法

名称	说明
Abort()	取消对 Internet 资源的请求
BeginGetRequestStream(AsyncCallback, Object)	开始对用来写入数据的 Stream 对象的异步请求
BeginGetResponse(AsyncCallback, Object)	开始对 Internet 资源的异步请求
EndGetRequestStream(IAsyncResult)	结束对用于写入数据的 Stream 对象的异步请求
EndGetResponse(IAsyncResult)	结束对 Internet 资源的异步请求

表 13.5 HttpWebRequest 类常用属性

名称	说明
Accept	获取或设置 Accept HTTP 标头的值
AllowReadStreamBuffering	获取或设置一个值，该值指示是否对从 Internet 资源读取的数据进行缓冲处理
ContentType	获取或设置 Content-type HTTP 标头的值
CookieContainer	指定与 HTTP 请求相关联的 CookieCollection 对象的集合
CreatorInstance	当在子类中重写时，获取从 IWebRequestCreate 类派生的工厂对象，该类用于创建为生成对指定 URI 的请求而实例化的 WebRequest
Credentials	当在子类中被重写时，获取或设置用于对 Internet 资源请求进行身份验证的网络凭据
HaveResponse	获取一个值，该值指示是否收到了来自 Internet 资源的响应
Headers	指定构成 HTTP 标头的名称/值对的集合
Method	获取或设置请求的方法
RequestUri	获取请求的原始统一资源标识符（URI）

3. HttpWebRequest 类和 WebClient 类的区别

HttpWebRequest 是个抽象类，所以无法 new 的，需要调用 HttpWebRequest. Create()。其 Method 指定了请求类型：GET 或 POST。其请求的 URI 必须是绝对地址；其请求是异步回调方式的，从 BeginGetResponse 开始，并通过 AsyncCallback 指定回调方法。另外，WebClient 类使用基于事件的异步编程模型，在 HTTP 响应返回时引发的 WebClient 回调是在 UI 线程中调用的，因此可用于更新 UI 元素的属性，例如把 HTTP 响应中的数据绑定到 UI 的指定控件上进行显示。HttpWebRequest 是基于后台进程运行的，回调不是 UI 线程，所以不能直接对 UI 进行操作，通常使用 Dispatcher. BeginInvoke()跟界面进行通信。

下面给出获取网站标题的示例：使用了 HttpWebRequest 类和 WebClient 类两种不同的方式来获取博客园网站的标题信息。

代码清单 13-1：获取网站标题(源代码：第 13 章\Examples_13_1)

MainPage. xaml 文件主要代码

```
<phone:PhoneApplicationPage
  ...
  Loaded = "PhoneApplicationPage_Loaded">
  ...
        <Grid x:Name = "ContentPanel" Grid.Row = "1" Margin = "12,0,12,0">
            <TextBlock Height = "38" HorizontalAlignment = "Left" Margin = "12,6,0,0" Name =
"webClientTextBlock" Text = "使用 webClient 获取网页内容" VerticalAlignment = "Top" Width =
"438" />
            <TextBlock Height = "44" HorizontalAlignment = "Left" Margin = "9,266,0,0" Name =
"httpWebRequestTextBlock" Text = "使用 httpWebRequest 获取网页内容" VerticalAlignment = "Top"
Width = "438" />
            <TextBox Height = "210" HorizontalAlignment = "Left" Margin = "9,50,0,0" Name =
"textBox1" Text = "" VerticalAlignment = "Top" Width = "438" />
```

```
            <TextBox Height = "239" HorizontalAlignment = "Left" Margin = "12,316,0,0" Name =
"textBox2" Text = "" VerticalAlignment = "Top" Width = "444" />
        </Grid>
    …
</phone:PhoneApplicationPage>
```

MainPage.xaml.cs 文件主要代码

```
private void PhoneApplicationPage_Loaded(object sender, RoutedEventArgs e)
{
    DoWebClient();
    DoHttpWebRequest();
}
private void DoWebClient()
{
    WebClient webClient = new WebClient();
    //在不阻止调用线程的情况下,从资源返回数据
    webClient.OpenReadAsync(new Uri("http://www.cnblogs.com/"));
    //异步操作完成时发生
    webClient.OpenReadCompleted += new OpenReadCompletedEventHandler(webClient_
OpenReadCompleted);
}
void webClient_OpenReadCompleted(object sender, OpenReadCompletedEventArgs e)
{
    using (StreamReader reader = new StreamReader(e.Result))
        string contents = reader.ReadToEnd();
        int begin = contents.ToString().IndexOf("<title>");
        int end = contents.ToString().IndexOf("</title>");
        textBox1.Text = contents.ToString().Substring(begin + 7, end - begin - 7);
    }
}
private void DoHttpWebRequest()
{
    string url = "http://www.cnblogs.com/";
    //创建 WebRequest 类
    WebRequest request = HttpWebRequest.Create(url);
    //返回异步操作的状态
    IAsyncResult result = (IAsyncResult)request.BeginGetResponse(ResponseCallback, request);
}
private void ResponseCallback(IAsyncResult result)
{
    //获取异步操作返回的信息
    HttpWebRequest request = (HttpWebRequest)result.AsyncState;
    //结束对 Internet 资源的异步请求
    WebResponse response = request.EndGetResponse(result);
    using (Stream stream = response.GetResponseStream())
    using (StreamReader reader = new StreamReader(stream))
    {
        string contents = reader.ReadToEnd();
        int begin = contents.ToString().IndexOf("<title>");
        int end = contents.ToString().IndexOf("</title>");
```

```
        //通过呼叫 UI Thread 来改变页面的显示
        Dispatcher.BeginInvoke(() => { textBox2.Text = contents.ToString().Substring
(begin + 7, end - begin - 7); });
    }
}
```

程序运行的效果如图 13.1 所示。

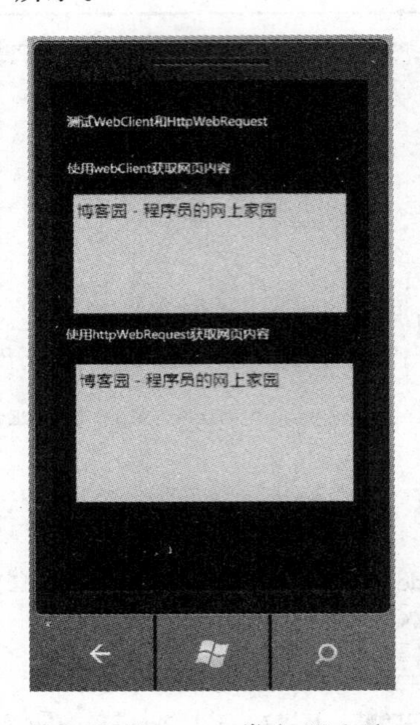

图 13.1　HttpWebRequest 类和 WebClient 类测试

13.1.2　天气预报应用

下面给出天气预报的示例：天气预报应用是通过 Http 协议异步调用 Google 天气 api (http://www.google.com/ig/api?weather=城市拼音)，对其返回的 xml 天气信息进行的数据解析，将其数据简单的展现在 Windows Phone 的客户端上。

代码清单 13-2：天气预报(源代码：第 13 章\Examples_13_2)

WeatherForecast.cs 文件代码：创建一个天气预报类来存储返回的天气预报信息，该类的属性对应接口 http://www.google.com/ig/api?weather=城市拼音返回的 XML 信息的节点类型

```
using System.ComponentModel;
namespace WeatherForecast
{
    /// <summary>
    ///天气预报绑定类
    /// </summary>
    public class ForecastPeriod : INotifyPropertyChanged
    {
```

```
private string day_of_week;     //星期
private int low;                //最低温度
private int high;               //最高温度
private string icon;            //图片地址
private string condition;       //天气情况
public event PropertyChangedEventHandler PropertyChanged;
public ForecastPeriod()
{
}
public string Day_of_week
{
    get
    {
        return day_of_week;
    }
    set
    {
        if (value != day_of_week)
        {
            this.day_of_week = value;
            NotifyPropertyChanged("Day_of_week");
        }
    }
}
public int Low
{
    get
    {
        return low;
    }
    set
    {
        if (value != low)
        {
            this.low = value;
            NotifyPropertyChanged("Low");
        }
    }
}
public int High
{
    get
    {
        return high;
    }
    set
    {
        if (value != high)
        {
            this.high = value;
```

```
                    NotifyPropertyChanged("High");
                }
            }
        }
        public string Icon
        {
            get
            {
                return icon;
            }
            set
            {
                if (value != icon)
                {
                    this.icon = value;
                    NotifyPropertyChanged("Icon");
                }
            }
        }
        public string Condition
        {
            get
            {
                return condition;
            }
            set
            {
                if (value != condition)
                {
                    this.condition = value;
                    NotifyPropertyChanged("Condition");
                }
            }
        }
        //属性改变事件
        private void NotifyPropertyChanged(string property)
        {
            if (PropertyChanged != null)
            {
                PropertyChanged(this, new PropertyChangedEventArgs(property));
            }
        }
    }
}
```

City.cs 文件代码：城市信息类，将城市的信息以面向对象的形式来存储

```
using System.ComponentModel;
namespace WeatherForecast
{
    ///<summary>
```

```
///城市绑定类
///</summary>
public class City : INotifyPropertyChanged
{
    private string cityPinyin;      //城市拼音
    private string province;        //省份
    private string cityName;        //城市名称
    public string CityPinyin
    {
        get
        {
            return cityPinyin;
        }
        set
        {
            if (value != cityPinyin)
            {
                cityPinyin = value;
                NotifyPropertyChanged("CityPinyin");
            }
        }
    }
    public string Province
    {
        get
        {
            return province;
        }
        set
        {
            if (value != province)
            {
                province = value;
                NotifyPropertyChanged("Province");
            }
        }
    }
    public string CityName
    {
        get
        {
            return cityName;
        }
        set
        {
            if (value != cityName)
            {
                cityName = value;
                NotifyPropertyChanged("CityName");
            }
```

```
            }
        }
        //定义属性改变事件
        public event PropertyChangedEventHandler PropertyChanged;
        ///<summary>
        ///构造 city 类
        ///</summary>
        public City(string cityPinyin, string province, string cityName)
        {
            CityPinyin = cityPinyin;
            Province = province;
            CityName = cityName;
        }
        ///<summary>
        ///用于绑定属性值改变触发的事件,动态改变
        ///</summary>
        private void NotifyPropertyChanged(string property)
        {
            if (PropertyChanged != null)
            {
                PropertyChanged(this, new PropertyChangedEventArgs(property));
            }
        }
    }
}
```

Cities.cs 文件代码：城市列表类，初始化需要显示天气预报的城市列表

```
using System.Collections.ObjectModel;
namespace WeatherForecast
{
    ///<summary>
    ///继承 ObservableCollection<City>用户数据绑定
    ///</summary>
    public class Cities : ObservableCollection<City>
    {
        public Cities() { }
        ///<summary>
        ///设置默认的绑定城市利用 App 类里面定义的静态变量 cityList
        ///</summary>
        public void LoadDefaultData()
        {
            App.cityList.Add(new City("Shenzhen", "广东省", "深圳市"));
            App.cityList.Add(new City("Beijing", "北京市", "北京市"));
            App.cityList.Add(new City("Shanghai", "上海市", "上海市"));
            App.cityList.Add(new City("Guangzhou", "广东省", "广州市"));
            App.cityList.Add(new City("Yangjiang", "广东省", "阳江市"));
        }
    }
}
```

Forecast.cs 文件代码：创建解析天气预报 XML 信息的天气预报类

```
using System;
using System.Net;
using System.Windows;
using System.ComponentModel;
using System.Collections.ObjectModel;
using System.IO;
using System.Linq;
using System.Xml.Linq;
namespace WeatherForecast
{
    ///< summary >
    ///天气类以及处理解析异步请求
    ///</summary >
    public class Forecast : INotifyPropertyChanged
    {
        //天气预报的城市
        private string city;
        //天气预报的时间
        private string forecast_date;
        public event PropertyChangedEventHandler PropertyChanged;
        //不同时间段的天气预报集合
        public ObservableCollection< ForecastPeriod > ForecastList
        {
            get;
            set;
        }
        public String City
        {
            get
            {
                return city;
            }
            set
            {
                if (value != city)
                {
                    city = value;
                    NotifyPropertyChanged("City");
                }
            }
        }
        public String Forecast_date
        {
            get
            {
                return forecast_date;
            }
            set
```

```
            {
                if (value != forecast_date)
                {
                    forecast_date = value;
                    NotifyPropertyChanged("Forecast_date");
                }
            }
        }
        public Forecast()
        {
            ForecastList = new ObservableCollection<ForecastPeriod>();
        }
        private void NotifyPropertyChanged(string property)
        {
            if (PropertyChanged != null)
            {
                PropertyChanged(this, new PropertyChangedEventArgs(property));
            }
        }
        ///<summary>
        ///获取 Forecast 类
        ///</summary>
        public void GetForecast(string cityPinyin)
        {
            UriBuilder fullUri = new UriBuilder("http://www.google.com/ig/api");
            fullUri.Query = "weather=" + cityPinyin;
            HttpWebRequest forecastRequest = (HttpWebRequest)WebRequest.Create(fullUri.Uri);
            ForecastUpdateState forecastState = new ForecastUpdateState();
            forecastState.AsyncRequest = forecastRequest;
            forecastRequest.BeginGetResponse(new AsyncCallback(HandleForecastResponse),
                forecastState);
        }
        ///<summary>
        ///异步获取信息
        ///</summary>
        ///<param name="asyncResult"></param>
        private void HandleForecastResponse(IAsyncResult asyncResult)
        {
            ForecastUpdateState forecastState = (ForecastUpdateState)asyncResult.AsyncState;
            HttpWebRequest forecastRequest = (HttpWebRequest)forecastState.AsyncRequest;
            forecastState.AsyncResponse = (HttpWebResponse)forecastRequest.EndGetResponse
(asyncResult);
            Stream streamResult;
            string city = "";
            string forecast_date = "";
            //创建一个临时的 ForecastPeriod 集合
            ObservableCollection<ForecastPeriod> newForecastList =
                new ObservableCollection<ForecastPeriod>();
            try
            {
```

```
                streamResult = forecastState.AsyncResponse.GetResponseStream();
                //加载 XML
                XElement xmlWeather = XElement.Load(streamResult);
                //解析 XML
                //http://www.google.com/ig/api?weather = Beijing
                //找到 forecast_information 节点获取 city 节点和 forecast_date 节点的信息
                XElement xmlCurrent = xmlWeather.Descendants("forecast_information").First();
                city = (string)(xmlCurrent.Element("city").Attribute("data"));
                forecast_date = (string)(xmlCurrent.Element("forecast_date").Attribute("data"));
                ForecastPeriod newPeriod;
                foreach (XElement curElement in xmlWeather.Descendants("forecast_conditions"))
                {
                    try
                    {
                        newPeriod = new ForecastPeriod();
                        newPeriod.Day_of_week = (string)(curElement.Element("day_of_
week").Attribute("data"));
                        newPeriod.Low = (int)(curElement.Element("low").Attribute("data"));
                        newPeriod.High = (int)(curElement.Element("high").Attribute("data"));
                        newPeriod.Icon = "http://www.google.com" + (string)(curElement.
Element("icon").Attribute("data"));
                        newPeriod.Condition = (string)(curElement.Element("condition").
Attribute("data"));
                        newForecastList.Add(newPeriod);
                    }
                    catch (FormatException)
                    {
                    }
                }
                Deployment.Current.Dispatcher.BeginInvoke(() =>
                {
                    //赋值 City Forecast_date
                    City = city;
                    Forecast_date = forecast_date;
                    ForecastList.Clear();
                    //赋值 ForecastList
                    foreach (ForecastPeriod forecastPeriod in newForecastList)
                    {
                        ForecastList.Add(forecastPeriod);
                    }
                });
            }
            catch (FormatException)
            {
                return;
            }
        }
    }
    public class ForecastUpdateState
    {
```

```
        public HttpWebRequest AsyncRequest { get; set; }
        public HttpWebResponse AsyncResponse { get; set; }
    }
}
```

MainPage.xaml 文件主要代码：城市列表显示界面

```
<Grid x:Name="ContentPanel" Grid.Row="1" Margin="12,0,12,0">
    <ListBox Height="646" HorizontalAlignment="Left" Margin="6,0,0,0" Name="CityList"
VerticalAlignment="Top" Width="474" SelectionChanged="CityList_SelectionChanged">
        <ListBox.ItemTemplate><!--数据绑定-->
            <DataTemplate>
                <StackPanel x:Name="stackPanelCityList" Orientation="Vertical" >
                    <TextBlock HorizontalAlignment="Left"
Foreground="{StaticResource PhoneForegroundBrush}" FontSize="40" Text="{Binding
CityPinyin}"/>
                    <StackPanel Orientation="Horizontal" HorizontalAlignment="Left">
                            <TextBlock Margin="0,0,10,0" FontSize="25" Text="省份:
" Foreground="LightBlue"/>
                        <TextBlock Margin="0,0,10,0" FontSize="25" Text="{Binding
Province}" Foreground="{StaticResource PhoneForegroundBrush}"/>
                            <TextBlock Margin="0,0,10,0" FontSize="25" Text="城市:
" Foreground="LightBlue"/>
                        <TextBlock Margin="0,0,10,0" FontSize="25" Text="{Binding
CityName}" Foreground="{StaticResource PhoneForegroundBrush}"/>
                    </StackPanel>
                </StackPanel>
            </DataTemplate>
        </ListBox.ItemTemplate>
    </ListBox>
</Grid>
```

MainPage.xaml.cs 文件代码

```
using System;
using System.Windows.Controls;
using Microsoft.Phone.Controls;
using System.Windows.Navigation;
namespace WeatherForecast
{
    public partial class MainPage : PhoneApplicationPage
    {
        public MainPage()
        {
            InitializeComponent();
            //绑定城市列表
            CityList.ItemsSource = App.cityList;
        }
        ///<summary>
        ///获取天气预报事件
        ///</summary>
```

```
        private void CityList_SelectionChanged(object sender, SelectionChangedEventArgs e)
        {
            //如果列被选中
            if (CityList.SelectedIndex != -1)
            {
                //获取当前选中的城市的绑定的类
                City curCity = (City)CityList.SelectedItem;
              //跳转向 ForecastPage.xaml 并传递参数 CityPinyin
                this.NavigationService.Navigate(new Uri("/ForecastPage.xaml?City=" +
                    curCity.CityPinyin, UriKind.Relative));
            }
        }
        /// <summary>
        /// 跳转到 ForecastPage.xaml 页面前执行该事件
        /// </summary>
        protected override void OnNavigatedFrom(NavigationEventArgs args)
        {
            //清空选中的列
            CityList.SelectedIndex = -1;
            CityList.SelectedItem = null;
        }
    }
}
```

App.xaml.cs 文件主要代码：初始化城市列表

```
//定义一个城市列表类的静态变量，绑定的城市列表
public static Cities cityList;
…
//在应用加载的时候进行初始化
private void Application_Launching(object sender, LaunchingEventArgs e)
{
    //创建城市列表
    if ( cityList == null)
    {
        cityList = new Cities();
        cityList.LoadDefaultData();
    }
}
```

ForecastPage.xaml 文件主要代码：天气预报信息显示页面

```
<ListBox Height="618" HorizontalAlignment="Left" Margin="0,5,0,0" Name="ForecastList"
  VerticalAlignment="Top" Width="474" Grid.RowSpan="2" SelectionChanged="ForecastList_
SelectionChanged" >
    <ListBox.ItemTemplate><!-- 数据绑定模板 -->
        <DataTemplate>
            <Grid >
                <Grid.RowDefinitions>
                    <RowDefinition Height="80"/>
                    <RowDefinition Height="80"/>
```

```
                    <RowDefinition />
                    <RowDefinition Height = " * " MinHeight = "80" />
                </Grid.RowDefinitions>
                <Grid.ColumnDefinitions>
                    <ColumnDefinition Width = "90" />
                    <ColumnDefinition Width = "70"/>
                    <ColumnDefinition Width = "180"/>
                    <ColumnDefinition Width = "90"/>
                </Grid.ColumnDefinitions>
                <TextBlock Text = "{Binding Day_of_week}" Foreground = "LightBlue" FontSize =
"40" Grid.Column = "0" Grid.Row = "0" Grid.ColumnSpan = "2"/>
                <Image Source = "{Binding Icon}" Grid.Column = "0" Grid.Row = "0"
VerticalAlignment = "Bottom" HorizontalAlignment = "Right" Grid.ColumnSpan = "2" />
                <TextBlock Text = "最低温度(K)" FontSize = "30" Foreground = "LightBlue"
Grid.Column = "0" Grid.Row = "1" Grid.ColumnSpan = "2"/>
                <TextBlock Text = "{Binding Low}" FontSize = "30" Foreground = "White" Grid.
Column = "1" Grid.Row = "1" Grid.ColumnSpan = "2" VerticalAlignment = "Bottom"
HorizontalAlignment = "Right"/>
                <TextBlock Text = "最高温度(K)" FontSize = "30" Foreground = "LightBlue"
Grid.Column = "0" Grid.Row = "2" Grid.ColumnSpan = "2"/>
                <TextBlock Text = "{Binding High}" FontSize = "30" Foreground = "White" Grid.
Column = "1" Grid.Row = "2" Grid.ColumnSpan = "2" VerticalAlignment =、"Bottom"
HorizontalAlignment = "Right"/>
                <TextBlock Text = "{Binding Condition}" FontSize = "25" Foreground = "White"
Grid.Column = "0" Grid.Row = "3" Grid.ColumnSpan = "4" TextWrapping = "Wrap"/>
            </Grid>
        </DataTemplate>
    </ListBox.ItemTemplate>
</ListBox>
```

ForecastPage.xaml.cs 文件主要代码

```
using System.Windows.Controls;
using Microsoft.Phone.Controls;
using System.Windows.Navigation;
namespace WeatherForecast
{
    public partial class ForecastPage : PhoneApplicationPage
    {
        Forecast forecast;
        public ForecastPage()
        {
            InitializeComponent();
        }
        //当该页面被链接打开时,会调用该事件
        protected override void OnNavigatedTo(NavigationEventArgs e)
        {
            //获取传过来的City值
            string cityPinyin = this.NavigationContext.QueryString["City"];
            forecast = new Forecast();
            //获取天气类
```

```
            forecast.GetForecast(cityPinyin);
            //设置页面数据绑定到 forecast
            DataContext = forecast;
            //设置 ForecastList 绑定到 forecast.ForecastList
            ForecastList.ItemsSource = forecast.ForecastList;
        }
        private void ForecastList_SelectionChanged(object sender, SelectionChangedEventArgs e)
        {
            ForecastList.SelectedIndex = -1;
            ForecastList.SelectedItem = null;
        }
    }
}
```

程序运行的效果如图 13.2 和图 13.3 所示。

图 13.2　天气预报城市列表页面

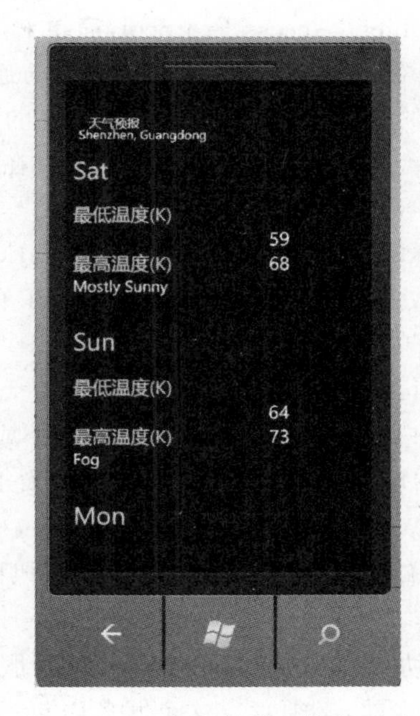

图 13.3　天气预报详细信息页面

13.2　使用 Web Service 进行网络编程

Web Service 是构建互联网分布式系统的基本部件。Web Service 正成为企业应用集成(Enterprise Application Integration)的有效平台。可以使用互联网中提供的 Web Service 构建应用程序，而不必考虑这些 Web Service 是怎样运行的。目前，互联网上有很多开发的 Web Service 接口，在开发 Windows Phone 7 手机客户端的时候可以直接来调用这些接口来实现它们提供的一些网络服务。

13.2.1 Web Service 简介

Web Service 是一种标准化的实现网络服务以及实现异构程序之间方法调用的机制，主要是为了使原来各孤立的站点之间的信息能够相互通信、共享而提出的一种接口。Web Service 也叫做 XML Web Service，是一种可以接纳从 Internet 或者 Intranet 上的其他设备中传送过去的请求，轻量级的独立的通信技术，是经过 SOAP 在 Web 上提供的软件服务，运用 WSDL 文件停止阐明，并经过 UDDI 停止注册。Web Service 所使用的是 Internet 上统一、开放的标准，如 HTTP、XML、SOAP（简单对象访问协议）、WSDL 等，所以 Web Service 可以在任何支持这些标准的环境（Windows、Linux）中使用。它通过 XML 格式的文件来描述方法、参数、调用和返回值，这种格式的 XML 文件称为 WSDL（Web Service Description Language，Web 服务描述语言）。Web Service 采用的通信协议是 SOAP（Simple Object Access Protocol，简单对象访问协议）。SOAP 协议是一个用于分散和分布式环境下网络信息交换的基于 XML 的通信协议。在此协议下，软件组件或应用程序能够通过标准的 HTTP 协议进行通信。它的设计目标就是简单性和扩展性，这有助于大量异构程序和平台之间的互操作性，从而使存在的应用程序能够被广泛的用户访问。

下面给出一些相关的术语的解释。

(1) XML：全称是 Extensible Markup Language，即扩展型可标志语言。面向短期的暂时数据处置、面向万维网络，是 SOAP 的根底。

(2) SOAP：全称是 Simple Object Access Protocol，即简单对象存取协议。是 XML Web Service 的通信协议。当用户经过 UDDI 找到 WSDL 描绘文档后，可以通过 SOAP 协议调用用户建立的 Web 服务中的一个或多个操纵。SOAP 是 XML 文档方式的调用方法的标准，它可以支撑不同的底层接口，像 HTTP(S)或 SMTP。

(3) WSDL：全称是 Web Services Description Language，即网络服务描述语言。WSDL 文件是一个 XML 文档，用于阐明一组 SOAP 音讯以及如何交流这些音讯。

(4) UDDI：全称是 Universal Description, Discovery, and Integration，即统一描述、发现和集成协议，是一个次要针对 Web 服务供应商和运用者的新项目。在用户可以调用 Web 服务之前，必须肯定这个服务内包括哪些商务办法，找到被调用的接口定义，还要在服务端来编制软件，UDDI 是一种依据描绘文档来指导设备查找相应服务的机制。UDDI 应用 SOAP 音讯机制（标准的 XML/HTTP）来公布、编辑、阅读以及查找注册音讯。它采用 XML 格式来封装各种不同类型的数据，并且发送到注册核心或由注册核心前往需求的数据。

13.2.2 在 Windows Phone 应用程序中调用 Web Service

本节会用一个手机号码归属地查询例子来演示 Windows Phone 7 的应用程序如何调用 Web Service 接口。在实例中将会使用到手机号码归属地查询 Web Service 接口：http://webservice.webxml.com.cn/WebServices/MobileCodeWS.asmx。这个 Web Service 接口是 http://www.webxml.com.cn/ 网站提供一个免费的 Web Service 接口，可

以在应用程序里面使用它实现一些功能。

1）接口的方法

getMobileCodeInfo 获得国内手机号码归属地省份、地区和手机卡类型信息。

2）输入参数

mobileCode ＝ 字符串（手机号码，最少前 7 位数字），userID ＝ 字符串（商业用户 ID）免费用户为空字符串；返回数据：字符串（手机号码：省份 城市 手机卡类型）。

3）返回的信息

如传入 mobileCode ＝13763324046 则会返回以下的 XML 信息：

```
<?xml version = "1.0" encoding = "UTF - 8"?>
< string xmlns = " http://WebXml. com. cn/" > 13763324046: 广东 广州 广东移动动感地带卡
</string >
```

代码清单 13-3：手机归属地查询（源代码：第 13 章\Examples_13_3）

下面来讲解在应用中调用 Web Service 的步骤。

1．在项目中引入 Web Service 服务

创建一个 Windows Phone 的项目工程，在工程中添加 webservice 的引用，将 web service 服务加入如图 13.4 所示，这时生成了上述 web 服务在本地的一个代理。

由于.net 平台内建了对 Web Service 的支持，包括 Web Service 的构建和使用，所以在 Windows Phone 项目中你不需要其他的工具或者 SDK 就可以完成 Web Service 的开发了。

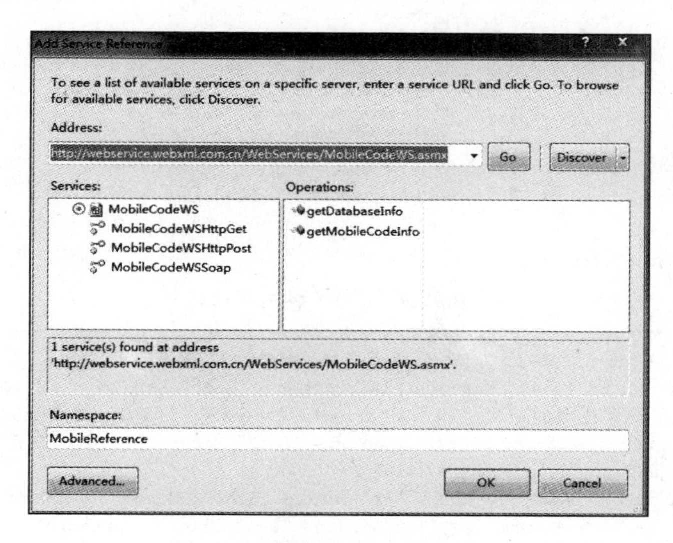

图 13.4　添加 Web Service 引用

添加 Web service 引用后，工程项目的文件目录如图 13.5 所示，可以看到在工程项目中多了 MobileReference 服务和 ServiceReferences. ClientConfig 文件。ServiceReferences. ClientConfig 文件是由编程工具自动生成的。

ServiceReferences. ClientConfig 文件代码

```
<configuration>
    <system.serviceModel>
        <bindings>
            <basicHttpBinding>
                <binding name = "MobileCodeWSSoap" maxBufferSize = "2147483647"
maxReceivedMessageSize = "2147483647">
                    <security mode = "None" />
                </binding>
            </basicHttpBinding>
        </bindings>
        <client>
            <endpoint address = "http://webservice.webxml.com.cn/WebServices/MobileCodeWS.asmx"
                binding = "basicHttpBinding" bindingConfiguration = "MobileCodeWSSoap"
                contract = "MobileReference.MobileCodeWSSoap" name = "MobileCodeWSSoap" />
        </client>
    </system.serviceModel>
</configuration>
```

图 13.5　Web Service 客户端项目工程目录

2. 调用 Web service

MainPage. xaml 文件主要代码

```
<Grid x:Name = "ContentPanel" Grid.Row = "1" Margin = "12,0,12,0">
    <TextBlock Height = "49" HorizontalAlignment = "Left" Margin = "12,66,0,0" Name = "des"
Text = "请输入你需要查询的手机号码" VerticalAlignment = "Top" Width = "284" />
    <TextBox Height = "72" HorizontalAlignment = "Left" Margin = "6,106,0,0" Name = "No" Text =
"" VerticalAlignment = "Top" Width = "415" />
    <Button Content = "查询" Height = "72" HorizontalAlignment = "Left" Margin = "12,184,0,0"
Name = "search" VerticalAlignment = "Top" Width = "160" Click = "search_Click" />
    <TextBlock Height = "211" HorizontalAlignment = "Left" Margin = "6,277,0,0" Name =
"information" Text = "" VerticalAlignment = "Top" Width = "444" />
</Grid>
```

MainPage. xaml. cs 文件代码

```
using System;
using System.Windows;
```

```
using Microsoft.Phone.Controls;
namespace WebServiceDemo
{
    public partial class MainPage : PhoneApplicationPage
    {
        public MainPage()
        {
            InitializeComponent();
        }
        private void search_Click(object sender, RoutedEventArgs e)
        {
            //实例化一个 web service 代理的对象
            MobileReference. MobileCodeWSSoapClient proxy = new MobileReference.
MobileCodeWSSoapClient();
            //getMobileCodeInfo 方法调用结束之后 触发的事件
            proxy. getMobileCodeInfoCompleted += new EventHandler < MobileReference.
getMobileCodeInfoCompletedEventArgs >(proxy_getMobileCodeInfoCompleted);
            //将调用信息包括方法名和参数加入到 soap 消息中通过 http 传送给 web service 服务端
            //这里对应的是调用了 web service 的 getMobileCodeInfo 方法
            proxy.getMobileCodeInfoAsync(No.Text, "");
        }
        void proxy_getMobileCodeInfoCompleted ( object sender, MobileReference.
getMobileCodeInfoCompletedEventArgs e)
        {
            if (e.Error == null)
            {
                //显示返回的结果
                information.Text = e.Result;
            }
        }
    }
```

程序运行的效果如图 13.6 所示。

图 13.6　手机号码查询页面

13.3 使用 WCF Service 进行网络编程

早在 Web Service 出现之前，已经有很多企业都实现了自己的 EAI（Enterprise Application Integration）企业应用整合，但此时并没有被大家所公认的技术规范，所以那时的集成方案比较分散，没有统一标准，尽管有些 EAI 做得还比较成功，苦于没有技术规范，很难得到推广。而出现 Web Service 后，由于更大厂商（包括 IBM、MS 等）的大力支持，SOAP 成为大家所公认的技术规范，很快就成为了解决这一难题的制胜法宝。MS 为响应这种变化，在.NET 平台中推出了自身的 Web Service 产品，也就是 ASP. NET Xml Web Service，但这个框架在通信安全和性能等方面存在着一些难以解决的问题，虽然后来又增加了 WSE 来弥补不足，但整体看来，ASP. NET Xml Web Service 的不足还是显而易见的。在安全方面，WSE 对 Xml Web Service 作了很大的改进，支持 WS 等网络服务的安全标准，但它作为框架的扩展，最新版本 3.0 还在 Beta 阶段，而且这个扩展框架鲜为人知，服务端如果使用了 WSE，那也要求客户端使用，而由于这个框架没有被广泛的推广，很可能给客户端开发人员增加开发和部署难度。在性能方面，.NET Remoting 技术相比 Web Service 略有优势，但学起来有一定难度，最致命的是它不能实现跨平台的操作，一个用.Net Remoting 写的 Service 很难用 Java 来调用，这就使得其实用性大打折扣，MSMQ 支持消息队列，但需要整合在其他系统之中，传统的 Enterprise Service 支持分布式事务，但同样没有 Web Service 的跨平台特性。此时，WCF 应运而生，它整合了 MS 历来最优秀的分布式系统开发技术，取其精华，弃其糟粕，是分布式应用程序开发技术的集大成者，它解决了跨平台的问题，同时支持安全通信和分布式事务。由于其简单易学，在推广上也必然比较顺利，可以说 WCF 将开辟分布式开发技术的一个新纪元，其影响应该不在当年 Web Service 之下。

13.3.1 WCF Service 简介

WCF（Windows Communication Foundation）是 Microsoft 为构建面向服务的应用提供的分布式通信编程框架，使用该框架，开发人员可以构建跨平台、安全、可靠和支持事务处理的企业级互联应用解决方案。WCF 是建立在 Microsoft .NET Framework 上类型的集合，整合了微软分布式应用程序开发中的众多成熟技术，如 Enterprise Services（COM＋）、.NET Remoting、Web Service（ASMX）、WSE 和 MSMQ 消息队列，并且存在于微软 Windows 操作系统上，在面向服务的世界和面向对象的世界里起着桥梁的作用。通常来说，与对象协作比在面向对象的世界里运行会更高效而且较低的错误，即便当这些对象发送、接受和处理面向对象消息的时候。WCF 给了用户可以在不同世界里工作的能力，但是它的目标是让用户可以在面向服务的世界里使用大家熟悉的知识编程。

（1）WCF 的通信范围：可以跨进程、跨机器、跨子网、企业网乃至于 Internet。

（2）WCF 的宿主：可以是 ASP. NET（IIS 或 WAS）、EXE、WPF、Windows Forms、NT Service、COM＋。

(3) WCF 的通信协议：TCP、HTTP、跨进程以及自定义。

下面介绍 WCF 的优点。

1) 统一性

WCF 是对于 ASMX、. NET Remoting、Enterprise Service、WSE、MSMQ 等技术的整合。由于 WCF 完全是由托管代码编写，因此开发 WCF 的应用程序与开发其他的. NET 应用程序没有太大的区别，仍然可以像创建面向对象的应用程序那样，利用 WCF 来创建面向服务的应用程序。

2) 互操作性

由于 WCF 最基本的通信机制是 SOAP，这就保证了系统之间的互操作性，即使是运行不同的上下文中。这种通信可以是基于. NET 到. NET 间的通信。可以跨进程、跨机器甚至于跨平台的通信，只要支持标准的 Web Service，例如，J2EE 应用服务器(如 WebSphere、WebLogic)。应用程序可以运行在 Windows 操作系统下，也可以运行在其他的操作系统，如 Sun Solaris、HP UNIX、Linux 等。

3) 安全与可信赖

WS Security、WS Trust 和 WS SecureConversation 均被添加到 SOAP 消息中，以用于用户认证，数据完整性验证，数据隐私等多种安全因素。在 SOAP 的 header 中增加了 WS ReliableMessaging 允许可信赖的端对端通信。而建立在 WS Coordination 和 WS AtomicTransaction 之上的基于 SOAP 格式交换的信息，则支持两阶段的事务提交(twophase commit transactions)。

上述的多种 WS Policy 在 WCF 中都给予了支持。对于 Messaging 而言，SOAP 是 Web Service 的基本协议，它包含了消息头(header)和消息体(body)。在消息头中，定义了 WS-Addressing 用于定位 SOAP 消息的地址信息，同时还包含了 MTOM(消息传输优化机制，Message Transmission Optimization Mechanism)。

4) 兼容性

WCF 充分的考虑到了与旧有系统的兼容性。安装 WCF 并不会影响原有的技术(如 ASMX 和. NET Remoting)。即使对于 WCF 和 ASMX 而言，虽然两者都使用了 SOAP，但基于 WCF 开发的应用程序，仍然可以直接与 ASMX 进行交互。

5) 易调试性

可以使用熟悉的调试方式对自承载环境中承载的 WCF 服务进行调试，而不必连接到单个应用程序来激活服务。

13.3.2 创建 WCF Service

本节演示如何搭建一个 WCF 服务，并简单地输出一个字符串。

代码清单 13-4：创建 WCF Service 服务(源代码：第 13 章\Examples_13_4)

下面来讲解在 Visual Studio 2010 中创建 WCF Service 的步骤。

1. 打开 Visual studio 2010 创建一个 WCF 服务应用程序如图 13.7 所示

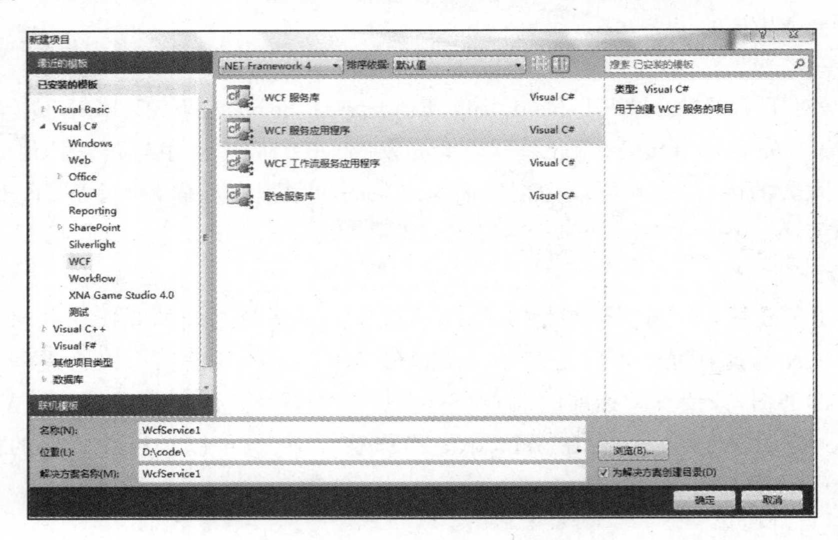

图 13.7　创建一个 WCF 项目

2. 修改 Service1.svc 和 IService1.cs 文件

创建好的项目默认的文件目录如图 13.8 所示。在 Service1.svc 中添加方法：

```
public string HelloWCF()
{
    return "Hello WCF";
}
```

在 IService1.cs 中添加接口：

```
[OperationContract]
string HelloWCF();
```

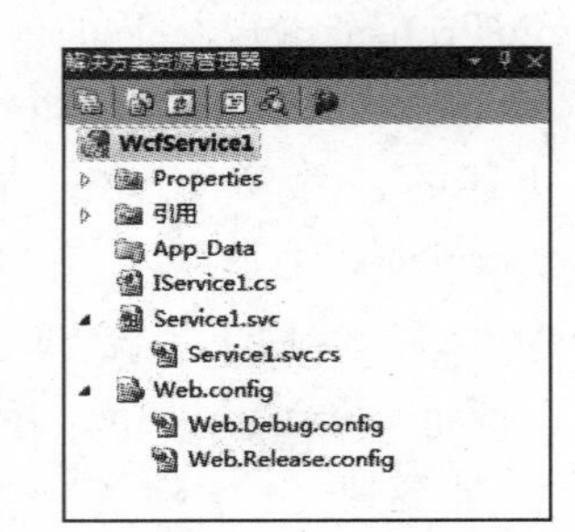

图 13.8　WCF 项目工程目录

3. 创建一个网站的虚拟目录指向 WCF 的项目工程

在控制面板中找到 Internet 信息管理器并打开，在 Default Web Site 节点下创建一个虚拟目录，命名为 wcf，路径指向本例子的 web 应用程序的代码，并单击确定，如图 13.9 所示。右键单击刚刚建好的虚拟目录 wcf，单击转换为应用程序，然后在 Windows 7 中需要设置 IIS 元数据库和 IIS 6 配置兼容性如图 13.10 所示。

4. 打开浏览器输入地址 http://localhost/wcf/Service1.svc

出现如图 13.11 所示的界面证明部署成功。

13.3.3　调用 WCF Service

本小节演示如何在 Windows Phone 的应用程序中调用 WCF 服务，代码清单 13-4：调用 WCF Service 服务(源代码：第 13 章\Examples_13_4)。步骤如下：

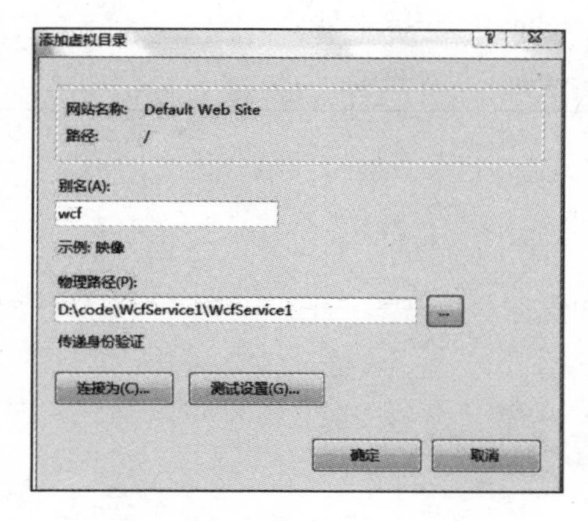

图 13.9　创建 WCF 服务程序

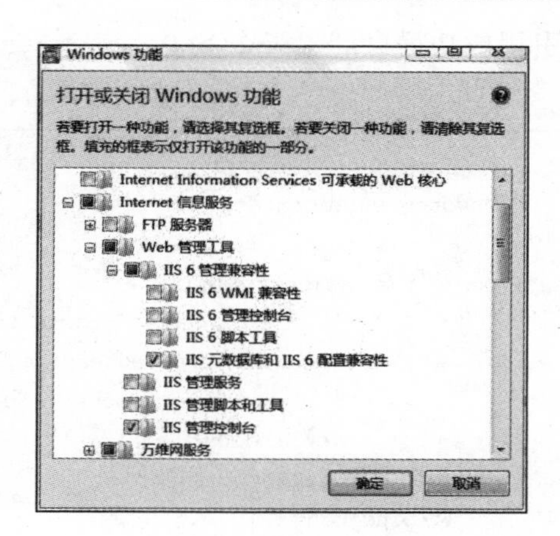

图 13.10　设置 IIS

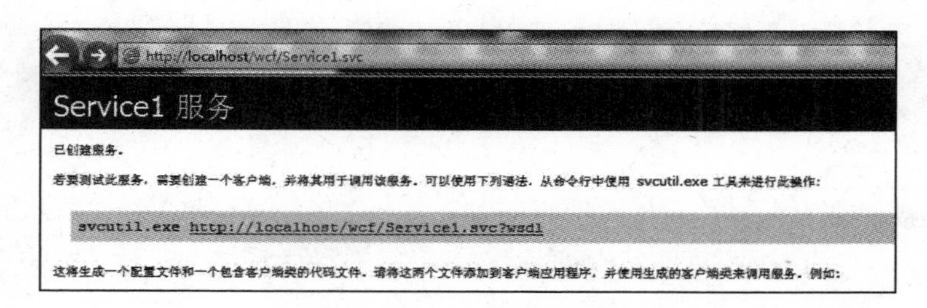

图 13.11　WCF 的网页浏览效果

1. 在 Windows Phone 项目中添加 WCF 服务引用，生成代理，如图 13.12 所示

图 13.12　在 Windows Phone 项目中引入 WCF 服务

引用成功后会产生一个 ServiceReferences. ClientConfig 文件。

ServiceReferences. ClientConfig 文件代码

```
<configuration>
    <system.serviceModel>
        <bindings>
            <basicHttpBinding>
                <binding name = "BasicHttpBinding_IService1" maxBufferSize = "2147483647"
                    maxReceivedMessageSize = "2147483647">
                    <security mode = "None" />
                </binding>
            </basicHttpBinding>
        </bindings>
        <client>
            <!-- http://localhost/wcf/Service1.svc 是指 WCF 服务的地址 -->
            <endpoint address = "http://localhost/wcf/Service1.svc" binding = "basicHttpBinding"
                bindingConfiguration = "BasicHttpBinding_IService1" contract = "WCFService.IService1"
                name = "BasicHttpBinding_IService1" />
        </client>
    </system.serviceModel>
</configuration>
```

2. 调用 WCF 服务

MainPage. xaml 文件主要代码

```
<Grid x:Name = "ContentPanel" Grid.Row = "1" Margin = "12,0,12,0">
    <TextBlock Height = "63" HorizontalAlignment = "Left" Margin = "60, 47, 0, 0" Name =
"textBlock1" Text = "" VerticalAlignment = "Top" Width = "249" />
</Grid>
```

MainPage. xaml. cs 文件主要代码

```
public MainPage()
{
    InitializeComponent();
    WCFDemo.WCFService.Service1Client proxy = new WCFService.Service1Client();
    proxy.HelloWCFCompleted += new EventHandler < WCFService.HelloWCFCompletedEventArgs >
(proxy_HelloWCFCompleted);
    proxy.HelloWCFAsync();
}
void proxy_HelloWCFCompleted(object sender, WCFService.HelloWCFCompletedEventArgs e)
{
    textBlock1.Text = e.Result.ToString();
}
```

程序运行效果如图 13.13 所示。

图 13.13 客户端运行效果

13.4 推送通知

移动互联网时代以前的手机，如果有事件发生要通知用户，会有一个窗口弹出，告诉用户正在发生什么。可能是未接电话的提示，日历的提醒，或者是一封新彩信。传统的 RSS 阅读器都是手动或以一定间隔自动抓取信息。手动需要你在想看的时候手动刷新内容，这么做的好处是省电。不足之处也很明显，麻烦，而且消息来的不及时。所以很多程序都以一定间隔自动进行刷新，比如十分钟上网抓取一次信息。这么做固然是方便了一些，但同样会带来问题，如果没有新内容，就白白耗费了不少的流量和电力。而且这个时间间隔长度本身也是一个问题，间隔太长就没有时效性，太短的话又过于费电费流量。那么推送通知就是专门为了解决这个问题的，它的原理很简单，第三方程序，比如微博客户端、聊天软件与推送通知的服务器保持连接，等有新的内容需要提供给手机之后，推送通知的服务器就会将数据推送到手机上。

推送通知是一个统一的通知服务。使用推送通知服务可以确保用户得到最新信息。很多类型的程序都可以使用这个服务。例如，体育程序可以在程序没有运行时更新关键的比赛信息。聊天程序可以显示会话中最新的回复。任务管理程序可以跟踪有多少任务用户还没有处理。推送通知服务对于移动设备可以说是一个非常搭配的技术功能。大部分重量级的运算都在服务器和通知服务器之间进行，和在后台运行的程序来说，对电池的影响更小，性能也更高。该服务会维持一个持久的 IP 连接从而在程序没有运行时也能通知用户。

13.4.1 推送通知简介

推送通知(Push Notificiation)是 Windows Phone 平台上内置特性，开发者可以利用 Windows Phone 提供的推送通知的服务，来实现网络的服务器端向手机的客户端程序推送一些通知或者消息，意思就是服务器端通过推送通知的服务主动地告诉手机客户端来了新的通知或者消息，这跟客户端调用 web service 去拉消息是两种不同原理的交互方式。

推送通知(Push Notificiation)的一个好处就是可以在应用程序没有执行的情况下，仍然可以将远端的消息传送到 Windows Phone 的客户端应用程序上，并且提醒这个消息的到来。下面来看一下推送通知的过程如图 13.14 所示。

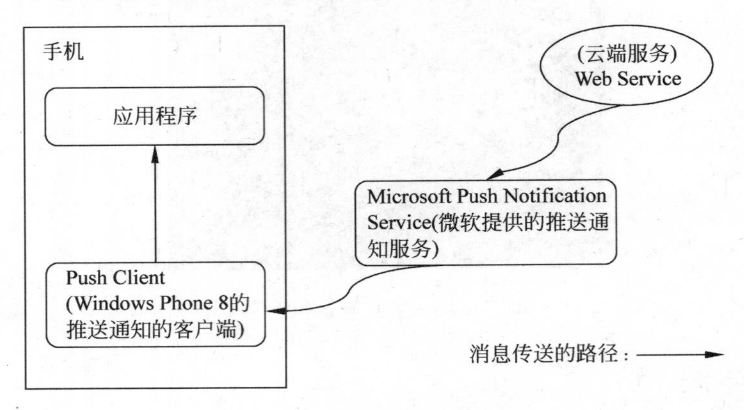

图 13.14 推送通知原理图

下面来介绍一下推送通知涉及的 3 个重要的服务。

1. Web Service（云端服务）

这是通知消息的出发点，也就是你要推送什么样的通知、什么内容的消息，就是从这里提供的。它怎么知道要推送到哪里呢？它怎么知道它的消息要传送到哪部手机哪个应用程序呢？这时候就出来了一个频道（Channel）的概念。使用推送通知的应用程序需要通过 Microsoft Push Notification Service 注册一个唯一的 Channel，然后把这个唯一的 Channel 告诉云端服务，这时候云端服务就可以将消息搭载这个唯一的 Channel，通过 Microsoft Push Notification Service 传送到手机的客户端应用程序。

2. Microsoft Push Notification Service（微软提供的推送通知服务）

这是推送通知的一个中介的角色，这是微软免费提供的一个服务，这个服务为手机客户端和服务器端的交流提供了一条特殊的通道。一种情况，它接受手机应用程序通过 Push Client 创建的 Channel 来作为整个推送通知过程的通道。另一种情况，它也接受云端服务所申请的 Service Name 来进行注册，让 Push Client 在建立 Channel 时指定云端服务所注册的 Service。

3. Push Client（Windows Phone 8 的推送通知的客户端）

这是推送通知在 Windows Phone 8 系统里面的客户端的支持，直接跟手机客户端打交道。Push Client 要取得资料的话，则需要向 Microsoft Push Notification Service 建立起独有的 Channel，因此 Push Client 会向 Microsoft Push Notification Service 送出询问是否存在指定的 Service Name 与专用的 Channel 名称。

整个推送通知的流程可以分解为下面的 4 步：

(1) Window Phone 客户端应用程序请求与微软推送通知服务建立通道连接，微软推送通知服务使用通道 URI 响应。

(2) Windows Phone 客户端应用程序向监视服务(Web Service 或 Cloud Application)发送包含推送通知服务通道 URI 以及负载的消息。

(3) 当监视服务检测到信息更改时(如航班取消、航班延期或天气警报),它会向微软推送通知服务发送消息。

(4) 微软推送通知服务将消息中继到 Windows Phone 设备,由 Window Phone 客户端应用程序处理收到的消息。

13.4.2　推送通知的分类

推送通知服务中有 3 种不同类型的通知,分别是原生通知(Raw Notification)、吐司通知(Toast Notification) 和瓷砖通知(Tile Notification)。这 3 种通知的表现形式和消息传送的格式都不一样,可以根据应用的具体情况来选择需要的通知的形式。

1. 原生通知(Raw Notification)

原生通知是一种只是针对于正在运行的应用程序而提供的通知,如果使用了原生通知的应用程序并没有运行,而服务器端又给应用程序发送了消息的情况下,那么这一条的原生通知将会被微软的推送通知服务所丢弃。原生通知一般是用于给正在运行的应用程序发送消息,比如聊天软件的好友上线通知等。

原生通知的特点:

(1) 可以发送任何格式的数据;

(2) 有效的载荷最大为 1KB;

(3) 只有在使用原生通知运行的情况下才能接收到消息;

(4) 允许在用户使用时更新用户界面。

原生通知的传送格式可以为任意的字符串格式,接收到消息的效果如图 13.15 所示。

图 13.15　原生通知(Raw Notification)运行效果

2. 吐司通知(Toast Notification)

吐司通知是一种直接在屏幕最上面弹出来的系统提示栏通知,总是显示在屏幕的最顶部,会有声音和振动的提示,十秒钟后会自动消失,当单击提示栏可以打开应用程序。例如,手机接收到新的短信的时候,在屏幕最顶端弹出来的消息就是吐司通知来的,单击进去就进去了短信的界面。吐司通知一般是用于一些比较重要的通知提示,比如短信提醒、恶劣天气提醒等。

吐司通知的特点:

(1) 发送的数据为指定的 XML 格式;

(2) 如果程序正在运行,内容发送到应用程序中;

(3) 如果程序不在运行,弹出 Toast 消息框显示消息；

(4) 会临时打断用户的操作；

(5) 消息的内容为 App 图标加上两个标题描述，标题为粗体字显示的字符串，副标题为非粗体字显示的字符串；

(6) 用户可以单击消息进行跟踪。

吐司通知的传送格式如下：

```
<?xml version = "1.0"encoding = "utf - 8"?>
    < wp:Notificationxmlns:wpwp:Notificationxmlns:wp = "WPNotification">
        < wp:Toast >
            < wp:Text1 >
                标题
                </wp:Text1 >
                < wp:Text2 >
                副标题
                </wp:Text2 >
         </wp:Toast >
</wp:Notification >
```

图 13.16 吐司通知(Toast Notification)运行的效果

接收到消息的效果如图 13.16 所示。

3. 瓷砖通知(Tile Notification)

瓷砖通知是一种针对于在启动屏幕(Start Screen)中的应用程序提供的通知，如果应用程序不在启动屏幕(Start Screen)中是不会接收到瓷砖通知通知的。瓷砖通知有 3 个元素：计数(Count)、标题(Title)和背景图像(Background)。计数(Count)是展现在图标的右上角的数字，标题(Title)是展示在图标的左下角的文本，背景图像(Background)则可以改变这个图标的背景。例如未读短信在短信图标的右上角显示就是瓷砖通知的表现形式。

瓷砖通知的特点：

(1) 发送的数据为指定的 XML 格式；

(2) 不会往应用程序进行发送；

(3) 可以改变启动屏幕(Start Screen)中的图标内容(图片，文字等)；

(4) 包含三个属性，背景，标题和计数器，每个属性都有固定的格式和位置，可以使用其中的属性，不一定三个属性一起使用。

瓷砖通知的传送格式如下：

```
<?xml version = "1.0" encoding = "utf - 8"?>
    < wp:Notificationxmlns:wpwp:Notificationxmlns:wp = "WPNotification">
        < wp:Tile >
            < wp:BackgroundImage >
              < backgroundimagepath >
              背景图像
              </wp:BackgroundImage >
              < wp:Count >
              计数
              < count >
```

```
            </wp:Count>
            <wp:Title>
            标题
            <title>
            </wp:Title>
      </wp:Tile>
</wp:Notification>
```

接收到消息的效果如图 13.17 所示。

图 13.17　瓷砖通知(Tile Notification)的运行效果

13.4.3　推送通知的实现

要在 Windows Phone 应用程序中实现推送通知的功能，需要从两个方面去实现，一个方面是 Windows Phone 应用客户端的对推送通知的调用；另一方面是 Web Service(云端服务)的设计和编码实现，通知的发送，通知的内容，通知的类型都需要在 Web Service 中去实现，然后通过微软的推送通知服务将消息传送到 Windows Phone 应用程序中去。下面通过一个实例来讲解如何实现推送通知的手机客户端和服务器端。

代码清单 13-5：推送通知(源代码：第 13 章\Examples_13_5)

实现的步骤如下。

1. 手机客户端的实现

1）应用页面的设计

MainPage.xaml 文件主要代码

```
<Grid x:Name="ContentPanel" Grid.Row="1" Margin="12,0,12,0">
    <Button Content="注册一个频道" Height="72" HorizontalAlignment="Left" Margin="101,
96,0,0" Name="button1" VerticalAlignment="Top" Width="221" Click="button1_Click" />
    <TextBlock Height="30" HorizontalAlignment="Left" Margin="9,321,0,0" Name=
"TextBlock1" Text="接收到的消息：" VerticalAlignment="Top" Width="144" />
    <TextBlock Height="30" HorizontalAlignment="Left" Margin="0,211,0,0" Name=
"textBlock2" Text="注册频道的状态：" VerticalAlignment="Top" />
    <TextBlock Height="68" HorizontalAlignment="Left" Margin="6,247,0,0" Name="state"
Text="" VerticalAlignment="Top" Width="414" />
    <TextBlock Height="216" HorizontalAlignment="Left" Margin="12,357,0,0" Name="msg"
Text="" VerticalAlignment="Top" Width="408" />
</Grid>
```

2）创建一个推送通知的频道

推送通知频道的类：HttpNotificationChannel。

HttpNotificationChannel 是一个在 Push Notification 服务和 Push Client 之间创建通知通道的类，它用来创建一个 raw，tile 和 toast 通知的订阅。该通道的创造流程是这样的：如果通道已经存在，则客户端应用程序应尝试重新打开它。试图重新创建已存在的通道将导致异常。如果通道没有打开，订阅通道事件，并尝试打开通道。一旦通道打开它会触发 ChannelUriUpdated 事件。此事件可能向客户端发送成功创建通道的信号。现有的通道可以根据名称找到，成功找到通道的情况下，通道将被重新激活，并可以在应用程序中使用。

整个过程是异步的。

MainPage. xaml. cs 文件实现创建推送通知频道的代码

```
HttpNotificationChannel httpChannel = null;
//频道的名字为 NotificationTest
string channelName = "NotificationTest";
//查找该频道是否已经存在,如果有则先将其关闭
httpChannel = HttpNotificationChannel.Find(channelName);
if (httpChannel != null)
{
    //关闭频道
    httpChannel.Close();
    //释放频道的资源
    httpChannel.Dispose();
}
//初始化一个名字为 NotificationTest 的频道
httpChannel = new HttpNotificationChannel(channelName);
```

3）订阅和实现频道注册成功事件

在注册成功事件里面,需要做的事情就是将注册成功的频道的 URL 告诉推送通知的服务器端也就是发送通知的 web service,服务端的 web service 的实现方式可以是 web 应用程序,也可以是各种方式的互联网的云服务,反正只有一个目的就是可以获取注册成功的频道的 URL。下面采用 web 应用程序的方式来接收这个 URL,这个 web 应用程序的实现在下面的推送通知的 Web Service 端的实现中会讲解。

MainPage. xaml. cs 文件订阅和实现频道注册成功事件的代码

```
//当频道注册成功的时候会触发这个事件
httpChannel. ChannelUriUpdated += new EventHandler < NotificationChannelUriEventArgs >
(httpChannel_ChannelUriUpdated);
 …
//处理频道注册成功的事件,获取频道注册成功的频道 URL,并告知云端服务该频道的具体地址
void httpChannel_ChannelUriUpdated(object sender, NotificationChannelUriEventArgs e)
{
        //当频道注册成功后会回传一个 URL,在服务器端就是通过这个 URL 要推送通知的
        WebClient webClient = new WebClient(); webClient. OpenReadAsync(new Uri("http://
localhost/pushnotification/send.aspx?url=" + e.ChannelUri));
        //异步操作完成时发生
          webClient. OpenReadCompleted += new OpenReadCompletedEventHandler(webClient_
OpenReadCompleted);
}
void webClient_OpenReadCompleted(object sender, OpenReadCompletedEventArgs e)
{
    if (e.Error != null)
    {
        state.Text = "注册失败!请检查 web 服务是否搭建成功。";
    }
    else
```

```
    {
        state.Text = "恭喜你,注册成功^_^";
    }
}
```

4）订阅和实现接收吐司通知(Toast Notification)的事件

虽然吐司通知在应用程序没有运行的情况下都会接收到提示的信息，但是当应用程序正在运行的情况下，吐司通知就不会直接在屏幕上方弹出来，这时候就需要在应用中捕获这个事件来实现消息的显示。

MainPage.xaml.cs 文件实现接收吐司通知(Toast Notification)的事件的代码

```
//接收到吐司通知时触发的 ShellToastNotificationReceived 事件
httpChannel.ShellToastNotificationReceived += new EventHandler<NotificationEventArgs>
(httpChannel_ShellToastNotificationReceived);
…
//吐司通知在应用程序里面的实现
void httpChannel_ShellToastNotificationReceived(object sender, NotificationEventArgs e)
{
    foreach (var key in e.Collection.Keys)
    {
        string pushmsg = e.Collection[key];
        Dispatcher.BeginInvoke(() =>
        {
            msg.Text += key + ": " + pushmsg + "\r\n";
        });
    }
}
```

5）订阅和实现接收原生通知(Raw Notification)的事件

原生通知是只有在应用程序运行的时候才能够接收到的，所以要接收原生通知必须要在应用程序里面实现其接收的事件。在现实接收通知中需要注意的是瓷砖通知(Tile Notification)不需要在程序里面去接收的。

MainPage.xaml.cs 文件实现接收原生通知(Raw Notification)的事件的代码

```
//接收到原生通知时触发的 HttpNotificationReceived 事件
httpChannel.HttpNotificationReceived += new EventHandler<HttpNotificationEventArgs>
(httpChannel_HttpNotificationReceived);
…
//原生通知在应用程序里面的实现
void httpChannel_HttpNotificationReceived(object sender, HttpNotificationEventArgs e)
{
    using (var reader = new StreamReader(e.Notification.Body))
    {
        string Rawmsg = reader.ReadToEnd();
        Dispatcher.BeginInvoke(() =>
        {
            msg.Text = Rawmsg;
        });
```

```
    }
}
```

6）实现处理频道的异常

MainPage. xaml. cs 文件实现处理频道的异常的代码

```
//频道的异常事件 ErrorOccurred,当频道发生异常或者错误的时候会触发该事件
httpChannel. ErrorOccurred += new EventHandler < NotificationChannelErrorEventArgs >
(httpChannel_ErrorOccurred);
…
void httpChannel_ErrorOccurred(object sender, NotificationChannelErrorEventArgs e)
{
    //发生错误提示错误信息
    Dispatcher.BeginInvoke(() =>
    {
        msg.Text = e.Message;
    });
}
```

7）打开频道并绑定推送通知的服务

MainPage. xaml. cs 文件实现打开频道并绑定推送通知的服务的代码

```
//打开频道
httpChannel.Open();
//绑定 Toast notification 推送服务
httpChannel.BindToShellToast();
//绑定 Tokens (tile) 推送服务
httpChannel.BindToShellTile();
```

程序运行的效果如图 13.18 所示。

2. 推送通知的 Web Service 端的实现

这里通过 Web 应用程序来实现通知的推送，为了简化 web 应用程序的搭建，本案例是将手机客户端传过来的 URL 保存到 Application 变量中，而没有保存到数据库里面。本案例的 web 应用程序是基于 IIS7 和 ASP. NET2. 0 的，所以在搭建环境的时候确保你已经安装了 IIS7 和 ASP. NET2. 0，下面讲解一下 Web 应用程序搭建的过程：

图 13.18　推送通知的主页面

（1）在控制面板中找到 Internet 信息管理器并打开。

（2）在 Default Web Site 节点下创建一个虚拟目录，命名为 pushNotification，路径指向本例子的 web 应用程序的代码，并单击确定，如图 13.19 所示。

（3）右键单击刚刚建好的虚拟目录 pushNotification，单击转换为应用程序。

（4）在浏览器中输入 http://localhost/pushnotification/send. aspx，若能正常显示说明环境搭建成功。

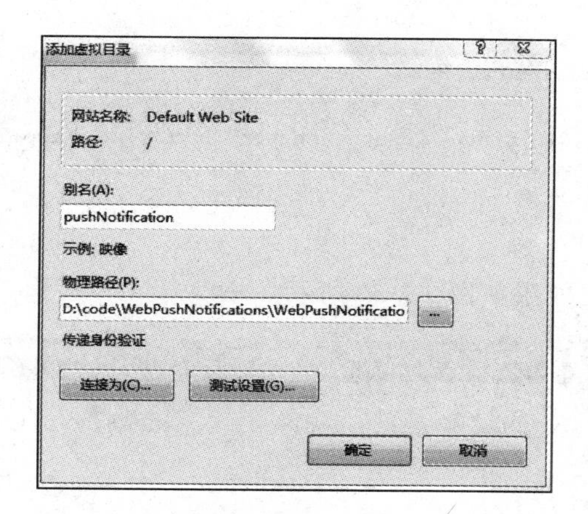

图 13.19　配置推送通知的 web 服务端

在该实例的推送通知服务器端中有 Send. aspx 和 Send. aspx. cs 两个页面，Send. aspx 是前端的网页页面，Send. aspx. cs 负责处理后台的通知推送功能。

Send. aspx 文件主要代码

```
<table style = "width: 100 % ; height: 274px;">
    <tr>
        <td class = "style2">
            URL 地址: </td>
        <td class = "style3">
            <asp:Label ID = "url" runat = "server" Text = ""></asp:Label>
        </td>
    </tr>
    <tr>
        <td class = "style1">
            请输入你要推送的消息: </td>
        <td>
            <asp:TextBox ID = "msg" runat = "server" Height = "36px" Width = "550px"></asp:TextBox>
        </td>
    </tr>
    <tr>
        <td class = "style1">
            推送的方式: </td>
        <td>
            <asp:Button ID = "SendToast" runat = "server"  Text = "发送 Toast 消息"
                Width = "113px" onclick = "SendToast_Click" />
            <asp:Button ID = "SendTile" runat = "server" Text = "发送 Tile 消息" Width = "128px"
                onclick = "SendTile_Click" />
            <asp:Button ID = "SendRaw" runat = "server" Text = "发送 Raw 消息" Width = "127px"
                onclick = "SendRaw_Click" />
        </td>
    </tr>
    <tr>
```

```
            <td class = "style1">
                发送状态：</td>
            <td>
             <asp:Label ID = "state" runat = "server" Text = ""></asp:Label>
            </td>
        </tr>
    </table>
```

Send. aspx 页面运行效果如图 13.20 所示。

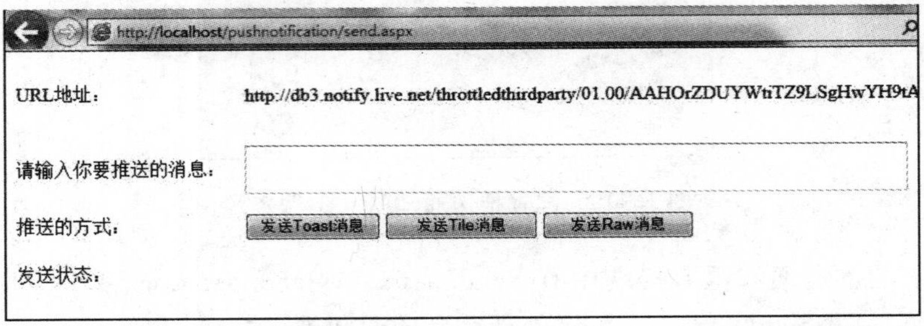

图 13.20　推送通知 web 服务端的运行效果

下面讲解在服务器端中如何实现将一个通知推送带手机客户端上的步骤。

1）创建一个 http 的 post 请求

这个 Post 请求是用来准备 Post 消息到手机应用对应的微软推送通知的 URL 地址。这个 URL 地址是在上面讲解的手机客户端的推送通知注册频道的时候产生的 URL，在实例中为了方便时保存到 Application["url"]应用程序变量中的，一般需要使用数据库来存储频道的信息，在实例中打开 send. aspx 页面时将取出来的 URL 地址放到 uil. Text 文本框中。

Send. aspx. cs 文件创建 http 的 post 请求的代码

```
HttpWebRequest sendNotificationRequest = (HttpWebRequest)WebRequest.Create(url.Text);
sendNotificationRequest.Method = WebRequestMethods.Http.Post;
```

2）准备消息头 HttpHeader

对于服务器端来说，发送不同的通知，都是以 Http 方式发出去的，但是在发送时，需要配置相应的参数，来告诉 PushNotificationService 所发送的类型是什么。

Send. aspx. cs 文件设置原生通知（Raw Notification）消息头的代码

```
//将消息头的 X - MessageID 设置成唯一的字符串
sendNotificationRequest.Headers["X - MessageID"] = Guid.NewGuid().ToString();
sendNotificationRequest.ContentType = "text/xml; charset = utf - 8";
//将 X - NotificationClass 设置为 3 表示立即发送消息，设置为 13 表示在 450 秒内传送出消息，设
//置为 23 表示在 900 秒内传送出消息
sendNotificationRequest.Headers.Add("X - NotificationClass", "3");
```

Send. aspx. cs 文件设置瓷砖消息（Tile Notification）消息头的代码

```
sendNotificationRequest.Headers["X - MessageID"] = Guid.NewGuid().ToString();
```

```
sendNotificationRequest.ContentType = "text/xml; charset = utf - 8";
//设置 X - WindowsPhone - Target 为 token 表示是瓷砖消息
sendNotificationRequest.Headers.Add("X - WindowsPhone - Target", "token");
//将 X - NotificationClass 设置为 1 表示立即发送消息,设置为 11 表示在 450 秒内传送出消息,设
//置为 21 表示在 900 秒内传送出消息
sendNotificationRequest.Headers.Add("X - NotificationClass", "1");
```

Send.aspx.cs 文件设置吐司通知(Toast Notification)消息头的代码

```
sendNotificationRequest.Headers = new WebHeaderCollection();
sendNotificationRequest.Headers["X - MessageID"] = Guid.NewGuid().ToString();
//设置 X - WindowsPhone - Target 为 toast 表示是吐司消息
sendNotificationRequest.Headers.Add("X - WindowsPhone - Target", "toast");
//将 X - NotificationClass 设置为 2 表示立即发送消息,设置为 12 表示在 450 秒内传送出消息,设
//置为 22 表示在 900 秒内传送出消息
sendNotificationRequest.Headers.Add("X - NotificationClass", "2");
```

3）创建消息的内容

Send.aspx.cs 文件创建消息的内容的代码

```
//原生通知(Raw Notification)的消息内容
byte[] strBytes = new UTF8Encoding().GetBytes(msg.Text);
…
//设置瓷砖消息(Tile Notification)消息头
string tileMessage = "<?xml version = \"1.0\" encoding = \"utf - 8\"?>" +
        "< wp:Notification xmlns:wp = \"WPNotification\">" +
           "< wp:Tile >" +
"< wp:BackgroundImage > ApplicationIcon.png </wp:BackgroundImage >" +
              "< wp:Count > 2 </wp:Count >" +
              "< wp:Title >" + msg.Text + "</wp:Title >" +
           "</wp:Tile > " +
        "</wp:Notification >";
byte[] strBytes = new UTF8Encoding().GetBytes(tileMessage);
…
//设置吐司通知(Toast Notification)消息头
string ToastPushXML = "<?xml version = \"1.0\" encoding = \"utf - 8\"?>" +
                         "< wp:Notification xmlns:wp = \"WPNotification\">" +
                           "< wp:Toast >" +
                             "< wp:Text1 >消息提醒</wp:Text1 >" +
                             "< wp:Text2 >{0}</wp:Text1 >" +
                           "</wp:Toast >" +
                         "</wp:Notification >";
string str = string.Format(ToastPushXML, msg.Text);
byte[] strBytes = new UTF8Encoding().GetBytes(str);
…
//将消息写入到请求的消息流里面
sendNotificationRequest.ContentLength = strBytes.Length;
using (Stream requestStream = sendNotificationRequest.GetRequestStream())
{
    requestStream.Write(strBytes, 0, strBytes.Length);
}
```

4）发送消息并获取发送的状态

Send. aspx. cs 文件发送消息并获取发送的状态的代码

```
//发送消息
HttpWebResponse response = (HttpWebResponse)sendNotificationRequest.GetResponse();
//获取通知状态
string notificationStatus = response.Headers["X-NotificationStatus"];
//获取频道状态
string notificationChannelStatus = response.Headers["X-SubscriptionStatus"];
//获取设备状态
string deviceConnectionStatus = response.Headers["X-DeviceConnectionStatus"];
```

13.5 WebBrowser

WebBrowser 控件是一个网页浏览控件，Windows Phone 的网络功能很强大，WebBrowser 控件可以直接加载网页，可以将其视为一个浏览器，但它不是一个完整的浏览器，因为它没有地址栏、收藏夹、选项卡等。可以把它当做 HTML 中的 iframe，但它提供了更丰富的界面。可以通过两个手指收缩（和双击）来进行缩放，平移和滚动是自动内置的，无须自己实现。WebBrowser 控件可以让用户浏览一个特定的网络上的网页，它的显示效果和浏览器差不多，但是手机必须得在连接到网络的状态。这个控件除了可以浏览网络上的网页，还可以加载本地的网页 HTML 文件的内容。利用这个特性我们通产可以将程序的帮助文档等，做成 HTML 的格式然后使用 WebBrowser 控件去进行加载和浏览。

WebBrowser 控件是在 Microsoft. Phone. Controls 空间下的控件，使用时需在 XAML 页面中加入空间"clr-namespace:Microsoft. Phone. Controls;assembly=Microsoft. Phone"的引用，然后再进行调用。WebBrowser 控件的常用方法、属性和事件分别如表 13.6～表 13.8 所示。

表 13.6　WebBrowser 控件的常用方法

名　　称	说　　明
InvokeScript	执行一个在当前加载的网页中存在的脚本函数，在当前加载的文档中定义；执行脚本的功能在 WebBrowser 控件中是默认关闭，即 IsScriptEnabled 属性默认为 false，如果要使用执行脚本的方法需要设置 IsScriptEnabled 属性为 true；InvokeScript (String)传入的参数是脚本的方法名称，InvokeScript(String, String[])传入的第二个参数是脚本方法的参数列表
Navigate	启动在 WebBrowser 控件所提供的 URI 地址；Navigate(Uri)参数为 Uri 网页地址，Navigate(Uri, Byte[], String)第二个参数是 post 过去的数据和 HTTP headers 信息
NavigateToString	注入 WebBrowser 控件 HTML 格式的字符串
SaveToString	返回 WebBrowser 控件中的网页的 HTML 内容
GetCookies	返回 WebBrowser 控件中当前网页的 Cookie 信息，返回值类型是 CookieCollection 类型

表 13.7 WebBrowser 控件的常用属性

名称	说明
Base	设置在手机应用程序独立存储中的路径位置
IsGeolocationEnabled	在 WebBrowser 控件中是否启用手机的定位服务
IsScriptEnabled	是否启动脚本执行，对下一个连接到的网页开始生效，而不是对当前的网页生效，默认值是 false，不启用脚本执行
Source	获取或设置 WebBrowser 控件的网页地址

表 13.8 WebBrowser 控件的常用事件

名称	说明
LoadCompleted	WebBrowser 控件加载网页完成后触发该事件
Navigated	WebBrowser 控件成功获取到网页的导航后触发该事件
Navigating	WebBrowser 控件成功正在导航到目标地址的过程中触发该事件
NavigationFailed	WebBrowser 控件成功正在导航到目标地址失败则触发该事件
ScriptNotify	当 Javascript 访问 window. external. notify(<data>)的时候触发该事件

下面给出 WebBrowser 控件打开并保存网页的示例：演示了如何在应用程序中使用 WebBrowser 控件，并且将 WebBrowser 控件的网页内容保存到应用程序的独立存储空间中。

代码清单 13-6：WebBrowser 控件打开并保存网页的示例(源代码：第 13 章\Examples_13_6)

MainPage. xaml 文件主要代码

```
<phone:PhoneApplicationPage
    ...
    xmlns:browser = "clr - namespace:Microsoft.Phone.Controls;assembly = Microsoft.Phone">
    ...
            <browser: WebBrowser  Margin  =  "  -  6,  6,  12,  332 "  Name  =  "  webBrowser1 "
HorizontalContentAlignment = "Left" />
            <TextBox Height = "90" HorizontalAlignment = "Left" Margin = "6,331,0,0" Name =
"textBox1" Text = "http://www.google.com.hk/?hl = en" VerticalAlignment = "Top" Width = "319" />
            <Button Content = "打开网页" Height = "70" HorizontalAlignment = "Right" Margin
= "0,331, - 12,0" Name = "button1" VerticalAlignment = "Top" Width = "160" Click = "button1_
Click"/>
            <Button Content = "把网页保存到本地" Height = "72" HorizontalAlignment = "Left"
Margin = "12,427,0,0" Name = "btnSave" VerticalAlignment = "Top" Width = "456" Click = "btnSave
_Click" />
            <Button Content = "加载本地保存的页面" Height = "72" HorizontalAlignment =
"Left" Margin = "12,505,0,0" Name = "btnLoad" VerticalAlignment = "Top" Width = "456" Click =
"btnLoad_Click" />
    ...
</phone:PhoneApplicationPage>
```

MainPage. xaml. cs 文件代码

```
using System;
using System.Windows;
using Microsoft.Phone.Controls;
using System.IO.IsolatedStorage;
using System.Windows.Resources;
using System.IO;
namespace WebBrowserTest
{
    public partial class MainPage : PhoneApplicationPage
    {
        public MainPage()
        {
            InitializeComponent();
        }
        //单击打开网页按钮事件处理
        private void button1_Click(object sender, RoutedEventArgs e)
        {
            //在控件中打开网页
            webBrowser1.Navigate(new Uri(textBox1.Text, UriKind.Absolute));
        }
        //保存网页到本地的独立存储
        private void SaveStringToIsoStore(string strWebContent)
        {
            //获取本地应用程序存储对象
            IsolatedStorageFile isoStore = IsolatedStorageFile.GetUserStoreForApplication();
            //清除之前保存的网页
            if (isoStore.FileExists("web.htm") == true)
            {
                isoStore.DeleteFile("web.htm");
            }
            //转化为流
            StreamResourceInfo sr = new StreamResourceInfo(new MemoryStream(System.Text.
Encoding.UTF8.GetBytes(strWebContent)), "html/text");
            using (BinaryReader br = new BinaryReader(sr.Stream))
            {
                byte[] data = br.ReadBytes((int)sr.Stream.Length);
                //保存文件到本地存储
                using (BinaryWriter bw = new BinaryWriter(isoStore.CreateFile("web.htm")))
                {
                    bw.Write(data);
                    bw.Close();
                }
            }
        }
        //把网页保存到本地按钮事件
        private void btnSave_Click(object sender, RoutedEventArgs e)
        {
            //获取网页的 html 代码
```

```
            string strWebContent = webBrowser1.SaveToString();
            SaveStringToIsoStore(strWebContent);
        }
        //加载本地保存的页面
        private void btnLoad_Click(object sender, RoutedEventArgs e)
        {
            //加载本地保存的页面
            webBrowser1.Navigate(new Uri("web.htm", UriKind.Relative));
        }
    }
}
```

程序运行的效果如图 13.21 所示。

图 13.21 WebBrowser 控件测试

第14章

异步编程与并行编程

Windows Phone 8 提供了一种新的异步编程和并行编程的模型，在 Windows Phone 8 中新增的耗时的 API 采用了这种异步的模型。异步编程主要是为了提供更好的用户体验和程序的效率，特别是对于耗时操作，这种异步编程的方式大大地提高了应用程序的交互体验。Windows Phone 8 中普遍地使用异步操作的主要作用是要促使开发者在编写应用程序上使用这种更加优越的开发方式，让 Windows Phone 8 的程序运行得更加流畅。本章会详细地讲解 Windows Phone 8 的这种新的异步编程和并行编程方式，为打造流畅的应用程序打下基础。

14.1 异步编程模式简介

异步编程模型是一种非常常用的编程方式，在.NET 里面这种编程的模式一直在发展，从之前的异步编程模型模式和基于事件的异步模式发展到目前 Windows Phone 8 的基于任务的异步模式。下面将介绍这三种异步编程模式的特点和如何去使用。

14.1.1 异步编程模型模式(APM)

异步编程模型模式(APM)(也称为 IAsyncResult 模式)，其中异步操作要求 Begin 和 End 方法(例如异步写操作的 BeginWrite 和 EndWrite)。下面来看一下异步编程模式的写法。

```
public class MyClass
{
     public IAsyncResult BeginRead(byte [ ] buffer, int offset, int count, AsyncCallback
callback, object state);
    public int EndRead(IAsyncResult asyncResult);
}
```

使用 IAsyncResult 设计模式的异步操作是通过名为“ Begin 操作名称”和“End 操作名称”的两个方法来实现的，这两个方法分别代表开始和结束异步操作。在调用“Begin 操作名称”后，应用程序可以继续在调用线程上执行指令，同时异步操作在另一个线程上执行。

每次调用“Begin 操作名称”时，应用程序还应调用“End 操作名称”来获取操作的结果。

1. 开始异步操作

Begin 方法开始异步操作，并返回实现 IAsyncResult 接口的对象。IAsyncResult 对象存储有关异步操作的信息，IAsyncResult 对象信息如表 14.1 所示。

表 14.1 IAsyncResult 对象信息

成员	说明
AsyncState	一个可选的应用程序特定的对象，包含有关异步操作的信息
AsyncWaitHandle	一个 WaitHandle，可用来在异步操作完成之前阻止应用程序执行
CompletedSynchronously	一个值，指示异步操作是否是在用于调用 Begin 操作名称的线程上完成，而不是在单独的 ThreadPool 线程上完成
IsCompleted	一个值，指示异步操作是否已完成

Begin 方法带有该方法的同步版本声明的参数和最后两个 Begin 方法规定的参数，Begin 方法中不包含任何输出参数。Begin 方法最后的两个参数，其中一个参数定义一个 AsyncCallback 委托，此委托引用在异步操作完成时调用的方法。如果调用方不希望在操作完成后调用方法，它可以指定 null。另外一个参数是一个用户定义的对象。此对象可用来向异步操作完成时调用的方法传递应用程序特定的状态信息。Begin 方法还可以带有其他一些操作特定的参数，不过 AsyncCallback 和应用程序状态对象一定是 Begin 方法中的最后两个参数。Begin 调用之后立即将控制权返回给调用线程。如果 Begin 方法引发异常，则会在开始异步操作之前引发异常，则意味着没有调用回调方法。

2. 结束异步操作

End 方法可结束异步操作，End 方法的返回值与其同步副本的返回值类型相同，并且是特定于异步操作的。例如，EndRead 方法返回从 FileStream 读取的字节数，EndGetHostByName 方法返回包含有关主机的信息的 IPHostEntry 对象。End 方法带有该方法同步版本中声明的所有输出参数或引用参数。除了来自同步方法的参数外，End 方法还包括 IAsyncResult 参数。如果调用 End 方法时 IAsyncResult 对象表示的异步操作尚未完成，则 End 方法将在异步操作完成之前阻止调用线程。

此异步编程模式的实施者通知调用方异步操作已完成，可以通过以下步骤实现：将 IsCompleted 设置为 true，调用异步回调方法（如果已指定一个回调方法），然后发送 AsyncWaitHandle 信号。对于访问异步操作的结果，应用程序开发人员有若干种设计选择。正确的选择取决于应用程序是否有可以在操作完成时执行的指令。如果应用程序在接收到异步操作结果之前不能进行任何其他工作，则必须先阻止该应用程序进行其他工作，等到获得这些操作结果后再继续进行。

3. 实例

下面给出异步编程模型模式的示例：使用异步编程模型模式实现一个网络请求调用和一个模拟的耗时请求调用。

代码清单 14-1：异步编程模型模式（源代码：第 14 章\Examples_14_1）

AsyncResultNoResult.cs 文件代码：异步返回对象，不带返回结果数据

```
using System;
using System.Threading;
namespace AsyncTaskDemo
{
    public class AsyncResultNoResult : IAsyncResult
    {
        private readonly AsyncCallback _asyncCallback;        //异步请求的返回方法
        private readonly object _asyncState;                  //异步请求的对象
        private ManualResetEvent _asyncWaitHandle;            //线程阻塞
        private Exception _exception;                         //操作异常
        private int _completedState;                          //完成状态
        private const int StateCompletedAsynchronously = 2; //异步完成
        private const int StateCompletedSynchronously = 1;  //同步完成
        private const int StatePending = 0;
        ///< summary >
        ///初始化
        ///</summary >
        ///< param name = "asyncCallback">异步返回方法</param >
        ///< param name = "state">异步调用对象</param >
        public AsyncResultNoResult(AsyncCallback asyncCallback, object state)
        {
            this._asyncCallback = asyncCallback;
            this._asyncState = state;
        }
        //结束任务
        public void EndInvoke()
        {
            if (!this.IsCompleted)                              //任务未完成
            {
                this.AsyncWaitHandle.WaitOne();
                this.AsyncWaitHandle.Close();
                this._asyncWaitHandle = null;
            }
            if (this._exception != null)
            {
                RethrowException(this._exception);
            }
        }
        private static void RethrowException(Exception ex)
        {
            throw ex;
        }
        ///< summary >
        ///完成
        ///</summary >
        ///< param name = "exception">异常</param >
```

```
        ///<param name = "completedSynchronously">是否同步完成</param>
        public void SetAsCompleted(Exception exception, bool completedSynchronously)
        {
            this._exception = exception;
            //如果是同步操作设置 completedSynchronously = 1
             if (Interlocked.Exchange(ref this._completedState, completedSynchronously ?
1 : 2) != 0)
            {
                //_completedState 等于 0 表示任务已经完成
                throw new InvalidOperationException("You can set a result only once", exception);
            }
            if (this._asyncWaitHandle != null)
            {
                this._asyncWaitHandle.Set();
            }
            if (this._asyncCallback != null)
            {
                this._asyncCallback(this);                    //调用异步请求的返回方法
            }
        }
        //放弃异步请求的对象
        public object AsyncState
        {
            get
            {
                return this._asyncState;
            }
        }
        //调节多个线程之间的同步,等待信号发生
        public WaitHandle AsyncWaitHandle
        {
            get
            {
                //处理多线程等待
                if (this._asyncWaitHandle == null)
                {
                    bool isCompleted = this.IsCompleted;
                    ManualResetEvent event2 = new ManualResetEvent(isCompleted);
                    //比较和交换操作按原子操作执行
                    if (Interlocked.CompareExchange<ManualResetEvent>(ref this._
asyncWaitHandle, event2, null) != null)
                    {
                        event2.Close();
                    }
                    else if (!isCompleted && this.IsCompleted)
                    {
                        //如果操作已经完成则释放阻塞
                        this._asyncWaitHandle.Set();
                    }
                }
```

```
                return this._asyncWaitHandle;
            }
        }
        //同步操作是否已经完成
        public bool CompletedSynchronously
        {
            get
            {
                return (this._completedState == 1);
            }
        }
        //是否已经完成
        public bool IsCompleted
        {
            get
            {
                return (this._completedState != 0);
            }
        }
    }
}
```

AsyncResult.cs 文件代码：继承 AsyncResultNoResult 类实现返回结果的异步对象

```
using System;
namespace AsyncTaskDemo
{
    public class AsyncResult<TResult> : AsyncResultNoResult
    {
        private TResult m_result;                                   //异步操作完成返回的对象
        public AsyncResult(AsyncCallback asyncCallback, object state)
            : base(asyncCallback, state)
        {
            //default 为泛型代码中的默认关键字
            this.m_result = default(TResult);
        }
        //停止异步返回对象
        public TResult EndInvoke()
        {
            base.EndInvoke();                                       //等待操作完成
            return m_result;                                        //返回结果
        }
        //操作完成
        public void SetAsCompleted(TResult result, bool completedSynchronously)
        public void SetAsCompleted(TResult result, bool completedSynchronously)
        {
            //保存异步操作的结果
            m_result = result;
            base.SetAsCompleted(null, completedSynchronously);
        }
```

```
    }
}
```

SimpleHttpGetRequest.cs 文件代码：异步封装网络请求

```
using System;
using System.IO;
using System.Net;
namespace AsyncTaskDemo
{
    internal class SimpleHttpGetRequest
    {
        private readonly Uri _url;                              //网络地址
        public SimpleHttpGetRequest(Uri url)
        {
            this._url = url;
        }
        //开始异步请求
        public virtual IAsyncResult BeginRequest(AsyncCallback asyncCallback, object state)
        {
            //创建异步返回的结果对象
            AsyncResult<Stream> asyncResult = new AsyncResult<Stream>(asyncCallback, state);
            //创建 web 请求
            HttpWebRequest webRequest = (HttpWebRequest)WebRequest.Create(this._url);
            webRequest.Method = "GET";
            AsyncCallback callback = delegate(IAsyncResult callBackResult)
            {
                try
                {
                    //设置异步操作完成
asyncResult.SetAsCompleted(((HttpWebResponse)webRequest.EndGetResponse(callBackResult)).
GetResponseStream(), callBackResult.CompletedSynchronously);
                }
                catch (Exception exception)
                {
                    //发生异常也需要完成异步操作
                    asyncResult.SetAsCompleted(exception, callBackResult.CompletedSynchronously);
                }
            };
            webRequest.BeginGetResponse(callback, null);
            return asyncResult;
        }
        // 结束异步请求
        public Stream EndRequest(IAsyncResult asyncResult)
        {
            return ((AsyncResult<Stream>)asyncResult).EndInvoke();
        }
    }
}
```

WebTask.cs 文件代码：网络异步任务

```
using System;
using System.IO;
using System.Threading;
namespace AsyncTaskDemo
{
    internal class WebTask
    {
        //网络请求
        private readonly SimpleHttpGetRequest _dataRequest;
        public WebTask(Uri url)
            : this(new SimpleHttpGetRequest(url))
        {
        }
        //创建一个网络任务
        public WebTask(SimpleHttpGetRequest httpPostRequest)
        {
            this._dataRequest = httpPostRequest;
        }
        //开始网络任务
        public IAsyncResult BeginWebTask(AsyncCallback asyncCallback, object state)
        {
            AsyncResult<string> asyncResult = new AsyncResult<string>(asyncCallback, state);
            this._dataRequest.BeginRequest(delegate(IAsyncResult o)
            {
                try
                {
                    //获取网络返回字符串
                    string result = WebResponseToString(_dataRequest.EndRequest(o));
                    asyncResult.SetAsCompleted(result, o.CompletedSynchronously);
                }
                catch (Exception exception)
                {
                    asyncResult.SetAsCompleted(exception, o.CompletedSynchronously);
                }
            }, null);
            return asyncResult;
        }
        //结束异步请求
        public string EndWebTask(IAsyncResult asyncResult)
        {
            return ((AsyncResult<string>)asyncResult).EndInvoke();
        }
        //处理网络请求的回应,转化为字符串
        private string WebResponseToString(Stream response)
        {
            using (StreamReader reader = new StreamReader(response))
            {
                return reader.ReadToEnd();
```

```
            }
        }
    }
}
```

TestTask.cs 文件代码：测试耗时任务

```
using System;
using System.Threading;
namespace AsyncTaskDemo
{
    public class TestTask
    {
        public TestTask(){}
        //开始异步任务
        public IAsyncResult BeginTestTask(string text, AsyncCallback asyncCallback, object state)
        {
            AsyncResult<string> asyncResult = new AsyncResult<string>(asyncCallback, state);
            ThreadPool.QueueUserWorkItem(DoTestTask, asyncResult);
            return asyncResult;
        }
        //结束异步任务
        public string EndTestTask(IAsyncResult asyncResult)
        {
            return ((AsyncResult<string>)asyncResult).EndInvoke();
        }
        //模拟耗时任务
        private void DoTestTask(Object asyncResult)
        {
            AsyncResult<string> ar = (AsyncResult<string>)asyncResult;
            try
            {
                Thread.Sleep(3000);
                ar.SetAsCompleted("测试完成", true);
            }
            catch (Exception e)
            {
                ar.SetAsCompleted(e, false);
            }
        }
    }
}
```

MainPage.xaml 文件主要代码：UI 界面

```
        <Grid x:Name="ContentPanel" Grid.Row="1" Margin="12,0,12,0">
            <Button Content="测试异步任务" Height="72" HorizontalAlignment="Left"
Margin="138,80,0,0" Name="button1" VerticalAlignment="Top" Width="219" Click="button1
_Click" />
```

```
        <Button Content = "测试网络任务" Height = "72" HorizontalAlignment = "Left"
Margin = "143, 280, 0, 0" Name = "button2" VerticalAlignment = "Top" Width = "224" Click =
"button2_Click" />
    </Grid>
```

MainPage.xaml.cs 文件主要代码

```
using System;
using System.Windows;
using Microsoft.Phone.Controls;
using Microsoft.Phone.Reactive;
namespace AsyncTaskDemo
{
    public partial class MainPage : PhoneApplicationPage
    {
        TestTask tt = new TestTask();                    //耗时任务
        //网络任务
        WebTask webTask = new WebTask(new Uri("http://www.cnblogs.com/linzheng"));
        IAsyncResult ia;
        public MainPage()
        {
            InitializeComponent();
        }
        //测试耗时任务的异步请求
        private void button1_Click(object sender, RoutedEventArgs e)
        {
            //用响应式编程创建一个异步模型
            Func<string, IObservable<string>> testResult = Observable.FromAsyncPattern
<string, string>(tt.BeginTestTask, tt.EndTestTask);
ObservableExtensions. Subscribe <string> ( Observable. ObserveOnDispatcher <string>
(testResult.Invoke("") ),
             delegate(string success) {MessageBox.Show(success);},
             delegate(Exception exception){MessageBox.Show(exception.Message);} );
        }
        //测试网络任务异步请求
        private void button2_Click(object sender, RoutedEventArgs e)
        {
            ia = webTask.BeginWebTask(delegate(IAsyncResult callBackResult)
            {
                string result = webTask.EndWebTask(callBackResult);
Deployment.Current.Dispatcher.BeginInvoke( delegate{MessageBox.Show(result);} );
            }, webTask);
        }
    }
}
```

程序的运行效果如图 14.1 所示。

14.1.2 基于事件的异步模式(EAP)

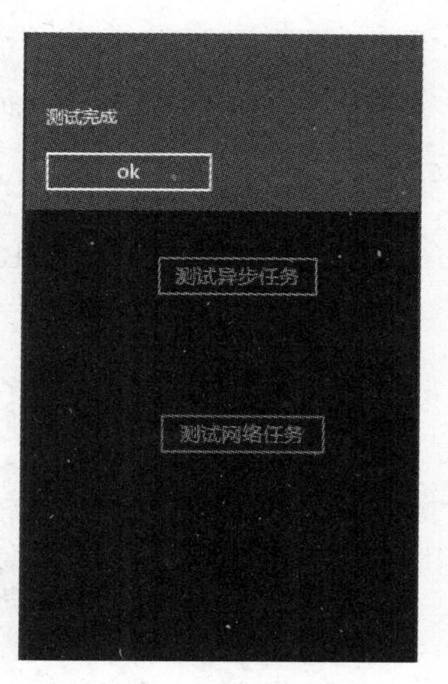

图 14.1 测试异步编程模型模式

基于事件的异步模式(EAP)需要一个具有 Async 后缀的方法,还需要一个或多个事件、事件处理程序、委托类型和 EventArg 派生的类型。下面来看一下基于事件的异步模式的写法。

```
public class MyClass
{
    public void ReadAsync(byte [] buffer, int offset, int count);
    public event ReadCompletedEventHandler ReadCompleted;
}
```

基于事件的异步模式具有多线程应用程序的优点,同时隐匿了多线程设计中固有的许多复杂问题。使用支持此模式的类,可以"在后台"执行耗时任务(例如下载和数据库操作),但不会中断应用程序,同时执行多个操作,每个操作完成时都会接到通知,等待资源变得可用,但不会停止或挂起的应用程序,并且可以使用熟悉的事件和委托模型与挂起的异步操作通信。

在 Windows Phone 里面的读取联系人操作的类 Contacts 就是支持基于事件的异步模式的组件。可以通过调用其 SearchAsync 方法来搜索联系人,然后再处理 SearchCompleted 事件。调用 SearchAsync 方法时,应用程序将继续运行,而搜索操作将在另一个线程上("在后台")继续。搜索操作完成时,将会调用事件处理程序,事件处理程序可以检查 ContactsSearchEventArgs 参数来获取搜索到的联系人。

基于事件的异步模式可以采用多种形式,具体取决于某个特定类支持的操作的复杂程度。最简单的类可能只有一个方法名称 Async 方法和一个对应的方法名称 Completed 事件。更复杂的类可能有若干个方法名称 Async 方法(每种方法都有一个对应的方法名称 Completed 事件),以及这些方法的同步版本。这些类分别支持各种异步方法的取消、进度报告和增量结果。异步方法可能还支持多个挂起的调用(多个并发调用),允许代码在此方法完成其他挂起的操作之前调用此方法任意多次。若要正确处理此种情况,必须让应用程序能够跟踪各个操作的完成。

下面给出异步编程模型模式的示例:使用异步编程模型模式测试网络的访问,异步返回网络测试的结果。

代码清单 14-2:基于事件的异步模式(源代码:第 14 章\Examples_14_2)

StateChangedEventArgs.cs 文件代码:网络事件状态参数

```
using System;
namespace EventDemo
{
```

```
    public class StateChangedEventArgs : EventArgs
    {
        public readonly string NewState;                        //网络状态值
        public readonly DateTime Timestamp;                     //事件
        public StateChangedEventArgs(string newstate)
        {
            this.NewState = newstate;
            this.Timestamp = DateTime.Now;
        }
    }
}
```

NetTask.cs 文件代码：异步网络任务

```
using System;
using System.Net;
using System.Threading;
using System.IO;
namespace EventDemo
{
    public class NetTask
    {
        public delegate void StateChanged(NetTask sender, StateChangedEventArgs args);//定义委托
        public event StateChanged OnStateChanged;               //定义事件
        public string NetTaskName = "";
        //开始网络任务
        public void StartNetTask(string url)
        {
            bool success = false;
            int attempt = 0;
            while (attempt < 3)
            {
                AsyncCallback callback = null;
                //开启线程等待
                ManualResetEvent webRequestWait = new ManualResetEvent(false);
                Uri targetUri = new Uri(url);
                HttpWebRequest request = (HttpWebRequest)WebRequest.Create(targetUri);
                request.Method = "POST";
                if (callback == null)
                {
                    callback = delegate(IAsyncResult asRequest)
                    {
                        try
                        {
                            success = true;
                            webRequestWait.Set();
                        }
                        catch
                        {
                            OnStateChanged(this, new StateChangedEventArgs("重试"));
                            webRequestWait.Set();
```

```
                    }
                };
            }
            request.BeginGetRequestStream(callback, request);
            //等待线程结束
            webRequestWait.WaitOne();
            if (success)
            {
                break;
            }
            attempt ++ ;
            Thread.Sleep(1000);
        }
        if (success)
        {
            OnStateChanged(this, new StateChangedEventArgs("成功"));
            Thread.Sleep(50);
        }
        else
        {
            OnStateChanged(this, new StateChangedEventArgs("失败"));
        }
    }
  }
}
```

MainPage.xaml 文件代码：UI 界面

```
    <Grid x:Name = "ContentPanel" Grid.Row = "1" Margin = "12,0,12,0">
        <Button Content = "测试网络" Height = "72" HorizontalAlignment = "Left" Margin =
"143,105,0,0" Name = "button1" VerticalAlignment = "Top" Width = "202" Click = "button1_Click" />
        <TextBlock Height = "50" HorizontalAlignment = "Left" Margin = "96,270,0,0" Name =
"textBlock1" Text = "网络的状态：" VerticalAlignment = "Top" Width = "126" />
        <TextBlock Height = "48" HorizontalAlignment = "Left" Margin = "34,326,0,0" Name =
"textBlock2" Text = "" VerticalAlignment = "Top" Width = "377" />
    </Grid>
```

MainPage.xaml.cs 文件代码

```
using System.Windows;
using Microsoft.Phone.Controls;
namespace EventDemo
{
    public partial class MainPage : PhoneApplicationPage
    {
        public MainPage()
        {
            InitializeComponent();
        }
        //开始网络异步任务
        private void button1_Click(object sender, RoutedEventArgs e)
        {
```

```
            NetTask netTask = new NetTask();
            netTask.OnStateChanged += OnStateChanged;
            netTask.NetTaskName = "测试网络";
            netTask.StartNetTask("http://www.cnblogs.com");
        }
        //处理状态变化事件
        public void OnStateChanged(object sender, StateChangedEventArgs e)
        {
            NetTask temp = sender as NetTask;
            textBlock2.Text = temp.NetTaskName + "," + e.NewState + "," + e.Timestamp.
ToLongTimeString();
        }
    }
}
```

程序的运行效果如图 14.2 所示。

14.1.3 基于任务的异步模式(TAP)

基于任务的异步模式（TAP），该模式使用一个方法表示异步操作的启动和完成。.NET Framework 4 中引入了 TAP，这是 .NET Framework 中异步编程的建议方法。基于任务的异步模式(TAP)是基于 System. Threading. Tasks 命名空间的 Task 和 Task<TResult>来实现，用于表示任意异步操作。TAP 是新开发的建议异步设计模式，也是 Windows Phone 8 新增的异步操作特性。下面来看一下基于任务的异步模式的写法。

图 14.2 测试事件使用

```
public class MyClass
{
    public Task < int > ReadAsync(byte [ ] buffer, int offset, int count);
}
```

1. 命名、参数和返回类型

基于任务的异步模式使用单个方法表示异步操作的开始和完成，这与异步编程模型模式和基于事件的异步模式不一样。TAP 方法返回 Task 或 Task<TResult>类型，基于相应的同步方法是否返回 void 或类型 TResult。TAP 方法的参数应与其同步副本的参数匹配，并应以相同顺序提供，但是 out 和 ref 参数不受此规则的限制，并应完全避免。将通过 out 或 ref 参数返回的任何数据均应改为作为 Task<TResult>返回的 TResult 的一部分，应使用元组或自定义数据结构来容纳多个值。

2. 初始化异步操作

基于任务的异步模式的异步方法可以同步完成少量工作，如在返回结果任务之前的验证参数和启动该异步操作应保持同步的最小权限，使该异步方法可以尽快返回。快速返回的原因包括以下几种：

(1) 异步方法可以从用户界面（UI）线程调用，因此，所有长期运行的同步工作可能会降低应用程序的响应能力。

(2) 多个异步方法可以同时启动。因此,在异步方法的同步部分中的任何长时间运行的工作可以延迟其他异步操作的启动,从而减少并发的优点。

(3) 在某些情况下,要求完成操作的工作量要比其将异步启动操作的工作量少。读取流时,按照在内存中已缓冲好的数据来满足该读取,这就是此类情形的一个示例。在这样的情况下,操作可能会同步完成,同时返回已完成的任务。

3. 异常

异步方法会引发异步方法调用的异常,以响应用法错误,但这类错误决不应该出现在生产代码中。例如,如果传递 null 引用作为某个方法的参数而导致错误状态(通常由 ArgumentNullException 异常表示),则可以修改调用代码确保绝对不传递 null 引用。对于所有其他错误,如在运行异步方法时发生异常,该异常应指定给返回的任务,即使是该异步方法碰巧在任务返回前以同步方式完成也如此。通常,任务最多包含一个异常。但是,如果任务表示多个操作,例如 WhenAll,那么多个异常可能与单个任务关联。

4. 目标环境

在实现 TAP 模式时,可以确定异步执行发生的地方,可选择在线程池上执行,可选择使用异步 I/O 实现它(不必绑定到大部分操作执行的线程)或可选择所需的特定线程上运行它(如 UI 线程)或使用其他上下文上。TAP 方法甚至没有要执行的异常,仅返回在系统的其他位置表示条件出现的 Task。TAP 方法的调用方可以会通过异步等待生成结果来阻止等待 TAP 方法完成,或可以在异步操作完成时运行其他操作。

5. 任务状态

Task 类用于异步操作时提供一个生命周期,该生命周期的状态由 TaskStatus 枚举表示,TaskStatus 枚举的成员如表 14.2 所示。Task 类提供了 Start 方法,用于支持从 Task 或者 Task＜TResult＞类型继承的自定义任务类,这些自定的任务类初始化之后的状态是 TaskStatus. Created,需要调用 Start 方法来启动,才会开始整个任务的生命周期。其他的任务都称为热任务,它们的状态都是 TaskStatus. Created 之外的状态值。如果所等待的 Task 或 Task＜TResult＞以 Canceled 关闭状态结束,则将引发 OperationCanceledException 异常。如果所等待的 Task 或 Task＜TResult＞以 Faulted 状态关闭,则将引发导致其出错的异常。

表 14.2　TaskStatus 枚举的成员

成员名称	说明
Created	该任务已初始化,但尚未被计划
WaitingForActivation	该任务正在等待 .NET Framework 基础结构在内部将其激活并进行计划
WaitingToRun	该任务已被计划执行,但尚未开始执行
Running	该任务正在运行,但尚未完成
WaitingForChildrenToComplete	该任务已完成执行,正在隐式等待附加的子任务完成
RanToCompletion	已成功完成执行的任务

续表

成员名称	说明
Canceled	该任务已通过对其自身的 CancellationToken 引发 OperationCanceled-Exception 对取消进行了确认，此时该标记处于已发送信号状态；或者在该任务开始执行之前，已向该任务的 CancellationToken 发出了信号
Faulted	由于未处理异常的原因而完成的任务

14.2 任务异步编程

任务的异步编程是 Windows Phone 8 里面最常见的异步编程模式，也是微软在新的 Windows 运行时框架下面力推的异步编程模式。下面将来详细地讲解任务的异步编程。

14.2.1 相关任务类介绍

对于任务异步编程的类在 System. Threading. Tasks 命名空间下，提供简化编写并发和异步代码的工作的类型。主要类型为 System. Threading. Tasks. Task（表示可以等待和取消的异步操作）和 System. Threading. Tasks. Task＜TResult＞（可以返回值的任务）。Factory 类提供用于创建和启动任务的静态方法，System. Threading. Tasks. TaskScheduler 类提供默认线程调度基础结构。

1. Task 类

Task 类表示一个异步操作，Task 类的主要方法和主要属性分别如表 14.3 和表 14.4 所示。Task 实例可以用各种不同的方式创建，最常见的方法是使用任务类型的 Factory 属性检索可用来创建用于多个用途的任务的 TaskFactory 实例。例如，要创建运行操作的 Task，可以使用工厂的 StartNew 方法：

```
var t = Task.Factory.StartNew(() => DoAction());
```

Task 类还提供了初始化任务但不计划执行任务的构造函数。出于性能方面的考虑，TaskFactory 的 StartNew 方法应该是创建和计划计算任务的首选机制，但是对于创建和计划必须分开的情况，可以使用构造函数，然后可以使用任务的 Start 方法计划任务在稍后执行。

表 14.3 Task 类的主要属性

名称	说明
AsyncState	获取在创建 Task 时提供的状态对象，如果未提供，则为 null
CreationOptions	获取用于创建此任务的 TaskCreationOptions
CurrentId	返回当前正在执行的 Task 的唯一 ID

续表

名　　称	说　　明
Exception	获取导致 Task 提前结束的 AggregateException，如果 Task 成功完成或尚未引发任何异常，则将返回 null
Factory	提供对用于创建 Task 和 Task<TResult> 实例的工厂方法的访问
Id	获取此 Task 实例的唯一 ID
IsCanceled	获取此 Task 实例是否由于被取消的原因而已完成执行
IsCompleted	获取此 Task 是否已完成
IsFaulted	获取 Task 是否由于未经处理异常的原因而完成
Status	获取此任务的 TaskStatus

表 14.4　Task 类的主要方法

名　　称	说　　明
ContinueWith(Action<Task>)	创建一个在目标 Task 完成时异步执行的延续任务
RunSynchronously()	对当前的 TaskScheduler 同步运行 Task
Start()	启动 Task，并将它安排到当前的 TaskScheduler 中执行
Delay(Int32)	创建将在时间延迟后完成的任务
Run(Action)	将在线程池上运行的指定工作排队，并返回该工作的任务句柄
Wait()	等待 Task 完成执行过程
WaitAll(Task[])	等待提供的所有 Task 对象完成执行过程
WaitAny(Task[])	等待提供的任一 Task 对象完成执行过程
WhenAll(IEnumerable<Task>)	所有提供的任务已完成时，创建将完成的任务
Yield	创建异步返回的上下文信息，是等待中的任务

2. TaskFactory 类

TaskFactory 类提供对创建和计划 Task 对象的支持，TaskFactory 类的主要方法和主要属性分别如表 14.5 和表 14.6 所示。TaskFactory 类将其中一些常用 Task 模式编码到获取默认设置的方法中，可以通过其构造函数进行配置。TaskFactory 的默认实例通过 Factory 属性获得。

表 14.5　TaskFactory 类的主要属性

名　　称	说　　明
CancellationToken	获取此 TaskFactory 的默认 CancellationToken
ContinuationOptions	获取此 TaskFactory 的 TaskContinuationOptions 值
CreationOptions	获取此 TaskFactory 的 TaskCreationOptions 值
Scheduler	获取此 TaskFactory 的 TaskScheduler

表 14.6　TaskFactory 类的主要方法

名　　称	说　　明
ContinueWhenAll(Task[], Action<Task[]>)	创建一个延续 Task,它将在提供的一组任务完成后马上开始
ContinueWhenAny(Task[], Action<Task>)	创建一个延续 Task,它将在提供的组中的任何任务完成后马上开始
FromAsync(IAsyncResult, Action<IAsyncResult>)	创建一个 Task,它在指定的 IAsyncResult 完成时执行一个结束方法操作
StartNew(Action)	创建并启动 Task

3. TaskScheduler 类

TaskScheduler 类表示一个处理将任务排队到线程中的低级工作的对象，TaskScheduler 类的主要方法和主要属性分别如表 14.7 和表 14.8 所示。TaskScheduler 充当所有可插入的日程排定逻辑的扩展点。这包括如何计划执行任务以及如何向调试器公开计划任务的机制。

表 14.7　TaskScheduler 类主要属性

名　　称	说　　明
Current	获取与当前正在执行的任务关联的 TaskScheduler
Default	获取由 .NET Framework 提供的默认 TaskScheduler 实例
Id	获取此 TaskScheduler 的唯一 ID
MaximumConcurrencyLevel	指示此 TaskScheduler 能够支持的最大并发级别

表 14.8　TaskScheduler 类主要方法

名　　称	说　　明
FromCurrentSynchronizationContext	创建一个与当前 System. Threading. SynchronizationContext 关联的 TaskScheduler
GetScheduledTasks	仅对于调试器支持,生成当前排队到计划程序中等待执行的 Task 实例的枚举
QueueTask	将 Task 排队到计划程序中
TryDequeue	尝试将以前排队到此计划程序中的 Task 取消排队
TryExecuteTask	尝试在此计划程序上执行提供的 Task
TryExecuteTaskInline	确定是否可以在此调用中同步执行提供的 Task,如果可以,将执行该任务

14.2.2　async 关键字和 await 关键字

async 关键字和 await 关键字是任务异步编程的标志性关符号,在任务异步之中到处都需要用到这两个符号,async 表示是一个异步的方法,await 表示等待一个异步方法的完成。下面来看一下 async 关键字和 await 关键字的详细含义和用法。

1. async 关键字

async 关键字指示编译器方法或 lambda 表达式是异步的，并且可以使用 await 来指定在方法中悬挂点。使用 async 关键字标识的方法、lambda 表达式或匿名方法是表示其是异步执行的，此类方法被称为异步方法。异步方法提供了一种简便的不会阻塞线程来调用可能会长时间运行的操作的方法。异步方法的调用方可以恢复其工作，而无须等待完成的异步方法。

下面的示例显示了一个异步方法的结构。按照约定，异步方法名称结尾"Async"。

```
public async Task < int > ExampleMethodAsync()
{
    //...
    //通过使用 await 运算符来将方法挂起等待，如果 AwaitedProcessAsync 未结束
    //那么将会返回到 ExampleMethodAsync 方法的调用方继续运行
    int exampleInt = await AwaitedProcessAsync();
    //...
    //这个返回值将会告诉 ExampleMethodAsync 方法的调用方异步的任务已经完成
    //并且获得了异步的返回值，挂起的方法可以继续运行
    return exampleInt;
}
```

通常情况下，一种方法使用 async 关键字，那么方法体里包含至少一个 await 表达式或语句。该方法同步运行，直到到达第一个 await 表达式中，此时它将一直暂停来等待的任务完成。同时，程序的控制返回给调用方的方法，相当于是告诉对调用该方法的地址，这里正在等待操作的完成，这段时间你可以选择去做其他的事情。如果该方法不包含 await 表达式或语句，然后它将同步执行，编译器会警告错误，提示对于不包含 await 表达式的异步方法会产生错误。

异步方法可以具有返回类型的 void、Task 或 Task <TResult>，但是该方法不能声明任何 ref 或 out，虽然它可以调用具有此类参数的方法。void 返回类型主要用于定义事件处理程序，其中 void 返回类型是必需的。在其他情况下，如果返回语句的方法指定了操作数的类型为 TResult，那么可以指定 Task<TResult>作为异步方法的返回类型。如果方法完成时没有返回值，那么可以使用 Task 作为返回值。将 Task 作为返回类型，这意味着是"Task<void>"，也就是说，调用该方法返回 Task，但当 Task 完成后，任何 await 正在等待的表达式 Task 的计算结果为 void。

2. await 关键字

await 关键字会告诉编译器在这个代码点上正在等待着用 async 标识着的异步方法的完成，然后程序可以继续去做其他的事情，当异步方法完成之后 await 下面的代码段将会继续执行。在逻辑上，当在程序中编写" await someObject;"这意味着当编译器将生成代码来检查 someObject 所代表的操作是否已完成，如果完成，将继续执行在 await 之后的代码，如果没有，则生成一个挂起的代理方法，当异步操作完成之后再触发执行这个方法。在异步方法完成之后，在 await 关键字的这个点上将继续进行操作，同时也可以获取到异步方法的返

回对象。

await 关键字应用于一个异步方法的任务挂起方法的执行，然后直到等待任务完成，这个任务表示当前正在进行的工作。await 使用的异步方法必须被 async 关键字修饰，这个异步方法使用 async 修饰符，并且方法体里面通常包含一个或多个 await 表达式。

```
//await 关键字用于一个返回 Task<TResult>类型的异步方法
TResult result = await AsyncMethodThatReturnsTaskTResult();
//await 关键字用于一个返回 Task 类型的异步方法
await AsyncMethodThatReturnsTask();
```

await 表达式不阻止它在其上执行的线程，相反，它导致编译器注册异步方法的其余部分为等待的任务继续，然后控制回异步方法的调用方，当任务完成时，将会调用其延续任务，并且，异步方法的执行恢复它将会停止的位置。

await 表达式只能够用于在用 async 修饰的方法、lambda 表达式或匿名方法的主体上，在同步的方法、lambda 表达式或匿名方法内不能够使用 await 表达式，在异常处理语句的 catch 或 finally 块内和在 lock 语句的块里面也不能够使用 await 表达式。

大多数异步方法返回 Task 或 Task<TResult>，返回的任务的属性包含了有关其状态和历史记录的信息，例如任务是否已完成，异步方法是否引发异常或已取消，以及最终结果是什么等等的信息。

下面看一段代码，演示在事件中使用 await 表达式。

```
//把事件标识为异步
private async void button1_Click(object sender, EventArgs e)
{
    //调用异步运行的方法
    string result = await WaitAsynchronouslyAsync();
    //调用同步运行的方法
    //string result = await WaitSynchronously ();
    //展示结果
    textBox1.Text += result;
}
//这个方法是异步运行的，运行之后即使还没有运行完成也不会阻塞 UI 线程
public async Task<string> WaitAsynchronouslyAsync()
{
    await Task.Delay(10000);
    return "Finished";
}
//这个方法虽然使用了 async 关键字，但是它还是一个同步的方法，当调用 Thread.Sleep
//的时候整个 UI 会被阻塞住，一直等到方法完成之后 UI 线程才会被释放
public async Task<string> WaitSynchronously()
{
    Thread.Sleep(10000);
    return "Finished";
}
```

14.2.3 创建 Task 任务

在 Windows Phone 8 里面有很多的 API 都是直接基于任务的异步编程方式，通过异步

的任务方式来使用这些 API 是理所当然的，也是必需的。那么当自定义处理同步方法，想让其通过这种异步任务的方式来运行，该怎么做呢？下面有几种常用的当时来实现自定义的 Task 任务的运行。

1. 将工作排队到池的代码

```
Task.Run(() =>
{
  //其他的同步处理逻辑
});
```

2. 将工作排队到池并等待完成的代码

```
await Task.Run(() =>
{
  //其他的同步处理逻辑
});
```

3. 创建长时间运行的工作项的代码

```
Task.Factory.StartNew(() =>
{
  //其他的同步处理逻辑
}, TaskCreationOptions.LongRunning);
```

下面给出任务模式的示例：使用三种方式调用一个模拟耗时的任务，第一种方式不等待任务完成，将会阻塞 UI 线程直到任务的完成，第二种方式使用 TaskScheduler 类实现异步的返回，不会阻塞 UI 线程，第三种方式等待任务完成，不会阻塞 UI 线程。

代码清单 14-3：任务模式(源代码：第 14 章\Examples_14_3)

MainPage.xaml 文件主要代码

```
<Grid x:Name="ContentPanel" Grid.Row="1" Margin="12,0,12,0">
    <StackPanel>
        <Button Content="同步调用" x:Name="btSync" Click="btSync_Click"/>
        <Button Content="异步调用 1" x:Name="btAsync1" Click="btAsync1_Click"/>
        <Button Content="异步调用 2" x:Name="btAsync2" Click="btAsync2_Click"/>
    </StackPanel>
</Grid>
```

MainPage.xaml.cs 文件主要代码

```
//同步调用
private void btSync_Click(object sender, RoutedEventArgs e)
{
    //创建一个任务
    var someTask = Task<int>.Factory.StartNew(() => LongTimeFun(1, 2));
    //该任务的运行将会一直阻塞 UI 线程
    MessageBox.Show("Result: " + someTask.Result.ToString());
}
//使用 TaskScheduler 实现异步调用
private void btAsync1_Click(object sender, RoutedEventArgs e)
```

```
{
    //获取 UI 线程的上下文
    var uiScheduler = TaskScheduler.FromCurrentSynchronizationContext();
    //创建并且开始一个任务
    var someTask = Task<int>.Factory.StartNew(() => LongTimeFun(1, 2));
    someTask.ContinueWith(x =>
    {
        MessageBox.Show("Result: " + someTask.Result.ToString());
    }, uiScheduler);
}
//使用任务等待实现异步任务
private async void btAsync2_Click(object sender, RoutedEventArgs e)
{
    //创建一个任务
    var someTask = Task<int>.Factory.StartNew(() => LongTimeFun(1, 2));
    //等待任务,任务不会占用 UI 线程
    await someTask;
    MessageBox.Show("Result: " + someTask.Result.ToString());
}
//模拟耗时任务
private int LongTimeFun(int a, int b)
{
    System.Threading.Thread.Sleep(10000);
    return a + b;
}
```

程序的运行效果如图 14.3 所示。

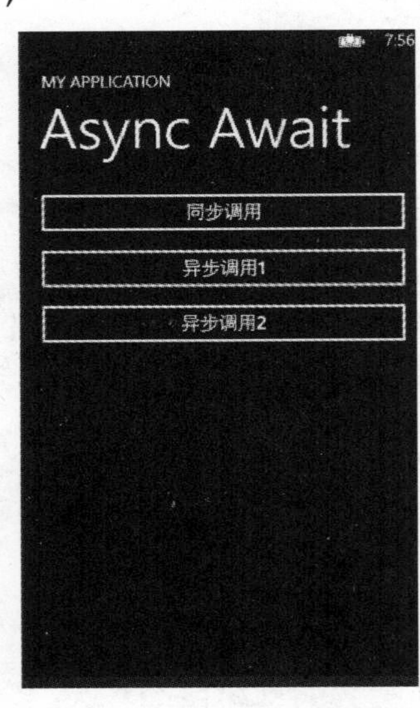

图 14.3　同步和异步调用

14.2.4　监视异步处理进度

当创建一个耗时时间较长的异步操作时，通常需要去跟踪这个耗时操作的完成情况，这时候就需要用到进度条了，异步任务提供了监视异步处理进度的实现方法，通过 IProgress＜T＞类结合异步任务的方法可以实现异步的操作过程与 UI 进度情况反映的汇报。Progress＜T＞类提供调用每个报告进度的值的回调的 IProgress＜T＞。IProgress＜T＞的常用方法事件如表 14.9 所示。

表 14.9　Progress＜T＞类的常用方法事件

名　称	说　明
构造函数 Progress＜T＞()	初始化 Progress＜T＞ 对象
构造函数 Progress＜T＞(Action＜T＞)	初始化使用指定的回调的 Progress＜T＞ 对象
方法 OnReport	报告进度更改
事件 ProgressChanged	引发为每个报告进度的值
显示方法实现 ProgressChanged	引发为每个报告进度的值

下面给出任务进度监控的示例：使 Progress<T>类对异步耗时任务的进度进行监控并且把进度的信息体现在界面上。

代码清单 14-4：任务进度监控（源代码：第 14 章\Examples_14_4）

MainPage. xaml 文件主要代码

```
<Grid x:Name = "ContentPanel" Grid.Row = "1" Margin = "12,0,12,0">
    <ProgressBar x:Name = "progressBar" HorizontalAlignment = "Left" Height = "23" Margin =
"30,103,0,0" VerticalAlignment = "Top" Width = "416"/>
    <Button Content = "测试任务进度监控" HorizontalAlignment = "Left" Margin = "94,191,0,0"
VerticalAlignment = "Top" Width = "238" Click = "Button_Click_1"/>
</Grid>
```

MainPage. xaml. cs 文件主要代码

```
//测试任务进度监控
private void Button_Click_1(object sender, RoutedEventArgs e)
{
    //创建 IProgress<T>对象来监控任务进度
    Progress<ProgressPartialResult> progress = new Progress<ProgressPartialResult>();
    //订阅进度改变事件处理
     progress.ProgressChanged += new EventHandler<ProgressPartialResult>(progress_
ProgressChanged);
    //启动异步任务并且使用进度回报
    DoSomething(progress);
}
//进度变化事件处理
void progress_ProgressChanged(object sender, ProgressPartialResult value)
{

    progressBar.Value = (float)value.Current / value.Total * 100;
}
//开始异步任务
private async void DoSomething(IProgress<ProgressPartialResult> progress)
{
    int total = 100;
    for (int i = 0; i <= total; i++)
    {
        await Task.Delay(20); //等待 2 秒钟
        if (progress != null)
        { // 报告任务进度情况
            progress.Report(new ProgressPartialResult() { Current = i + 1, Total = total });
        }
    }
    progress.Report(new ProgressPartialResult() { Current = 0, Total = total });
}
//进度返回的对象数据
public class ProgressPartialResult
{
    public int Current { get; set; }
    public int Total { get; set; }
}
```

程序的运行效果如图 14.4 所示。

图 14.4　任务进度

14.3　多线程与并行编程

Windows Phone 8 是一个支持多处理器的操作系统，并行编程的目的就是为了充分地利用多个处理器并发处理的性能，来打造出高效的应用程序。Windows Phone 8 里面对并行编程提供了很好的支持，提供了任务并行库不仅仅可以处理任务并行，也可以处理数据并行，功能非常强大。并行编程是在多线程的基础上，进化发展而来，学习并行编程，还需要对多线程编程有一定的了解和认识，才能够编写出更加强壮的并行编程的代码。

14.3.1　多线程介绍

什么是进程呢？当一个程序开始运行时，它就是一个进程，进程所指包括运行中的程序和程序所使用到的内存和系统资源。而一个进程又是由多个线程所组成的，线程是程序中的一个执行流，每个线程都有自己的专有寄存器（栈指针、程序计数器等），但代码区是共享的，即不同的线程可以执行同样的函数。多线程是指程序中包含多个执行流，即在一个程序中可以同时运行多个不同的线程来执行不同的任务，也就是说允许单个程序创建多个并行执行的线程来完成各自的任务。

1. 线程的生命周期

线程是一个动态执行的过程，它也有一个从产生到死亡的过程，这就是所谓的生命周期。一个线程在它的生命周期内有 5 种状态：

1）新建（new Thread）

当创建 Thread 类的一个实例（对象）时，此线程进入新建状态（未被启动），例如：

```
Thread t1 = new Thread();
```

2）就绪(runnable)

线程已经被启动，正在等待被分配给 CPU 时间片，也就是说此时线程正在就绪队列中排队等候得到 CPU 资源，例如：

```
t1.start();
```

3）运行(running)

线程获得 CPU 资源正在执行任务(run()方法)，此时除非此线程自动放弃 CPU 资源或者有优先级更高的线程进入，线程将一直运行到结束。

4）死亡(dead)

当线程执行完毕或被其他线程杀死，线程就进入死亡状态，这时线程不可能再进入就绪状态等待执行。自然终止：正常运行 run()方法后终止。异常终止：调用 stop()方法让一个线程终止运行。

5）堵塞(blocked)

由于某种原因导致正在运行的线程让出 CPU 并暂停自己的执行，即进入堵塞状态。正在睡眠：用 sleep(long t) 方法可使线程进入睡眠方式。一个睡眠着的线程在指定的时间过去可进入就绪状态。

2. 开启多线程的优点和缺点

优点：提高界面程序响应速度。通过使用线程，可以将需要大量时间完成的流程在后台启动单独的线程完成，提高前台界面的相应速度。充分利用系统资源，提高效率。通过在一个程序内部同时执行多个流程，可以充分利用 CPU 等系统资源，从而最大限度的发挥硬件的性能。

缺点：当程序中的线程数量比较多时，系统将花费大量的时间进行线程的切换，这反而会降低程序的执行效率。但是，相对于优势来说，劣势还是很有限的，所以现在的项目开发中，多线程编程技术得到了广泛的应用。

14.3.2　线程

System. Threading. Thread 类是用来创建并控制线程，线程的属性和方法分别如表 14.10 和表 14.11 所示，下面简单地介绍一下 Thread 类的相关的方法的使用。

通过调用 Thread. Sleep，Thread. Suspend 或者 Thread. Join 可以暂停/阻塞线程。调用 Sleep()和 Suspend()方法意味着线程将不再得到 CPU 时间。这两种暂停线程的方法是有区别的，Sleep()使得线程立即停止执行，但是在调用 Suspend()方法之前，公共语言运行时必须到达一个安全点。一个线程不能对另外一个线程调用 Sleep()方法，但是可以调用 Suspend()方法使得另外一个线程暂停执行。对已经挂起的线程调用 Thread. Resume()方法会使其继续执行。不管使用多少次 Suspend()方法来阻塞一个线程，只需一次调用 Resume()方法就可以使得线程继续执行。已经终止的和还没有开始执行的线程都不能使用挂起。Thread. Sleep(int x)使线程阻塞 x 毫秒。只有当该线程是被其他的线程通过调用 Thread. Interrupt()或者 Thread. Abort()方法，才能被唤醒。

表 14.10 线程的属性

属　性	值
IsAlive	如果线程处于活动状态，则包含值 True
IsBackground	获取或设置一个布尔值，该值表示一个线程是否是或是否应当是后台线程；后台线程与前台线程类似，但后台线程不阻止进程停止；一旦某个进程的所有前台线程都停止，公共语言运行时就会对仍处于活动状态的后台线程调用 Abort 方法，从而结束该进程
Name	获取或设置线程的名称。通常用于在调试时发现各个线程
Priority	获取或设置操作系统用于确定线程调度优先顺序的值
ApartmentState	获取或设置用于特定线程的线程模型。线程模型在线程调用非托管代码时很重要
ThreadState	包含描述线程状态的值

表 14.11 单个线程的方法

方　法	操　作
Start	使线程开始运行
Sleep	使线程暂停指定的一段时间
Suspend	在线程到达安全点时，使其暂停
Abort	在线程到达安全点时，使其停止
Resume	重新启动挂起的线程
Join	使当前线程一直等到另一线程完成，在与超时值一起使用时，如果该线程在分配的时间内完成，此方法将返回 True

14.3.3 线程池

线程池是可以用来在后台执行多个任务的线程集合。这使主线程可以自由地异步执行其他任务。ThreadPool 类提供一个线程池，该线程池可用于发送工作项、处理异步 I/O、代表其他线程等待以及处理计时器。线程池通过为应用程序提供一个由系统管理的辅助线程池，可以更为有效地使用线程。当一个等待操作完成时，线程池中的一个辅助线程就会执行，对应的每个进程都有一个线程池。线程池的默认大小为：每个处理器 250 个辅助线程，再加上 1000 个 I/O 完成线程。

ThreadPool 类是一个静态类，不能也不必要生成它的对象，而且一旦使用该方法在线程池中添加了一个项目，那么该项目将是没有办法取消的。在这里无须自己建立线程，只需把要做的工作写成函数，然后作为参数传递给 ThreadPool. QueueUserWorkItem()方法就行了，传递的方法就是依靠 WaitCallback 代理对象，而线程的建立、管理、运行等工作都是由系统自动完成的，无须考虑那些复杂的细节问题，线程池的优点也就在这里体现出来了，就好像你是公司老板——只需要安排工作，而不必亲自动手。许多应用程序创建的线程都要在休眠状态中消耗大量时间，以等待事件发生。其他线程可能进入休眠状态，只被定期唤醒以轮询更改或更新状态信息。当创建 Task 或 Task＜TResult＞ 对象来异步执行某些任务时，默认情况下，计划任务将在线程池线程上运行。

下面来看一个使用线程池运行多线程的代码示例：

```
public void DoWork()
{
    //启动一个线程池的线程任务
    System.Threading.ThreadPool.QueueUserWorkItem(
        new System.Threading.WaitCallback(SomeLongTask));
    //启动另外一个线程池的任务
    System.Threading.ThreadPool.QueueUserWorkItem(
        new System.Threading.WaitCallback(AnotherLongTask));
}

private void SomeLongTask(Object state)
{
    //处理在这个线程中的任务
}

private void AnotherLongTask(Object state)
{
    //处理在这个线程中的任务
}
```

14.3.4　线程锁

在应用程序中使用多个线程的一个好处是每个线程都可以异步执行，将耗时的任务放在后台执行，而使应用程序 UI 保持响应，保持了应用程序的高效运作。然而，线程的异步特性意味着必须协调对资源（如文件句柄、网络连接和内存）的访问，否则，两个或更多的线程可能在同一时间访问相同的资源，而每个线程都不知道其他线程的操作。结果将产生不可预知的数据损坏。

对于整数数据类型的简单操作，可以用 Interlocked 类的成员来实现线程同步。对于其他所有数据类型和非线程安全的资源，可以用 lock 语句来实现线程同步。使用 Interlocked 类的方法来避免在多个线程尝试同时更新或比较同一个值时可能出现的问题。使用这个类的方法可以安全地递增、递减、交换和比较任何线程中的值。lock 语句可以用来确保代码块完成运行，而不会被其他线程中断。这是通过在代码块运行期间为给定对象获取互斥锁来实现的。lock 语句有一个作为参数的对象，在该参数的后面还有一个一次只能由一个线程执行的代码块。

例如：

```
public class TestThreading
{
    private System.Object lockThis = new System.Object();
```

```
    public void Process()
    {
        lock (lockThis)
        {
            //访问线程的公共资源
        }
    }
}
```

提供给 lock 关键字的参数必须为基于引用类型的对象，该对象用来定义锁的范围。在上面的示例中，锁的范围限定为此函数，因为函数外不存在任何对对象 lockThis 的引用。如果确实存在此类引用，锁的范围将扩展到该对象。严格地说，提供的对象只是用来唯一地标识由多个线程共享的资源，所以它可以是任意类实例。然而，实际上，此对象通常表示需要进行线程同步的资源。例如，如果一个容器对象将被多个线程使用，则可以将该容器传递给 lock，而 lock 后面的同步代码块将访问该容器。只要其他线程在访问该容器前先锁定该容器，则对该对象的访问将是安全同步的。

通常，最好避免锁定 public 类型或锁定不受应用程序控制的对象实例。例如，如果该实例可以被公开访问，则 lock(this) 可能会有问题，因为不受控制的代码也可能会锁定该对象。这可能导致死锁，即两个或更多个线程等待释放同一对象。出于同样的原因，锁定公共数据类型(相比于对象)也可能导致问题。

14.3.5 同步事件和等待句柄

使用锁对于防止同时执行区分线程的代码块很有用，但是这些构造不允许一个线程向另一个线程传达事件。这需要同步事件，它是有两个状态(终止和非终止)的对象，可以用来激活和挂起线程。让线程等待非终止的同步事件可以将线程挂起，将事件状态更改为终止可以将线程激活。如果线程尝试等待已经终止的事件，则线程将继续执行，而不会延迟。

同步事件有两种：AutoResetEvent 和 ManualResetEvent。它们之间唯一的不同在于，无论何时，只要 AutoResetEvent 激活线程，它的状态将自动从终止变为非终止。相反，ManualResetEvent 允许它的终止状态激活任意多个线程，只有当它的 Reset 方法被调用时才还原到非终止状态。

可以通过调用 WaitOne、WaitAny 或 WaitAll 等中的某个等待方法使线程等待事件。WaitHandle.WaitOne() 使线程一直等待，直到单个事件变为终止状态；WaitHandle.WaitAny() 阻止线程，直到一个或多个指示的事件变为终止状态；WaitHandle.WaitAll() 阻止线程，直到所有指示的事件都变为终止状态。当调用事件的 Set 方法时，事件将变为终止状态。

AutoResetEvent 和 ManualResetEvent 的构造函数中，都有 bool 变量来指明线程的终止状态和非终止状态。true 表示终止状态，false 表示非终止状态。

下面用两个示例代码来说明 AutoResetEvent 和 ManualResetEvent 的区别。

示例 1：使用 AutoResetEvent

```
AutoResetEvent _autoResetEvent = new AutoResetEvent(false);
private void TestAutoResetEvent()
{
    Thread t1 = new Thread(this.Thread1);
    t1.Start();
    Thread t2 = new Thread(this.Thread2);
    t2.Start();
    Thread.Sleep(3000);
    _autoResetEvent.Set();
}
void Thread1()
{
    _autoResetEvent.WaitOne();
    MessageBox.Show("线程 1 完成");
}
void Thread2()
{
    _autoResetEvent.WaitOne();
    MessageBox.Show("线程 2 完成");
}
```

该段代码运行的效果是，过 3 秒后，要么弹出“线程 1 完成”，要么弹出“线程 2 完成”，不会两个都弹出，其中一个进行将会结束，而另一个线程永远不会结束。

示例 2：使用 ManualResetEvent

```
ManualResetEvent _menuRestEvent = new ManualResetEvent(false);
private void Test ManualResetEvent()
{
    Thread t1 = new Thread(this.Thread1);
    t1.Start();
    Thread t2 = new Thread(this.Thread2);
    t2.Start();
    Thread.Sleep(3000);
    _menuRestEvent.Set();
}
void Thread1()
{
    _menuRestEvent.WaitOne();
    MessageBox.Show("线程 1 完成");
}
void Thread2()
{
    _menuRestEvent.WaitOne();
    MessageBox.Show("线程 2 完成");
}
```

该段代码运行的效果是，过 3 秒后，“线程 1 完成”和“线程 2 完成”，两个都被弹出。也就是说，两个线程都结束了。这个对比说明了 AutoResetEvent 只会给一个线程发送信号，而不会给多个线程发送信号。在需要同步多个线程的时候，就只能采用 ManualResetEvent

了。深层次的原因是，AutoResetEvent 在 set()之后，会将线程状态自动置为 false，而 ManualResetEvent 在 Set()后，线程的状态就变为 true 了，必须手动 ReSet()之后，才会重新将线程置为 false。这也就是为什么他们的名字一个为 Auto，一个为 Manual 的原因。

14.3.6 数据并行

数据并行是指对源集合或数组中的元素同时（即并行）执行相同操作的情况。System. Threading. Tasks. Parallel 类中 For 和 ForEach 方法的若干重载支持使用强制性语法的数据并行。在数据并行操作中，将对源集合进行分区，以便多个线程能够同时对不同的片段进行操作。并行编程模型支持通过 System. Threading. Tasks. Parallel 类实现的数据并行，此类提供 for 和 foreach 循环基于方法的并行实现，为 Parallel. For 或 Parallel. ForEach 循环编写循环逻辑与编写顺序循环非常类似，不必创建线程或队列工作项。在基本的循环中，不必采用锁，并行编程模型会处理所有低级别工作。下面的代码示例演示一个简单的 foreach 循环及其并行等效项。

```
//非并行编程编程的写法
foreach (var item in sourceCollection)
{
    Process(item);
}
//并行编程的写法
Parallel.ForEach(sourceCollection, item => Process(item));
```

Parallel. For 和 Parallel. ForEach 方法都有若干重载，利用这些重载可以停止或中断循环执行、监视其他线程上循环的状态、维护线程本地状态、完成线程本地对象、控制并发程度，等等。启用此功能的帮助器类型包括 ParallelLoopState、ParallelOptions 以及 ParallelLoopResult、CancellationToken 和 CancellationTokenSource。

通过使用并行循环，将会产生对源集合进行分区和同步工作线程的开销。计算机上的处理器数进一步限制了并行化的优点。在仅仅一个处理器上运行多个主要进行计算的线程时，速度并不会得到提升。因此，必须小心不要对循环进行过度并行化。在嵌套循环中，最有可能发生过度并行化的情况。在多数情况下，除非满足以下一个或多个条件，否则最好仅对外部循环进行并行化。

14.3.7 任务并行

在 Windows Phone 8 编程里面，任务就是一种基于并行编程的异步操作，在某些方面它类似于创建新线程或 ThreadPool 工作项，但抽象级别较高。任务对系统资源的使用效率更高，任务是用于编写多线程，异步和并行代码的首选 API。在前面的异步编程里面已经介绍过任务的创建和监视等内容，这里将主要介绍多个任务并行的协作编程。

Task 和 Task＜TResult＞类提供了相关的方法来操作多个任务的互相协作，如 WhenAll、WhenAny、Delay、FromResult＜TResult＞、WaitAll 和 WaitAny 方法，使用这些

方法可以实现多个任务的等待延时等相关的操作。

1. 等待所有任务完成的任务：Task.WhenAll 方法

Task.WhenAll 方法异步等待多个 Task 或 Task<TResult>对象完成。

下面的基本示例使用 Task.WhenAll 创建表示其他三个任务的完成的任务。

```
//开始多个任务
Task taskA = Task.Run(() => Debug.WriteLine("Hello from taskA."));
Task taskB = Task.Run(() => Debug.WriteLine("Hello from taskB."));
Task taskC = Task.Run(() => Debug.WriteLine("Hello from taskC."));
Task joinTask = Task.WhenAll(new Task[] { taskA, taskB, taskC });
//从 joinTask 的线程中输出消息
Debug.WriteLine("Hello from the joining thread.");
//等待 taskA,taskB,taskC 任务的完成
joinTask.Wait();
/* 输出:
    Hello from the joining thread.
    Hello from taskB.
    Hello from taskA.
    Hello from taskC.
*/
```

2. 等待一个任务完成的任务：Task.WhenAny 方法

Task.WhenAny 方法异步等待多个 Task 或 Task<TResult>对象中任何一个任务的完成，只要其中一个任务完成了，任务就完成了。WhenAny 策略特别用于以下情况。

(1) 冗余操作。考虑在许多方面可以执行的算法或运算。可以使用 WhenAny 方法选择最先完成取消剩余操作的操作。

(2) 交错的操作。可以启动均必须完成和使用 WhenAny 方法处理结果的多操作。在一个操作完成后，可以开始一个或多个其他任务。

(3) 将限制的操作。可以使用 WhenAny 方法通过限制并发数扩展以上方案。

(4) 过期的操作。可以使用 WhenAny 方法选择一个或多个任务以及在特定时间后完成的任务之间，如使用 Delay 方法返回的任务。

下面的基本示例使用 Task.WhenAny 方法。

```
//开始多个任务
Task<int> taskA = Task.Run(() => 88);
Task<int> taskB = Task.Run(() => 42);
Task<int> taskC = Task.Run(() => 99);
Task<Task<int>> first = Task.WhenAny(new Task<int>[] { taskA, taskB, taskC });
Debug.WriteLine("First task to finish returns {0}.", (await first).Result);
/* 输出:
    First task to finish returns 88.
*/
```

3. 任务的延迟：Task.Delay 方法

Task.Delay 方法使在指定时间内完成多个任务之中的一个任务，便终止任务。可以

使用此方法生成为数据偶尔轮询，引入了超时，延迟在一个排序时用户输入处理，以此类推循环。

下面的基本示例创建完成的延续任务，当两个长操作之一完成，或另一种超时过期。

```
//开始两个任务。一个任务的时间是1500毫秒，另一个是2500毫秒
    var taskA = Task.Run(() => Thread.Sleep(1500));
    var taskB = Task.Run(() => Thread.Sleep(2500));

//等待taskA,taskB,Task.Delay(1000)三个任务的最快完成的任务
//这种方法常常会用来作为检查超时操作,Task.Delay方法来定义超时时间
    var result = Task.Factory.ContinueWhenAny(
    new Task[] { taskA, taskB, Task.Delay(1000) },
    first =>
    {
        if (first == taskA)
            Debug.WriteLine("Task A finishes first.");
        else if (first == taskB)
            Debug.WriteLine("Task B finishes first.");
        else
            Debug.WriteLine("Task A and Task B both timed-out.");
    });

//等待任务完成
result.Wait();

/* 输出:
    Task A and Task B both timed-out.
 */
```

4. 有结果的任务：Task.FromResult<TResult>方法

使用 Task.FromResult<TResult> 方法可以创建包含一个预先计算结果的 Task<TResult>对象，也就是意味着直接返回了结果，例如，该方法可以用于对缓存的操作，如果缓存存在则使用 Task.FromResult<TResult> 方法返回结果，若不存在则开始真正的异步操作。

下面的基本示例模拟了 Task.FromResult<TResult> 方法的使用。

```
public string Resut = "";
public Task<string> GetResultAsync()
{
    //如果结果值已经获取将直接返回结果的任务
    if (Resut! = "")
    {
        return Task.FromResult<string>( Resut );
    }
    //如果结果值不存在,则开始计算结果
    return Task.Run(async () =>
    {
        Resut = await GetResult();
```

```
        return Resut;
    });
}
```

5. 任务等待：Task. WaitAll 和 Task. WaitAny 方法

单个 Task 和 Task<TResult>任务可以使用 Task. Wait 和 Task<TResult>. Wait 方法来等待任务完成，多个任务可以使用 Task. WaitAll 方法来等待所有的任务完成，使用 Task. WaitAny 方法来等待多个任务中任一个任务的完成。通常，会出于以下某个原因等待任务：主线程依赖于任务计算的最终结果；必须处理可能从任务引发的异常。

下面的示例演示等待多个任务完成的基本模式。

```
Task[] tasks = new Task[3]
{
    Task.Factory.StartNew(() => MethodA()),
    Task.Factory.StartNew(() => MethodB()),
    Task.Factory.StartNew(() => MethodC())
};
//等待上面 3 个任务的完成
Task.WaitAll(tasks);
```

6. 任务的延续：Task.ContinueWith 方法和 Task<TResult>.ContinueWith 方法

Task. ContinueWith 和 Task<TResult>. ContinueWith 方法使在前面的任务完成之后再启动另外的一个任务。可以在 Result 属性中将用户定义的值从前面的任务传递到其延续任务，以便前面的任务的输出可以作为延续任务的输入。

下面的示例演示 Task. ContinueWith 方法的使用，下面的示例会先完成任务 A，然后把 A 的结果传到任务 B，完成任务 B 之后把 B 的结果传到任务 C，组后完成任务 C。

```
Task<int> taskA = Task.Run(() => 1);
Task<int> taskB = taskA.ContinueWith(x => (int)x.Result + 1);
Task<int> taskC = taskB.ContinueWith(y => (int)y.Result + 2);
taskA.Start();
//或者
Task<string> reportData2 = Task.Factory.StartNew(() => 1)
                                        .ContinueWith((x) => (int)x.Result + 1)
                                       .ContinueWith((y) => (int)y.Result + 2);
```

第15章

联系人和日程安排

Windows Phone 手机允许应用程序读取一些手机以及用户相关的信息，如联系人、日程安排与设备信息。在之前的 Windows Phone 7 版本手机里面对于联系人的操作只能够进行读取，不过在 Windows Phone 8 版本里面，手机对应用程序开放了更大的权限，允许应用程序可以创建手机的联系人，并可以实现增删改的功能。本章将讲解系统联系人查询，日程安排查询和应用程序联系人的增删改等技术知识。

15.1 系统联系人

系统联系人是手机上必备的信息，也是手机里面最原始的特性，在应用程序中获取了手机系统的联系人，然后可以在应用程序里面利用这些联系人的数据，例如打电话、发短信等。

15.1.1 Contacts 类与 Contact 类

Contacts 类提供用于与用户的联系人数据交互的方法和事件。Contacts 类的主要属性事件方法如表 15.1 所示。

表 15.1 Contacts 类的主要属性事件方法

名称	说明
属性 Accounts	获取可从中获取联系人的账户
属性 BaseUri	获取用于指示数据上下文的字符串
方法 ExecuteRequest<T>	在数据上下文中执行搜索对象
方法 SearchAsync	在用户的联系人数据中异步搜索联系人
事件 SearchCompleted	当搜索联系人完成时发生

Contact 类包含有关单个联系人的所有可用的信息。所有单个 Contact 对象都来自 Contacts 对象的 SearchAsync 方法。Contact 类的主要属性方法分别如表 15.2 和表 15.3 所示。

表 15.2 Contact 类的属性

名 称	说 明
Accounts	获取与此联系人关联的数据源
Addresses	获取与此联系人关联的地址
Birthdays	获取与此联系人关联的生日
Children	获取与此联系人关联的孩子
Companies	获取与此联系人关联的公司
CompleteName	获取与此联系人关联的全称
DisplayName	获取此联系人的显示名称
EmailAddresses	获取与此联系人关联的电子邮件地址
IsPinnedToStart	获取一个值,该值指示此联系人是否固定到“开始”屏幕
Notes	获取与此联系人关联的备注
PhoneNumbers	获取与此联系人关联的电话号码
SignificantOthers	获取与此联系人关联的重要他人
Websites	获取与此联系人关联的网站

表 15.3 Contact 类的主要方法

名 称	说 明
GetHashCode	用作联系人类型的哈希函数
GetPicture	获取此联系人的照片
ToString	返回一个字符串,它表示当前的联系人

15.1.2 聚合数据源

Windows Phone 提供跨用户不同账户的用户联系人数据的聚合视图。信息可以来自诸如在手机自身中输入的数据、社交网络站点或其他数据服务提供商之类的源。并非来自所有服务提供商的数据都通过此 API 公开显示。StorageKind 枚举列出了数据的可能来源。StorageKind 枚举的成员如表 15.4 所示,StorageKind 枚举对应的提供商数据如表 15.5 所示。

表 15.4 StorageKind 枚举的成员

成员名称	说 明
Phone	数据来自手机本身
WindowsLive	数据来自 Windows Live 账户
Outlook	数据来自 Microsoft Outlook 账户
Facebook	数据来自 Facebook 账户
Other	数据来自其他位置

表 15.5　StorageKind 枚举对应的数据供应商

数据提供商	联系人姓名	联系人照片	其他联系人数据	日历约会	StorageKind 枚举
Windows Phone 设备	是	是	是	是	Phone
Windows Live 社交	是	是	是	是	WindowsLive
Windows Live Rolodex	是	是	是	是	WindowsLive
Exchange 账户（联系人仅来自本地地址簿，而不是全局地址列表）	是	是	是	是	Outlook
移动运营商地址簿	是	是	是	否	Other
Facebook	是	是	否	否	Facebook
Windows Live 聚合网络（Twitter、LinkedIn 等）	否	否	否	否	None

15.1.3　联系人搜索

可以通过使用内置的搜索筛选器在应用程序中搜索联系人数据，可以搜索到所有联系人或固定到“开始”屏幕的联系人，也可以按姓名、电话号码或电子邮件地址进行搜索。在使用到联系人搜索的 API 的程序里面在需要在 WMAppManifest. xml 文件里面添加 CAPABILITY 元素<Capability Name="ID_CAP_CONTACTS" />，表示允许应用程序获取系统的联系人。

访问联系人数据的常规过程是获取对 Contacts 对象的引用，通过调用 SearchAsync 方法对其执行异步搜索，然后在 SearchCompleted 事件处理程序中捕获 Contact 对象集合形式的结果。有内置的搜索筛选器，可用于按姓名、电话号码或电子邮件地址搜索联系人。针对基础数据库，快速建立这些搜索的索引。也可以搜索所有联系人，然后使用 LINQ 查询结果集合，但与使用内置搜索相比，速度较慢。使用内置筛选器执行的各种搜索的示例如表 15.6 所示。

表 15.6　各种搜索的示例

筛选器种类	示例搜索	说明
None	SearchAsync(String. Empty, FilterKind. None, "State String 1")	搜索所有联系人
PinnedToStart	SearchAsync(String. Empty, FilterKind. PinnedToStart, "State String 2")	搜索固定到“开始”屏幕的所有联系人
DisplayName	SearchAsync("A", FilterKind. DisplayName, "State String 3")	按显示名称进行搜索
EmailAddress	SearchAsync("Chris@example. com", FilterKind. EmailAddress, "State String 4")	按电子邮件地址进行搜索
PhoneNumber	SearchAsync("555-0004", FilterKind. PhoneNumber, "State String 5")	按电话号码进行搜索

下面给出查询手机联系人的示例：查询手机里面所有的联系人信息。

代码清单 15-1：查询手机联系人(源代码：第 15 章\Examples_15_1)

MainPage.xaml 文件主要代码

```
        <Grid x:Name = "ContentPanel" Grid.Row = "1" Margin = "12,0,12,0">
            <ListBox Name = "ContactResultsData" ItemsSource = "{Binding}" >
                <ListBox.ItemTemplate>
                    <DataTemplate>
                        <StackPanel>
                            <TextBlock Text = "{Binding DisplayName}" />
                            <! -- 电话号码 -->
                            <ListBox ItemsSource = "{Binding PhoneNumbers}">
                                <ListBox.ItemTemplate>
                                    <DataTemplate>
                                        <StackPanel Orientation = "Horizontal">
                                            <TextBlock Text = "{Binding PhoneNumber}" />
                                        </StackPanel>
                                    </DataTemplate>
                                </ListBox.ItemTemplate>
                            </ListBox>
                            <! -- 邮箱 -->
                            <ListBox ItemsSource = "{Binding EmailAddresses}">
                                <ListBox.ItemTemplate>
                                    <DataTemplate>
                                        <StackPanel Orientation = "Horizontal">
                                            <TextBlock Text = "{Binding EmailAddress}"/>
                                        </StackPanel>
                                    </DataTemplate>
                                </ListBox.ItemTemplate>
                            </ListBox>
                            <! -- 地址信息 -->
                            <ListBox ItemsSource = "{Binding Addresses}">
                                <ListBox.ItemTemplate>
                                    <DataTemplate>
                                        <StackPanel Orientation = "Horizontal">
                                            <StackPanel>
                                                <TextBlock Text = "{Binding PhysicalAddress.
AddressLine1}"/>
                                                <TextBlock Text = "{Binding PhysicalAddress.
City}"/>
                                            </StackPanel>
                                        </StackPanel>
                                    </DataTemplate>
                                </ListBox.ItemTemplate>
                            </ListBox>
                        </StackPanel>
                    </DataTemplate>
                </ListBox.ItemTemplate>
```

```
        </ListBox>
    </Grid>
```

MainPage. xaml. cs 文件主要代码

```
private Contacts contacts; //手机联系人数据对象
public MainPage()
{
    InitializeComponent();
    contacts = new Contacts();
    contacts.SearchCompleted += contacts_SearchCompleted;
    contacts.SearchAsync("", FilterKind.DisplayName, "查询联系人");
}
//通讯录搜索完成事件处理
void contacts_SearchCompleted(object sender, ContactsSearchEventArgs e)
{
    ContactResultsData.DataContext = e.Results;
}
```

程序的运行效果如图 15.1 所示。

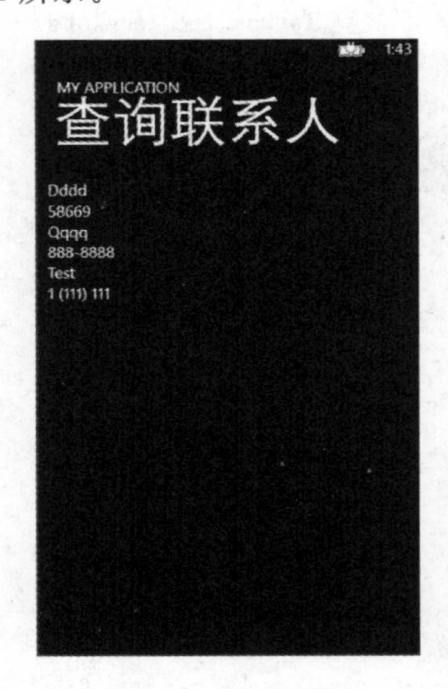

图 15.1　查询联系人

15.2　日程安排

日程安排是指 Windows Phone 手机内置日历里面的记事记录和日历信息。在第三方程序里面可以获取到用户在日历上的记事信息和事件日程的安排。

15.2.1　Appointments 类与 Appointment 类

Appointments 类提供用于与用户的日程安排数据交互的方法和事件。Appointments

类的主要成员如表 15.7 所示。Appointment 类包含有关单个约会的所有可用的信息，Appointment 类属性如表 15.8 所示。

表 15.7　Appointments 类的主要成员

名　称	说　明
属性 Accounts	获取约会所来自的数据源
属性 BaseUri	获取用于指示数据上下文的字符串
方法 ExecuteRequest＜T＞	在数据上下文中执行搜索对象
方法 SearchAsync(DateTime，DateTime，Object)	异步搜索在指定的开始日期和时间与结束日期和时间之间发生的约会
方法 SearchAsync(DateTime，DateTime，Account，Object)	从指定的数据源中，异步搜索在指定的开始日期和时间与结束日期和时间之间发生的约会
方法 SearchAsync(DateTime，DateTime，Int32，Object)	异步搜索在指定的开始日期和时间与结束日期和时间之间发生的约会，只返回不超过指定数量的约会
方法 SearchAsync(DateTime，DateTime，Int32，Account，Object)	从指定的数据源中，异步搜索在指定的开始日期和时间与结束日期和时间之间发生的约会，只返回不超过指定数量的约会
事件 SearchCompleted	当搜索约会完成时发生
字段 DefaultMaximumItems	表示默认情况下从一个搜索返回的约会的最大数量

表 15.8　Appointment 类属性

名　称	说　明
Account	获取与此约会关联的数据源
Attendees	获取与此约会关联的参与者
Details	获取约会的详细说明
EndTime	获取约会结束的日期和时间
IsAllDayEvent	获取一个值，该值指示此约会是否为全天事件
IsPrivate	获取一个值，该值指示此约会是否为私人约会
Location	获取约会的地点
Organizer	获取约会的组织者
StartTime	获取约会开始的日期和时间
Status	获取有关如何处理此约会的时间块的信息，如忙碌或外出
Subject	获取约会的主题

15.2.2　日程安排查询

访问日历数据的过程是获取对 Appointments 对象的引用，对其执行异步搜索，然后捕获 Appointment 对象集合形式的结果。

下面给出查询日程安排的示例：查询手机里面所有的日程安排信息。

代码清单 15-2：查询日程安排(源代码：第 15 章\Examples_15_2)

MainPage. xaml 文件主要代码

```
<Grid x:Name = "ContentPanel" Grid.Row = "1" Margin = "12,0,12,0">
    <ListBox x:Name = "lbApp"> </ListBox>
</Grid>
```

MainPage. xaml. cs 文件主要代码

```
private Appointments appointments;                          //手机日程安排数据对象
public MainPage()
{
    InitializeComponent();
    appointments = new Appointments();
    appointments.SearchCompleted += contacts_SearchCompleted;
    //查询所有的日程安排
    appointments.SearchAsync(DateTime.MinValue, DateTime.MaxValue, this);
}
//日程安排搜索完成事件处理
void contacts_SearchCompleted(object sender, AppointmentsSearchEventArgs e)
{
    foreach(var appointment in e.Results)
    {
        lbApp.Items.Add(new TextBlock { Text = "标题：" + appointment.Subject +
"内容：" + appointment.Details});
    }
}
```

程序的运行效果如图 15.2 所示。

图 15.2　日程安排查询

15.3 程序联系人存储

程序联系人存储是第三方的应用程序创建的联系人数据，这些联系人的数据也可以在手机的通讯录里面进行显示，但是它们是由创建这些联系人数据的第三方应用程序所管理的。联系人数据的归属应用程序可以设置这些联系人数据的系统和其他程序的访问权限，对属于它自己的联系人具有增删改的权限，并且一旦用户卸载了联系人数据归属应用程序，这些联系人也会被删除掉。程序联系人存储的 API 在空间 Windows. Phone. PersonalInformation 下，下面来看一下如何去使用这些 API 来操作联系人。

15.3.1 ContactStore 类和 StoredContact 类

ContactStore 类表示一个 Windows Phone 应用程序的自定义联系人存储，它是应用程序存储的一个管理者，负责管理应用程序所创建的联系人。ContactStore 类的主要成员如表 15.9 所示。StoredContact 类表示一个应用程序自定义的联系人存储，它继承了 IContactInformation 接口，所有由应用程序创建的联系人都是一个 StoredContact 类的对象。StoredContact 类的主要成员如表 15.10 所示。

表 15.9 ContactStore 类的主要成员

成　　员	说　　明
public ulong RevisionNumber { get; }	联系人存储的版本号
public ContactQueryResult CreateContactQuery()	创建一个默认的联系人查询，返回 ContactQueryResult 对象，包含了存储中的联系人
public ContactQueryResult CreateContactQuery (ContactQueryOptions options)	创建一个自定义的联系人查询，返回 ContactQueryResult 对象，包含了存储中的联系人
public static IAsyncOperation<ContactStore> CreateOrOpenAsync()	异步方法创建或者打开应用程序的自定义联系人存储，假如存储不存在将创建一个存储
public static IAsyncOperation<ContactStore> CreateOrOpenAsync (ContactStoreSystemAccessMode access ContactStoreApplicationAccessMode sharing)	异步方法创建或者打开应用程序的自定义联系人存储，假如存储不存在将创建一个存储，返回当前的联系人存储对象 access：联系人是否可以在手机系统通讯录里面进行编辑还是只能在应用程序中创建 sharing：是否存储的联系人所有属性都可以在另外的应用程序里面进行访问
public IAsyncAction DeleteAsync()	异步方法删除应用程序的联系人存储
public IAsyncAction DeleteContactAsync(string id)	异步方法通过联系人的 ID 删除应用程序里面存储的联系人
public IAsyncOperation < StoredContact > FindContactByIdAsync(string id)	异步方法通过 ID 查找应用程序的联系人，返回 StoredContact 对象
public IAsyncOperation < StoredContact > FindContactByRemoteIdAsync(string id)	异步方法通过 remote ID 查找应用程序的联系人，返回 StoredContact 对象

续表

成　员	说　明
public IAsyncOperation < IReadOnlyList < ContactChangeRecord >> GetChangesAsync (ulong baseREvisionNumber)	异步方法通过联系人的版本号获取联系人改动记录
public IAsyncOperation<IDictionary<string，object>> LoadExtendedPropertiesAsync()	异步方法加载应用程序联系人的扩展属性 Map 表
public IAsyncAction SaveExtendedPropertiesAsync (IReadOnlyDictionary<string，object> data)	异步方法保存应用程序联系人的扩展属性 Map 表

表 15.10　StoredContact 类的主要成员

成　员	说　明
public StoredContact(ContactStore store)	通过当前应用程序的 ContactStore 来初始化一个 StoredContact 对象
public StoredContact (ContactStore store，ContactInformation contact)	通过 ContactStore 对象和 ContactInformation 对象来创建一个 StoredContact 对象，StoredContact 对象的信息由 ContactInformation 对象来提供
public string DisplayName { get；set；}	获取或设置一个存储联系人的显示名称
public IRandomAccessStreamReference DisplayPicture { get；}	获取一个存储联系人的图片
public string FamilyName { get；set；}	获取或设置一个存储联系人的家庭名
public string GivenName { get；set；}	获取或设置一个存储联系人的名字
public string HonorificPrefix { get；set；}	获取或设置一个存储联系人的尊称前缀
public string HonorificSuffix { get；set；}	获取或设置一个存储联系人的尊称后缀
public string Id { get；}	获取应用程序存储联系人的 ID
public string RemoteId { get；set；}	获取或设置应用程序联系人的 RemoteId
public ContactStore Store { get；}	获取当前应用程序联系人所在的联系存储对象
public IAsyncOperation<IRandomAccessStream> GetDisplayPictureAsync()	获取一个存储联系人的图片
public IAsyncOperation<System. Collections. Generic. IDictionary < string， object >> GetExtendedPropertiesAsync()	异步方法获取联系人的扩展属性 Map 表
public IAsyncOperation<System. Collections. Generic. IDictionary < string， object >> GetPropertiesAsync()	异步方法获取联系人的已知属性 Map 表
public IAsyncAction ReplaceExistingContactAsync(string id)	异步方法使用当前的联系人来替换联系人存储中某个 ID 的联系人
public IAsyncAction SaveAsync()	异步方法保存当前的联系人到联系人存储中
public IAsyncAction SetDisplayPictureAsync (IInputStream stream)	异步方法保存当前的联系人的图片
public IAsyncOperation<IRandomAccessStream> ToVcardAsync()	异步方法把当前的联系人转化为 VCard 信息流

15.3.2　程序联系人的新增

新增程序联系人需要先创建或者打开程序的联系人存储 ContactStore，并且可以设置该程序联系人存储的被访问权限。创建的代码如下：

```
ContactStore conStore = await ContactStore.CreateOrOpenAsync();
```

联系人存储对于系统通讯和其他程序的都有权限的限制，ContactStoreSystemAccessMode 枚举表示手机系统通讯录对应用程序联系人的访问权限，有 ReadOnly 只读权限和 ReadWrite 读写两个权限，ContactStoreApplicationAccessMode 枚举表示第三方应用程序对应用程序联系人的访问权限类型，有 LimitedReadOnly 限制只读权限和 ReadOnly 只读权限。上面的代码创建联系人存储的代码是默认用了最低的访问权限来创建联系人存储，即联系人对于系统通讯录是只读的权限，对于其他程序的访问权限是限制只读权限。下面来看一下自定义权限的创建联系人存储。

```
//创建一个系统通讯可以读写和其他程序只读的联系人存储
ContactStore conStore = await ContactStore.CreateOrOpenAsync(ContactStoreSystemAccessMode.
ReadWrite, ContactStoreApplicationAccessMode.ReadOnly);
```

接下来看一下如何创建一个联系人。

1. 第一种方式直接通过联系人存储创建联系人

```
//创建或者打开联系人存储
ContactStore conStore = await ContactStore.CreateOrOpenAsync();
//保存联系人
StoredContact storedContact = new StoredContact(conStore);
//设置联系人的展示名称
storedContact.DisplayName = "展示名称";
//保存联系人
await storedContact.SaveAsync();
```

2. 第二种方式通过 ContactInformation 类对象创建联系人

ContactInformation 类表示一个非系统存储中联系人的联系人信息。ContactInformation 类的主要成员如表 15.11 所示。

```
//创建一个 ContactInformation 类
ContactInformation conInfo = new ContactInformation();
// 获取 ContactInformation 类的属性 map 表
var properties = await conInfo.GetPropertiesAsync();
//添加电话属性
properties.Add(KnownContactProperties.Telephone, "123456");
//添加名字属性
properties.Add(KnownContactProperties.GivenName, "名字");
//创建或者打开联系人存储
ContactStore conStore = await ContactStore.CreateOrOpenAsync();
//保存联系人
StoredContact storedContact = new StoredContact(conStore, conInfo);
//保存联系人
```

```
await storedContact.SaveAsync();
```

表 15.11 ContactInformation 类的主要成员

成　　员	说　　明
ContactInformation()	初始化
string DisplayName	联系人的显示名字
IRandomAccessStreamReference DisplayPicture	联系人图片信息
string FamilyName	联系人姓
string GivenName	名字
string HonorificPrefix	尊称前缀
string HonorificSuffix	尊称后缀
IAsyncOperation<IRandomAccessStream> GetDisplayPictureAsync()	异步获取联系人图片
IAsyncOperation < System. Collections. Generic. IDictionary <string, object>> GetPropertiesAsync()	异步获取联系人的 Map 表信息
IAsyncOperation < ContactInformation > ParseVcardAsync (IInputStream vcard)	异步解析一个 VCard 数据流为 ContactInformation 对象
IAsyncAction SetDisplayPictureAsync(IInputStream stream)	异步设置联系人的图片通过图片数据流
IAsyncOperation<IRandomAccessStream> ToVcardAsync()	异步转换 ContactInformation 对象为一个 VCard 信息流

15.3.3 程序联系人的查询

联系人查询也需要创建联系人存储，创建联系人存储的形式和联系人新增是一样的。联系人查询是通过 ContactStore 的 CreateContactQuery 方法来创建一个查询，可以查询的参数 ContactQueryOptions 来设置查询返回的结果和排序的规则，创建的查询时 ContactQueryResult 类型。可以通过 ContactQueryResult 类的 GetContactsAsync 异步方法获取联系人存储中的联系人列表和通过 GetCurrentQueryOptions 方法获取当前的查询条件。下面来看创建联系人查询的代码如下：

```
conStore = await ContactStore.CreateOrOpenAsync();
ContactQueryResult conQueryResult = conStore.CreateContactQuery();
uint count = await conQueryResult.GetContactCountAsync();
IReadOnlyList<StoredContact> conList = await conQueryResult.GetContactsAsync();
```

15.3.4 程序联系人的编辑

联系人编辑删除也需要创建联系人存储，创建联系人存储的形式和联系人新增是一样的。联系人的编辑需要首先要获取要编辑的联系人，获取编辑的联系人可以通过联系人的 id 或者 remoteid 来获取，获取到的联系人是一个 StoredContact 对象，通过修改该对象的属性，然后再调用 SaveAsync 保存方法就可以实现编辑联系人了。删除联系人可以分为删除一个联系人和删除所有的联系人，删除一个联系人可以通过联系人的 id 然后调用

ContactStore 的 DeleteContactAsync 方法来进行删除，如果要删除所有的联系人那么就要调用 ContactStore 的 DeleteAsync 方法。联系人的新增，编辑和删除都会有相关的操作记录，GetChangesAsync 方法来获取联系人的修改记录。下面来看一下修改一个联系人的代码：

```
ContactStore conStore = await ContactStore.CreateOrOpenAsync();
StoredContact storCon = await conStore.FindContactByIdAsync(id);
var properties = await storCon.GetPropertiesAsync();
properties[KnownContactProperties.Telephone] = "12345678";
await storCon.SaveAsync();
```

15.3.5 程序联系人的删除

删除联系人的可以分为删除一个联系人和删除所有的联系人，删除一个联系人的可以通过联系人的 id 然后调用 ContactStore 的 DeleteContactAsync 方法来进行删除，如果要删除所有的联系人那么就要调用 ContactStore 的 DeleteAsync 方法。联系人的新增，编辑和删除都会有相关的操作记录，GetChangesAsync 方法来获取联系人的修改记录。下面来看一下删除一个联系人的代码：

```
ContactStore conStore = await ContactStore.CreateOrOpenAsync();
await conStore.DeleteContactAsync (id);
await conStore.DeleteAsync ();
```

15.3.6 实例演示联系人存储的使用

下面给出查询日程安排的示例：查询手机里面所有的日程安排信息。

代码清单 15-3：联系人存储的增删改(源代码：第 15 章\Examples_15_3)

MainPage.xaml 文件主要代码：联系人新增页面

```
<Grid x:Name = "ContentPanel" Grid.Row = "1" Margin = "12,0,12,0">
    <StackPanel >
        <TextBlock HorizontalAlignment = "Left" Text = "名字" FontSize = "30"/>
        <TextBox HorizontalAlignment = "Left" x:Name = "name" Height = "85" Text =
"" Width = "296"/>
        <TextBlock HorizontalAlignment = "Left" Text = "电话" FontSize = "30" />
        <TextBox HorizontalAlignment = "Left" x:Name = "phone" Height = "85" Text =
"" Width = "296"/>
        <Button Content = "保存" HorizontalAlignment = "Left" Width = "308" Height =
"91" Click = "Button_Click_1"/>
        <Button Content = "查询应用存储的联系人" HorizontalAlignment = "Left" Width =
"308" Height = "91" Click = "Button_Click_2"/>
    </StackPanel >
</Grid >
```

MainPage.xaml.cs 文件主要代码

```
//新增一个联系人
```

```
private async void Button_Click_1(object sender, RoutedEventArgs e)
{
    if (name.Text != "" && phone.Text != "")
    {
        try
        {
            //创建一个联系人的信息对象
            ContactInformation conInfo = new ContactInformation();
            //获取联系人的属性字典
            var properties = await conInfo.GetPropertiesAsync();
            //添加联系人的属性
            properties.Add(KnownContactProperties.Telephone, phone.Text);
            properties.Add(KnownContactProperties.GivenName, name.Text);
            //创建或者打开联系人存储
            ContactStore conStore = await ContactStore.CreateOrOpenAsync();
            StoredContact storedContact = new StoredContact(conStore, conInfo);
            //保存联系人
            await storedContact.SaveAsync();
            MessageBox.Show("保存成功");
        }
        catch (Exception ex)
        {
            MessageBox.Show("保存失败,错误信息：" + ex.Message);

        }
    }
    else
    {
        MessageBox.Show("名字或电话不能为空");
    }
}
//跳转到联系人列表页面
private void Button_Click_2(object sender, RoutedEventArgs e)
{
    NavigationService.Navigate(new Uri("/ContactsList.xaml", UriKind.Relative));
}
```

ContactsList.xaml 文件主要代码：联系人列表页面

```
<Grid x:Name="ContentPanel" Grid.Row="1" Margin="12,0,12,0">
  <StackPanel>
      <ListBox x:Name="conListBox" ItemsSource="{Binding}" >
          <ListBox.ItemTemplate >
            <DataTemplate>
              <StackPanel>
                <TextBlock Text="{Binding Name}" />
                <TextBlock Text="{Binding Id}" />
                <TextBlock Text="{Binding Phone}" />
                <Button Content="删除" HorizontalAlignment="Left" Width="308"
Height="91" Click="Button_Click_1"/>
                <Button Content="编辑" HorizontalAlignment="Left" Width="308"
```

```
Height = "91" Click = "Button_Click_2"/>
                        <TextBlock Text = " ----------------------------- " />
                    </StackPanel>
                </DataTemplate>
            </ListBox.ItemTemplate>
        </ListBox>
    </StackPanel>
</Grid>
```

ContactsList.xaml.cs 文件主要代码

```
private ContactStore conStore;                          //联系人存储
public ContactsList()
{
    InitializeComponent();
}
//进入页面事件
protected override void OnNavigatedTo(System.Windows.Navigation.NavigationEventArgs e)
{
    GetContacts();
    base.OnNavigatedTo(e);
}
//获取联系人列表
async private void GetContacts()
{
    conStore = await ContactStore.CreateOrOpenAsync();
    ContactQueryResult conQueryResult = conStore.CreateContactQuery();
    //查询联系人
    IReadOnlyList<StoredContact> conList = await conQueryResult.GetContactsAsync();
    List<Item> list = new List<Item>();
    foreach (StoredContact storCon in conList)
    {
        var properties = await storCon.GetPropertiesAsync();
        list.Add(
            new Item
            {
                Name = storCon.FamilyName + storCon.GivenName,
                Id = storCon.Id,
                Phone = properties[KnownContactProperties.Telephone].ToString()
            });
    }
    conListBox.ItemsSource = list;
}
//删除联系人事件处理
private async void Button_Click_1(object sender, RoutedEventArgs e)
{
    Button deleteButton = sender as Button;
    Item deleteItem = deleteButton.DataContext as Item;
    await conStore.DeleteContactAsync(deleteItem.Id);
    GetContacts();
}
```

```
        //跳转到编辑联系人页面
        private void Button_Click_2(object sender, RoutedEventArgs e)
        {
            Button deleteButton = sender as Button;
            Item editItem = deleteButton.DataContext as Item;
            NavigationService.Navigate(new Uri("/EditContact.xaml?Id=" + editItem.Id,
UriKind.Relative));
        }
    }
    //自定义绑定的联系人数据对象
    class Item
    {
        public string Name { get; set; }
        public string Id { get; set; }
        public string Phone { get; set; }
    }
```

EditContact.xaml 文件主要代码：联系人编辑页面

```
        <Grid x:Name="ContentPanel" Grid.Row="1" Margin="12,0,12,0">
          <StackPanel>
            <TextBlock HorizontalAlignment="Left" Text="名字" FontSize="30"/>
            <TextBox HorizontalAlignment="Left" x:Name="name" Height="85" Text=
"" Width="296"/>
            <TextBlock HorizontalAlignment="Left" Text="电话" FontSize="30" />
            <TextBox HorizontalAlignment="Left" x:Name="phone" Height="85" Text=
"" Width="296"/>
            <Button Content="保存" HorizontalAlignment="Left" Width="308" Height="91"
Click="Button_Click_1"/>
          </StackPanel>
        </Grid>
```

EditContact.xaml.cs 文件主要代码

```
    private string conId = "";
    private ContactStore conStore;                              //联系人数据存储
    private StoredContact storCon;                              //联系人对象
    private IDictionary<string, object> properties;             //联系人属性字典
    public EditContact()
    {
        InitializeComponent();
    }

    //进入页面事件处理
    protected override void OnNavigatedTo(System.Windows.Navigation.NavigationEventArgs e)
    {
        //通过联系人的 id 获取联系人的信息
        if (NavigationContext.QueryString.ContainsKey("Id"))
        {
            conId = NavigationContext.QueryString["Id"].ToString();
            GetContact(conId);
        }
```

```
        base.OnNavigatedTo(e);
    }
    //保存编辑的联系人
    private async void Button_Click_1(object sender, RoutedEventArgs e)
    {
        if (name.Text != "" && phone.Text != "")
        {
            storCon.GivenName = name.Text;
            properties[KnownContactProperties.Telephone] = phone.Text;
            await storCon.SaveAsync();//保存联系人
            NavigationService.GoBack(); //返回上一个页面
        }
        else
        {
            MessageBox.Show("名字或者电话不能为空");
        }
    }
    //获取需要编辑的联系人信息
    async private void GetContact(string id)
    {
        conStore = await ContactStore.CreateOrOpenAsync();
        storCon = await conStore.FindContactByIdAsync(id);
        properties = await storCon.GetPropertiesAsync();
        name.Text = storCon.GivenName;
        phone.Text = properties[KnownContactProperties.Telephone].ToString();
    }
        phone.Text = properties[KnownContactProperties.Telephone].ToString();
    }
```

程序的运行效果如图 15.3～图 15.5 所示。

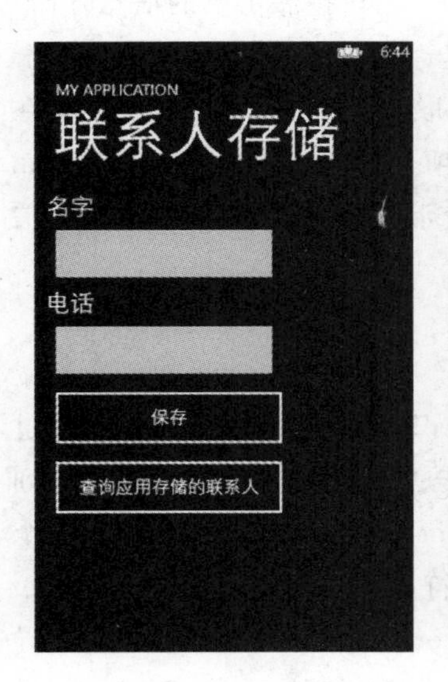

图 15.3　联系人存储

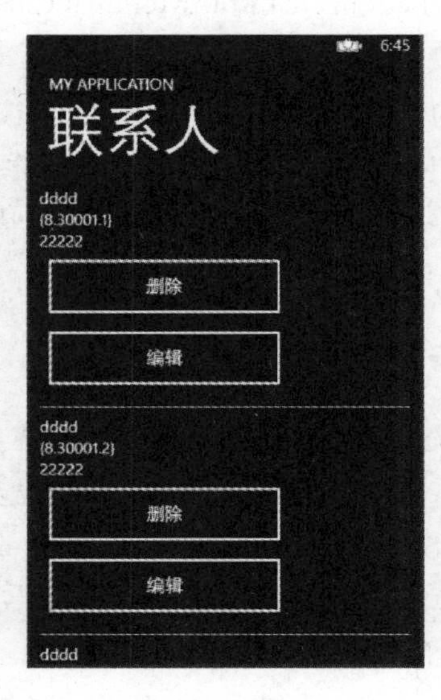

图 15.4　联系人

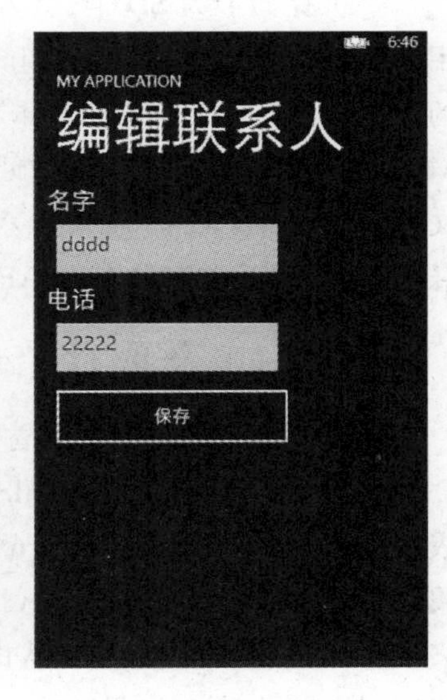

图 15.5　编辑联系人

第16章

手机文件数据读写

在 Windows Phone 里面的手机文件数据包括手机的图片音频，存储卡的文件和应用程序自身存储的文件。在 Windows Phone 8 里面新增了存储卡的特性，并提供了读取存储卡信息的相关 API，不过并没有提供写入存储卡的相关 API。对于图片音频文件，Windows Phone 8 系统给第三方应用程序提供了读写的权限，应用程序里可以调用手机本地的图片音频并进行保存。在应用程序内建立的文件也是属于手机文件的数据，Windows Phone 8 提供了两种访问的协议，开发者可以根据应用的自身情况进行选取。下面来看一下手机文件数据读写的内容。

16.1 手机存储卡数据

手机存储卡为 SD 卡（全名为 Secure Digital Memory Card），用户扩展手机的信息存储，用户可以很方便地进行更换和安装。在 Windows Phone 8 里面第三方程序只有对存储卡的读取权限，包括读取存储卡的文件目录和文件的信息。读取存储卡信息的相 API 都在 Microsoft. Phone. Storage 空间下，读取存储卡的信息需要在项目的 WMAppManifest. xml 文件中添加 ID_CAP_REMOVABLE_STORAGE 的能力，表示程序需要读取存储卡的信息，否则调用读取存储卡的 API 将会引发异常。

16.1.1 获取存储卡文件夹

要读取存储卡文件夹首先需要去识别存储卡设备，识别存储卡设备可以通过 ExternalStorage. GetExternalStorageDevicesAsync()方法异步去获取一个存储卡设备的列表，存储卡设备对象用 ExternalStorageDevice 类表示。ExternalStorageDevice 类包含两个属性，一个是 ExternalStorageID 表示扩展存储的唯一 ID 字符串，另外一个是 RootFolder 表示存储卡的根目录，是默认的最顶层的文件夹。ExternalStorageDevice 类还包含两个异步方法，一个是 GetFileAsync 方法通过文件的路径来获取文件的信息，一个是 GetFolderAsync 方法通过文件夹的路径来获取文件夹的信息。

ExternalStorageFolder 类表示一个存储卡的文件夹类，里面包含了文件夹的一些详细的信息，以及文件夹的相关方法，ExternalStorageFolder 类的成员如表 16.1 所示。

表 16.1　ExternalStorageFolder 类的成员

名　称	说　明
DateModified	文件修改的时间
Name	文件夹的名字
Path	文件夹的路径
GetFilesAsync()	获取该文件夹的所有文件
GetFolderAsync(string name)	获取该文件夹目录下该名字文件夹的信息
GetFoldersAsync()	获取该文件夹下面所有的文件夹

下面开看一下获取存储卡文件目录的代码示例：

```
async void GetFolder()
{
    //获取扩展的存储卡列表
    IEnumerable <ExternalStorageDevice>  deviceList  =  await  ExternalStorage.
GetExternalStorageDevicesAsync();
    //遍历存储卡列表
    foreach (ExternalStorageDevice device in deviceList)
    {
        //遍历存储卡根目录
        foreach (ExternalStorageFolder folder in await device.RootFolder.GetFoldersAsync())
        {    //遍历存储卡里面的文件夹
            foreach (ExternalStorageFolder folder2 in await folder.GetFoldersAsync())
            {
                //获取文件夹的信息 如名字 folder2.Name 等
            }
        }
    }
}
```

16.1.2　获取存储卡文件

ExternalStorageFile 类表示一个存储卡的文件类，里面包含了文件的一些详细的信息，以及打开文件的方法，ExternalStorageFile 类的成员如表 16.2 所示。要读取存储卡的文件，必须要先读取出文件夹之后，然后再通过 ExternalStorageFolder 类的 GetFilesAsync() 的方法来获取该文件夹下的所有文件，如果知道文件的存储路径也可以使用 ExternalStorageDevice 存储卡设备类的 GetFileAsync(string filePath) 方法来获取具体路径的文件。

表 16.2　ExternalStorageFile 类的成员

名　称	说　明
DateModified	文件修改的时间
Name	文件夹的名字
Path	文件夹的路径
Task<IO.Stream> OpenForReadAsync()	打开文件夹的方法

下面来看一下获取存储卡根目录文件的代码示例：

```
async void GetFile()
{
    //获取扩展的存储卡列表
    IEnumerable <ExternalStorageDevice> deviceList = await ExternalStorage.
GetExternalStorageDevicesAsync();
    //遍历存储卡列表
    foreach (ExternalStorageDevice device in deviceList)
    {    //遍历存储卡根目录的文件
        foreach (ExternalStorageFile file in await device.RootFolder.GetFilesAsync())
        {
            //处理文件的信息
        }
    }
}
```

16.1.3　实例：读取存储卡信息

下面给出读取存储卡信息的示例：在手机应用程序里面读取手机的存储卡的信息，按照文件夹目录一层层来展示存储卡的文件。

代码清单 16-1：读取存储卡信息(源代码：第 16 章\Examples_16_1)

MainPage.xaml 文件主要代码

```
<Grid x:Name="ContentPanel" Grid.Row="1" Margin="12,0,12,0">
    <StackPanel>
        <Button Content="获取 SD 卡的文件信息" x:Name="btGetFile" Margin="0,0,
0,80" Click="btGetFile_Click"/>
        <TextBlock Text="文件夹列表："/>
        <ListBox x:Name="lbFolder"></ListBox>
        <Button Content="打开选中的文件夹" Click="Button_Click_1" />
        <TextBlock Text="文件列表：" Margin="0,50,0,0"/>
        <ListBox x:Name="lbFile"></ListBox>
    </StackPanel>
</Grid>
```

MainPage.xaml.cs 文件代码

```
using System.Collections.Generic;
using System.Windows;
```

```
using System.Windows.Controls;
using Microsoft.Phone.Controls;
using Microsoft.Phone.Storage;
namespace SDCardDemo
{
    public partial class MainPage : PhoneApplicationPage
    {
        IEnumerable<ExternalStorageDevice> deviceList;    //存储卡设备列表
        public MainPage()
        {
            InitializeComponent();
        }
        //打开存储卡的根目录
        private async void btGetFile_Click(object sender, System.Windows.RoutedEventArgs e)
        {
            //获取存储卡设备列表
            deviceList = await ExternalStorage.GetExternalStorageDevicesAsync();
            lbFolder.Items.Clear();
            foreach (ExternalStorageDevice device in deviceList)
            {
                //把获取到的设备根目录信息添加到ListBox控件里面
                ListBoxItem item = new ListBoxItem();
                item.Content = "根目录" + device.ExternalStorageID;
                item.DataContext = device.RootFolder;
                lbFolder.Items.Add(item);
            }
        }
        //打开文件夹获取文件夹里面的文件夹和文件
        private async void Button_Click_1(object sender, System.Windows.RoutedEventArgs e)
        {
            if (lbFolder.SelectedIndex == -1)
            {
                MessageBox.Show("请选择一个文件夹");
            }
            else
            {
                //获取选中的文件夹
                ListBoxItem item = lbFolder.SelectedItem as ListBoxItem;
                ExternalStorageFolder    externalStorageDevice    =    item.DataContext    as
ExternalStorageFolder;
                //获取文件夹中的文件夹
                IEnumerable< ExternalStorageFolder > folderList = await externalStorageDevice.
GetFoldersAsync();
                lbFolder.Items.Clear();
                foreach (ExternalStorageFolder folder in folderList)
                {
                    //把获取到的文件夹信息添加到ListBox控件里面
                    ListBoxItem item2 = new ListBoxItem();
                    item2.Content = "文件夹: " + folder.Name;
                    item2.DataContext = folder;
```

```
                        lbFolder.Items.Add(item2);
                    }
                    //获取文件夹中的文件
                    IEnumerable<ExternalStorageFile> fileList = await externalStorageDevice.
GetFilesAsync();
                    lbFile.Items.Clear();
                    foreach (ExternalStorageFile file in fileList)
                    {
                        //把获取到的文件信息添加到 ListBox 控件里面
                        ListBoxItem item3 = new ListBoxItem();
                        item3.Content = "文件：" + file.Name;
                        item3.DataContext = file;
                        lbFile.Items.Add(item3);
                    }
                }
            }
        }
    }
```

程序的运行效果如图 16.1 所示。

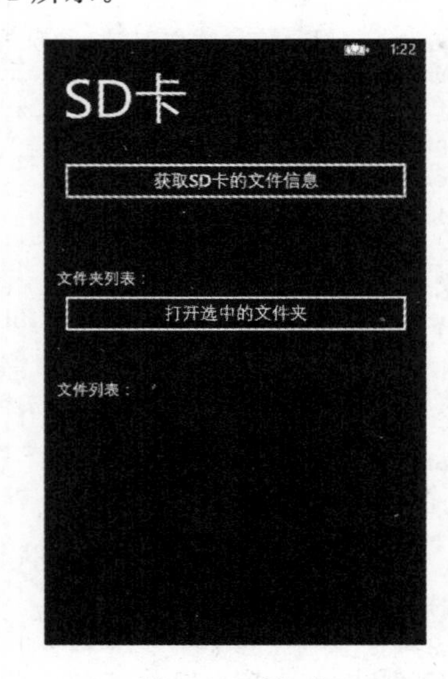

图 16.1　SD 卡

16.2　图片音频数据

图片和音频数据是手机里面非常常用也是非常重要的数据，在 Windows Phone 手机里面，我们可以通过应用程序来获取手机的图片和音频文件，也可以把图片和音频文件保存到手机里面。手机的图片有很多用途，与手机相关的图片应用程序也五花八门的，例如图片美化，图片备份，图片分享等。手机音频通常会是手机里面的音乐数据，应用程序也可以利用这些音频的数据来开发音乐播放器等。

16.2.1 获取手机图片和音频数据

Windows Phone 里面的图片和音频数据是存在在手机的多媒体库里面的，并不是像SD存储卡里面的文件一样，存储在一个文件夹目录分明的存储区域里面。在应用程序里面访问手机的图片和音频数据，需要通过手机的媒体库来获取。MediaLibrary 类是 Windows Phone 中的媒体库，它包含了影音、图片，是所有媒体的集合，该类在 Microsoft. Xna. Framework. Media 空间名下，是 XNA 的 API 的一部分，MediaLibrary 类的主要属性如表 16.3 所示。在应用程序里面可以直接通过 Pictures 属性来获取到手机里面的图片文件的集合和通过 Songs 属性来获取到手机的音频的集合。示例代码如下：

```
//创建媒体库
MediaLibrary media = new MediaLibrary();
//获得媒体库中所有的图片
PictureCollection pics = media.Pictures;
if (pics.Count > 1)
{
    Picture pic = pics[0];
    //pic 是多媒体库里面的一个图片对象
}
SongCollection scs = media.Songs;
if (scs.Count > 1)
{
    Song s = sc[0];
    //s 是多媒体库里面的一个音频对象
}
```

表 16.3 MediaLibrary 类的主要属性

名　称	说　明
Albums	获取在媒体库中包含的所有专辑
Artists	获取在媒体库中包含所有的艺术家
Pictures	获取包含在媒体库中的所有图片
Playlists	获取包含在媒体库中的播放列表
SavedPictures	获取媒体库中保存的图片
Songs	获取在媒体库中的所有歌曲

16.2.2 保存图片到手机

在 Windows Phone 应用程序里面可以将图片保存到图片中心的“保存的图片”或“本机拍照”相册中。为此，分别使用 Microsoft. Xna. Framework. Media 命名空间中的 SavePicture 或 SavePictureToCameraRoll 方法。SavePicture(String, Byte[]), SavePicture(String, Stream)保存图片到图片中心的“相册”，第一个参数是保存到媒体库的图片名称，第二个参数是需要被保存的图片数据。SavePictureToCameraRoll(String，Byte[])，SavePictureToCameraRoll(String，Stream)——保存图片到图片中心的“本机拍照”，第一个参数是保存到媒体库的图片名称，第二个参数是需要被保存的图片数据。下面的代码示例如保存图片到手机：

```
//从独立存储里读出的图片文件,picPath为图片在独立存储中位置
IsolatedStorageFileStream myFileStream = myStore.OpenFile(picPath, FileMode.Open, FileAccess.Read);
//把图片保存到手机的媒体库
Picture pic = library.SavePicture(picPath, myFileStream);
myFileStream.Close();
```

16.2.3 保存和删除手机音频

Windows Phone 8 可以保存和删除手机的音频，保存和删除手机的音频操作的 API 在 Microsoft.Xna.Framework.Media.PhoneExtensions 空间下，通过 MediaLibraryExtensions 类的 SaveSong(this MediaLibrary library，Uri filename，SongMetadata songMetadata，SaveSongOperation operation)方法实现保存音频到手机上，library 是指手机的媒体库，filename 为音频的路径，songMetadata 是音频的元数据，operation 表示操作的类型，调用 Delete(this MediaLibrary library，Song song)方法实现删除手机的音频，library 是指手机的媒体库，song 为手机的音频对象。代码示例删除手机的一个音频：

```
//创建媒体库
MediaLibrary media = new MediaLibrary();
SongCollection scs = media.Songs;
if (scs.Count > 1)
{
    Song s = scs[0];
    //删除音频
    MediaLibraryExtensions.Delete(media,s);
}
```

16.3 应用程序本地数据

应用程序本身也有一个文件目录的结构，存放的是应用程序的文件，包括在应用程序下载的文件，应用程序新建的文件，应用程序在程序包里面添加的文件等。下面来看一下应用程序本地数据的操作。

16.3.1 应用程序本地文件夹和文件

应用程序本地的根目录文件夹可以通过 ApplicationData 类的 LocalFolder 属性来获取到，ApplicationData 类是一个单例模式结构的类，所以通过 ApplicationData.Current.LocalFolder 便可以取得应用程序的根目录文件夹。应用程序内部文件夹的类为 StorageFolder 类，表示操作文件夹及其内容，并提供有关它们的信息，StorageFolder 类的主要成员如表 16.4 所示。StorageFile 类为应用程序里面的文件类它的主要成员如表 16.5

所示。应用程序会有一些默认的文件夹,在程序建立的时候就已经存在了,并且这些默认的文件夹都有一些特殊的作用,如表 16.6 所示。

表 16.4 StorageFolder 类的主要成员

名 称	说 明
DateCreated	获取创建文件夹的日期和时间
Name	获取存储文件夹的名称
Path	获取存储文件夹的路径
CreateFileAsync(string desiredName)	在文件夹或文件组中创建一个新文件;desiredName:要创建的文件的所需名称,返回作为 StorageFile 的新文件
CreateFolderAsync(string desiredName)	在当前文件夹中创建新的文件夹;desiredName:要创建的文件夹的所需名称,返回新文件作为 StorageFolder
DeleteAsync();	删除当前文件夹或文件组
GetFileAsync(string name)	从当前文件夹获取指定文件;name:要检索的文件的名称,返回表示文件的 StorageFile
GetFilesAsync();	在当前文件夹中获取文件;返回文件夹中的文件列表(类型 IReadOnlyList<StorageFile >),列表中的每个文件均由一个 StorageFile 对象表示
GetFolderAsync(string name)	从当前文件夹获取指定文件夹;name:要检索的文件夹的名称,返回表示子文件夹的 StorageFolder
RenameAsync(string desiredName)	重命名当前文件夹;desiredName:当前文件夹所需的新名称

表 16.5 StorageFile 类的主要成员

名 称	说 明
DateCreated	获取创建文件夹的日期和时间
Name	获取存储文件夹的名称
Path	获取存储文件夹的路径
CopyAndReplaceAsync(IStorageFile fileToReplace)	将指定文件替换为当前文件的副本;fileToReplace:要替换的文件
CopyAsync(IStorageFolder destinationFolder)	在指定文件夹中创建文件的副本;destinationFolder:从中创建副本的目标文件夹,返回表示副本的 StorageFile
CopyAsync (IStorageFolder destinationFolder, string desiredNewName)	使用所需的名称,在指定文件夹中创建文件的副本;destinationFolder:从中创建副本的目标文件夹;desiredNewName:副本的所需名称;如果在已经指定 desiredNewName 的目标文件夹中存在现有文件,则为副本生成唯一的名称,返回表示副本的 StorageFile
DeleteAsync()	删除当前文件
GetFileFromPathAsync(string path)	获取 StorageFile 对象以代表指定路径中的文件;path:表示获取 StorageFile 的文件路径,返回作为 StorageFile 的文件
RenameAsync(string desiredName)	重命名当前文件,desiredName:当前项所需的新名称

表 16.6　应用程序初始化默认的文件夹

路　径	用　途	备　注
Shared	用于托管 Media、ShellContent 和 Transfers 文件夹	该文件夹在应用程序安装时创建，但可以将其删除
Shared/Media	专辑封面：应用程序可以在播放背景音频时使用此文件夹在通用音量控件（UVC）中显示专辑封面	该文件夹在应用程序安装时创建，但可以将其删除
Shared/ShellContent	应用程序磁贴：磁贴可以显示每个磁贴正面和背面上的背景图像；背景图像可以存储在独立存储中，但是必须位于此文件夹或其子文件夹中	该文件夹在应用程序安装时创建，但可以将其删除
Shared/Transfers	后台文件传输：即使当应用程序不再在前台中运行时，应用程序仍可以将文件保存到后台中的独立存储中	该文件夹在应用程序安装时创建，但可以将其删除

16.3.2　实例演示本地文件和文件夹的操作

下面给出本地文件和文件夹操作的示例：该实例通过目录的形式展示出应用程序里面的所有文件夹和文件，并演示了在文件夹里面添加文件和删除文件的操作。

代码清单 16-2：本地文件和文件夹操作（源代码：第 16 章\Examples_16_2）

MainPage.xaml 文件的主要代码

```
        <Grid x:Name = "ContentPanel" Grid.Row = "1" Margin = "12,0,12,0">
            <StackPanel >
                <Button Content = "获取程序的文件目录" x:Name = "btGetFile" Margin = "0,0,
0,0" Click = "btGetFile_Click"/>
                <TextBlock Text = "文件夹列表：" VerticalAlignment = "Top"/>
                <ListBox x:Name = "lbFolder" VerticalAlignment = "Top">
                </ListBox>
                <Button Content = "打开选中的文件夹" x:Name = "open" Click = "open_Click" />
                <TextBlock Text = "文件列表：" Margin = "0,0,0,0"/>
                <ListBox x:Name = "lbFile">
                </ListBox>
                <Button Content = "在选中文件夹下创建测试文件" x:Name = "create" Click =
"create_Click" />
                <Button Content = "删除选中的文件" x:Name = "delete" Click = "delete_Click" />
            </StackPanel>
        </Grid>
```

MainPage.xaml.cs 文件的主要代码

```
using System;
using System.Windows;
using System.Windows.Controls;
using Microsoft.Phone.Controls;
using Windows.Storage;
namespace LocalFileDemo
```

```
{
    public partial class MainPage : PhoneApplicationPage
    {
        public MainPage()
        {
            InitializeComponent();
        }
        //打开应用程序的根目录
        private async void btGetFile_Click(object sender, System.Windows.RoutedEventArgs e)
        {
            //清理文件夹列表
            lbFolder.Items.Clear();
            //获取根目录
            StorageFolder localFolder = ApplicationData.Current.LocalFolder;
            //添加遍历根目录的文件夹添加到文件夹列表
            foreach (StorageFolder folder in await localFolder.GetFoldersAsync())
            {
                ListBoxItem item = new ListBoxItem();
                item.Content = "应用程序目录" + folder.Name;
                item.DataContext = folder;
                lbFolder.Items.Add(item);
            }
            //清理文件列表
            lbFile.Items.Clear();
            //添加遍历根目录的文件添加到文件列表
            foreach (StorageFile file in await localFolder.GetFilesAsync())
            {
                ListBoxItem item3 = new ListBoxItem();
                item3.Content = "文件：" + file.Name;
                item3.DataContext = file;
                lbFile.Items.Add(item3);
            }
        }
        //打开选中的文件夹

        private async void open_Click(object sender, System.Windows.RoutedEventArgs e)
        {
            if (lbFolder.SelectedIndex == -1)
            {
                MessageBox.Show("请选择一个文件夹");
            }
            else
            {
                ListBoxItem item = lbFolder.SelectedItem as ListBoxItem;
                //获取选中的文件夹
                StorageFolder folder = item.DataContext as StorageFolder;
                //清理文件夹列表
                lbFolder.Items.Clear();
                //添加遍历到的文件夹添加到文件夹列表
                foreach (StorageFolder folder2 in await folder.GetFoldersAsync())
```

```
                {
                    ListBoxItem item2 = new ListBoxItem();
                    item2.Content = "文件夹：" + folder2.Name;
                    item2.DataContext = folder;
                    lbFolder.Items.Add(item2);
                }
                //清理文件列表
                lbFile.Items.Clear();
                //添加遍历到的文件添加到文件列表
                foreach (StorageFile file in await folder.GetFilesAsync())
                {
                    ListBoxItem item3 = new ListBoxItem();
                    item3.Content = "文件：" + file.Name;
                    item3.DataContext = file;
                    lbFile.Items.Add(item3);
                }
            }
        }
        //在选中的文件中新建一个文件
        private async void create_Click(object sender, RoutedEventArgs e)
        {
            if (lbFolder.SelectedIndex == -1)
            {
                MessageBox.Show("请选择一个文件夹");
            }
            else
            {
                ListBoxItem item = lbFolder.SelectedItem as ListBoxItem;

                //获取选中的文件
                StorageFolder folder = item.DataContext as StorageFolder;
                //在文件夹中创建一个文件
                StorageFile file = await folder.CreateFileAsync(DateTime.Now.Millisecond.
ToString() + ".txt");
                //添加到文件列表中
                ListBoxItem item3 = new ListBoxItem();
                item3.Content = "文件：" + file.Name;
                item3.DataContext = file;
                lbFile.Items.Add(item3);
                MessageBox.Show("创建文件成功");
            }
        }
        //删除选中的文件
        private async void delete_Click(object sender, RoutedEventArgs e)
        {
            if (lbFile.SelectedIndex == -1)
            {
                MessageBox.Show("请选择一个文件");
            }
            else
```

```
                {
                    ListBoxItem item = lbFile.SelectedItem as ListBoxItem;
                    //获取选中的文件
                    StorageFile file = item.DataContext as StorageFile;
                    //删除文件
                    await file.DeleteAsync();
                    lbFile.Items.Remove(item);
                    MessageBox.Show("删除成功");
                }
            }
        }
    }
```

程序的运行效果如图 16.2 所示。

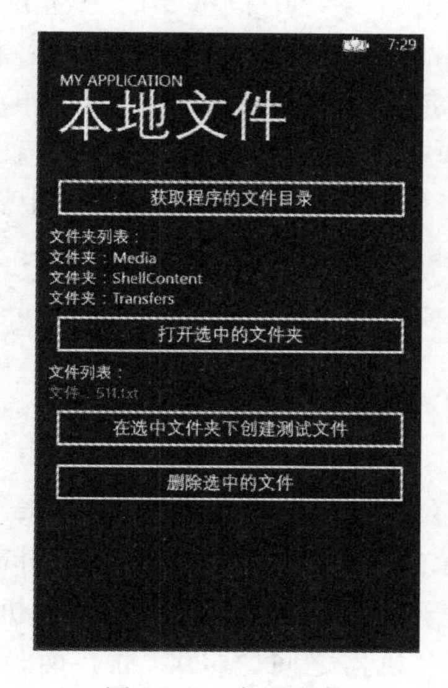

图 16.2 本地文件

16.3.3 获取安装包下的文件夹和文件

安装包下的文件是指在项目中添加的文件，可以在应用程序里面获取到安装包下的文件，不过编译的文件如源代码文件和资源类文件是获取不到。获取安装包的文件可以先通过 Windows.ApplicationModel.Package 类的 InstalledLocation 属性来获取安装包的文件夹，语法如下：

```
StorageFolder localFolder = Windows.ApplicationModel.Package.Current.InstalledLocation;
```

然后就可以通过查看文件夹和操作文件的方式来对安装包的文件进行操作了，编程的方式和对应用程序的文件的编程方式一样。

第17章

Socket编程

Socket 编程是 Windows Phone 系统中很重要的一部分，从 Windows Phone 7.1 版本开始支持 Socket 编程，到 Windows Phone 8 版本增加了新的基于 Windows 运行时的 Socket API。使用 Socket 编程可以实现那些基于 Socket 相关协议的网络编程以及蓝牙编程、近场通信编程等。蓝牙编程和近场通信编程也使用到了 Windows 运行时的 Socket 消息传输机制，蓝牙编程和近场通信编程将会在后续的章节中做详细的介绍，本章将会重要讲解在 Windows Phone 中的 Socket 编程，包括基于 Windows Phone 7.1 的 Socket 以及基于 Windows 运行时的 Socket。

17.1 Socket 编程介绍

Socket 通常也称作套接字，用于描述 IP 地址和端口，是一个通信链的句柄。应用程序通常通过 Socket 向网络发出请求或者应答网络请求。Socket 是一种用于表达两台机器之间连接"终端"的软件抽象。对于一个给定的连接，在每台机器上都有一个 Socket，可以想象一个虚拟的"电缆"工作在两台机器之间，"电缆"插在两台机器的 Socket 上。当然，物理硬件和两台机器之间的"电缆"这些连接装置都是未知的，抽象的所有目的就是为了让用户不必了解更多的细节。

网络上的两个程序通过一个双向的通信连接实现数据的交换，这个双向链路的一端称为一个 Socket。Socket 通常用来实现客户方和服务方的连接。Socket 是 TCP/IP 协议的一个十分流行的编程界面，一个 Socket 由一个 IP 地址和一个端口号唯一确定。但是，Socket 所支持的协议种类也不只 TCP/IP 一种，因此两者之间没有必然的联系。简单地说，一台机器上的 Socket 同另一台机器通话创建一个通信信道，程序员可以用这个信道在两台机器之间发送数据。发送数据时，TCP/IP 协议栈的每一层都给数据添加适当的报头。Socket 像电话听筒一样在电话的任意一端——通过一个专门的信道来进行通话和接听。会话将一直进行下去直到挂断电话，除非我们挂断电话，否则各自的电话线路都会占线。

通过使用 Socket 编程的 API 可以使开发人员方便地创建出包括 FTP、电子邮件、聊天系统和流媒体等网络应用。Windows Phone 平台提供了一些虽然简单但是相当强大的高层抽象以使创建和使用 Socket 更加容易一些。

17.1.1　Socket 的相关概念

1. 端口

网络中可以被命名和寻址的通信端口，是操作系统可分配的一种资源。按照 OSI 七层协议的描述，传输层与网络层在功能上的最大区别是传输层提供进程通信能力。从这个意义上讲，网络通信的最终地址就不仅仅是主机地址了，还包括可以描述进程的某种标识符。为此，TCP/IP 协议提出了协议端口（protocol port，简称端口）的概念，用于标识通信的进程。

端口是一种抽象的软件结构（包括一些数据结构和 I/O 缓冲区）。应用程序（即进程）通过系统调用与某端口建立连接（binding）后，传输层传给该端口的数据都被相应进程所接收，相应进程发给传输层的数据都通过该端口输出。在 TCP/IP 协议的实现中，端口的操作类似于一般的 I/O 操作，进程获取一个端口，相当于获取本地唯一的 I/O 文件，可以用一般的读写原语访问。

类似于文件描述符，每个端口都拥有一个叫做端口号（port number）的整数型标识符，用于区别不同端口。由于 TCP/IP 传输层的两个协议 TCP 和 UDP 是完全独立的两个软件模块，因此各自的端口号也相互独立，如 TCP 有一个 255 号端口，UDP 也可以有一个 255 号端口，二者并不冲突。

端口号的分配是一个重要问题。有两种基本分配方式：一种叫全局分配，这是一种集中控制方式，由一个公认的中央机构根据用户需要进行统一分配，并将结果公布于众；另一种是本地分配，又称动态连接，即进程需要访问传输层服务时，向本地操作系统提出申请，操作系统返回一个本地唯一的端口号，进程再通过合适的系统调用将自己与该端口号联系起来（绑扎）。TCP/IP 端口号的分配中综合了上述两种方式。TCP/IP 将端口号分为两部分，少量的作为保留端口，以全局方式分配给服务进程。因此，每一个标准服务器都拥有一个全局公认的端口（即 well-known port），即使在不同的机器上，其端口号也相同。剩余的为自由端口，以本地方式进行分配。TCP 和 UDP 均规定，小于 256 的端口号才能作保留端口。

2. 地址

网络通信中通信的两个进程分别在两台地址不同的机器上。在互连网络中，两台机器可能位于不同位置的网络上，这些网络通过网络互联设备（网关、网桥、路由器等）连接。因此需要三级寻址：

(1) 某一主机可与多个网络相连，必须指定一个特定网络地址；

(2) 网络上每一台主机应有其唯一的地址；

(3) 每一主机上的每一进程应有在该主机上的唯一标识符。

通常主机地址由网络 ID 和主机 ID 组成，在 TCP/IP 协议中用 32 位整数值表示；TCP

和 UDP 均使用 16 位端口号标识用户进程。

3. IPv4 和 IPv6

IPv4 是互联网协议(Internet Protocol,IP)的第 4 版,也是第一个被广泛使用,构成现今互联网技术的基石的协议。1981 年 Jon Postel 在 RFC791 中定义了 IP,Ipv4 可以运行在各种各样的底层网络上,比如端对端的串行数据链路(PPP 协议和 SLIP 协议)、卫星链路等。局域网中最常用的是以太网。IPv6 是 Internet Protocol Version 6 的缩写,其中 Internet Protocol 译为"互联网协议"。IPv6 是 IETF(Internet Engineering Task Force,互联网工程任务组)设计的用于替代现行版本 IP 协议(IPv4)的下一代 IP 协议。目前,IP 协议的版本号是 4(IPv4),它的下一个版本就是 IPv6。

4. 广播

广播是指在一个局域网中向所有的网上节点发送信息。广播有一个广播组,即只有一个广播组内的节点才能收到发往这个广播组的信息。什么决定了一个广播组呢？就是端口号,局域网内一个节点,如果设置了广播属性并监听了端口号 A 后,那么就加入了 A 组广播,这个局域网内所有发往广播端口 A 的信息他都收的到。在广播的实现中,如果一个节点想接受 A 组广播信息,那么就要先将他绑定给地址和端口 A,然后设置这个 Socket 的属性为广播属性。如果一个节点不想接受广播信息,而只想发送广播信息,那么不用绑定端口,只需要先为 Socket 设置广播属性后,向广播地址的 A 端口发送信息即可。

5. TCP 协议

TCP 是 Transfer Control Protocol 的简称,是一种面向连接的保证可靠传输的协议。通过 TCP 协议传输,得到的是一个顺序的无差错的数据流。发送方和接收方的成对的两个 Socket 之间必须建立连接,以便在 TCP 协议的基础上进行通信,当一个 Socket(通常都是 Server Socket)等待建立连接时,另一个 Socket 可以要求进行连接,一旦这两个 Socket 连接起来,它们就可以进行双向数据传输,双方都可以进行发送或接收操作。

6. UDP 协议

UDP 是 User Datagram Protocol 的简称,是一种无连接的协议,每个数据报都是一个独立的信息,包括完整的源地址或目的地址,它在网络上以任何可能的路径传往目的地,因此能否到达目的地,到达目的地的时间以及内容的正确性都是不能被保证的。

7. TCP 协议和 UDP 协议的比较

1) UDP 的特性

(1) 每个数据报中都给出了完整的地址信息,因此无须建立发送方和接收方的连接。

(2) UDP 传输数据时是有大小限制的,每个被传输的数据报必须限定在 64KB 之内。

(3) UDP 是一个不可靠的协议,发送方所发送的数据报并不一定以相同的次序到达接收方。

(4) TCP 在网络通信上有极强的生命力,例如,远程连接(Telnet)和文件传输(FTP)都需要不定长度的数据被可靠地传输。但是可靠的传输是要付出代价的,对数据内容正确性的检验必然占用计算机的处理时间和网络的带宽,因此 TCP 传输的效率不如 UDP 高。

2）TCP 的特性

(1) 面向连接的协议，在 Socket 之间进行数据传输之前必然要建立连接，所以在 TCP 中需要连接时间。

(2) TCP 传输数据大小限制，一旦连接建立起来，双方的 Socket 就可以按统一的格式传输大的数据。

(3) TCP 是一个可靠的协议，它确保接收方完全正确地获取发送方所发送的全部数据。

(4) UDP 操作简单，而且仅需要较少的监护，因此通常用于局域网高可靠性的分散系统中 Client/Server 应用程序。如视频会议系统，并不要求音频视频数据绝对的正确，只要保证连贯性就可以了，这种情况下显然使用 UDP 会更合理一些。

8. Socket 连接与 HTTP 连接的比较

由于通常情况下 Socket 连接就是 TCP 连接，因此 Socket 连接一旦建立，通信双方即可开始相互发送数据内容，直到双方连接断开。但在实际网络应用中，客户端到服务器之间的通信往往需要穿越多个中间节点，如路由器、网关、防火墙等。大部分防火墙默认会关闭长时间处于非活跃状态的连接而导致 Socket 连接断连，因此需要通过轮询告诉网络，该连接处于活跃状态。而 HTTP 连接使用的是“请求—响应”的方式，不仅在请求时需要先建立连接，而且需要客户端向服务器发出请求后，服务器端才能回复数据。很多情况下，需要服务器端主动向客户端推送数据，保持客户端与服务器数据的实时与同步。此时若双方建立的是 Socket 连接，服务器就可以直接将数据传送给客户端；若双方建立的是 HTTP 连接，则服务器需要等到客户端发送一次请求后才能将数据传回给客户端，因此，客户端定时向服务器端发送连接请求，不仅可以保持在线，同时也是在“询问”服务器是否有新的数据，如果有就将数据传给客户端。

17.1.2 Socket 通信的过程

Socket 通信在客户端和服务器进行，建立 Socket 连接至少需要一对 Socket，其中一个运行于客户端，称为 ClientSocket，另一个运行于服务器端，称为 ServerSocket。

Socket 之间的连接过程分为 3 个步骤：服务器监听、客户端请求、连接确认。

(1) 服务器监听：服务器端 Socket 并不定位具体的客户端 Socket，而是处于等待连接的状态，实时监控网络状态，等待客户端的连接请求。

(2) 客户端请求：指客户端的 Socket 提出连接请求，要连接的目标是服务器端的 Socket。为此，客户端的 Socket 必须首先描述它要连接的服务器端 Socket，指出服务器端 Socket 的地址和端口号，然后就向服务器端 Socket 提出连接请求。

(3) 连接确认：当服务器端 Socket 监听到或者说接收到客户端 Socket 的连接请求时，就响应客户端 Socket 的请求，建立一个新的线程，把服务器端 Socket 的描述发给客户端，一旦客户端确认了此描述，双方就正式建立连接。而服务器端 Socket 继续处于监听状态，继续接收其他客户端 Socket 的连接请求。

整个 Windows Phone 应用程序 Socket 通信的过程包括了七个步骤，如图 17.1 所示。

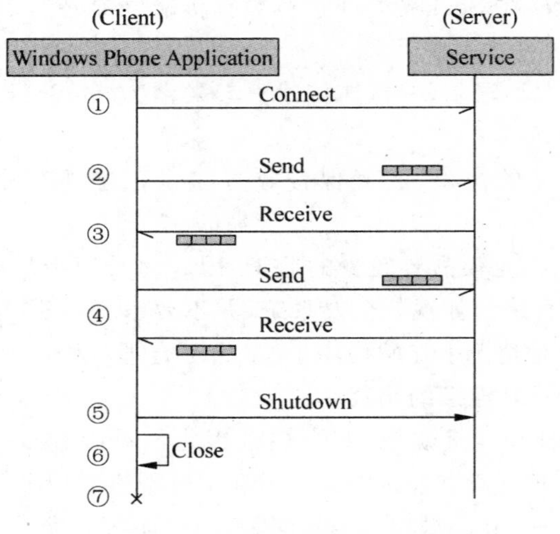

图 17.1 Socket 通信的步骤

(1) 创建一个客户端和服务器端的 Socket 连接。

(2) 客户端发送消息的过程，客户端向服务器发送消息，服务器端接收客户端发过来的消息。

(3) 客户端接收消息的过程，客户端接收服务端返回来的消息。

(4) 连接继续保持，将可以不断地重复(2)和(3)的发送消息和接收消息的动作。

(5) 关闭发送接收通道，可以只关闭发送通道或者接收通道，也可以两者同时关闭。

(6) 关闭 Socket 连接。

(7) 整个通信过程到此终止。

17.2 .NET 框架的 Socket 编程

在.NET 框架下的 Socket 编程也就是指 Windows Phone 7.1 版本的 Socket 编程，如果应用程序既要在 Windows Phone 8 的手机上运行，要同时要支持久的 Windows Phone 7 手机，那么建议选择.NET 框架下的 Socket 编程，若不再兼容 Windows Phone 7 的手机那么可以选择 Windows 运行时下的 Socket 编程。

17.2.1 Windows Phone 7.1 中的 Socket API

Windows Phone 7.1 版本新增的 Socket 编程的类在 System.Net.Sockets 命名空间下，Socket 类实现 Socket 编程的接口，SocketAsyncEventArgs 类表示异步 Socket 操作，SocketException 类发生 Socket 错误时引发的异常。System.Net.Sockets 命名空间下的枚举类型如表 17.1 所示。

表 17.1 Sockets 命名空间下的枚举类型

枚举	说明
AddressFamily	指定 Socket 类的实例可以使用的寻址方案
ProtocolType	指定 Socket 类支持的协议
SocketAsyncOperation	最近使用此对象执行的异步 Socket 操作的类型
SocketClientAccessPolicyProtocol	指定用于下载 Socket 类的实例将使用的策略文件的方法
SocketError	定义 Socket 类的错误代码
SocketShutdown	定义 Socket.Shutdown 方法使用的常量
SocketType	指定 Socket 类的实例表示的 Socket 类型

Socket 类为网络通信提供了一组方法和属性分别如表 17.2 和表 17.3 所示。SocketAsyncEventArgs 类表示异步 Socket 操作,提供由 Socket 应用程序使用的异步模式,SocketAsyncEventArgs 类的属性如表 17.4 所示。在 Socket 类中,异步 Socket 操作由分配的可重用 SocketAsyncEventArgs 对象描述并由应用程序维护。应用程序可以根据自身需要创建任意多的 SocketAsyncEventArgs 对象。例如,如果 Windows Phone 的应用程序需要同时有 10 个未完成的 Socket 发送操作,则可以为此目的提前分配 10 个可重用的 SocketAsyncEventArgs 对象。

表 17.2 Socket 类的常用属性

名称	说明
AddressFamily	获取 Socket 的地址族
Connected	获取 Socket 连接的状态,该值指示 Socket 是否在上次操作后连接到了远程主机,true 表示连接状态,false 表示断开状态
NoDelay	获取或设置 Socket 是否使用 Nagle 算法
OSSupportsIPv4	获取当前的主机是否支持 IPv4 协议并且已启用,如果支持返回 true,否则为 false
OSSupportsIPv6	获取当前的主机是否支持 IPv6 协议并且已启用,如果支持返回 true,否则为 false
ProtocolType	获取 Socket 的协议类型
ReceiveBufferSize	获取或设置接收缓冲区的大小
RemoteEndPoint	获取远程端点
SendBufferSize	获取或设置发送缓冲区的大
Ttl	获取或设置 Socket 的 IP 包的生存时间(Time To Live)

表 17.3 Socket 类的常用方法

名称	说明
CancelConnectAsync	取消一个活动的 socket 连接
Close	关闭 Socket 连接并释放所有相关资源
ConnectAsync(SocketAsyncEventArgs) ConnectAsync(SocketType, ProtocolType, SocketAsyncEventArgs)	开始一个远程主机连接的异步请求
Dispose Dispose(Boolean)	释放由 Socket 使用的非托管资源

续表

名称	说明
ReceiveAsync	开始接收 Socket 异步请求的数据，从一个连接 Socket 的主机接收
ReceiveFromAsync	开始接收 Socket 异步请求的数据，从一个指定的 Socket 的主机接收
SendAsync	发送数据到一个连接 Socket 的主机上
SendToAsync	发送数据到一个指定的 Socket 的主机上
Shutdown	禁止 Socket 数据的发送和接收，完成所有挂起的发送操作，并用信号通知远程终结点应关闭连接。如果指定了 Send，在远程主机关闭其连接端（表现为接收 0 字节）之前，可能仍会收到数据

表 17.4　SocketAsyncEventArgs 类的常用属性

名称	说明
Buffer	获取要用于异步 Socket 方法的数据缓冲区
BufferList	获取或设置一个要用于异步 Socket 方法的数据缓冲区数组
BytesTransferred	获取在 Socket 操作中传输的字节数
ConnectByNameError	当使用 DnsEndPoint 时，在出现连接故障的情况下获取异常
ConnectSocket	成功完成 ConnectAsync 方法后创建和连接的 Socket 对象
Count	获取可在异步操作中发送或接收的最大数据量（以字节为单位）
LastOperation	获取最近使用此对象执行的 Socket 操作的类型
Offset	获取 Buffer 属性引用的数据缓冲区的偏移量（以字节为单位）
RemoteEndPoint	获取或设置异步操作的远程 IP 或 DNS 终结点
SocketClientAccessPolicyProtocol	获取或设置用于下载 Socket 类的实例将使用的策略文件的方法
SocketError	获取或设置异步 Socket 操作的结果
UserToken	获取或设置与此异步 Socket 操作关联的用户或应用程序对象

执行异步 Socket 操作的模式包含以下步骤：

（1）分配一个新的 SocketAsyncEventArgs 对象，或者从应用程序池中获取一个这样的空闲对象。

（2）针对 SocketAsyncEventArgs 对象设置即将执行的操作所需的属性（例如，附加到 Completed 事件的事件处理程序以及 ConnectAsync 方法的 RemoteEndPoint 属性）。

（3）调用适当的 Socket 方法以启动异步操作。

（4）如果异步 Socket 方法返回 true，则说明 I/O 操作处于挂起状态。操作完成时，将引发传递给 Socket 方法的 SocketAsyncEventArgs 对象的 SocketAsyncEventArgs.Completed 事件。该事件在事件处理程序中引发时，将查询 SocketAsyncEventArgs 属性来获取完成状态和 Socket 操作结果（例如，缓冲区中为 ReceiveAsync 方法调用接收的数据量）。

(5) 如果异步 Socket 方法返回 false，则说明操作是同步完成的。可以查询 SocketAsyncEventArgs 属性来获取完成状态和 Socket 操作结果。

(6) 将 SocketAsyncEventArgs 重用于另一个操作，将它放回到应用程序池中，或者将它丢弃。

异步 Socket 操作中使用的新的 SocketAsyncEventArgs 对象的生存期由应用程序代码中的引用和异步 I/O 引用决定。在对 SocketAsyncEventArgs 对象的引用作为一个参数提交给某个异步 Socket 方法之后，应用程序不必保留该引用。在完成回调返回之前将一直引用该对象。但是，应用程序保留对 SocketAsyncEventArgs 对象的引用是有好处的，这样该引用就可以重用于将来的异步 Socket 操作。当数据发送和数据接收完成之后，可使用 Shutdown 方法来禁用 Socket。在调用 Shutdown 并(可选)接收到确认信息(远程终结点通过获取 0 字节的接收操作也已关闭了连接)后，调用 Close 方法来释放与 Socket 关联的所有资源。

17.2.2 Socket 示例：实现手机客户端和计算机服务器端的通信

下面给出 Socket 实现手机客户端和计算机服务器端的通信的示例：使用 Windows Phone 的客户端程序作为 Socket 的客户端，Windows 的控制台程序作为服务器端，实现了客户端和服务器端之间的信息发送和接收。

代码清单 17-1：Socket 实现手机客户端和计算机服务器端的通信(源代码：第 17 章\Examples_17_1)

1. Windows Phone 客户端客户端的实现

MainPage.xaml 文件主要代码

```
        <Grid x:Name = "ContentPanel" Grid.Row = "1" Margin = "12,0,12,0">
            <TextBlock FontSize = " 30 " Text = " 主机 IP:" Margin = " 12, 23, 0, 540 "
HorizontalAlignment = "Left" Width = "99" />
            <TextBox x:Name = "Host" InputScope = "Digits" HorizontalAlignment = "Stretch"
Text = "192.168.1.102" Margin = "110,6,0,523" />
            <TextBlock FontSize = "30" Text = "端口号:" Margin = "9,102,345,451" />
            <TextBox x:Name = "Port" InputScope = "Digits"
                     HorizontalAlignment = "Stretch"
                     Text = "8888" Margin = "110,90,0,433" />
            <TextBlock FontSize = "30" Text = "发送的消息内容:" Margin = "6,180,157,374" />
                <TextBox x:Name = "Message"
                     HorizontalAlignment = "Stretch" Margin = " - 6,226,6,300" />
            <Button x:Name = "SendButton"
                     Content = "发送"
                     Click = "SendButton_Click" Margin = "0,509,12,6" />
            <ListBox Height = "190" HorizontalAlignment = "Left" Margin = " - 4,313,0,0" Name =
"listBox1" VerticalAlignment = "Top" Width = "460" />
        </Grid>
```

MainPage.xaml.cs 文件代码

```
using System;
using System.Net;
using System.Windows;
using Microsoft.Phone.Controls;
using System.Text;
using System.Net.Sockets;
namespace SocketTest
{
    public partial class MainPage : PhoneApplicationPage
    {
        public MainPage()
        {
            InitializeComponent();
        }
        private void SendButton_Click(object sender, RoutedEventArgs e)
        {
            //判断是否已经输入了IP地址和端口
            if (string.IsNullOrEmpty(Host.Text) || string.IsNullOrEmpty(Port.Text))
            {
                MessageBox.Show("麻烦输入以下主机IP和端口号,谢谢!");
                return;
            }
            //主机IP地址
            string host = Host.Text.Trim();
            //端口号
            int port = Convert.ToInt32(Port.Text.Trim());
            //建立一个终结点对象
            DnsEndPoint hostEntry = new DnsEndPoint(host, port);
            //创建一个Socket对象
            Socket sock = new Socket(AddressFamily.InterNetwork, SocketType.Stream,
ProtocolType.Tcp);
            //创建一个Socket异步事件参数
            SocketAsyncEventArgs socketEventArg = new SocketAsyncEventArgs();
            //将消息内容转化为发送的byte[]格式
            byte[] buffer = Encoding.UTF8.GetBytes(Message.Text);
            //将发送内容的信息存放进Socket异步事件参数中
            socketEventArg.SetBuffer(buffer, 0, buffer.Length);
            //注册Socket完成事件
            socketEventArg.Completed += new EventHandler<SocketAsyncEventArgs>
(SocketAsyncEventArgs_Completed);
            //设置Socket异步事件参数的Socket远程终结点
            socketEventArg.RemoteEndPoint = hostEntry;
            //将定义好的Socket对象赋值给Socket异步事件参数的运行实例属性
            socketEventArg.UserToken = sock;
            try
            {
                //运行Socket
                sock.ConnectAsync(socketEventArg);
```

```
            }
            catch (SocketException ex)
            {
                throw new SocketException((int)ex.ErrorCode);
            }
        }
        private void SocketAsyncEventArgs_Completed(object sender, SocketAsyncEventArgs e)
        {
            //检查是否发送出错
            if (e.SocketError != SocketError.Success)
            {
                if (e.SocketError == SocketError.ConnectionAborted)
                {
                    Dispatcher.BeginInvoke(() => MessageBox.Show("连接超时请重试! " + e.
SocketError));
                }
                else if (e.SocketError == SocketError.ConnectionRefused)
                {
                    Dispatcher.BeginInvoke(() => MessageBox.Show("服务器端启动"
                        + e.SocketError));
                }
                else
                {
                    Dispatcher.BeginInvoke(() => MessageBox.Show("出错了"
                        + e.SocketError));
                }
                //关闭连接清理资源
                if (e.UserToken != null)
                {
                    Socket sock = e.UserToken as Socket;
                    sock.Shutdown(SocketShutdown.Both);
                    sock.Close();
                }
                return;
            }
            //检查 Socket 的当前最后的一个操作
            switch (e.LastOperation)
            {
                //如果最后的一个操作是连接,那么下一步处理就是发送消息
                case SocketAsyncOperation.Connect:
                    if (e.UserToken != null)
                    {
                        //获取运行中的 Socket 对象
                        Socket sock = e.UserToken as Socket;
                        //开始发送
                        bool completesAsynchronously = sock.SendAsync(e);
                        //检查 socket 发送是否被挂起,如果被挂起将继续进行处理
                        if (!completesAsynchronously)
                        {
                            SocketAsyncEventArgs_Completed(e.UserToken, e);
```

```
                    }
                };
                break;
            //如果最后的一个操作是发送,那么显示刚才发送成功的消息,然后开始下一步
            //处理就是接收消息
            case SocketAsyncOperation.Send:
                //将已成功发送的消息内容绑定到 listBox1 控件中
                Dispatcher.BeginInvoke(() =>
                    {
                        listBox1.Items.Add("客户端" + DateTime.Now.ToShortTimeString()
+ "发送的消息 :" + Message.Text);
                    }
                    );
                if (e.UserToken != null)
                {
                    //获取运行中的 Socket 对象
                    Socket sock = e.UserToken as Socket;
                    //开始接收服务器端的消息
                    bool completesAsynchronously = sock.ReceiveAsync(e);
                    //检查 socket 发送是否被挂起,如果被挂起将继续进行处理
                    if (!completesAsynchronously)
                    {
                        SocketAsyncEventArgs_Completed(e.UserToken, e);
                    }

                }
                break;
            //如果是最后的一个操作时接收,那么显示接收到的消息内容,并清理资源
            case SocketAsyncOperation.Receive:
                if (e.UserToken != null)
                {
                    //获取接收到的消息,并转化为字符串
                        string dataFromServer = Encoding.UTF8.GetString(e.Buffer, 0, e.
BytesTransferred);
                    //获取运行中的 Socket 对象
                    Socket sock = e.UserToken as Socket;
                    //将接收到的消息内容绑定到 listBox1 控件中
                    Dispatcher.BeginInvoke(() =>
                    {
                        listBox1.Items.Add("服务器端" + DateTime.Now.ToShortTimeString()
+ "传过来的消息:" + dataFromServer);
                    });
                }
                break;
        }
    }
}
}
```

2. Socket 服务器端的实现，使用 Windows 的控制台程序

SocketServer.cs 文件代码

```
using System;
using System.Linq;
using System.Net;
using System.Net.Sockets;
using System.Text;
using System.Threading;
namespace SocketServer
{
    static class Program
    {
        private static AutoResetEvent _flipFlop = new AutoResetEvent(false);
        static void Main(string[] args)
        {
            //创建 socket,使用的是 TCP 协议,如果用 UDP 协议,则要用 SocketType.Dgram 类型的 Socket
            Socket listener = new Socket(AddressFamily.InterNetwork,
                SocketType.Stream,
                ProtocolType.Tcp);
            //创建终结点 EndPoint 取当前主机的 IP
            IPHostEntry ipHostInfo = Dns.Resolve(Dns.GetHostName());
            //把 IP 和端口转化为 IPEndpoint 实例,端口号取 8888
            IPEndPoint localEP = new IPEndPoint(ipHostInfo.AddressList.First(), 8888);
            Console.WriteLine("本地的 IP 地址和端口是 : {0}", localEP);
            try
            {
                //绑定 EndPoint 对象(8888 端口和 IP 地址)
                listener.Bind(localEP);
                //开始监听
                listener.Listen(10);
                //一直循环接收客户端的消息
                while (true)
                {
                    Console.WriteLine("等待 Windows Phone 客户端的连接...");
                    //开始接收下一个连接
                    listener.BeginAccept(AcceptCallback, listener);
                    //开始线程等待
                    _flipFlop.WaitOne();
                }
            }
            catch (Exception e)
            {
                Console.WriteLine(e.ToString());
            }
        }
        //接收返回事件处理
        private static void AcceptCallback(IAsyncResult ar)
        {
            Socket listener = (Socket)ar.AsyncState;
            Socket socket = listener.EndAccept(ar);
```

```
        Console.WriteLine("连接到 Windows Phone 客户端。");
        _flipFlop.Set();
        //开始接收,传递 StateObject 对象过去
        var state = new StateObject();
        state.Socket = socket;
        socket.BeginReceive(state.Buffer,
            0,
            StateObject.BufferSize,
            0,
            ReceiveCallback,

            state);
    }
    private static void ReceiveCallback(IAsyncResult ar)
    {
        StateObject state = (StateObject)ar.AsyncState;
        Socket socket = state.Socket;
        //读取客户端 socket 发送过来的数据
        int read = socket.EndReceive(ar);
        //如果成功读取了客户端 socket 发送过来的数据
        if (read > 0)
        {
            //获取客户端的消息,转化为字符串格式
            string chunk = Encoding.UTF8.GetString(state.Buffer, 0, read);
            state.StringBuilder.Append(chunk);
            if (state.StringBuilder.Length > 0)
            {
                string result = state.StringBuilder.ToString();
                Console.WriteLine("收到客户端传过来的消息：{0}", result);
                //发送数据信息给客户端
                Send(socket, result);
            }
        }
    }
    //返回客户端数据
    private static void Send(Socket handler, String data)
    {
        //将消息内容转化为发送的 byte[]格式
        byte[] byteData = Encoding.UTF8.GetBytes(data);
        //发送消息到 Windows Phone 客户端
        handler.BeginSend(byteData, 0, byteData.Length, 0,
            new AsyncCallback(SendCallback), handler);
    }
    private static void SendCallback(IAsyncResult ar)
    {
        try
        {
            Socket handler = (Socket)ar.AsyncState;
            // 完成发送消息到 Windows Phone 客户端
            int bytesSent = handler.EndSend(ar);
            if (bytesSent > 0)
            {
```

```
                Console.WriteLine("发送成功!");
            }
        }

        catch (Exception e)
        {
            Console.WriteLine(e.ToString());
        }
    }
}
public class StateObject
{
    public Socket Socket;
    public StringBuilder StringBuilder = new StringBuilder();
    public const int BufferSize = 10;
    public byte[] Buffer = new byte[BufferSize];
    public int TotalSize;
}
}
```

运行服务器端的程序，显示如图 17.2 所示，运行 Windows Phone 的客户端程序如图 17.3 所示。用 Windows Phone 客户端向服务器发送一条消息，注意，IP 地址要填写当前计算机主机网络的 IP 地址，因为服务器端自动去了当前计算机主机的 IP 地址，发送的显示效果如图 17.4 所示，服务器端的接收和自动发送消失效果如图 17.5 所示。

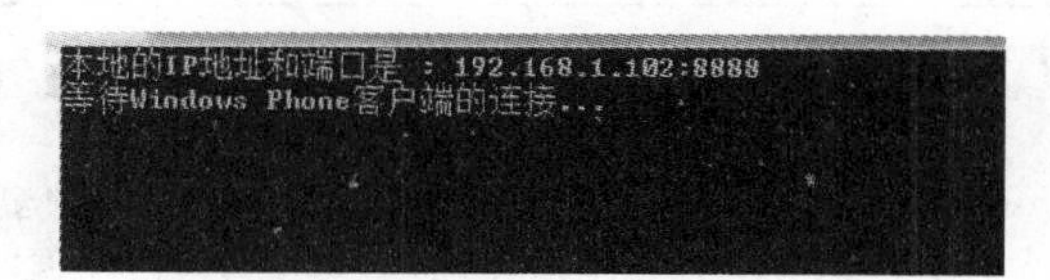

图 17.2　打开 Socket 服务器端的效果

图 17.3　打开 Socket 客户端的效果

图 17.4　Socket 发送测试

图 17.5 Socket 服务器端的接收信息和发送信息

17.3 Windows 运行时的 Socket 编程

在.NET 框架下的 Socket 编程的 API 采用的是基于事件的异步模式，而 Windows 运行时的 Socket 编程的 API 是基于任务的异步模式。Windows 运行时下的 Socket 编程可以在 Windows 8 和 Windows Phone 8 两个平台公用，也就是这两个平台使用的时同一套的 Socket 编程 API，可以通过 Windows 运行时组件来共用代码，实现跨平台。

17.3.1 StreamSocket 简介以及 TCP Socket 编程步骤

StreamSocket 类在 Windows. Networking. Sockets 空间下，表示对象连接到网络资源，以使用异步方法发送数据，StreamSocket 类的成员如表 17.5 所示。

表 17.5 StreamSocket 类成员

名　称	说　明
StreamSocket()	创建新的 StreamSocket 对象
StreamSocketControl Control	获取 StreamSocket 对象上的套接字控件数据，返回某一 StreamSocket 对象上的套接字控件数据
StreamSocketInformation Information	获取 StreamSocket 对象上的套接字信息，返回该 StreamSocket 对象的套接字信息
IInputStream InputStream	获取要从 StreamSocket 对象上的远程目标读取的输入流，返回要从远程目标读取的有序字节流
IOutputStream OutputStream	获取 StreamSocket 对象上写入远程主机的输出流，返回要写入远程目标的有序字节流
IAsyncAction ConnectAsync (EndpointPair endpointPair)	启动 StreamSocket 对象连接到被指定为 EndpointPair 对象的远程网络目标的异步操作。endpointPair：指定本地主机名或 IP 地址、本地服务名或 UDP 端口、远程主机名或远程 IP 地址，以及远程网络目标的远程服务名或远程 TCP 端口的 EndpointPair 对象。返回 StreamSocket 对象的异步连接操作
IAsyncAction ConnectAsync (EndpointPair endpointPair, SocketProtectionLevel protectionLevel)	启动 StreamSocket 对象连接到被指定为 EndpointPair 对象和 SocketProtectionLevel 枚举的远程网络目标的异步操作，protectionLevel：表示 StreamSocket 对象的完整性和加密的保护级别，返回 StreamSocket 对象的异步连接操作

续表

名　称	说　明
IAsyncAction ConnectAsync (HostName remoteHostName, string remoteServiceName)	在 StreamSocket 上启动远程主机名和远程服务名所指定的远程网络目标的连接操作，remoteHostName：远程网络目标的主机名或 IP 地址，remoteServiceName：远程网络目标的服务名称或 TCP 端口号，返回 StreamSocket 对象的异步连接操作
IAsyncAction ConnectAsync (HostName remoteHostName, string remoteServiceName, SocketProtectionLevel protectionLevel)	启动 StreamSocket 对象连接远程主机名、远程服务名以及 SocketProtectionLevel 所指定的远程目标的异步操作
IAsyncAction UpgradeToSslAsync (SocketProtectionLevel protectionLevel, HostName validationHostName)	启动 StreamSocket 对象将连接的套接字升级到使用 SSL 的异步操作，protectionLevel：表示 StreamSocket 对象上完整性和加密的保护级别，validationHostName：在升级到 SSL 时用于验证的远程网络目标的主机名，返回 StreamSocket 对象升级到使用 SSL 的异步操作

使用 StreamSocket 类进行 TCP Socket 编程的步骤如下：

(1) 使用 StreamSocket 类创建 TCP 套 Socket 字。

(2) 使用 StreamSocket.ConnectAsync 方法之一建立与 TCP 网络服务器的网络连接。

(3) 使用 Streams.DataWriter 对象将数据发送到服务器，该对象允许程序员在任何流上写入常用类型(例如整数和字符串)。

(4) 使用 Streams.DataReader 对象从服务器接收数据，该对象允许程序员在任何流上读取常用类型(例如整数和字符串)。

17.3.2　连接 Socket

连接 Socket 是使用 Socket 编程的第一步，创建 StreamSocket 并连接到服务器，在这个步骤里面还会定义主机名和服务名(TCP 端口)以连接到服务器。连接的过程是采用异步任务的模式，当连接成功的时候将会继续执行到 await 连接服务器后面的代码，如果连接失败将 ConnectAsync 方法会抛出异常，表示无法与网络服务器建立 TCP 连接。连接失败，需要捕获异常的信息来获取是什么类型的异常，然后再进行判断该怎么操作，是否需要重新连接还是释放掉资源等等。连接 Socket 示例代码如下：

```
async void Connect()
{
    //创建一个 StreamSocket 对象
    StreamSocket clientSocket = new StreamSocket();
    try
    {
        //创建一个主机名字
        HostName serverHost = new HostName(serverHostname);
        //开始链接
```

```
        await clientSocket.ConnectAsync(serverHost, serverPort);
        //链接成功
    }
    catch (Exception exception)
    {
        //获取错误的类型 SocketError.GetStatus(exception.HResult)
        //错误消息 exception.Message;
        //如果关闭 Socket 则释放资源
        clientSocket.Dispose();
        clientSocket = null;
    }
}
```

17.3.3 发送和接收消息

Socket 连接成功之后便可以发送和接收 Socket 消息了，发送消息需要使用 Streams.DataWriter 对象将数据发送到服务器，接收消息使用 Streams.DataReader 对象从服务器接收数据。发送接收消息的示例代码如下：

```
async void Send()
{
    try
    {   //发送消息
        string sendData = "测试消息字符串";
        //创建一个 DataWriter 对象
        DataWriter writer = new DataWriter(clientSocket.OutputStream);
        //获取 UTF-8 字符串的长度
        Int32 len = writer.MeasureString(sendData);
        //存储数据到输出流里面
        await writer.StoreAsync();
        //发送成功释放资源
        writer.DetachStream();
        writer.Dispose();
    }
    catch (Exception exception)
    {
        //获取错误的类型 SocketError.GetStatus(exception.HResult)
        //错误消息 exception.Message;
        //如果关闭 Socket 则释放资源
        clientSocket.Dispose();
        clientSocket = null;
    }
    try
    {   //接收数据
        DataReader reader = new DataReader(clientSocket.InputStream);
        //设置接收数据的模式,这里选择了部分
        reader.InputStreamOptions = InputStreamOptions.Partial;
        await reader.LoadAsync(reader.UnconsumedBufferLength);
        //接收成功
    }
```

```
        catch (Exception exception)
        {
            // …
        }
    }
```

17.3.4　启动 Socket 监听

Socket 监听表示是服务器端程序对客户端的 Socket 连接和发送消息的监听，在 Windows Phone 8 里面可以通过 Windows.Networking.Sockets 空间下的 StreamSocketListener 类来实现监听的操作，StreamSocketListener 类的成员如表 17.6 所示。Socket 监听示例代码如下：

```
async void Listene()
{
    //创建一个监听对象
    StreamSocketListener listener = new StreamSocketListener();
    listener.ConnectionReceived += OnConnection;
    //开始监听操作
    try
    {
        await listener.BindServiceNameAsync(localServiceName);
    }
    catch (Exception exception)
    {
        //处理异常
    }
}
//监听的连接事件处理
private async void OnConnection(StreamSocketListener sender,
StreamSocketListenerConnectionReceivedEventArgs args)
{
    DataReader reader = new DataReader(args.Socket.InputStream);
    try
    {
        //循环接收数据
        while (true)
        {
            //读取监听到的消息
            uint stringLength = reader.ReadUInt32();
            uint actualStringLength = await reader.LoadAsync(stringLength);
            //通过 reader 去读取监听到的消息内容
            reader.ReadString(actualStringLength)
        }
    }
    catch (Exception exception)
    {
        //异常处理
    }
}
```

表 17.6 StreamSocketListener 类成员

名 称	说 明
StreamSocketListener()	创建新的 StreamSocketListener 对象
StreamSocketListenerControl Control	获取 StreamSocketListener 对象上的套接字控件数据，返回某一 StreamSocketListener 对象上的套接字控件数据
StreamSocketListenerInformation Information	获取该 StreamSocketListener 对象的套接字信息，返回该 StreamSocketListener 对象的套接字信息
event TypedEventHandler < StreamSocketListener, StreamSocketListenerConnectionReceivedEventArgs > ConnectionReceived	指示在 StreamSocketListener 对象上收到连接的事件
IAsyncAction BindEndpointAsync（HostName localHostName）IAsyncAction BindEndpointAsync（HostName localHostName）	启动 StreamSocketListener 本地主机名和本地服务名的绑定操作。localHostName：用于绑定 StreamSocketListener 对象的本地主机名或 IP 地址；localServiceName：用于绑定 StreamSocketListener 对象的本地服务名称或 TCP 端口号，返回 StreamSocketListener 对象的异步绑定操作
IAsyncAction BindServiceNameAsync（string localServiceName）	启动 StreamSocketListener 本地服务名的绑定操作；localServiceName：用于绑定 StreamSocketListener 对象的本地服务名称或 TCP 端口号，返回 StreamSocketListener 对象的异步绑定操作

17.3.5 实例：模拟 Socket 通信过程

下面给出模拟 Socket 通信过程的示例：使用 Windows 运行时的 Socket 模拟收发消息和监听消息。

代码清单 17-2：查询日程安排（源代码：第 17 章\Examples_17_2）

MainPage.xaml 文件主要代码

```
<Grid x:Name="ContentPanel" Grid.Row="1" Margin="12,0,12,0">
    <StackPanel>
        <Button Content = "开始监听" x: Name = " btStartListener " Click =
"btStartListener_Click" />
        <Button Content = "连接 socket" x: Name = " btConnectSocket " Click =
"btConnectSocket_Click" />
        <TextBox Text="hello" x:Name="tbMsg"/>
        <Button Content="发送消息" x:Name="btSendMsg" Click="btSendMsg_Click" />
        <Button Content="关闭" x:Name="btClose" Click="btClose_Click" />
        <TextBlock Text="收到的消息："/>
        <ListBox x:Name="lbMsg">
        </ListBox>
    </StackPanel>
</Grid>
```

MainPage.xaml.cs 文件主要代码

```
using System;
using System.Windows;
using System.Windows.Controls;
using Microsoft.Phone.Controls;
using Windows.Networking;
using Windows.Networking.Sockets;
using Windows.Storage.Streams;
namespace StreamSocketDemo
{
    public partial class MainPage : PhoneApplicationPage
    {
        StreamSocketListener listener;                    //监听器
        StreamSocket socket;                              //Socket 数据流对象
        DataWriter writer;                                //输出流的写入数据对象
        public MainPage()
        {
            InitializeComponent();
        }
        //监听
        private async void btStartListener_Click(object sender, RoutedEventArgs e)
        {
            if (listener != null)
            {
                MessageBox.Show("监听已经启动了");
                return;
            }
            listener = new StreamSocketListener();
            listener.ConnectionReceived += OnConnection;
            //开始监听操作
            try
            {
                await listener.BindServiceNameAsync("22112");
                MessageBox.Show("正在监听中");
            }
            catch (Exception exception)
            {
                listener = null;
                //未知错误
                if (SocketError.GetStatus(exception.HResult) == SocketErrorStatus.Unknown)
                {
                    throw;

                }
            }
        }
        ///<summary>
        ///监听成功后会出发的连接事件处理
        ///</summary>
```

```
        ///<param name="sender">连接的监听者</param>
        ///<param name="args">连接的监听到的数据参数</param>
        private async void OnConnection(StreamSocketListener sender,
    StreamSocketListenerConnectionReceivedEventArgs args)
        {
            DataReader reader = new DataReader(args.Socket.InputStream);
            try
            {
                //循环接收数据
                while (true)
                {
                    //读取数据前面的4个字节,代表的是接收到的数据的长度
                    uint sizeFieldCount = await reader.LoadAsync(sizeof(uint));
                    if (sizeFieldCount != sizeof(uint))
                    {
                        //在socket被关闭之前才可以读取全部的数据
                        return;
                    }
                    //读取字符串
                    uint stringLength = reader.ReadUInt32();
                    uint actualStringLength = await reader.LoadAsync(stringLength);
                    if (stringLength != actualStringLength)
                    {
                        //在socket被关闭之前才可以读取全部的数据
                        return;
                    }
                    string msg = reader.ReadString(actualStringLength);
                    //通知到界面监听到的消息
                    Deployment.Current.Dispatcher.BeginInvoke(() =>
                    lbMsg.Items.Add(new TextBlock { Text = msg })
                    );
                }
            }
            catch (Exception exception)
            {
                //未知异常
                if (SocketError.GetStatus(exception.HResult) == SocketErrorStatus.Unknown)

                {
                    throw;
                }
            }
        }
        //连接Socket
        private async void btConnectSocket_Click(object sender, RoutedEventArgs e)
        {
            if (socket != null)
            {
                MessageBox.Show("已经连接了Socket");
                return;
```

```
        }
        HostName hostName;
        try
        {
            hostName = new HostName("localhost");
        }
        catch (ArgumentException)
        {
            MessageBox.Show("主机名不可用");
            return;
        }
        socket = new StreamSocket();
        try
        {
            //连接到 socket 服务器
            await socket.ConnectAsync(hostName, "22112");
            MessageBox.Show("连接成功");
        }
        catch (Exception exception)
        {
            //未知异常
            if (SocketError.GetStatus(exception.HResult) == SocketErrorStatus.Unknown)
            {
                throw;
            }
        }
    }
    //发送消息
    private async void btSendMsg_Click(object sender, RoutedEventArgs e)
    {
        if (listener == null)

        {
            MessageBox.Show("监听未启动");
            return;
        }
        if (socket == null)
        {
            MessageBox.Show("未连接 Socket");
            return;
        }
        if (writer == null)
        {
            writer = new DataWriter(socket.OutputStream);
        }
        //先写入数据的长度,长度的类型为 UInt32 类型然后再写入数据
        string stringToSend = tbMsg.Text;
        writer.WriteUInt32(writer.MeasureString(stringToSend));
        writer.WriteString(stringToSend);
        //把数据发送到网络
```

```
            try
            {
                await writer.StoreAsync();
                MessageBox.Show("发送成功");
            }
            catch (Exception exception)
            {
                if (SocketError.GetStatus(exception.HResult) == SocketErrorStatus.Unknown)
                {
                    throw;
                }
            }
        }
        //关闭连接,清空资源
        private void btClose_Click(object sender, RoutedEventArgs e)
        {
            if (writer != null)
            {
                writer.DetachStream();
                writer.Dispose();
                writer = null;
            }
            if (socket != null)
            {
                socket.Dispose();
                socket = null;
            }
            if (listener != null)
            {
                listener.Dispose();
                listener = null;
            }
        }
    }
}
```

程序的运行效果如图 17.6 所示。

图 17.6 Socket 模拟运行

第18章

墓碑机制与后台任务

智能手机屏幕越来越大、性能越来越强、机身越来越薄，但电池技术一直没有突破性进展，所以智能手机系统对于多任务的运行一直都处于谨慎的态度。为了确保创建一个快速响应的用户体验以及为了优化手机上的电源使用，Windows Phone 一次仅允许在前台中运行一个应用程序。Windows Phone 会通过一个墓碑机制来实现前台程序的伪多任务，可以让程序暂时挂起来，返回的时候再重新激活进行运行。Windows Phone 也为应用留出一些自由，这就是这些后台任务，这些任务可以在手机后台运行，即使在应用程序不是活动的前台应用程序时，仍然允许应用程序执行操作，比如音乐应用可以保持后台播放等，不过这些后台任务都会有比较严格的限制和规定。在 Windows Phone 8 之中这些后台的任务得到了进一步的加强和优化，例如后台传输文件由之前最高值为 5 个提高到了 25 个。

18.1 墓碑机制

墓碑机制(Tombstone)是 Windows Phone 系统中的一个程序运行规则，也是一个伪多任务的特性。简单来说，这个机制就是手机上一个任务被迫中断时，系统记录下当前应用程序的状态后，将程序暂停。当需要恢复时，系统再根据记录恢复到中断前的状态。在 Windows Phone 7.5 的版本中其“墓碑”的数量被限定在了 5 个，在 Windows Phone 8 增加到 7 个，用户能够在这些应用之间进行切换，如果超过，则之前的记录将被挤掉。

18.1.1 执行模式概述

Windows Phone 执行模型控制着 Windows Phone 上运行的应用程序的生命周期，该过程从启动应用程序开始，直至应用程序终止。该执行模型旨在始终为用户提供快速响应的体验。为此，在任何给定时间内，Windows Phone 仅允许一个应用程序在前台运行。当应用程序不再运行于前台时，操作系统将使应用程序进入休眠状态。如果可用于前台应用程序的设备内存不足，无法提供出色的用户体验，则操作系统将开始终止休眠的应用程序(首先终止最早使用的应用程序)。在应用程序里面可以通过相关的编程技巧来管理应用程序的状态。这有助于创建一种在用户看来应用程序保持单一实例的用户体验，即使应用程序已终止或已重新激活。

该执行模型还为用户提供了应用程序间一致的导航体验。在 Windows Phone 中，用户可通过从已安装程序列表或从“开始”中的磁贴启动应用程序来向前导航，除此之外，还可以通过其他方式向前导航，如点按与应用程序相关的 Toast 通知。用户还可以使用硬件“返回”按钮向后导航各个运行应用程序的页面，或者向后导航先前运行的应用程序的堆栈。Windows Phone 添加了通过按住硬件“返回”按钮来切换到先前运行的应用程序的功能。

18.1.2 应用程序的生命周期

图 18.1 演示了 Windows Phone 应用程序的生命周期。在该图中，圆圈表示应用程序的状态，矩形显示应用程序应管理其状态的应用程序级别或页面级别的事件。当启动一个应用程序的时候会先触发 Application 的 Launching 事件，然后跳转到默认的首页，调用该页面的 OnNavigatedTo 方法，程序进入运行的状态，当离开页面的时候将会调用该页面的 OnNavigatedFrom 方法。程序离开运行状态的时候有两种去向，一种是关闭应用应用程序，这种情况下会触发 Application 的 Closing 事件，程序处于关闭的状态；另外一种情况是进入休眠的状态，这时候会触发 Application 的 Deactivated 事件，程序处于休眠的状态。当手机的内存被耗尽，程序将会由休眠的状态转为了墓碑的状态，表示程序之前存储在内存的数据都被清空掉了，只剩下一个状态。程序从休眠状态或者墓碑状态恢复到运行状态的时候，将会触发 Application 的 Activated 事件，然后调用当前页面的 OnNavigatedTo 方法进

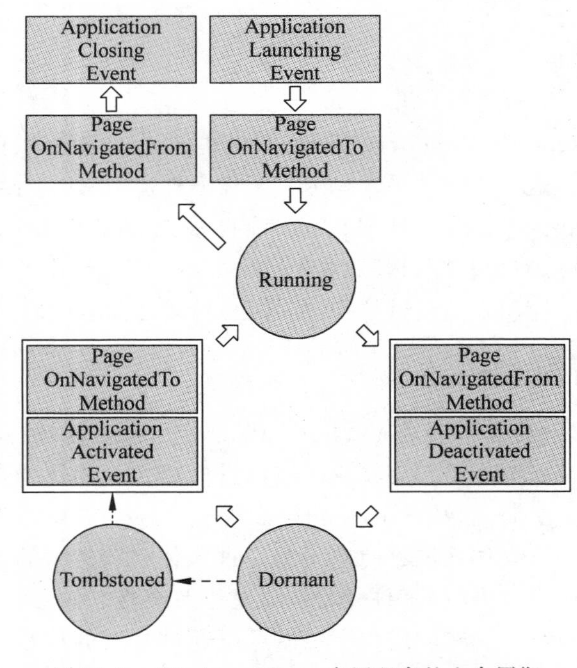

图 18.1　Windows Phone 应用程序的生命周期

入程序的页面。

在 Windows Phone 应用生命周期中需要注意一些问题，在 Launching 事件中执行非常少量的代码，不要执行资源密集型操作，例如，访问独立存储。在 Windows Phone 的 silverlight 框架程序中，退出程序的唯一方法就是按 Back 键，当 Back Stack 中不存在页面时程序就会退出，目前系统没有提供任何 Exit 方法来以代码的方式结束程序。在程序退出时会触发 Application 中最后一个事件，Closing。在这个事件中我们可以释放一些使用的资源，保存数据等，要注意的是，如果一个程序从墓碑状态被结束，是不会触发此事件的。关闭程序的时间被限制在 10s，超过这个时间程序会被终止掉。

18.1.3 休眠状态和墓碑状态处理

休眠状态和墓碑状态这两个状态容易让人混淆，休眠状态时，程序停止运行，但不同的是此时整个进程还是存在于内存中。当恢复这个程序时，就不需要创建一个新的实例。这样就加快了程序恢复和切换的速度，并且从休眠状态恢复时不需要去恢复墓碑数据。一般情况下按 home 键，程序会进入休眠状态，当前台程序在运行过程中，内存不足或者不足以让程序流畅运行时，系统会执行一些操作来释放内存，此时程序就可能从休眠状态变换为的墓碑状态。一个程序进入到墓碑状态时，其进程被终止掉，但是程序的回退栈中的信息，以及保存的一些信息会保留在内存中。

程序恢复时会触发 Application 类中的 Activated 事件，可以通过检查 IsApplicationInstance Preserved 参数来判断程序是从休眠状态还是墓碑状态返回的，在此方法中可以用来恢复之前在 Deactivated 事件中保存的数据。

```
private void Application_Activated(object sender, ActivatedEventArgs e)
{
    if (e.IsApplicationInstancePreserved)
    {
      //从休眠状态恢复
    }
    else
    {
      //从墓碑状态恢复
      //这时候程序在内存中的信息已经被清除掉了,需要在这里处理如何恢复之前的数据
    }
}
```

在 Windows Phone 中，只要有足够的内存可供前台应用程序顺畅运行，应用程序就会在用户导航离开时置于休眠状态。当应用程序休眠后还原时，UI 状态会自动保留。若要验证页面状态在逻辑删除后是否正确还原，需要启用调试器中的自动逻辑删除功能。

通过从“Project Properties”菜单中选择“[应用程序名称] Properties…”，或者通过右键单击“Project”中的项目，然后选择“Properties”，打开“Solution Explorer”。在“Debug”标签中，选中标有“Tombstone upon deactivation while debugging”的复选框，如图 18.2 所示。选中此复选框后，可以开始调试应用程序，单击 Start 按钮将立即逻辑删除应用程序让应用

程序进入墓碑状态，然后按 Back 键返回应用程序，并验证页面状态是否正确保存和还原。

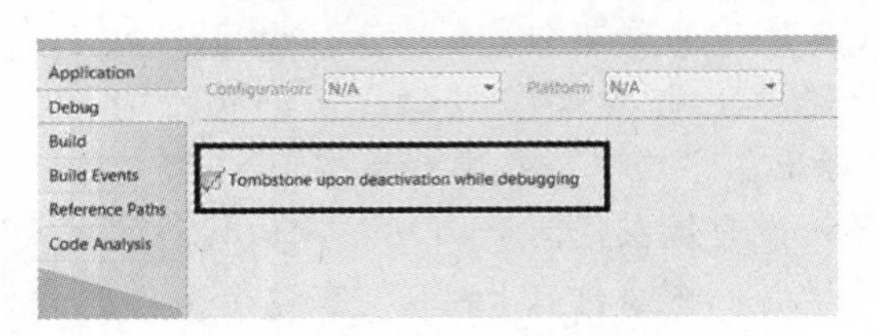

图 18.2　设置墓碑机制调试状态

从休眠状态恢复到程序的时候，如果这时候程序正在发送 http 请求的话，有可能这个请求会被取消掉，这时候需要在程序中来捕获这个异常并进行处理。处理代码如下：

```
private void GetSomeResponse(IAsyncResult MyResultAsync)
{
    HttpWebRequest request = (HttpWebRequest)MyResultAsync.AsyncState;
    try
    {
        HttpWebResponse response = (HttpWebResponse)request.EndGetResponse(MyResultAsync);
        if (response.StatusCode == HttpStatusCode.OK && response.ContentLength > 0)
        {
            using (StreamReader sr = new StreamReader(response.GetResponseStream()))
            {
              //…处理网络返回信息操作
            }
        }
    }
    catch (WebException e)
    {
    if(e.Status == WebExceptionStatus.RequestCanceled)
            //这里有可能就是从休眠和墓碑状态返回来的
            //可以在这里处理休眠和墓碑状态的恢复,例如重新发送 http 请求…
    else
        {
            //其他的异常
            using (HttpWebResponse response = (HttpWebResponse)e.Response)
            {
                MessageBox.Show(response.StatusCode.ToString());
            }
        }
    }
}
```

18.2　后台文件传输

后台文件传输是指应用程序能够对一个或多个使用文件上传或下载操作进行排队，这些操作将在后台执行，即使当应用程序不再在前台运行时也是如此。在应用程序里面我们也可以使用用于启动文件传输的 API 来查询现有传输的状态，从而能够为最终用户提供进度指示器。

18.2.1　后台文件传输概述

后台传输服务可以使用 BackgroundTransferService 在后台上传和下载文件。仅支持使用 HTTP 和 HTTPS 的传输，不支持 FTP，GET HTTP 方法支持下载文件，POST 方法支持下载或上传文件。所有后台传输都需要本地文件路径。下载需要目标路径，该路径指定将保存下载的文件的位置。上传需要源路径，该路径指定将从中上传文件的位置。后台传输的所有本地路径都必须位于应用程序的独立存储中，在名为“/shared/transfers”的根目录中。该目录是操作系统在安装应用程序时创建的，但如果应用程序删除或重命名该目录，则必须在启动任何文件传输之前重新创建该目录。可以在“/shared/transfers”根目录下创建选择的任何其他目录结构，并且可以在传输完成之后复制或删除文件，目的是确保后台传输服务不修改文件，但尝试使用“/shared/transfers”目录之外的路径启动传输将引发异常。

18.2.2　后台传输策略

操作系统对于与文件大小、连接速度以及设备资源有关的后台传输强制很多限制。

1. 传输大小的限制

通过后台传输服务上传最大文件大小为 5MB。通过手机网络连接的最大下载大小为 20MB，如果超过此限制，则传输的 TransferPreferences 属性将自动更改为 AllowBattery。通过 Wi-Fi 而没有外部电源的最大下载大小为 100MB，大于 100MB 的文件必须将传输的 TransferPreferences 属性设置为 None，否则传输将失败。如果不知道传输文件的大小，则其有可能会超出该限制，应将值设置为 None。

2. 数量的限制

每个应用程序的队列中未完成的最大请求数(这包括活动和挂起的请求)为 25 个，设备上所有应用程序的最大并发传输数为 2 个，设备上所有应用程序的最大排队传输数为 500，每个请求的最大 HTTP 标头数位 15，HTTP 标头的最大大小为每个 16KB。

3. 传输策略的限制

后台传输服务在 3G 以及更高版本的网络上运行，在 2G、EDGE 和标准 GPRS 的网络上不运行后台传输服务。服务器端所需的文件内容长度或文件范围标头大于 5MB，服务器应该始终在响应中返回内容长度，不这样做可能会导致严重降低传输的性能。

4. 慢速传输的限制

如果设备的网络连接速度 3G 网络低于 50Kbps，Wi-Fi/USB 网络低于 100Kbps，则会暂停传输并重试。

18.2.3 后台传输的 API

可以在 Microsoft.Phone.BackgroundTransfer 命名空间中找到后台传输服务 API，主要编程元素为 BackgroundTransferRequest 和 BackgroundTransferService 类。BackgroundTransferRequest 对象表示单个传输请求，包括目标和目标文件路径、传输方法以及传输的当前状态，BackgroundTransferRequest 类的主要成员如表 18.1 所示。BackgroundTransferService 对象用于启动新传输以及查询或删除现有文件传输，BackgroundTransferService 的主要成员如表 18.2 所示。BackgroundTransferRequest 对象的 Headers 属性用于设置传输请求的 HTTP 标头。当使用 Add(BackgroundTransferRequest) 方法排列传输请求时，向 Headers 集合中添加以下标头之一将引发 NotSupportedException：If-Modified-Since、If-None-Match、If-Range、Range 和 Unless-Modified-Since。保留以上标头以供系统使用，不能通过调用应用程序使用。

表 18.1 BackgroundTransferRequest 类的主要成员

名　称	说　明
BytesReceived	获取已为请求下载的字节数
BytesSent	获取已为请求上传的字节数
DownloadLocation	获取或设置所请求的文件将下载到的位置
Headers	获取请求 HTTP 标头的词典
Method	获取或设置请求的 HTTP 方法
RequestId	获取与请求关联的唯一标识符
RequestUri	获取与请求关联的目标 URI
StatusCode	请求的 HTTP 状态代码
Tag	获取或设置与请求关联的其他数据
TotalBytesToReceive	获取将为请求下载的字节总数
TotalBytesToSend	获取将为请求上传的字节总数
TransferError	获取与失败的后台传输请求关联的异常，传输请求可能具有 Completed 的 TransferStatus，指示传输是否成功完成，如果传输成功完成，则 TransferError 将为 null
TransferPreferences	获取或设置启用传输的首选条件
TransferStatus	获取请求的状态
UploadLocation	获取或设置将从中上传所请求文件的位置
事件 TransferProgressChanged	当传输的进度发生更改时发生
事件 TransferStatusChanged	当请求的 TransferStatus 属性发生更改时发生

表 18.2 BackgroundTransferService 类主要成员

名　称	说　明
属性 Requests	获取与正在调用的应用程序关联的所有活动传输请求的枚举
方法 Add	将后台传输请求添加到队列中
方法 Find	尝试返回具有指定 ID 的后台传输请求
方法 Remove	尝试删除具有指定 ID 的后台传输请求

18.2.4　后台传输编程步骤

后台传输编程可以分为以下3个步骤。

1. 应用程序启动之后启动传输服务

使用BackgroundTransferService对象的Requests属性循环访问应用程序的所有后台传输。对于每个请求，为TransferStatusChanged事件注册一个处理程序以便应用程序可以响应该应用程序正在运行时发生的状态更改。对于每个请求，为TransferProgressChanged事件注册一个处理程序。这对于通知用户有关活动传输进度的信息非常有用。对于每个请求，检查TransferStatus属性，以确定该应用程序处于非活动状态时任何传输的状态是否为已完成或已更改。使用前面的回调更新应用程序UI，而不是使用计时器或其他一些轮询服务更改的机制。

2. 处理添加新的传输请求

使用BackgroundTransferService对象的Add(BackgroundTransferRequest)方法添加传输请求。首先，查看应用程序是否已达到每个应用程序25个并发传输的限制。如果已达到，则可以通知用户，提示用户等待现有传输完成或取消现有传输。也可以将有关新传输的信息存储在独立存储中，然后在将来的某些时间文件传输队列低于此限制时加载该信息并启动传输。将对Add的调用放置在一个try块中并捕获任何异常。尝试创建新的传输失败时，应该向用户提供一个可使用的消息。

使用"/shared/transfers"作为所有传输操作的本地根目录。对于文件下载，需要将DownloadLocation属性设置为该目录中的某个文件名。这就是传输完成时文件所在的位置。对于文件上传，需要将UploadLocation属性设置为该目录中的某个文件名以指定要上传的文件。使用RequestUri属性指定文件传输的远程服务器地址。后台传输服务使用Uri的OriginalString属性。因此，应该使用Uri.EscapeUriString方法转义Uri中的任何特殊字符(如果Uri尚未转义)。如果要传输的文件大于100 MB，请将传输的TransferPreferences的属性设置为None，否则传输将失败。如果不知道传输文件的大小，则其有可能会超出该限制，也应将该属性设置为None。

3. 处理传输的完成

当传输完成时TransferStatus属性的值是Completed。通过检查包含从目标服务器返回的HTTP状态代码的StatusCode属性确定传输是否成功。根据服务器配置，如果传输成功，则该值将为200或206。其他任何状态代码都指示服务器错误。处理服务器错误的方式取决于你的应用程序。检查TransferError属性以帮助确定失败的传输未成功的原因。如果文件下载成功，可能希望将该文件从"/shared/transfers"目录移动到独立存储中的某个新位置，以便不向将来的后台传输操作公开该文件。通过调用BackgroundTransferService的Remove(BackgroundTransferRequest)方法从队列中删除传输请求。

18.2.5 后台文件传输实例

下面给出后台文件传输的示例：演示使用后台传输服务进行下载文件。

代码清单 18-1：后台文件传输(源代码：第 18 章\Examples_18_1)

MainPage.xaml 文件主要代码：添加后台文件下载的界面

```
<Grid x:Name="ContentPanel" Grid.Row="1" Margin="12,0,12,0">
    <StackPanel VerticalAlignment="Top">
        <TextBlock Margin="10" Text="输入文件的网络地址:"/>
        <TextBox x:Name="fileUrl" Text=" "></TextBox>
        <Button x:Name="downloadButton" Click="downloadButton_Click" Content=
"提交后台下载" />
        <Button Click="Button_Click" Content="查看下载的文件" />
    </StackPanel>
</Grid>
```

MainPage.xaml.cs 文件主要代码

```
//进入页面事件处理
protected override void OnNavigatedTo(System.Windows.Navigation.NavigationEventArgs e)
{
    //确保在独立存储中的传输文件保存目录存在
    using (IsolatedStorageFile isoStore = IsolatedStorageFile.
GetUserStoreForApplication())
    {
        if (!isoStore.DirectoryExists("/shared/transfers"))
        {
            isoStore.CreateDirectory("/shared/transfers");
        }
    }
    base.OnNavigatedTo(e);
}
//添加一个后台下载的任务
private void downloadButton_Click(object sender, RoutedEventArgs e)
{
    //后台传输的文件数量不能超过 5 个文件
    if (BackgroundTransferService.Requests.Count() >= 25)
    {
        MessageBox.Show("正在传输的文件数不能超过 25 个");
        return;
    }
    if (fileUrl.Text == "")
    {
        MessageBox.Show("文件地址不能为空");
        return;
    }
    //传输的文件网络路径
    string transferFileName = fileUrl.Text;
    Uri transferUri;
    try
```

```
            {
                transferUri = new Uri(Uri.EscapeUriString(transferFileName), UriKind.
RelativeOrAbsolute);
            }
            catch (Exception ex)
            {
                MessageBox.Show("文件地址不符合格式");
                return;
            }

            //创建一个后台文件传输请求
              BackgroundTransferRequest transferRequest = new BackgroundTransferRequest
(transferUri);
            //设置传输的方法为 GET 请求
            transferRequest.Method = "GET";
            //获取文件的名称
            string downloadFile = transferFileName.Substring(transferFileName.LastIndexOf
("/") + 1);
            Uri downloadUri = new Uri ( " shared/transfers/" + downloadFile, UriKind.
RelativeOrAbsolute);
            transferRequest.DownloadLocation = downloadUri;
            //添加请求的 Tag 属性,Tag 属性不能超过 4000 个字符
            transferRequest.Tag = downloadFile;
            //使用 BackgroundTransferService 添加文件传输请求
            try
            {
                BackgroundTransferService.Add(transferRequest);
            }
            catch (Exception ex)
            {
                MessageBox.Show("无法添加请求:" + ex.Message);
            }
        }
        //跳转到下载文件列表
        private void Button_Click(object sender, RoutedEventArgs e)
        {
            NavigationService.Navigate(new Uri("/BackgroundTransferList.xaml", UriKind.
RelativeOrAbsolute));
        }
```

BackgroundTransferList.xaml 文件主要代码：后台文件下载和已下载文件列表界面

```
<phone:Pivot>
    <phone:PivotItem Header="正在下载文件">
        <Grid Margin="12,0,12,0">
            <ListBox Name="TransferListBox"><!-- 正在下载文件的 ListBox 绑定模板 -->
                <ListBox.ItemTemplate>
                    <DataTemplate>
                        <Grid Background="Transparent" Margin="0,0,0,30">
                            <Grid.ColumnDefinitions>
                                <ColumnDefinition Width="380"/>
```

```
                                <ColumnDefinition Width="50"/>
                            </Grid.ColumnDefinitions>
                            <Grid Grid.Column="0">
                                <StackPanel Orientation="Vertical">
                                    <TextBlock Text = "{Binding Tag}" Foreground =
"{StaticResource PhoneAccentBrush}" FontWeight="Bold"/>
                                    <StackPanel Orientation="Horizontal">
                                        <TextBlock Text="status: "/>
                                        <TextBlock Text = "{Binding TransferStatus}"
HorizontalAlignment="Right"/>
                                    </StackPanel>
                                    <StackPanel Orientation="Horizontal">
                                        <TextBlock Text="bytes received: "/>
                                        <TextBlock Text = "{Binding BytesReceived}"
HorizontalAlignment="Right"/>
                                    </StackPanel>
                                    <StackPanel Orientation="Horizontal">
                                        <TextBlock Text="total bytes: "/>
                                        <TextBlock Text="{Binding TotalBytesToReceive}"
HorizontalAlignment="Right"/>
                                    </StackPanel>
                                </StackPanel>
                            </Grid>
                            <Grid Grid.Column="1">
                              <!--取消下载按钮-->
                                <Button Tag = " { Binding RequestId }" Click =
"CancelButton_Click" Content = " X" BorderBrush = " Red" Background = " Red" Foreground =
"{StaticResource PhoneBackgroundBrush}" VerticalAlignment="Top" BorderThickness="0" Width
="50" Padding="0,0,0,0"></Button>
                            </Grid>
                        </Grid>
                    </DataTemplate>
                </ListBox.ItemTemplate>
            </ListBox>
        </Grid>
    </phone:PivotItem>
    <phone:PivotItem Header="文件列表">
        <Grid Margin="12,0,12,0">
            <ListBox Name="FileListBox">
                <ListBox.ItemTemplate><!--已下载文件的 ListBox 绑定模板-->
                    <DataTemplate>
                        <Grid Margin="5,5,5,5">
                            <Grid.RowDefinitions>
                                <RowDefinition />
                                <RowDefinition Height="1" />
                                <RowDefinition />
                                <RowDefinition />
                            </Grid.RowDefinitions>
                            <Grid.ColumnDefinitions>
                                <ColumnDefinition />
```

```
                                    </Grid.ColumnDefinitions>
                                    <Grid Grid.Row = "0">
                                        <Grid.ColumnDefinitions>
                                            <ColumnDefinition Width = "Auto" />
                                            <ColumnDefinition Width = " * " />
                                        </Grid.ColumnDefinitions>
                                        <TextBlock Grid. Column = " 1 " Text = " { Binding
EntryName}" FontSize = "28" FontWeight = "ExtraBold" />
                                    </Grid>
                                    <TextBlock Grid. Row = " 2 " Text = " { Binding Detail }"
FontSize = "20" />
                                </Grid>
                            </DataTemplate>
                        </ListBox.ItemTemplate>
                    </ListBox>
                </Grid>
            </phone:PivotItem>
        </phone:Pivot>
```

BackgroundTransferList.xaml.cs 文件代码

```
using System;
using System.Collections.Generic;
using System.Windows;
using System.Windows.Controls;
using Microsoft.Phone.Controls;
using Microsoft.Phone.BackgroundTransfer;
namespace BackgroundTransferDownLoadDemo
{
    public partial class BackgroundTransferList : PhoneApplicationPage
    {
        //正在进行中的后台传输列表
        IEnumerable<BackgroundTransferRequest> transferRequests;
        //传输服务的条件限制
        bool WaitingForExternalPower;
        bool WaitingForExternalPowerDueToBatterySaverMode;
        bool WaitingForNonVoiceBlockingNetwork;
        bool WaitingForWiFi;
        public BackgroundTransferList()
        {

            InitializeComponent();
            DirectoryEntry[] directoryEntries = IsolatedStorageHelper.GetDirectoryEntries
("/shared/transfers");
             if (directoryEntries != null)
             {
                 FileListBox.ItemsSource = directoryEntries;
             }
        }
        //进入页面的初始化工作
        protected override void OnNavigatedTo(System.Windows.Navigation.NavigationEventArgs e)
        {
```

```
            //初始化所有的限制为 false
            WaitingForExternalPower = false;
            WaitingForExternalPowerDueToBatterySaverMode = false;
            WaitingForNonVoiceBlockingNetwork = false;
            WaitingForWiFi = false;
            //刷新传输的服务
            InitialTansferStatusCheck();
            UpdateUI();
        }
        private void InitialTansferStatusCheck()
        {
            UpdateRequestsList();
            foreach (var transfer in transferRequests)
            {
                transfer.TransferStatusChanged += new EventHandler<BackgroundTransferEventArgs>
(transfer_TransferStatusChanged);
                transfer.TransferProgressChanged += new EventHandler<BackgroundTransferEventArgs>
(transfer_TransferProgressChanged);
                ProcessTransfer(transfer);
            }
            if (WaitingForExternalPower)
            {
                MessageBox.Show("电力不足!传输服务不能启动。");
            }
            if (WaitingForExternalPowerDueToBatterySaverMode)
            {
                MessageBox.Show("省电模式限制!传输服务不能启动。");
            }
            if (WaitingForNonVoiceBlockingNetwork)
            {
                MessageBox.Show("等待网络连接!");
            }

            if (WaitingForWiFi)
            {
                MessageBox.Show("等待 WiFi 网络连接!");
            }
        }
        //传输状态的改变
        void transfer_TransferStatusChanged(object sender, BackgroundTransferEventArgs e)
        {
            //处理不同情况的传输状态
            ProcessTransfer(e.Request);
            //更新 UI 界面
            UpdateUI();
        }
        //处理不同情况的传输状态
        private void ProcessTransfer(BackgroundTransferRequest transfer)
        {
            switch (transfer.TransferStatus)
            {
                case TransferStatus.Completed:
```

```
                //当传输的状态是 200 或者 206 表示传输成功
                if (transfer.StatusCode == 200 || transfer.StatusCode == 206)
                {
                    //需要主动移除已经传输成功的请求,传输服务不会注意移除
                    RemoveTransferRequest(transfer.RequestId);
                }
                else
                {
                    //传输失败,移除传输的服务
                    RemoveTransferRequest(transfer.RequestId);
                    if (transfer.TransferError != null)
                    {
                        MessageBox.Show("失败的信息: " + transfer.TransferError);
                    }
                }
                break;
            case TransferStatus.WaitingForExternalPower:
                WaitingForExternalPower = true;
                break;
            case TransferStatus.WaitingForExternalPowerDueToBatterySaverMode:
                WaitingForExternalPowerDueToBatterySaverMode = true;
                break;

            case TransferStatus.WaitingForNonVoiceBlockingNetwork:
                WaitingForNonVoiceBlockingNetwork = true;
                break;
            case TransferStatus.WaitingForWiFi:
                WaitingForWiFi = true;
                break;
        }
    }
    //传输进度的改变
    void transfer_TransferProgressChanged(object sender, BackgroundTransferEventArgs e)
    {
        UpdateUI();
    }
    //更新 UI 界面
    private void UpdateUI()
    {
        //更新请求列表
        UpdateRequestsList();
        //更新绑定的 UI 源
        TransferListBox.ItemsSource = transferRequests;
    }
    //移除后台文件传输请求
    private void RemoveTransferRequest(string transferID)
    {
        //在后台文件传输服务中寻找到传输 ID 对应的传输请求
        BackgroundTransferRequest transferToRemove = BackgroundTransferService.
Find(transferID);
        //移除文件传输请求
        if (transferToRemove != null)
```

```
                {
                    BackgroundTransferService.Remove(transferToRemove);
                }
            }
            //取消文件传输
            private void CancelButton_Click(object sender, EventArgs e)
            {
                //获取 ListBox 中绑定的传输请求的 ID,Button 里面的 Tag 属性
                string transferID = ((Button)sender).Tag as string;
                //移除传输请求
                RemoveTransferRequest(transferID);
                //刷新文件传输 UI
                UpdateUI();
            }

            private void UpdateRequestsList()
            {
                //销毁旧的请求,从新赋值新的文件传输队列
                if (transferRequests != null)
                {
                    foreach (var request in transferRequests)
                    {
                        request.Dispose();
                    }
                }
                transferRequests = BackgroundTransferService.Requests;
            }
        }
    }
```

程序的运行效果如图 18.3～图 18.5 所示。

图 18.3　文件下载

图 18.4　正在下载文件

图 18.5　文件列表

18.3　后台代理

后台代理是指 Windows Phone 的应用程序允许在系统的后台去执行一些计划任务，即使应用程序关闭后这些计划任务依然可以继续运行。在手机系统的后台任务列表里面，可以找到那些程序使用后台代理的功能，以及这些后台代理的表述，并且可以看到当前那些后台代理正在运行中那些已经关闭，用户也可以手动去关闭和打开这些程序的后台代理任务。后台代理为应用程序提供了更加好的用户体验，例如 Windows Phone 的新浪微博客户端可以在磁贴上提示未读消息条数就是使用了后台代理的功能去实现了这种在关闭了应用程序之后还可以把最重要的消息告诉给用户的功能。

18.3.1　后台代理简介

后台代理有两个计划任务的类型，分别为定期代理(PeriodicTask)和资源密集型代理(ResourceIntensiveTask)。定期代理以定期重复的间隔运行一小段时间。这种类型的任务的典型方案包括上传设备的位置以及执行少量的数据同步。资源密集型代理在手机符合与处理器活动、电源以及网络连接有关的一组要求时运行，运行的时间相对较长。这种类型任务的典型方案是，在用户没有主动使用手机时将大量数据同步到手机。

一个应用程序可能只有一个后台代理，可以将该代理注册为 PeriodicTask、ResourceIntensiveTask 或两者。运行代理的计划取决于注册的任务类型，一次只能运行代理的一个实例。代理的代码由应用程序在从 BackgroundAgent 继承的类中实现。代理启动时，操作系统调用 OnInvoke(ScheduledTask)。在该方法中，应用程序可以确定它以哪种 ScheduledTask 类型运行，并执行相应的操作。代理完成其任务之后，它应该调用 NotifyComplete()或 Abort() 以让操作系统知道它已完成。如果任务成功，则应该使用 NotifyComplete。如果代理无法执行其任务(如所需的服务器不可用)，则代理应该调用 Abort，这将导致 IsScheduled 属性设置为 false。前台应用程序可以在其运行时检查该属性，以确定是否调用了 Abort。

18.3.2　实现后台代理的 API

ScheduledTaskAgent 类是表示是后台代理，该类包含一个方法 OnInvoke，只要执行计划任务就会调用该方法。BackgroundAgent 类是后台代理 ScheduledTaskAgent 类的基类，主要包含了 3 个方法：NotifyComplete 方法，通知操作系统，代理已完成其目标任务；OnCancel 方法由操作系统调用，用于警告将要置于休眠状态或终止状态的后台代理；Abort 方法通知操作系统，代理无法执行其目标任务并且在前台应用程序解决阻止问题并重新启用该代理之前不能再次启动代理。在后台代理编程中必须要重写 ScheduledTaskAgent 类，在后台运行的代码放置在从 ScheduledTaskAgent 派生的类中。

PeriodicTask 类表示在短时间内定期运行的计划任务，它允许应用程序在主应用程序不在前台中时也执行处理，该类用于在较短的时间内需要定期运行并且允许非常有限的设备资源的任务。ResourceIntensiveTask 类表示偶尔运行并且允许使用大量设备资源的计划任务。ResourceIntensiveTask，它允许应用程序在主应用程序不在前台中时也执行处理，该类用于可能不需要定期运行，但需要运行较长时间并且使用更多设备资源的任务。PeriodicTask 类和 ResourceIntensiveTask 类派生自 ScheduledTask 类，ScheduledTask 类表示一个将定期运行或者在手机处于支持资源密集型处理的状态时将运行的任务，该类从 ScheduledAction 继承，是无法直接实例化的抽象类。ScheduledTask 派生自 ScheduledAction 类，ScheduledTask 的属性如表 18.3 所示。

表 18.3 ScheduledTask 类的属性

名称	说明
BeginTime	获取或设置开始操作计划的时间
ExpirationTime	获取或设置操作计划的结束时间
IsEnabled	在当前版本中未使用该值
IsScheduled	获取操作的计划状态
Name	获取计划操作的名称
Description	获取或设置 ScheduledTask 的说明，该字符串在设备的“后台任务”设置页面中向用户显示
LastExitReason	获取代理在上次运行时退出的原因
LastScheduledTime	获取上次计划运行代理的时间，采用设备的本地时间

18.3.3 后台代理不支持运行的 API

有一组 API 不能在由计划任务执行的代码中使用。其中一些 API 由开发工具进行检测，并且在运行时或编译时将引发错误。其他 API 受 Windows Phone 商城应用程序提交过程的限制。不支持后台代理的 API 如表 18.4 所示，用于后台代理时需要注意的 API 如表 18.5 所示。

表 18.4 不支持后台代理的空间和类

命名空间	不支持的 API
Microsoft. Devices	Camera
Microsoft. Devices	VibrateController
Microsoft. Devices	NowPlaying
Microsoft. Devices. Radio	不支持此命名空间中的所有 API
Microsoft. Devices. Sensors	不支持此命名空间中的所有 API
Microsoft. Phone. BackgroundAudio	BackgroundAudioPlayer
Microsoft. Phone. BackgroundTransfer	Add(BackgroundTransferRequest)
Microsoft. Phone. Controls	WebBrowser

续表

命名空间	不支持的 API
Microsoft. Phone. Info	IsKeyboardDeployed
Microsoft. Phone. Notification	不支持此命名空间中的所有 API
Microsoft. Phone. Scheduler	Add(ScheduledAction) Remove(String) Replace(ScheduledAction)
Microsoft. Phone. Shell	不支持除以下 API 之外的所有 API： ShellToast 类 ShellTile 类的 Update(ShellTileData) 方法 ShellTile 类的 Delete()()()() 方法 ShellTile 类的 ActiveTiles 属性
Microsoft. Phone. Tasks	不支持此命名空间中的所有 API
Microsoft. Xna. *	不支持所有 XNA Framework 命名空间中的所有 API
System. Windows	MessageBox
System. Windows	Clipboard
System. Windows. Controls	MediaElement
System. Windows. Controls	MultiScaleImage
System. Windows. Media	LicenseAcquirer
System. Windows. Media	A/V Capture
System. Windows. Navigation	不支持此命名空间中的所有 API

表 18.5　用于后台代理时需要注意的 API

类　名	说　明
GeoCoordinateWatcher	此 API 用于获取设备的地理坐标，支持在后台代理中使用，但它使用缓存的位置值而不是实时数据；设备每 15 分钟更新缓存的位置值
Mutex 类	应该使用 Mutex 类同步对在前台应用程序和后台代理之间共享的资源（如独立存储中的文件）的访问
ShellToast 类	该类可以用于从正在运行的后台代理弹出 Toast 通知
ShellTile 类的 Update(ShellTileData) 方法 ShellTile 类的 Delete() 方法 ShellTile 类的 ActiveTiles 属性	这些方法可以用于修改正在运行的后台代理中的 shell 磁贴；应注意，不能在后台代理中创建 shell 磁贴
HttpWebRequest 类	该类允许从正在运行的后台代理进行 Web 请求

18.3.4　后台代理的限制

在 Windows Phone 中使用后台代理有着严格的限制，除了上一节列出的不支持的 API 之外，后台代理还有内存，时间和程序稳定性的限制。

(1) 定期代理和资源密集型代码任何时候都不能使用超过 6MB 的内存。音频代理限制为 15MB。如果超过此内存上限，则立即终止。在调试器下运行时，不作内存和超时限制。可以使用 ApplicationMemoryUsageLimit API 查询前台应用程序和后台代理的内存限制。

(2) 使用 ScheduledTask 对象的 ExpirationTime 属性设置不再运行任务的时间。当使用 Add(ScheduledAction) 方法计划操作时，该值必须设置为两周之内的某个时间。当与该任务关联的应用程序在前台运行时，可能会重新计划该任务并将过期时间重置为从当前时间起最多两周。

(3) 如果由于超过内存配额或任何其他未处理的异常而连续两次退出，则取消定期代理和资源密集型代理的计划。代理必须由前台应用程序重新计划。

定期代理和资源密集型代理还在代理的计划、持续时间和常规限制上有着不同的限制，定期代理的计划、持续时间和常规限制如表 18.6 所示，资源密集型代理的计划、持续时间和常规限制如表 18.7 所示。

表 18.6　定期代理的计划、持续时间和常规限制

限　制	说　明
计划间隔：30 分钟	定期代理通常每隔 30 分钟运行一次。若要优化电池使用时间，定期代理的运行可以与其他后台进程一致，因此执行时间可能最多漂移 10 分钟
计划持续时间：25 秒	定期代理通常运行 25 秒；其他限制可能会导致代理提取终止
节电模式可能会阻止执行	节电模式是一个选项，用户可以在设备上启用该选项以指示应该优先考虑电池使用时间；如果启用此模式，则定期代理可能不运行，即使间隔已过也是如此
每个设备的定期代理限制	为了帮助最大程度地提高设备的电池使用时间，对手机上可以计划的定期代理数量进行了硬性限制；它因每个设备配置而异并且可以低到 6；还有另一个低于硬性限制的限制，超过该限制之后会警告用户他们正在运行多个后台代理，因此电池消耗可能比较快 警告： 当超出设备限制时，如果尝试添加定期后台代理，则对 Add(ScheduledAction) 的调用将引发 InvalidOperationException；由于每个设备的定期代理限制非常低，因此应用程序很可能会遇到此异常；出于此原因，在添加定期代理时捕获此异常非常重要，这样应用程序就不会崩溃

表 18.7　资源密集型代理的计划、持续时间和常规限制

限　制	说　明
持续时间：10 分钟	资源密集型代理通常运行 10 分钟；其他限制可能会导致代理提取终止
需要外部电源	除非设备连接到外部电源，否则不运行资源密集型代理
需要非手机网络连接	除非设备通过 Wi-Fi 或 PC 连接建立网络连接，否则资源密集型代理不运行

续表

限　制	说　明
最低电池电量	除非设备的电池电量高于 90%，否则资源密集型代理不运行
需要设备屏幕锁定	除非设备屏幕锁定，否则资源密集型代理不运行
非活动手机呼叫	当手机呼叫处于活动状态时，资源密集型代理不运行
不能将网络更改为手机网络	如果资源密集型代理尝试调用指定 MobileBroadbandGSM()或 MobileBroadbandCDMA()的 AssociateToNetworkInterface(Socket, NetworkInterfaceInfo)，则该方法调用失败

18.3.5 后台任务实例

下面给出后台任务的示例：演示使用后台代理服务实现一个后台运行的任务。

代码清单 18-2：后台任务(源代码：第 18 章\Examples_18_2)

ScheduledAgent.cs 文件主要代码：后台任务的处理逻辑

```
protected override void OnInvoke(ScheduledTask task)
{
    string toastTitle = "PeriodicTask";
    string toastMessage = "测试 PeriodicTask 后台任务";
    //通过吐司通知来提示后台代理的运行
    ShellToast toast = new ShellToast();
    toast.Title = toastTitle;
    toast.Content = toastMessage;
    toast.Show();
    //设置测试的时间
    ScheduledActionService.LaunchForTest(task.Name, TimeSpan.FromSeconds(60));
    NotifyComplete();
}
```

MainPage.xaml 文件主要代码

```
<Grid x:Name = "ContentPanel" Grid.Row = "1" Margin = "12,0,12,0">
    <Button Content = "启动任务" Height = "72" HorizontalAlignment = "Left" Margin = "90,60,0,
0" Name = "button1" VerticalAlignment = "Top" Width = "160" Click = "button1_Click" />
    <TextBlock Height = "50" HorizontalAlignment = "Left" Margin = "112,169,0,0" Name =
"textBlock1" Text = "" VerticalAlignment = "Top" Width = "193" />
</Grid>
```

MainPage.xaml.cs 文件代码

```
using System;
using System.Windows;
using Microsoft.Phone.Controls;
using Microsoft.Phone.Scheduler;
namespace PeriodicAgentDemo
{
    public partial class MainPage : PhoneApplicationPage
```

```
    {
        PeriodicTask periodicTask;                                   //后台代理对象
        string periodicTaskName = "PeriodicAgent";                   //后台代理的名字
        public MainPage()
        {
            InitializeComponent();
        }
        //开始启动后台任务的按钮事件处理
        private void button1_Click(object sender, RoutedEventArgs e)
        {
            StartPeriodicAgent();
        }
        //启动后台任务
        private void StartPeriodicAgent()
        {
            //通过名字查找后台代理服务
            periodicTask = ScheduledActionService.Find(periodicTaskName) as PeriodicTask;
            //如果查找到了相同名字的代理服务,必须先移除掉再添加才能更新代理
            if (periodicTask != null)
            {
                ScheduledActionService.Remove(periodicTaskName);
            }
            periodicTask = new PeriodicTask(periodicTaskName);
            //这里的描述可以在手机的后台服务别表里面看到
            periodicTask.Description = "这是一个测试的后台服务";
            try
            {
                ScheduledActionService.Add(periodicTask);
                //测试后台任务
                ScheduledActionService.LaunchForTest(periodicTaskName, TimeSpan.FromSeconds(60));
            }
            catch (InvalidOperationException exception)
            {
                if (exception.Message.Contains("BNS Error: The action is disabled"))
                {
                    MessageBox.Show("该后台代理服务已经被用户禁止了。");
                }
                if ( exception. Message. Contains ( " BNS Error: The maximum number of
ScheduledActions of this type have already been added."))
                {
                    MessageBox.Show("已经达到最大的后台代理服务数量了。");
                }

            }
            catch (SchedulerServiceException)
            {
            }
            textBlock1.Text = "任务已经启动";
        }
    }
}
```

程序的运行效果如图 18.6 和图 18.7 所示。

图 18.6 后台任务

图 18.7 设置页面

18.4 后台音频

后台音频是指在系统后台播放音频，即使播放音频所属的应用程序已经关闭了，音频还可以继续在手机上播放。这意味着，即使用户已通过“返回”或“开始”按钮离开应用程序，应用程序也可以继续播放音频。

18.4.1 后台音频概述

Windows Phone 上的所有媒体都是通过 Zune 媒体队列播放的。后台音频应用程序通过向 Zune 媒体队列发送命令来设置当前播放堆栈、开始播放、暂停、快进、后退等。通过在 BackgroundAudioPlayer 类中调用方法来完成这些操作，然后与 Zune 媒体队列通信以操作音频的播放。

通用音量控制（UVC）是一个音频控件，当正在播放音频或当用户按音量控制开关时手机会显示这个控件。UVC 还可以作 Zune 媒体队列当随后从应用程序开始播放某些内容时，可以使用 UVC 控制音频。UVC 向应用程序中的 AudioPlayerAgent 发送事件，从而允许实现播放列表逻辑。

有两种类型的后台音频应用程序。一种类型是实现简单的播放列表并将一个包含媒体文件地址的 URI 传递给 Zune 媒体队列以设置当前曲目。URI 可以是手机的本地或远程地址。在任何一种情况下，音频需要是 Windows Phone 支持的类型才能播放。另一种类型的后台音频应用程序使用 MediaStreamSource 实现音频流以向播放系统输送音频示例。此流的格式可以是任何格式，因为需要在程序中实现一个从 MediaStreamSource 派生的类来处理音频的流处理以及对音频的解码。实现 MediaStreamSource 不在本文的讨论范围内。

18.4.2 后台音频的 API

AudioPlayerAgent 类是 BackgroundAgent 的实现，专门设计为在后台播放音频，主要有 3 个方法，OnError 方法当播放出现错误（如未正确下载音频曲目）时调用；OnPlayStateChanged 方法当播放状态发生更改时调用，错误状态除外；OnUserAction 方法当用户使用一些应用程序提供的 UI 或通用音量控制（UVC）请求操作并且应用程序已请求该操作的通知时调用。

BackgroundAudioPlayer 类提供对音频播放功能（如播放、暂停、快进和后退）的后台访问。这是用于执行播放选项和注册 PlayStateChanged 事件的主类，与当前正在播放的播放列表无关。它由前台应用程序和后台代理使用。后台代理依赖回调，而不是依赖事件。BackgroundAudioPlayer 类的属性和方法分别如表 18.8 和表 18.9 所示。

表 18.8 BackgroundAudioPlayer 类属性

名称	说明
BufferingProgress	为媒体内容完成的缓冲量
CanPause	获取一个值，该值指示当调用 Pause() 方法时是否可以暂停媒体
CanSeek	获取一个值，该值指示是否可以通过设置 Position 属性的值来重新放置媒体
Error	播放当前的 AudioTrack 时发生的最后一个错误（如果有）
Instance	返回 BackgroundAudioPlayer 的实例，如果此应用程序已分配后台音频播放资源，则返回的 BackgroundAudioPlayer 将包含对这些资源的引用
PlayerState	获取播放器的当前 PlayState
Position	获取或设置当前 Track 中的当前位置
Track	获取或设置此应用程序的当前曲目以及该曲目当前是否正在播放
Volume	在 0～1 之间的线性标尺上所表示的媒体音量，默认值为 0.85

表 18.9 BackgroundAudioPlayer 类方法

名称	说明
Close	关闭播放器并移除为其保留的所有资源，包括当前的 AudioTrack
FastForward	开始快进当前的 AudioTrack
Pause	暂停当前位置上的播放
Play	播放或继续播放当前 AudioTrack 的当前位置
Rewind	开始后退当前的 AudioTrack
SkipNext	跳至下一个曲目
SkipPrevious	跳至上一个曲目
Stop	停止媒体并将其重设为从头播放

AudioTrack 类表示单一音频曲目。该类使用 IEditableObject 来启用属性更改请求的批处理。例如，当某个曲目更改时，应用程序可能希望一次更新元数据的几个片段，如曲目名称、艺术家、专辑等，AudioTrack 类的属性如表 18.10 所示。使用 AudioTrack 构造函数之一创建曲目之后，必须使用 BeginEdit() 和 EndEdit() 方法更新对象。使用 BeginEdit 方法将曲目对象设置为编辑模式，使用 EndEdit 方法结束编辑操作，将所有属性更改保存到基础存储中。

表 18.10 AudioTrack 类的属性

名　称	说　明
Album	要作为曲目的专辑显示的文本
AlbumArt	曲目的专辑封面的路径
Artist	要作为曲目的艺术家显示的文本
Duration	曲目的长度
PlayerControls	确定在系统用户界面中启用了哪些播放控件
Source	曲目的 URI 路径
Tag	与此曲目关联的任意字符串，用于存储特定于应用程序的状态
Title	要作为曲目的标题显示的文本

18.4.3 后台音乐实例

下面给出后台音乐的示例：演示使用后台音乐服务播放音乐。

代码清单 18-3：后台音乐(源代码：第 18 章\Examples_18_3)

App.xaml.cs 文件主要代码：启动程序的时候把程序的文件复制到独立存储里面

```
public App()
{
    CopyToIsolatedStorage();                          //把文件复制到独立存储里面去
    …
}
…
private void CopyToIsolatedStorage()
{
    using (IsolatedStorageFile storage = IsolatedStorageFile.GetUserStoreForApplication())
    {
        if (!storage.FileExists("Ring01.wma"))
        {
            string _filePath = "Audio/Ring01.wma";
            // 获取程序中的音频
            StreamResourceInfo resource = Application.GetResourceStream(new Uri(_
filePath, UriKind.Relative));
            using (IsolatedStorageFileStream file = storage.CreateFile("Ring01.wma"))
            {
```

```
                    int chunkSize = 4096;
                    byte[ ] bytes = new byte[chunkSize];
                    int byteCount;
                    //把数据写入独立存储文件
                    while ((byteCount = resource.Stream.Read(bytes, 0, chunkSize)) > 0)
                    {
                        file.Write(bytes, 0, byteCount);
                    }
                }
            }
        }
    }
```

MainPage.xaml 文件主要代码

```
<Grid x:Name = "ContentPanel" Grid.Row = "1" Margin = "12,0,12,0">
        < Button Content = "播放" FontSize = "50" HorizontalAlignment = "Center"
VerticalAlignment = "Center" Click = "OnPlayButtonClick" />
</Grid>
```

MainPage.xaml.cs 文件主要代码

```
//使用后台音乐服务进行播放
private void OnPlayButtonClick(object sender, RoutedEventArgs e)
{
        AudioTrack audioTrack = new AudioTrack ( new Uri ( "Ring01.wma", UriKind.
Relative),"Ringtone 1","艺术家", "专辑", null);
      BackgroundAudioPlayer.Instance.Track = audioTrack;
}
```

程序的运行效果如图 18.8 所示。

图 18.8　播放后台音乐

18.5　计划通知

计划通知也是一种后台任务的类型，它有两种类型：一种是警报，另外一种是提醒，都是在一定的计划时间中提示的通知。在 Windows Phone 中可以利用计划通知的功能去开发一些闹铃应用程序和提醒类的应用程序。

18.5.1　计划通知简介

计划通知是一个在指定时间在屏幕上弹出的对话框，它类似于手机的内置应用程序所显示的通知。该对话框向用户显示一些可自定义的文本信息并允许用户关闭该通知或将其推迟到稍后的时间。如果用户点按通知，则会启动创建该通知的应用程序。在应用程序里面可以将通知配置为启动一次，或配置为按定期计划启动多次。请注意，这些通知的计划仅在一分钟的范围内是准确的。换言之，在计划之后多达一分钟才能启动通知。

有两种类型的计划通知：警报(Alarm)和提醒(Reminder)。警报允许指定启动通知时要播放的声音文件。当创建提醒时，可以指定一个指向应用程序中某个页面的相对导航URI，URI 可以包含查询字符串参数。向用户显示提醒并且用户点击提醒按钮时，会启动应用程序并且传递指定的 URI。在应用程序里面的警报或提醒是有限制的，每个应用程序一次只能有 50 个警报或提醒。

警报和提醒拥有不同的向用户显示的用户界面并且拥有不同的应用程序自定义选项。

1. 警报 UI 属性

如图 18.9 所示显示创建警报(Alarm)的应用程序的名称；始终将文本“警报(Alarm)”显示为 UI 标题；显示一个包含应用程序所提供文本的内容区域；可以使用由应用程序设置的警报声音，该声音开始时非常安静，但会随着时间的推移越来越大；当用户在“推迟(snooze)”和“消除(dismiss)”按钮之外点按警报(Alarm)时，会启动应用程序并且显示初始页面，就像用户从开始页面或应用程序列表中启动了该应用程序一样。

2. 提醒 UI 属性

如图 18.10 所示显示创建提醒的应用程序的名称；显示由应用程序提供的标题；显示一个包含应用程序所提供文本的内容区域；使用设备的默认通知声音；当用户在“推迟(snooze)”和“消除(dismiss)”按钮之外点按提醒(Reminder)时，会启动应用程序。应用程序可以选择提供在启动时将导航到的导航 URI 和查询字符串参数。

图 18.9　警报界面

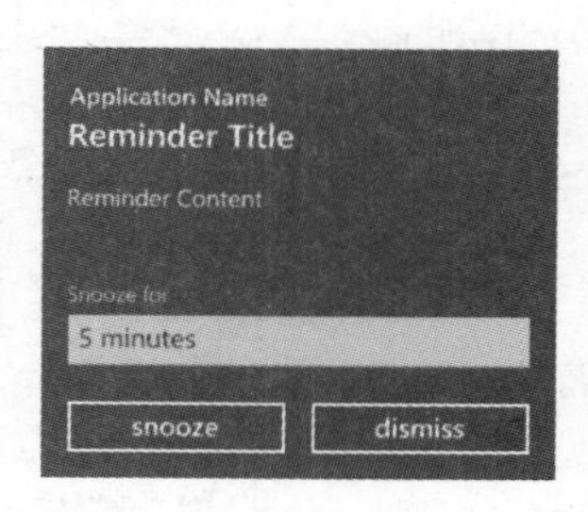

图 18.10　提醒界面

18.5.2 计划通知的 API

Alarm 类为警报类可以为其指定自定义声音的通知，在警报对话框中显示的标题始终为“Alarm”。Reminder 类为提醒类可以将上下文信息传递到父应用程序的通知。当调用提醒并向用户显示时，用户可以单击通知以启动创建它的应用程序。当采用这种方式启动时，应用程序可以查询 NavigationUri 属性以获得创建提醒时设置的上下文信息。Alarm 类和 Reminder 类都从 ScheduledNotification 和 ScheduledAction 继承，Alarm 类和 Reminder 类的主要属性如表 18.11 所示。

表 18.11 Alarm 与 Reminder 的属性

类别	属性	说明
Reminder	NavigationUri	获取或设置当 Reminder 执行时，用户点击 snooze 后前往应用程式指定的 Navigation Uri
Reminder	Title	取得通知信息的标题
Alarm/Reminder	Content	获取或设置启动通知时向用户显示的通知的文本内容
Alarm/Reminder	BeginTime	获取或设置开始操作计划的时间
Alarm/Reminder	ExpirationTime	获取或设置操作计划的结束时间
Alarm/Reminder	RecurrenceType	获取或设置通知的 RecurrenceType 类型
Alarm	Sound	获取或设置当 Alarm 执行要播放的声音文件
Alarm/Reminder	IsScheduled	获取操作的计划状态
Alarm/Reminder	Name	获取计划操作的名称

18.5.3 计划通知实例

下面给出计划通知的示例：测试警报通知和提醒通知。

代码清单 18-4：计划通知(源代码：第 18 章\Examples_18_4)

MainPage.xaml 文件主要代码

```
<Grid x:Name = "ContentPanel" Grid.Row = "1" Margin = "12,0,12,0">
  <StackPanel>
          <StackPanel Orientation = "Horizontal">
               <RadioButton Content = "提醒" Name = "reminderRadioButton" GroupName =
"ReminderOrAlarm" IsChecked = "True"></RadioButton>
               <RadioButton Content = "警报" Name = "alarmRadioButton" GroupName =
"ReminderOrAlarm" ></RadioButton>
          </StackPanel>
          <TextBlock Height = "30" HorizontalAlignment = "Left" Name = "titleLabel"
Text = "标题" VerticalAlignment = "Top" />
          <TextBox Height = "72" HorizontalAlignment = "Left" Name = "titleTextBox"
Text = "" VerticalAlignment = "Top" Width = "460" MaxLength = "63"/>
```

```
            <TextBlock Height = "30" HorizontalAlignment = "Left" Name = "contentLabel"
Text = "内容" VerticalAlignment = "Top" />
            <TextBox Height = "160" HorizontalAlignment = "Left" Name = "contentTextBox"
Text = "" VerticalAlignment = "Top" Width = "460" TextWrapping = "Wrap" MaxLength = "256"
AcceptsReturn = "True" />
            <Button x:Name = "Save" Content = "保存" Click = "Save_Click"/>
        </StackPanel>
    </Grid>
```

MainPage.xaml.cs 文件主要代码

```
    //保存警报提醒的按钮事件处理
    private void Save_Click(object sender, RoutedEventArgs e)
    {
        if (titleTextBox.Text == "" || contentTextBox.Text == "")
        {
            MessageBox.Show("标题和内容不能为空");
            return;
        }
        if ((bool)reminderRadioButton.IsChecked)

        {
            Reminder reminder = new Reminder(titleTextBox.Text);
            reminder.Title = titleTextBox.Text;
            reminder.Content = contentTextBox.Text;
            reminder.BeginTime = DateTime.Now;
            reminder.ExpirationTime = DateTime.Now.Add(new TimeSpan(0, 1, 0));
            reminder.RecurrenceType = RecurrenceInterval.None;  //只提醒一次
            reminder.NavigationUri = new Uri("/Page1.xaml", UriKind.Relative); ;
            ScheduledActionService.Add(reminder);  //添加计划通知服务
        }
        else
        {
            Alarm alarm = new Alarm(titleTextBox.Text);
            alarm.Content = contentTextBox.Text;
            alarm.Sound = new Uri("/Ringtones/Ring01.wma", UriKind.Relative);
            alarm.BeginTime = DateTime.Now;
            alarm.ExpirationTime = DateTime.Now.Add(new TimeSpan(0, 3, 0));
            alarm.RecurrenceType = RecurrenceInterval.None;
            ScheduledActionService.Add(alarm);  //添加计划通知服务
        }
    }
```

程序的运行效果如图 18.11 所示。

图 18.11　播放后台音乐

18.6　后台定位

Microsoft 定位服务允许为 Windows Phone 创建位置感知应用程序。该服务从获得来源(如 GPS、Wi-Fi 和 Cellular)获取位置数据。它可以使用一个或多个来源推导出 Windows Phone 的位置，从而根据应用程序的需要平衡性能和电能利用。通过事件驱动的托管代码接口向应用程序公开位置。

18.6.1　定位服务概述

定位服务使用多个来源确定设备的位置，但偶尔会有位置数据不可用的时刻。应用程序应该正常处理这种缺少数据的情况。当定位服务的状态发生更改时引发 StatusChanged 事件。应用程序可以使用该事件的处理程序，以让用户知道位置数据是否可用以及相应地修改应用程序的行为。当定位服务能够获取位置数据时，初始化并获取第一个读数通常需要 15 秒的时间，但也可能需要最多 120 秒的时间。在设计应用程序时应牢记这一点，如果在等待服务启动时应用程序没有响应，则应该警告用户。由于较低的精度设置不等待 GPS 返回数据，因此通常在 GPS 尚未激活时速度更快。如果其他应用程序正在使用 GPS 并且正在检索数据，则较高精度设置返回数据的速度将与较低精度设置一样快或更快。

由于移动设备中的 GPS 硬件没有天线，因此传感器通常设计得非常敏感。这种灵敏度可能会导致信号中有少量来自表面反射以及其他环境影响的噪音。主定位服务类 Geolocator 显示 MovementThreshold 属性。该属性指定在引发 PositionChanged 事件之前必须进行的位置方面的最小更改。如果将 MovementThreshold 属性设置为一个非常低的值，则可能会导致应用程序接收实际上是由信号噪音所导致的事件。为了平滑信号以便仅表示位置中的重大更改，请将 MovementThreshold 属性设置为至少 20m。这也会使应用程序的耗电量降低。将移动阈值设置为 0m 将导致频繁引发事件，一秒钟一个事件。该设置对于导航应用程序可能非常有用。

18.6.2 后台运行事件

使用 Geolocator 类进行定位服务可以运行在系统后台，当应用程序开始定位服务之后，单击开始键，将会执行后台运行事件，后台运行事件需要在 App. xaml. cs 中添加 Application_RunningInBackground(object sender, RunningInBackgroundEventArgs args) 事件的处理。

18.6.3 跟踪位置变化实例

下面给出跟踪位置变化的示例：使用 Geolocator 类进行定位，并且判断定位服务是否在后台服务。

代码清单 18-5：跟踪位置变化（源代码：第 18 章\Examples_18_5）

App. xaml. cs 文件主要代码

```
public static bool InBackground { get; set; }                    //后台运行标识符
…
//后台运行事件
private void Application_RunningInBackground(object sender, RunningInBackgroundEventArgs args)
{
    App.InBackground = true;
}
```

MainPage. xaml 文件主要代码

```
        <Grid x:Name = "ContentPanel" Grid.Row = "1" Margin = "12,0,12,0">
            <StackPanel>
                <Button x:Name = "TrackLocationButton" Click = "TrackLocation_Click" Content
= "开始跟踪位置变化"/>
                <TextBlock x:Name = "LatitudeTextBlock"/>
                <TextBlock x:Name = "LongitudeTextBlock"/>
                <TextBlock x:Name = "StatusTextBlock"/>
            </StackPanel>
        </Grid>
```

MainPage. xaml. cs 文件主要代码

```
using System.Windows;
using Microsoft.Phone.Controls;
using Windows.Devices.Geolocation;
namespace ContinuousLocationDemo
{
    public partial class MainPage : PhoneApplicationPage
    {
        Geolocator geolocator = null;
```

```
        bool tracking = false;
        public MainPage()
        {
            InitializeComponent();
        }
        //跟踪位置按钮点击事件处理
        private void TrackLocation_Click(object sender, RoutedEventArgs e)
        {
            if (!tracking)
            {
                geolocator = new Geolocator();
                geolocator.DesiredAccuracy = PositionAccuracy.High;
                geolocator.MovementThreshold = 100;
                geolocator.StatusChanged += geolocator_StatusChanged;
                geolocator.PositionChanged += geolocator_PositionChanged;
                tracking = true;
                TrackLocationButton.Content = "停止跟踪位置变化";
            }
            else
            {
                geolocator.PositionChanged -= geolocator_PositionChanged;
                geolocator.StatusChanged -= geolocator_StatusChanged;
                geolocator = null;
                tracking = false;
                TrackLocationButton.Content = "开始跟踪位置变化";
                StatusTextBlock.Text = "已停止";
            }
        }
        //位置变化事件处理
        void geolocator_PositionChanged(Geolocator sender, PositionChangedEventArgs args)
        {
            //只有当程序在前台运行时才更新界面数据
            if (App.InBackground == false)
            {
                Dispatcher.BeginInvoke(() =>
                {
                     LatitudeTextBlock.Text = args.Position.Coordinate.Latitude.ToString
("0.00000000");
                    LongitudeTextBlock.Text = args.Position.Coordinate.Longitude.ToString
("0.00000000");
                });
            }
        }
        //状态变化事件处理
        void geolocator_StatusChanged(Geolocator sender, StatusChangedEventArgs args)
        {
            string status = "";
            switch (args.Status)
            {
                case PositionStatus.Disabled:
```

```
                status = "位置服务已被禁止";
                break;
            case PositionStatus.Initializing:
                status = "正在初始化";
                break;
            case PositionStatus.NoData:
                status = "没有数据";
                break;
            case PositionStatus.Ready:
                status = "准备中";
                break;
            case PositionStatus.NotAvailable:
                status = "不可用";
                break;
            case PositionStatus.NotInitialized:
                status = "未进行初始化";
                break;
        }
        Dispatcher.BeginInvoke(() =>
        {
            StatusTextBlock.Text = status;
        });
    }
  }
}
```

程序的运行效果如图 18.12 所示。

图 18.12 定位

第19章

蓝牙和近场通信

蓝牙和近场通信技术都是手机的近距离无线传输的技术，Windows Phone 7 系统手机仅支持蓝牙耳机功能，并不支持蓝牙文件信息传输和近场通信技术，而在 Windows Phone 8 手机里面将全面支持蓝牙和近场通信技术，并且提供了相关的 API 来给开发者使用。开发者可以利用蓝牙和近场通信的相关 API 来创建应用程序，在应用程序里面使用手机的蓝牙或者近场通信技术来进行近距离的文件传输和发送接收消息，创造出更加有趣和方便的应用软件。

19.1 蓝牙

Windows Phone 8 的配置符合蓝牙技术联盟的标准，并且支持智能手机之间的文件传输。蓝牙是手机上面一种很常见的技术，在非智能手机上蓝牙功能也很普遍，蓝牙通常会用来在近距离的两台手机之间传输文件。在 Windows Phone 8 里面提供了相关的蓝牙 API 给开发者去开发一些蓝牙相关的应用，例如通过蓝牙交换联系人等。

19.1.1 蓝牙原理

蓝牙是一种低成本、短距离的无线通信技术。蓝牙技术支持设备短距离通信(一般 10m 内)，能在包括移动电话、PDA、无线耳机、笔记本电脑、相关外设等众多设备之间进行无线信息交换。利用“蓝牙”技术，能够有效地简化移动通信终端设备之间的通信，也能够成功地简化设备与 Internet 之间的通信，从而数据传输变得更加迅速高效，为无线通信拓宽道路。蓝牙采用分散式网络结构以及快跳频和短包技术，支持点对点及点对多点通信，工作在全球通用的 2.4GHz ISM(即工业、科学、医学)频段。其数据速率为 1Mbps。采用时分双工传输方案实现全双工传输。

蓝牙技术是一种无线数据与语音通信的开放性全球规范，它以低成本的近距离无线连接为基础，为固定与移动设备通信环境建立一个特别连接。其程序写在一个 9mm×9mm 的微芯片中。

蓝牙协议栈允许采用多种方法，包括 RFCOMM 和 Object Exchange(OBEX)，在设备之间发送和接收文件。图19.1所示显示了协议栈的细节。

栈的最底层是HCI（Host Controller Interface），即主机控制器接口。这一层顾名思义就是主机（计算机）和控制器（蓝牙设备）之间的接口。可以看到，其他所有的层都要经过HCI。HCI上面的一层是L2CAP，即逻辑链接控制器适配协议（Logical Link Controller Adaptation Protocol）。这一层充当其他所有层的数据多路复用器。接下来一层是BNEP，即蓝牙网络封装协议（Bluetooth Network Encapsulation Protocol）。使用BNEP，可以在蓝牙上运行其他网络协议，例如IP、TCP和UDP。RFCOMM称作虚拟串口协议（virtual serial port protocol），因为它允许蓝牙设备模拟串口的功能。OBEX协议层是在RFCOMM层上面实现的，如果想把数据以对象（例如文件）的形式传输，那么OBEX很有用。SDP是服务发现协议（Service Discovery Protocol）层，用于在远程蓝牙设备上寻找服务。最后两层是AVCTP和AVDTP，用于蓝牙上音频和视频的控制和发布。AVCTP和AVDTP是蓝牙协议中增加的相对较新的层；如果想控制媒体播放器的功能或者想以立体声播放音频流，则要使用它们。

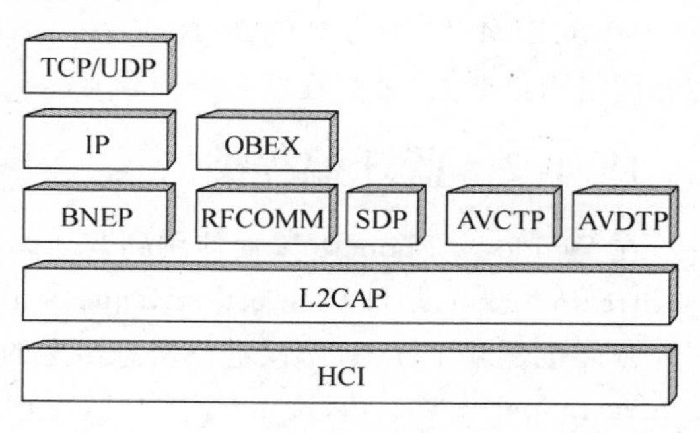

图19.1 蓝牙协议栈

19.1.2 Windows Phone蓝牙技术

在Windows Phone 8里面可以在应用程序里面利用蓝牙进行通信，使用蓝牙相关的API，可以让应用程序连接到另外的一个应用程序，也可以让应用程序连接到一个设备上。Windows Phone 8的蓝牙技术支持两个蓝牙方案：一个是应用程序到应用程序的通信，另外一个是应用程序到设备的通信。

1. 应用程序到应用程序的通信

应用程序到应用程序的通信的过程是，应用程序使用蓝牙去查找正在广播蓝牙服务的对等的应用程序，如果在应用程序提供服务的范围内发现一个应用程序，那么该应用程序可以发起连接请求。当这两个应用程序接受连接，它们之间就可以进行通信了，通信的过程是使用socket的消息发送接收机制。在Windows Phone 8中使用到应用程序到应用程序的蓝牙通信技术，需要在项目的WMAppManifest.xml文件中添加ID_CAP_PROXIMITY的功能选项，表示支持临近的设备通信能力，否则程序会出现异常。

2. 应用程序到设备的通信

在应用程序到设备的通信过程时，应用程序使用蓝牙去查找提供服务的设备，如果提供的服务范围之内发现一个可以连接的蓝牙设备，那么该应用程序可以发起连接请求。当应用程序和设备同时接受该连接，它们之间就可以进行通信了，通信的过程也是使用socket的消息发送接收机制，类似于应用程序到应用程序的通信。在Windows Phone 8中使用到应用程序到设备的蓝牙通信技术，需要在项目的 WMAppManifest. xml 文件中添加 ID_CAP_PROXIMITY 和 ID_CAP_NETWORKING 的功能选项，表示支持临近的设备通信能力和网络通信能力，否则程序会出现异常。

19.1.3 蓝牙编程类

在 Windows Phone 8 里面使用到蓝牙编程主要会用到 PeerFinder 类，PeerInformation 类，StreamSocket 类和 ConnectionRequestedEventArgs 类，这些类的说明如表 19.1 所示。因为蓝牙也是基于 TCP 协议进行消息传递的，所以需要用到 Socket 的相关的编程知识，以及 StreamSocket 类。PeerFinder 类是蓝牙查找类，它的主要成员如表 19.2 所示。

表 19.1 蓝牙编程类的说明

类　　名	说　　明
PeerFinder	用于去查找附近的设备是否有运行和当前应用程序相同的应用程序，并且可以在两个应用程序之间建立起 socket 连接，从而可以进行通信；对等应用程序是在其他设备上运行的应用程序的另一个实例
PeerInformation	包含对等应用程序或设备的识别信息
StreamSocket	支持使用一个 TCP 的 Socket 流的网络通信
ConnectionRequestedEventArgs	表示传递到一个应用程序的 ConnectionRequested 事件的属性

表 19.2 PeerFinder 类的成员

成　　员	说　　明
bool AllowBluetooth	指定 PeerFinder 类的此实例是否可以通过使用 Bluetooth 来连接 ProximityStreamSocket 对象；如果 PeerFinder 的此实例可以通过使用 Bluetooth 来连接 ProximityStreamSocket 对象，则为 true；否则为 false。默认为 true
bool AllowInfrastructure	是否使用 TCP/IP 协议连接到 StreamSocket
bool AllowWiFiDirect	指定 PeerFinder 类的此实例是否可以通过使用 Wi-Fi Direct 来连接 ProximityStreamSocket 对象；如果 PeerFinder 的此实例可以通过使用 Wi-Fi Direct 来连接 ProximityStreamSocket 对象，则为 true；否则为 false。默认为 true
IDictionary<string, string> AlternateIdentities	获取要与其他平台上的对等应用程序匹配的备用 AppId 值列表；返回要与其他平台的对等类应用程序匹配的备用 AppId 值列表
string DisplayName	获取或设置标识计算机到远程对等类的名称
PeerDiscoveryTypes SupportedDiscoveryTypes	获取一个值，该值指示哪些发现选项可与 PeerFinder 类一同使用

续表

成　　员	说　　明
event TypedEventHandler<object, ConnectionRequestedEventArgs> ConnectionRequested	远程对等类使用 ConnectAsync 方法请求连接时发生
event TypedEventHandler<object, TriggeredConnectionStateChangedEventArgs> TriggeredConnectionStateChanged	在远程对等类的轻击笔势期间发生
IAsyncOperation< StreamSocket> ConnectAsync(PeerInformation peerInformation)	连接已发现了对 FindAllPeersAsync 方法的调用的对等类；peerInformation 表示连接到的对等类的对等类信息对象；返回通过使用所提供的临近 StreamSocket 对象连接远程对等类的异步操作
IAsyncOperation<IReadOnlyList<PeerInformation>> FindAllPeersAsync()	适用于无线范围内运行相同应用程序的对等计算机的异步浏览；返回通过使用 Wi-Fi 直连技术浏览对等类的异步操作
void Start(string peerMessage)	向临近设备上的对等类应用程序传递消息
void Stop()	停止查找对等类应用程序或广播对等类连接的过程

19.1.4　查找蓝牙设备和对等项

查找在服务范围内的蓝牙设备和对等项是蓝牙编程的第一步，查找蓝牙设备和对等项中会使用到 PeerFinder 类的 FindAllPeersAsync 方法去进行查找，然后以异步的方式返回查找到的对等项列表的信息 IReadOnlyList<PeerInformation>，注意要使查找对等的应用程序时，在调用 FindAllPeersAsync 方法前必须先调用 PeerFinder 类的 Start 方法，主要的目的是启动广播服务，让对方的应用程序也能查找到自己。PeerInformation 包含三个属性：一个是 DisplayName 表示对等项的名字，这个名字一般都是由对方的设备的名称或者查找到的应用程序自身设置的现实名字，一个是 HostName 表示主机名字或者 IP 地址，还有一个属性是 ServiceName 表示服务名称或者 TCP 协议的端口号。然后可以利用查找到的 PeerInformation 信息进行连接和通信。

查找对等的应用程序的代码示例：

```
async void AppToApp()
{
    //启动查找服务
    PeerFinder.Start();
    //开始查找
    ObservableCollection<PeerInformation> peers = await PeerFinder.FindAllPeersAsync();
    if (peers.Count == 0)
    {
        //未找到任何的对等项
    }
```

```
    else
    {
        //处理查找到的对等项,可以使用 PeerFinder 类的 ConnectAsync 方法来连接选择的要进行
        //通信的对等项
    }
}
```

查找蓝牙设备的代码示例：

```
private async void AppToDevice()
{
    //设置查找所匹配的蓝牙设备
    PeerFinder.AlternateIdentities["Bluetooth:Paired"] = "";
    //开始查找
    ObservableCollection<PeerInformation> pairedDevices = await PeerFinder.FindAllPeersAsync();
    if (pairedDevices.Count == 0)
    {
        //没有找到可用的蓝牙设备
    }
    else
    {
        //处理查找到的蓝牙设备,可以新建一个 StreamSocket 对象,然后使用 StreamSocket 类的
        //ConnectAsync 方法通过 HostName 和 ServiceName 来连接蓝牙设备
    }
}
```

19.1.5 蓝牙发送消息

蓝牙编程的发送消息机制使用的是 TCP 的 StreamSocket 的方式，原理与 Socket 的一致。在蓝牙连接成功后，可以获取到一个 StreamSocket 类的对象，然后我们使用该对象的 OutputStream 属性来初始化一个 DataWriter 对象，通过 DataWriter 对象来进行发送消息。OutputStream 属性表示的是 Socket 的输出流，用于发送消息给对方。下面来看一下发送消息的示例：

```
async void SendMessage(string message)
{
    //连接选中的对等项,selectedPeer 为查找到的 PeerInformation 对象
    StreamSocket _socket = = await PeerFinder.ConnectAsync(selectedPeer);
    //创建 DataWriter
    DataWriter _dataWriter = new DataWriter(_socket.OutputStream);
    //先写入发送消息的长度
     _dataWriter.WriteInt32(message.Length);
     await _dataWriter.StoreAsync();
    //最后写入发送消息的内容
     _dataWriter.WriteString(message);
     await _dataWriter.StoreAsync();
}
```

19.1.6 蓝牙接收消息

蓝牙编程的接收消息机制同样使用的是 TCP 的 StreamSocket 的方式，原理与 Socket 的一致。在蓝牙连接成功后，可以获取到一个 StreamSocket 类的对象，然后我们使用该对象的 InputStream 属性来初始化一个 DataReader 对象，通过 DataReader 对象来进行接收消息。InputStream 属性表示的是 Socket 的输入流，用于接收对方的消息。下面来看一下接收消息的示例：

```
async Task<string> GetMessage()
{
    //连接选中的对等项,selectedPeer 为查找到的 PeerInformation 对象
    StreamSocket _socket = = await PeerFinder.ConnectAsync(selectedPeer);
    //创建 DataReader
    DataReader _dataReader = new DataReader(_socket.InputStream);
    //先读取消息的长度
    await _dataReader.LoadAsync(4);
    uint messageLen = (uint)_dataReader.ReadInt32();
    //最后读取消息的内容
    await _dataReader.LoadAsync(messageLen);
    return _dataReader.ReadString(messageLen);
}
```

19.1.7 实例：实现蓝牙程序对程序的传输

下面给出蓝牙程序对程序传输的示例：通过使用蓝牙功能查找周边也要使用该应用的手机，互相建立起连接和发送测试消息。

代码清单 19-1：蓝牙程序对程序传输(源代码：第 19 章\Examples_19_1)

MainPage.xaml 文件主要代码

```
        <Grid x:Name="ContentPanel" Grid.Row="1" Margin="12,0,12,0">
          <StackPanel>
             <Button x:Name="btFindBluetooth" Content="通过蓝牙查找该应用设备" Click=
"btFindBluetooth_Click"/>
             <ListBox x:Name="lbBluetoothApp" ItemsSource="{Binding}" >
               <ListBox.ItemTemplate >
                  <DataTemplate>
                    <StackPanel>
                       <TextBlock Text="{Binding DisplayName}" />
                       <TextBlock Text="{Binding ServiceName}" />
                       <Button Content = "连接" HorizontalAlignment = "Left" Width = "308"
Height="91" Click="btConnect_Click"/>
                    </StackPanel>
                  </DataTemplate>
               </ListBox.ItemTemplate>
             </ListBox>
          </StackPanel>
        </Grid>
```

MainPage.xaml.cs 文件代码

```
using System;
using System.Windows;
using System.Windows.Controls;
using Microsoft.Phone.Controls;
using Windows.Networking.Proximity;
using Windows.Networking.Sockets;
using Windows.Storage.Streams;
namespace BluetoothDemo
{
    public partial class MainPage : PhoneApplicationPage
    {
        private StreamSocket _socket = null;                  //Socket 数据流对象
        private DataWriter _dataWriter;                       //数据写入对象
        private DataReader _dataReader;                       //数据读取对象
        public MainPage()
        {
            InitializeComponent();
            Loaded += MainPage_Loaded;                        //页面加载事件
        }
        //查找蓝牙对等项按钮事件处理
        private async void btFindBluetooth_Click(object sender, RoutedEventArgs e)
        {
            try
            {
                //开始查找对等项
                PeerFinder.Start();
                //等待找到的对等项
                var peers = await PeerFinder.FindAllPeersAsync();
                if (peers.Count == 0)
                {
                    MessageBox.Show("没有发现对等的蓝牙应用");
                }
                else
                {
                    //把对等项目绑定到列表中
                    lbBluetoothApp.ItemsSource = peers;
                }
            }
            catch(Exception ex)
            {
                if ((uint)ex.HResult == 0x8007048F)
                {

                    MessageBox.Show("Bluetooth 已关闭请打开手机的蓝牙开关");
                }
                else
                {
                    MessageBox.Show(ex.Message);
```

```
            }
        }
    }
    //连接蓝牙对等项的按钮事件处理
    private async void btConnect_Click(object sender, RoutedEventArgs e)
    {
        Button deleteButton = sender as Button;
        PeerInformation selectedPeer = deleteButton.DataContext as PeerInformation;
        //连接到选择的对等项
        _socket = await PeerFinder.ConnectAsync(selectedPeer);
        //使用输出输入流建立数据读写对象
        _dataReader = new DataReader(_socket.InputStream);
        _dataWriter = new DataWriter(_socket.OutputStream);
        //开始读取消息
        PeerFinder_StartReader();
    }
    //读取消息
    async void PeerFinder_StartReader()
    {
        try
        {
            uint bytesRead = await _dataReader.LoadAsync(sizeof(uint));
            if (bytesRead > 0)
            {
                //获取消息内容的大小
                uint strLength = (uint)_dataReader.ReadUInt32();
                bytesRead = await _dataReader.LoadAsync(strLength);
                if (bytesRead > 0)
                {
                    String message = _dataReader.ReadString(strLength);
                    MessageBox.Show("获取到消息：" + message);
                    //开始下一条消息读取
                    PeerFinder_StartReader();
                }
                else
                {
                    MessageBox.Show("对方已关闭连接");
                }
            }

            else
            {
                MessageBox.Show("对方已关闭连接");
            }
        }
        catch (Exception e)
        {
            MessageBox.Show("读取失败：" + e.Message);
        }
    }
```

```
        //页面加载事件处理
        void MainPage_Loaded(object sender, RoutedEventArgs e)
        {
            //订阅连接请求事件
            PeerFinder.ConnectionRequested += PeerFinder_ConnectionRequested;
        }
        //连接请求事件处理
        void PeerFinder_ConnectionRequested(object sender, ConnectionRequestedEventArgs args)
        {
            //连接并且发送消息
            ConnectToPeer(args.PeerInformation);
        }
        //连接并发送消息给对方
        async void ConnectToPeer(PeerInformation peer)
        {
            _socket = await PeerFinder.ConnectAsync(peer);
            _dataReader = new DataReader(_socket.InputStream);
            _dataWriter = new DataWriter(_socket.OutputStream);
            string message = "测试消息";
            uint strLength = _dataWriter.MeasureString(message);
            _dataWriter.WriteUInt32(strLength);              //写入消息的长度
            _dataWriter.WriteString(message);                //写入消息的内容
            uint numBytesWritten = await _dataWriter.StoreAsync();
        }
    }
}
```

程序的运行效果如图 19.2 所示。

图 19.2　查找蓝牙应用

19.1.8 实例：实现蓝牙程序对设备的连接

下面给出蓝牙程序对设备连接的示例：查找蓝牙设备，并对找到的第一个蓝牙设备进行连接。

代码清单 19-2：蓝牙程序对设备连接(源代码：第 19 章\Examples_19_2)

MainPage.xaml 文件主要代码

```
<Grid x:Name = "ContentPanel" Grid.Row = "1" Margin = "12,0,12,0">
    <StackPanel>
        <Button x:Name = "btFindBluetooth" Content = "连接周围的蓝牙设备" Click =
"btFindBluetooth_Click"/>
    </StackPanel>
</Grid>
```

MainPage.xaml.cs 文件主要代码

```
//查找蓝牙设备事件处理
private async void btFindBluetooth_Click(object sender, RoutedEventArgs e)
{
    try
    {
        //配置 PeerFinder 蓝牙服务的 GUID 去搜索设备
        PeerFinder.AlternateIdentities["Bluetooth:SDP"] = "5bec6b8f - 7eba - 4452 -
bf59 - 1a510745e99d";
        var peers = await PeerFinder.FindAllPeersAsync();
        if (peers.Count == 0)
        {
            Debug.WriteLine("没发现蓝牙设备");
        }

        else
        {
            //连接找到的第一个蓝牙设备
            PeerInformation selectedPeer = peers[0];
            StreamSocket socket = new StreamSocket();
            await socket.ConnectAsync(selectedPeer.HostName, selectedPeer.ServiceName);
            MessageBox.Show("连接上了 HostName:" + selectedPeer.HostName +
"ServiceName:" + selectedPeer.ServiceName);
        }
    }
    catch (Exception ex)
    {
        if ((uint)ex.HResult == 0x8007048F)
        {
            MessageBox.Show("Bluetooth is turned off");
        }
    }
}
```

程序的运行效果如图 19.3 所示。

图 19.3　查找蓝牙设备

19.2　近场通信

近场通信(NFC)技术已经逐渐走入了生活，在各大主流平台中也开始普遍采用，而 Windows Phone 8 也将 NFC 技术完美应用其中。Windows Phone 8 支持近场通信芯片，这些芯片能够帮助用户与其他附近的设备进行通信，而且还能够让用户与好友共享诸如图片或视频等内容，另外还可以与其他支持 NFC 技术的设备进行数据传输。

19.2.1　近场通信的介绍

近场通信又称近距离无线通信，其英文简称为 NFC，是 Near Field Communication 的缩写，是一种短距离的高频无线通信技术，允许电子设备之间进行非接触式点对点数据传输，交换数据。近场通信是一种短距高频的无线电技术，在 13.56MHz 频率运行于 20 厘米距离内。其传输速度有 106kbit/s、212kbit/s 或者 424kbit/s 三种。可以在移动设备、消费类电子产品、PC 和智能控件工具间进行近距离无线通信。NFC 提供了一种简单、触控式的解决方案，可以让消费者简单直观地交换信息、访问内容与服务。

近场通信技术是由 Nokia、Philips、Sony 合作制定的标准，在 ISO 18092、ECMA 340 和 ETSI TS 102 190 框架下推动标准化，同时也兼容应用广泛的 ISO 14443 、Type-A、ISO 15693、B 以及 Felica 标准非接触式智能卡的基础架构。近场通信标准详细规定近场通信设备的调制方案、编码、传输速度与 RF 接口的帧格式，以及主动与被动近场通信模式初始化过程中数据冲突控制所需的初始化方案和条件，此外还定义了传输协议，包括协议启动和数据交换方法等。

近场通信是基于RFID技术发展起来的一种近距离无线通信技术。与RFID一样,近场通信信息也是通过频谱中无线频率部分的电磁感应耦合方式传递,但两者之间还是存在很大的区别。近场通信的传输范围比RFID小,RFID的传输范围可以达到0～1m,但由于近场通信采取了独特的信号衰减技术,相对于RFID来说近场通信具有成本低、带宽高、能耗低等特点。

近场通信技术的主要特征如下:

(1) 用于近距离(10cm以内)安全通信的无线通信技术。

(2) 射频频率:13.56MHz。

(3) 射频兼容:ISO 14443,ISO 15693,Felica标准。

(4) 数据传输速度:106kbit/s,212kbit/s,424kbit/s。

19.2.2 近场通信编程类

在Windows Phone 8里面使用到蓝牙编程主要会用到PeerFinder类、PeerInformation类、ConnectionRequestedEventArgs类、TriggeredConnectionStateChangedEventArgs类和ProximityDevice类,这些类的说明如表19.3所示。

表19.3 蓝牙编程的主要类

类名	说明
PeerFinder	用于去查找附近的设备是否有运行和当前应用程序相同的应用程序,并且可以在两个应用程序之间建立起socket连接,从而可以进行通信;对等应用程序是在其他设备上运行的应用程序的另一个实例
PeerInformation	包含对等应用程序或设备的识别信息
ConnectionRequestedEventArgs	表示传递到一个应用程序的ConnectionRequested事件的属性
ProximityDevice	可以通过轻轻地碰击来使应用程序能够以3～4厘米的大致范围内的其他设备进行通信和进行数据交换
ProximityMessage	表示从订阅收到的消息
TriggeredConnectionStateChangedEventArgs	表示传递到一个应用程序的TriggeredConnectionStateChanged事件的属性

19.2.3 发现近场通信设备

发现近场通信设备需要通过注册PeerFinder类的TriggeredConnectionStateChanged事件来监听是否有近场通信设备的连接,然后调用Start方法开始查找和被查找。在TriggeredConnectionStateChanged事件的实现中可以通过TriggeredConnectionStateChangedEventArgs的State来判断连接的状态和情况。示例代码如下:

```
//获取近场通信
```

```
ProximityDevice device = ProximityDevice.GetDefault();
//确认设备支持近场通信
if (device! = null)
{
    //注册触发事件
    PeerFinder.TriggeredConnectionStateChanged += OnTriggeredConnectionStateChanged;
    //开始查找对等项,同时也让自己被其他对等项目发现
    PeerFinder.Start();
}
//TriggeredConnectionStateChanged 事件处理
void OnTriggeredConnectionStateChanged(object sender, TriggeredConnectionStateChangedEventArgs args)
{
    switch (args.State)
    {
        case TriggeredConnectState.Listening:
            //正在监听,作为热点来进行连接
            break;
        case TriggeredConnectState.PeerFound:
            //接近手势完成后,用户可以让他们的设备,通过其他传输协议,如 TCP / IP 或蓝牙,
            //来建立起连接
            break;
        case TriggeredConnectState.Connecting:
            //正在连接
            break;
        case TriggeredConnectState.Completed:
            //连接完毕,获取到一个 StreamSocket 对象 args.Socket 来进行通信
            break;
        case TriggeredConnectState.Canceled:
            //连接被取消
            break;
        case TriggeredConnectState.Failed:
            //连接失败
            break;
    }
}
```

19.2.4 近场通信发布消息

发布消息可以通过 ProximityDevice 类的 PublishMessage 方法来实现，PublishMessage 中有两个参数，一个是发送给订阅用户的消息的类型，该类型会再接收消息中用到，作为接收消息的类型，一个是发送给订阅用户的消息，发送消息之后将会返回一个消息唯一的发布 ID，如果需要停止发送消息可以使用 StopPublishingMessage 方法来阻止消息的发送。下面来看一下发布消息的示例：

```
ProximityDevice device = ProximityDevice.GetDefault();
//确认设备支持近场通信
if (device! = null)
{
```

```
    //发布消息
    long Id = device.PublishMessage("Windows.SampleMessageType", "Hello World!");
}
```

19.2.5　近场通信订阅消息

订阅消息可以通过 ProximityDevice 类的 SubscribeForMessage 方法为指定的消息类型创建订阅，SubscribeForMessage 中有两个参数，一个是发送本订阅的消息类型，该类型会匹配到发送消息的类型，一个是接收消息的处理方法，发送消息之后将会返回一个消息唯一的订阅 ID，如果需要停止订阅消息可以使用 StopSubscribingForMessage 方法来阻止消息的订阅。下面来看一下订阅消息的示例：

```
ProximityDevice device = ProximityDevice.GetDefault();
//确认设备支持近场通信
if (device!= null)
{
    //开始订阅消息
    long Id = device.SubscribeForMessage ("Windows.SampleMessageType", messageReceived);
}
//接收消息的处理
private void messageReceived(ProximityDevice sender, ProximityMessage message)
{
    //接收消息成功，设备的 id 为 sender.DeviceId，消息的内容为 message.DataAsString
}
```

19.2.6　实例：实现近场通信的消息发布订阅

下面给出近场通信的消息发布订阅的示例：实现了近场通信的消息发布和订阅的功能。

代码清单 19-3：近场通信的消息发布订阅(源代码：第 19 章\Examples_19_3)

MainPage.xaml 文件主要代码

```
<Grid x:Name="ContentPanel" Grid.Row="1" Margin="12,0,12,0">
 <StackPanel>
      <TextBlock Text="发送的消息："/>
      <TextBox x:Name="tbSendMessage"/>
         <Button x:Name="btSend" Content="发送测试消息" Click="btSend_Click_1"/>
     <Button x:Name="btReceive" Content="接收消息" Click="btReceive_Click_1"/>
         <TextBlock Text="接受到的消息："/>
         <TextBlock x:Name="tbReceiveMessage"/>
    </StackPanel>
</Grid>
```

MainPage.xaml.cs 文件主要代码

```
using System;
using System.Windows;
```

```
using Microsoft.Phone.Controls;
using Windows.Networking.Proximity;
namespace NFCMessageDemo
{
    public partial class MainPage : PhoneApplicationPage
    {
        private ProximityDevice _proximityDevice;              //邻近设备类对象
        private long _publishedMessageId = -1;                 //发布消息的 ID
        private long _subscribedMessageId = -1;                //订阅消息的 ID
        public MainPage()
        {
            try
            {
                InitializeComponent();
                _proximityDevice = ProximityDevice.GetDefault();
                //确认手机是否支持 NFC 功能
                if (_proximityDevice == null)
                {
                    MessageBox.Show("你的手机不支持 NFC 功能.");
                }
            }
            catch (Exception ex)
            {
                MessageBox.Show("你的手机不支持 NFC 功能.");
            }

        }
        //发布消息按钮事件处理
        private void btSend_Click_1(object sender, RoutedEventArgs e)
        {
            if (_proximityDevice == null)
            {
                MessageBox.Show("你的手机不支持 NFC 功能.");
                return;
            }
            if (_publishedMessageId == -1)
            {
                String publishText = tbSendMessage.Text;
                tbSendMessage.Text = "";
                if (publishText.Length > 0)
                {
                    //发布消息并获取消息的 ID
                    _publishedMessageId = _proximityDevice.PublishMessage("Windows.
SampleMessageType", publishText);
                    MessageBox.Show("消息已经发送,可以接触其他的设备来进行接收消息.");
                }
                else
                {
                    MessageBox.Show("发送的消息不能为空.");
                }
```

```
            }
            else
            {
                MessageBox.Show("消息已经发送,请接收.");
            }
        }
        //订阅消息按钮事件处理
        private void btReceive_Click_1(object sender, RoutedEventArgs e)
        {
            if (_proximityDevice == null)
            {
                MessageBox.Show("你的手机不支持 NFC 功能.");
                return;
            }
            if (_subscribedMessageId == -1)
            {
                _subscribedMessageId = _proximityDevice.SubscribeForMessage ( " Windows.
SampleMessageType", MessageReceived);
                MessageBox.Show("订阅 NFC 接收的消息成功.");

            }
            else
            {
                MessageBox.Show("已订阅 NFC 接收的消息.");
            }
        }
        //接收到消息事件的处理
        void MessageReceived(ProximityDevice proximityDevice, ProximityMessage message)
        {
            Dispatcher.BeginInvoke(() =>
        {
            tbReceiveMessage.Text = message.DataAsString;
        });
        }
        //离开页面停止消息的发送和订阅
        protected override void OnNavigatedFrom(System.Windows.Navigation.NavigationEventArgs e)
        {
            if (_proximityDevice != null)
            {
                if (_publishedMessageId != -1)
                {
                    _proximityDevice.StopPublishingMessage(_publishedMessageId);
                    _publishedMessageId = -1;
                }
                if (_subscribedMessageId != -1)
                {
                    _proximityDevice.StopSubscribingForMessage(_subscribedMessageId);
                    _subscribedMessageId = 1;
                }
            }
```

```
            base.OnNavigatedFrom(e);
        }
    }
}
```

程序的运行效果如图 19.4 所示。

图 19.4　NFC 消息传递

响应式编程

响应式编程是一种更加优美的异步编程模式，它可以对所在运行环境中的变化做出响应，从而达到一种高效简洁的编程模式，而传统上的处理方式是使用锁和事件句柄来协调这些变化，实现起来就显得低效和烦琐。在 Windows Phone 的程序里面可以通过.NET 的 Reactive Extension（响应式扩展框架）来实现响应式编程，Reactive Extension 提供了强大的类和方法并且使用了 LINQ 的语法形式来支持响应式编程，给 Windows Phone 的程序提供了更加简洁和高效的编程模式，在 Windows Phone 的应用开发中应该多去使用这种先进的编程模式来实现更加优美的程序。响应式编程是基于观察者设计模式理论来实现的一种编程模式，所以对观察者设计模式理论的深刻理解是掌握响应式编程的前提。本章首先介绍观察者设计模式理论和编程方式以及 LINQ 编程的一些基础语法，然后介绍.NET 的 Reactive Extension 和在 Windows Phone 中使用响应式编程来编写程序。

20.1 观察者模式

观察者（Observer）模式，有时又称为发布-订阅（Publish-Subscribe）模式、模型-视图（Model-View）模式、源-收听者（Source-Listener）模式或从属者（Dependents）模式，是软件设计模式的一种。在观察者模式中，一个目标物件管理所有相依于它的观察者物件，并且在它本身的状态改变时主动发出通知，一般会通过呼叫各观察者所提供的方法来实现。Windows Phone 中的响应式编程框架 Reactive Extension 正是以这种设计模式为出发点而设计的一种编程的方式，对观察者模式的深入了解可以更好地学习响应式编程和理解其中的编程思想。

20.1.1 观察者模式理论

观察者模式有很多实现方式，从根本上说，该模式必须包含两个角色：观察者和被观察对象。观察者和被观察者之间存在“观察”的逻辑关联，当被观察者发生改变的时候，观察者就会观察到这样的变化，并且做出相应的响应。

实现观察者模式有很多形式，比较直观的一种是使用一种“注册—通知—撤销注册”的形式。下面描述了观察者模式的3种过程。

1．观察者

观察者将自己注册到被观察对象(Subject)中，被观察对象将观察者存放在一个容器(Container)里。

2．被观察对象

被观察对象发生了某种变化，从容器中得到所有注册过的观察者，将变化通知观察者。

3．撤销观察

观察者告诉被观察者要撤销观察，被观察者从容器中将观察者去除。观察者将自己注册到被观察者的容器中时，被观察者不应该过问观察者的具体类型，而是应该使用观察者的接口。这样的优点是：假定程序中还有别的观察者，那么只要这个观察者也是相同的接口实现即可。一个被观察者可以对应多个观察者，当被观察者发生变化的时候，他可以将消息一一通知给所有的观察者。基于接口，而不是具体的实现——这一点为程序提供了更大的灵活性。

20.1.2 观察者模式的实现

使用观察者模式在独立的对象(主体)中维护一个对主体感兴趣的依赖项(观察器)列表。让所有观察器各自实现公共的Observer接口，以取消主体和依赖性对象之间的直接依赖关系，如图20.1所示。

同样，如果对视图类进行了更改，则模型很可能受到影响。包含许多紧耦合类的应用程序往往是脆弱的和难于维护的，因为一个类中的更改可能影响所有的紧耦合类。

在与依赖性对象相关的客户端中发生状态更改时，ConcreteSubject会调用Notify()方法。Subject超类用于维护所有感兴趣观察器组成的列表，以便让Notify()方法能够遍历所有观察器列表，并调用每个已注册的观察器上的Update()方法。观察器通过调用Subject上的subscribe()和unsubscribe()方法，来注册到更新和取消注册。ConcreteObserver的一个或多个实例可能也会访问ConcreteSubject以获取详细信息，因此通常依赖于ConcreteSubject类。但是，如图20.2所示，ConcreteSubject类既不直接也不间接依赖于ConcreteObserver类。

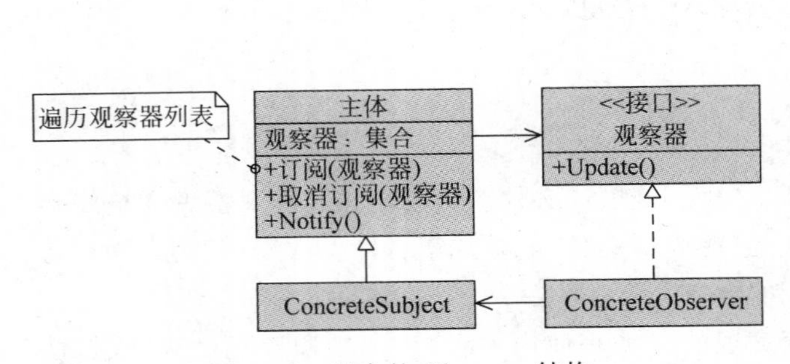

图20.1 基本的Observer结构

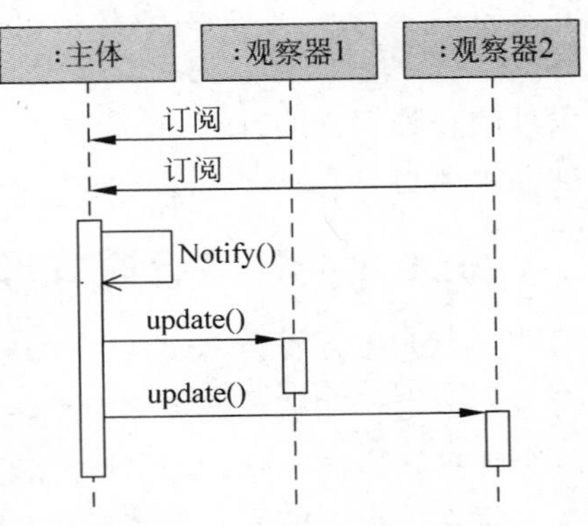

图20.2 基本的Observer交互

下面来看一下一个简单的观察者示例的实现代码：

```
using System;
//主体
class Observable
{
    public event EventHandler SomethingHappened;
    public void DoSomething()
    {
        EventHandler handler = SomethingHappened;
        if (handler != null)
        {
            handler(this, EventArgs.Empty);
        }
    }
}
//观察器 1
class Observer1
{
    public void HandleEvent(object sender, EventArgs args)
    {
        //处理观察器 1 需要处理的事情
    }
}
//观察器 2
class Observer2
{
    public void HandleEvent(object sender, EventArgs args)
    {
        //处理观察器 2 需要处理的事情
    }
}
class Test
{
    static void Main()
    {
        Observable observable = new Observable();                  //创建观察者主体
        Observer1 observer1 = new Observer1();                     //创建观察器 1
        Observer2 observer2 = new Observer2();                     //创建观察器 2
        observable.SomethingHappened += observer1.HandleEvent;     //订阅
        observable.SomethingHappened += observer2.HandleEvent;     //订阅
        observable.DoSomething();                                  //开始处理相关的事情
    }
}}
```

20.1.3 观察者模式的优缺点

1. 观察者模式的优点

(1) 具体主题和具体观察者是松耦合关系。由于主体(Subject)接口仅仅依赖于观察者

(Observer)接口，因此具体主体只是知道它的观察者是实现观察者接口的某个类的实例，但不需要知道具体是哪个类。同样，由于观察者仅仅依赖于主体接口，因此具体观察者只是知道它依赖的主题是实现主体接口的某个类的实例，但不需要知道具体是哪个类。

(2) 观察模式满足“开-闭原则”。主体接口仅仅依赖于观察者接口，这样就可以让创建具体主题的类也仅仅是依赖于观察者接口，因此如果增加新的实现观察者接口的类，不必修改创建具体主体的类的代码。同样，创建具体观察者的类仅仅依赖于主体接口，如果增加新的实现主体接口的类，也不必修改创建具体观察者类的代码。

2. 观察者模式的缺点

(1) 如果一个被观察者对象有很多直接和间接的观察者的话，将所有的观察者都通知到会花费很多时间。

(2) 如果在被观察者之间有循环依赖的话，给观察者会触发它们之间进行循环调用，导致系统崩溃。在使用观察者模式时要特别注意这一点。

(3) 虽然观察者模式可以随时使观察者知道所观察的对象发生了变化，但是观察者模式没有相应的机制使观察者知道所观察的对象是怎么发生变化的。

20.1.4 观察者模式的使用场景

适合使用观察者模式的情景：

(1) 当一个对象的数据更新时需要通知其他对象，但这个对象又不希望和被通知的那些对象形成紧耦合。

(2) 当一个对象的数据更新时，这个对象需要让其他对象也各自更新自己的数据，但这个对象不知道具体有多少对象需要更新数据。

20.2 LINQ语法

语言集成查询(Language INtegrated Query，LINQ)是一组用于C#和Visual Basic语言的扩展，它允许编写C#或者Visual Basic代码以查询数据库相同的方式操作内存数据。

LINQ定义了一组标准查询操作符用于在所有基于.NET平台的编程语言中更加直接地声明跨越、过滤和投射操作的统一方式，标准查询操作符允许查询作用于所有基于IEnumerable<T>接口的源，并且允许适合于目标域或技术的第三方特定域操作符来扩大标准查询操作符集，更重要的是，第三方操作符可以用它们自己提供附加服务的实现来自由地替换标准查询操作符，根据LINQ模式的习俗，这些查询喜欢采用与标准查询操作符相同的语言集成和工具支持。

LINQ通过提供一种跨各种数据源和数据格式使用数据的一致模型。在LINQ查询中，始终会用到数据集合对象，可以使用相同的基本编码模式来查询和转换XML文档、SQL数据库、ADO.NET数据集、.NET集合中的数据以及对其有LINQ提供程序可用的任何其他格式的数据。

20.2.1 LINQ查询的组成

LINQ语句会包含如下几个主要元素。

1. 数据源

数据源表示LINQ查询将从哪里查找数据，它通常是一个或多个数据集，每个数据集包含一系列的元素。数据集是一个类型为IEnumerable或IQueryable的对象，可以对它进行枚举，遍历每一个元素。此外，它的元素可以是任何数据类型，所以可以表示任何数据的集合。

2. 目标数据

数据源中的元素并不一定是查询所需要的结果。目标数据用来指定查询的具体是什么数据。在LINQ中，它定义了查询结果数据集中元素的具体类型。

3. 筛选条件

筛选条件定义了对数据源中元素的过滤条件。只有满足条件的元素才作为查询结果返回。筛选条件可以是简单的逻辑表达式，也可以用具有复杂逻辑的函数来表示。

4. 附加操作

附加操作表示一些其他的具体操作，比如对查询结果进行分组等。其中，数据源和目标数据库是LINQ查询的必备元素，筛选条件和附加操作是可选元素。

20.2.2 LINQ的标准查询操作符

LINQ操作符有多种类型，如表20.1所示展示了所有LINQ操作符和其说明。许多标准查询操作符在对序列执行运算时都使用Func委托来处理单个元素。Lambda表达式可与标准查询操作符结合使用以代表委托。lambda表达式是创建委托实现的简略表达形式，并可用于匿名委托适用的所有场合。

下面来看看如何对一个整数数组使用Single操作符。这个整数数组的每个元素代表2的1～10次方。先创建此数组，然后使用Single操作符来检索满足Lambda表达式中指定条件的单个整数元素：

```
int[] nums = { 1, 2, 4, 8, 16, 32, 64, 128, 256, 512, 1024 };
int singleNum = nums.Single(x => x > 16 && x < 64);
```

Lambda表达式包含多个关键部分。Lambda表达式首先定义传入委托的变量。在以上代码示例中，x(在=>操作符左侧声明)是参数，代表传递给它的nums数组中的每个元素。Lambda表达式的剩余部分代表数组中每个元素的评估逻辑。可使用匿名委托轻松地重新编写以上表达式，如下所示：

```
int singleNum = nums.Single<int>(
  delegate(int x) {return (x > 16 && x < 64); }
);
```

表 20.1 LINQ 操作符

聚合操作符	说　明
Aggregate	对序列执行一个自定义方法
Average	计算数值序列的平均值
Count	返回序列中的项目数(整数)
LongCount	返回序列中的项目数(长型)
Min	查找数字序列中的最小数
Max	查找数字序列中的最大数
Sum	汇总序列中的数字
连接操作符	**说　明**
Concat	将两个序列连成一个序列
转换操作符	**说　明**
Cast	将序列中的元素转换成指定类型
OfType	筛选序列中指定类型的元素
ToArray	从序列返回一个数组
ToDictionary	从序列返回一个字典
ToList	从序列返回一个列表
ToLookup	从序列返回一个查询
ToSequence	返回一个 IEnumerable 序列
元素操作符	**说　明**
DefaultIfEmpty	为空序列创建默认元素
ElementAt	返回序列中指定索引的元素
ElementAtOrDefault	返回序列中指定索引的元素，或者如果索引超出范围，则返回默认值
First	返回序列中的第一个元素
FirstOrDefault	返回序列中的第一个元素，或者如果未找到元素，则返回默认值
Last	返回序列中的最后一个元素
LastOrDefault	返回序列中的最后一个元素，或者如果未找到元素，则返回默认值
Single	返回序列中的单个元素
SingleOrDefault	返回序列中的单个元素，或者如果未找到元素，则返回默认值
相等操作符	**说　明**
SequenceEqual	比较两个序列看其是否相等
生成操作符	**说　明**
Empty	生成一个空序列
Range	生成一个指定范围的序列
Repeat	通过将某个项目重复指定次数来生成一个序列
分组操作符	**说　明**
GroupBy	按指定分组方法对序列中的项目进行分组

续表

联接操作符	说　明
GroupJoin	通过归组将两个序列联接在一起
Join	将两个序列从内部联接起来
排序操作符	**说　明**
OrderBy	以升序按值排列序列
OrderByDescending	以降序按值排列序列
ThenBy	升序排列已排序的序列
ThenByDescending	降序排列已排序的序列
Reverse	颠倒序列中项目的顺序
分区操作符	**说　明**
Skip	返回跳过指定数目项目的序列
SkipWhile	返回跳过不满足表达式项目的序列
Take	返回具有指定数目项目的序列
TakeWhile	返回具有满足表达式项目的序列
投影操作符	**说　明**
Select	创建部分序列的投影
SelectMany	创建部分序列的一对多投影
限定符操作符	**说　明**
All	确定序列中的所有项目是否满足某个条件
Any	确定序列中是否有任何项目满足条件
Contains	确定序列是否包含指定项目
限制操作符	**说　明**
Where	筛选序列中的项目
设置操作符	**说　明**
Distinct	返回无重复项目的序列
Except	返回代表两个序列差集的序列
Intersect	返回代表两个序列交集的序列
Union	返回代表两个序列交集的序列

20.2.3　IEnumerable 和 IEnumerator 的理解

在 C# 中，凡是实现了 IEnumerator 接口的数据类型都可以用 foreach 语句进行迭代访问，可是，对于自定义类型如何实现这个接口以支持 foreach 的迭代呢？

要实现这个功能，先来看看 IEnumerable 和 IEnumerator 接口的定义：

```
public interface IEnumerable
{
        //IEnumerable 只有一个方法，返回可循环访问集合的枚举数
```

```
        IEnumerator GetEnumerator() ;
    }
    public interface IEnumerator
    {
        //方法
        //移到集合的下一个元素。如果成功则返回为 true; 如果超过集合结尾,则返回 false
        bool MoveNext();
        //将集合设置为初始位置,该位置位于集合中第一个元素之前
        void Reset();
        //属性：获取集合中的当前元素
        object Current { get; }
    }
```

IEnumerator 是所有枚举数的基接口。枚举数只允许读取集合中的数据。枚举数无法用于修改基础集合。最初，枚举数被定位于集合中第一个元素的前面。Reset 也将枚举数返回到此位置。在此位置，调用 Current 会引发异常。因此，在读取 Current 的值之前，必须调用 MoveNext 将枚举数提前到集合的第一个元素。在调用 MoveNext 或 Reset 之前，Current 返回同一对象。MoveNext 将 Current 设置为下一个元素。在传递到集合的末尾之后，枚举数放在集合中最后一个元素后面，且调用 MoveNext 会返回 false。如果最后一次调用 MoveNext 返回 false，则调用 Current 会引发异常。若要再次将 Current 设置为集合的第一个元素，可以调用 Reset，然后再调用 MoveNext。

只要集合保持不变，枚举数就将保持有效。如果对集合进行了更改(例如添加、修改或删除元素)，则该枚举数将失效且不可恢复，并且下一次对 MoveNext 或 Reset 的调用将引发 InvalidOperationException。如果在 MoveNext 和 Current 之间修改集合，那么即使枚举数已经无效，Current 也将返回它所设置成的元素。

枚举数没有对集合的独占访问权，因此，枚举一个集合在本质上不是一个线程安全的过程。甚至在对集合进行同步处理时，其他线程仍可以修改该集合，这会导致枚举数引发异常。若要在枚举过程中保证线程安全，可以在整个枚举过程中锁定集合，或者捕捉由于其他线程进行的更改而引发的异常。

下面再来总结一下 IEnumerable 和 IEnumerator 的区别：

(1) 一个 Collection 要支持 foreach 方式的遍历，必须实现 IEnumerable 接口(亦即，必须以某种方式返回 IEnumerator 对象)。

(2) IEnumerator 对象具体实现了 iterator(通过 MoveNext()，Reset()，Current)。

(3) 从这两个接口的用词选择上，也可以看出其不同：IEnumerable 是一个声明式的接口，声明实现该接口的 class 是“可枚举(enumerable)”的，但并没有说明如何实现枚举器(iterator)；IEnumerator 是一个实现式的接口，IEnumerator 对象人就是一个 iterator。

(4) IEnumerable 和 IEnumerator 通过 IEnumerable 的 GetEnumerator()方法建立了连接，客户端可以通过 IEnumerable 的 GetEnumerator()得到 IEnumerator 对象，在这个意义上，将 GetEnumerator()看作 IEnumerator 对象的工厂方法也未尝不可。

20.3 .NET 的响应式框架

在C#中,程序员在指定异步操作回调或事件处理程序的时候使用响应式编程。在异步操作完成或者事件触发的时候,就会调用方法并作为对该事件的响应。.NET 的响应式框架就是为了实现和优化这种情景模式而诞生的,在 Windows Phone 中可以使用这种优秀的编程模式来实现应用的相关逻辑。

20.3.1 响应式框架概述

.NET 的响应式框架(Reactive Extensions for .NET Framework)是一个托管库,它提供用于编写响应式应用程序的 API。响应式应用程序是由其环境驱动的,在响应式模型中,数据流、异步请求以及事件都表示为可观察序列。应用程序可以订阅这些可观察序列,以在新数据到达时接收异步消息。响应式框架允许应用程序使用查询运算符组合这些序列。

如果应用程序与多个数据源交互,则管理所有这些交互的便利方法是为每个数据流实现单独的处理程序。在这些处理程序中,必须提供代码以在所有不同的数据流之间进行协调并将该数据处理为可使用的形式。响应式框架可以编写一个查询,该查询将所有这些数据流组合成触发单个处理程序的单个流。筛选、同步和转换数据等工作由响应式框架查询执行,以便处理程序只需对接收的数据进行反应并对该数据进行某些处理。

响应式框架把事件驱动 UI 与 LINQ、并发性和异步调用结合起来。响应式框架尝试解决从基于事件的 UI 异步访问数据的问题。标准的迭代器模式以及它的基本接口 IEnumerable 和 IEnumerator 对于异步操作是不足够的,因此响应式框架通过引入观察者模式来解决这个问题,这个模式包含两个主要的接口:IObservable 和 IObserver。不是客户端一步一步地迭代数据集合,而是集合把数据作为异步调用的结果推送给客户端,终结了调用循环。响应式框架的强大之处在于它内置了 LINQ 和并发性的支持。因此,用户并不仅能以异步的方式获得一组数据,而且可以在一组数据上产生异步 LINQ 查询,然后并行地运行它们。

20.3.2 IObserver＜T＞和 IObservable＜T＞

IObserver＜T＞ 和 IObservable＜T＞ 接口为基于推送的通知提供通用机制,也称为观察者设计模式。IObservable＜T＞ 接口表示发送通知(提供程序)的类;IObserver＜T＞ 接口表示接收通知(观察器)的类。T 表示提供通知信息的类。在某些基于推送的通知中,IObserver＜T＞ 实现和 T 可以表示相同的类型。

该提供程序必须实现一个方法 Subscribe,指出某个观察器要接收基于推送的通知。该方法的调用方传递观察器的实例。方法返回 IDisposable 实现,其可让观察程序在提供程序已停止发送它们的任意时间之前取消通知。

在任何给定时间，给定的提供程序可能具有零个、一个或多个观察器。该提供程序负责存储对观察器的引用，并且在发送通知之前确保它们有效。IObservable<T>接口对于观察器的个数或发送通知的顺序不作任何假设。

程序可以通过调用IObserver<T>的相关方法向观察器发送以下三种通知：

(1) 当前数据：提供程序可以调用IObserver<T>.OnNext方法来为观察器传递包含当前数据、更改的数据或全新数据的T对象。

(2) 一个异常情况：提供程序可以调用IObserver<T>.OnError方法来通知观察器出现了某种错误情况。

(3) 没有进一步的数据：提供程序可以调用IObserver<T>.OnCompleted方法来通知观察器它已经完成通知的发送。

20.3.3 IObservable和IEnumerable

在.NET响应式框架中，LINQ to Object使用IEnumerable和IEnumerator两个接口来迭代数据集。枚举器的MoveNext()方法用于从前一个元素枚举到下一个元素，而Current属性则用于检索个别的元素。这种机制采取广泛使用的简洁易懂的“拉(pull)”过程。IEnumerable有对偶特性：一个是可以从集合中提取数据，同时可以把数据推进同样的集合。这意味着另一种响应式编程的方式。理论上来说，IEnumerable接口可对集合添加对象，但是由于它的阻塞性所以不能在异步操作中进行。那就是为什么会诞生两个新的接口：IObservable和IObserver。可以把IObserver赋给数据集并把它作为IObservable订阅。当一项新的数据可用时，就可以通过传递IObservable来把它压进集合，接着再传给IObserver。要遍历IObservable，需要做的就是执行与IEnumerable相反的操作。创建IObserver，把它赋给IObservable，接着IObservable通过调用自身的方法来把数据压进IObserver。在IObservable对Observer调用OnUpdate方法的时候，就相当于IEnumerable方法使用关键字yield向IEnumerable传递信息。类似地，在IObservable对Observer调用OnCompleted方法的时候，就相当于IEnumerable使用关键字break来表示没有数据一样。

20.4 在Windows Phone上实践响应式编程

响应式框架背后的理念是将应用程序中所触发的一系列事件塑造成一个数据流(也称为可观察序列，observable sequence)。然后应用程序可以订阅一个可观察序列，并在添加新项时收到通知。要在应用程序中使用响应式框架，必须先添加对Microsoft.Phone.Reactive和System.Observable程序集的引用。完成后，即可创建一个Observable集合。一旦拥有了合适的Observable集合，即可调用Subscribe方法来对其进行订阅(侦听更改)。可将Subscribe视为一种关联事件处理程序的方法，每当有新数据被添加到序列中时，就令调用事件处理程序。本节将会通过Windows Phone上的实例来演示如何在Windows Phone应用程序上使用响应式编程。

20.4.1　事件联动模拟用户登录实例

下面给出模拟用户登录的实例：一般的用户登录窗口都会设置成，当用户输入了用户名和密码之后登录的按钮才会从不可点状态变成可点状态，当用户删除了用户名或者密码之后，登录按钮又从可点状态变成不可点状态，那么这个示例就是要用响应式编程实现这个业务的逻辑。

代码清单 20-1：模拟用户登录(源代码：第 20 章\Examples_20_1)

MainPage.xaml 文件主要代码

```
<Grid x:Name = "ContentPanel" Grid.Row = "1" Margin = "12,0,12,0">
    <StackPanel x:Name = "CredentialsPannel">
        <TextBox Name = "SignInUserTextBox" InputScope = "EmailUserName" IsTabStop = "True"
MaxLength = "255" />
        <PasswordBox Name = "SignInPasswordBox" MaxLength = "16" IsTabStop = "True" />
        <Button Name = "SignInButton" Content = "登录" IsEnabled = "False"/>
    </StackPanel>
</Grid>
```

MainPage.xaml.cs 文件主要代码

```
using System;
using System.Windows;
using System.Windows.Controls;
using Microsoft.Phone.Controls;
using Microsoft.Phone.Reactive;
namespace ObservableDemo
{
    public partial class MainPage : PhoneApplicationPage
    {
        public MainPage()
        {
            InitializeComponent();
            //将可观察序列的每个值投射到新表格中
            IObservable< RoutedEventArgs > observable = Observable.Select < IEvent <
TextChangedEventArgs >, RoutedEventArgs >
                (
                //返回一个可观察序列,该序列包含基础 .NET Framework 事件的值
                Observable.FromEvent< TextChangedEventHandler, TextChangedEventArgs >
                    (h => new TextChangedEventHandler(h.Invoke),
                    delegate(TextChangedEventHandler h)
                    {
                        this.SignInUserTextBox.TextChanged += h;
                    },
                    delegate(TextChangedEventHandler h)
                    {
                        this.SignInUserTextBox.TextChanged -= h;
                    }),
                evt => new RoutedEventArgs()
```

```
            );
        IObservable < RoutedEventArgs > observable2 = Observable. Select < IEvent <
RoutedEventArgs >, RoutedEventArgs >
            //返回一个可观察序列,该序列包含基础 .NET Framework 事件的值
            (Observable. FromEvent < RoutedEventHandler, RoutedEventArgs >( h = > new
RoutedEventHandler(h. Invoke), delegate(RoutedEventHandler h)
            {
                this. SignInPasswordBox. PasswordChanged += h;
            }, delegate(RoutedEventHandler h)
            {
                this. SignInPasswordBox. PasswordChanged -= h;
            }), evt = > new RoutedEventArgs());
        //监听事件观察者对象源

        ObservableExtensions. Subscribe(
            //返回一个可观察序列,它只包含独特的连续值
           // Observable. DistinctUntilChanged(
            Observable. Select(
            //将可观察序列的可观察序列合并到一个可观察序列中
            Observable. Merge < RoutedEventArgs >(observable, observable2),
            evt = > new User { Password = this. SignInPasswordBox. Password, Login =
this. SignInUserTextBox. Text }
            ),
           // ),
            user = > this. SignInButton. IsEnabled = ! string. IsNullOrWhiteSpace( user.
Login) && ! string. IsNullOrWhiteSpace(user. Password)
            );
    }
  }
  public class User
  {
      public string Password { get; set; }
      public string Login { get; set; }
  }
}
```

程序的运行效果如图 20.3 所示。

图 20.3　用户名密码输入

20.4.2　网络请求实例

下面给出网络请求的示例：该示例演示了使用普通的网络请求方式和响应式编程的网络请求方式。

代码清单 20-2：网络请求(源代码：第 20 章\Examples_20_2)

MainPage. xaml 文件主要代码

```
< Grid x:Name = "ContentPanel" Grid. Row = "1" Margin = "12,0,12,0">
    < Button Content = "普通模式获取网络响应" Height = "72" HorizontalAlignment = "Left"
Margin = "79,97,0,0" Name = "button1" VerticalAlignment = "Top" Width = "333" Click = "button1_
Click" />
```

```
    <Button Content="观察者模式获取网络响应" Height="72" HorizontalAlignment="Left" Margin="79,205,0,0" Name="button2" VerticalAlignment="Top" Width="333" Click="button2_Click" />
</Grid>
```

MainPage.xaml.cs 文件主要代码

```
//普通模式
private void button1_Click(object sender, RoutedEventArgs e)
{
    HttpWebRequest request = (HttpWebRequest)WebRequest.Create("http://www.cnblogs.com/linzheng");
    request.Method = "GET";
    request.BeginGetResponse(new AsyncCallback(GetResponse), request);
}
void GetResponse(IAsyncResult res)
{
    HttpWebRequest request = (HttpWebRequest)res.AsyncState;
    HttpWebResponse response = (HttpWebResponse)request.EndGetResponse(res);
    string content = "";
    using (StreamReader reader = new StreamReader(response.GetResponseStream(), Encoding.UTF8))
    {
        content = reader.ReadToEnd();
    }
    Deployment.Current.Dispatcher.BeginInvoke(delegate
    {
        MessageBox.Show(content);
    });
}
//观察者模式
private void button2_Click(object sender, RoutedEventArgs e)
{
    HttpWebRequest request = (HttpWebRequest)WebRequest.Create("http://www.cnblogs.com/linzheng");
    request.Method = "GET";
    Observable.FromAsyncPattern<WebResponse>(request.BeginGetResponse, request.EndGetResponse)().Subscribe(HandleResult);
}
void HandleResult(WebResponse res)
{
    string content = "";
    using (StreamReader reader = new StreamReader(res.GetResponseStream(), Encoding.UTF8))
    {
        content = reader.ReadToEnd();
    }
    Deployment.Current.Dispatcher.BeginInvoke(delegate
    {
        MessageBox.Show(content);
    });
}
```

程序的运行效果如图 20.4 所示。

图 20.4　测试网络请求

20.4.3　响应式线程实例

下面给出响应式线程的示例：该示例演示了在响应式编程中如何使用后台线程和 UI 线程进行处理的方法。

代码清单 20-3：响应式线程(源代码：第 20 章\Examples_20_3)

MainPage. xaml 文件主要代码

```
<Grid x:Name = "ContentPanel" Grid.Row = "1" Margin = "12,0,12,0">
    <Button Content = "测试请求" Height = "88" HorizontalAlignment = "Left" Margin = "120,129,
0,0" Name = "button1" VerticalAlignment = "Top" Width = "233" Click = "button1_Click" />
</Grid>
```

MainPage. xaml. cs 文件主要代码

```
private void button1_Click(object sender, RoutedEventArgs e)
{
    button1.IsEnabled = false;
    WebClient wc = new System.Net.WebClient();
    //后台线程
    var o = Observable.FromEvent<DownloadStringCompletedEventArgs>(wc, "DownloadStringCompleted")
                      .ObserveOn(Scheduler.ThreadPool)
                      .Select(evt => evt.EventArgs.Result);
    //UI 线程
    o.ObserveOn(Scheduler.Dispatcher).Subscribe(s => MessageBox.Show(s));
    //开始请求
    wc.DownloadStringAsync(new Uri("http://www.cnblogs.com/linzheng"));
}
```

程序的运行效果如图 20.5 和图 20.6 所示。

图 20.5　线程首页

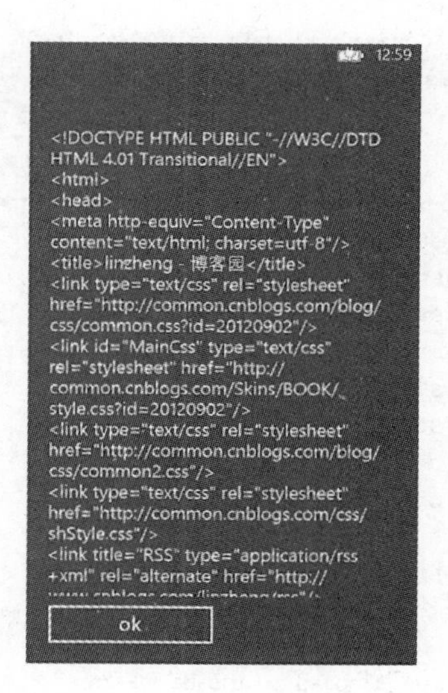

图 20.6　请求返回页面

20.4.4　豆瓣搜索实例

下面给出的示例：该示例演示了在响应式编程中如何使用后台线程和 UI 线程进行处理的方法。

代码清单 20-4：响应式线程(源代码：第 20 章\Examples_20_4)

MainPage.xaml 文件主要代码

```
<StackPanel x:Name="TitlePanel" Grid.Row="0" Margin="12,17,0,28">
    <ProgressBar IsIndeterminate="True" x:Name="LoadingIndicator" Visibility="Collapsed">
</ProgressBar>
    <TextBlock x:Name="ApplicationTitle" Text="豆瓣搜书" Style="{StaticResource
PhoneTextNormalStyle}"/>
    <TextBox x:Name="searchTextBox" Text=""/>
</StackPanel>
<Grid x:Name="ContentPanel" Grid.Row="1" Margin="12,0,12,0">
    <ListBox x:Name="searchResults"
     Grid.Row="1">
        <ListBox.ItemTemplate>
            <DataTemplate>
                <StackPanel Orientation="Horizontal">
                    <StackPanel Orientation="Vertical">
                        <TextBlock Text="{Binding Num}" FontSize="30"/>
                        <TextBlock Text="{Binding Title}" TextWrapping="Wrap" />
                        <TextBlock FontWeight="Bold" Text="{Binding Author}"/>
                        <TextBlock Text="{Binding Id}" FontSize="30"/>
                    </StackPanel>
                </StackPanel>
            </DataTemplate>
```

```
            </ListBox.ItemTemplate>
        </ListBox>
    </Grid>
```

MainPage.xaml.cs 文件主要代码

```
using System;
using System.Collections.Generic;
using System.Linq;
using System.Net;
using System.Windows;
using System.Windows.Controls;
using Microsoft.Phone.Controls;
using Microsoft.Phone.Reactive;
using System.Xml.Linq;
using System.IO;
namespace DouBanRxDemo
{
    public partial class MainPage : PhoneApplicationPage
    {
        private static string _atomNamespace = "http://www.w3.org/2005/Atom";
        //书籍实体节点
        private static XName _entryName = XName.Get("entry", _atomNamespace);
        //书籍标题节点
        private static XName _titleName = XName.Get("title", _atomNamespace);
        //书籍 ID 节点
        private static XName _idName = XName.Get("id", _atomNamespace);
        //书籍作者节点
        private static XName _publishedName = XName.Get("author", _atomNamespace);
        //书籍名字节点
        private static XName _nameName = XName.Get("name", _atomNamespace);
        public MainPage()
        {
            InitializeComponent();
            Func<string, IObservable<string>> search = searchText =>
            {
                var request = (HttpWebRequest)HttpWebRequest.Create(new Uri(string.Format
("http://api.douban.com/book/subjects?tag={0}&max-results=20",searchText)));
                var bookSearch = Observable.FromAsyncPattern<WebResponse>(request.
BeginGetResponse, request.EndGetResponse);
                return bookSearch().Select(res => WebResponseToString(res));
            };
            Observable.FromEvent<TextChangedEventArgs>(searchTextBox, "TextChanged")
                            //设置观察的对象为 searchTextBox 控件的 TextChanged 事件
                    .Select(e => ((TextBox)e.Sender).Text)
                            //选择文本的字符串为新的观察序列
                    .Where(text => text.Length > 2)
                            //文本长度大于 2
                    .Do(s => searchResults.Opacity = 0.5)
                            //设置 ListBox 控件为半透明
                    .Throttle(TimeSpan.FromMilliseconds(400))
```

```
                    //设置观察最短时间为 400 毫秒
        .ObserveOnDispatcher()
                    //在当前的 UI 线程运行
        .Do(s => LoadingIndicator.Visibility = Visibility.Visible)
                    //展示进度条
        .SelectMany(txt => search(txt))
                    //搜索书籍,返回网络获取的 xml 文件
        .Select(searchRes => ParseSearch(searchRes))
                    //解析 xml 文件为书的枚举序列
        .ObserveOnDispatcher()
                    //在当前的 UI 线程运行
        .Do(s => LoadingIndicator.Visibility = Visibility.Collapsed)
                    //隐藏进度条
        .Do(s => searchResults.Opacity = 1)
                    //设置 ListBox 控件可见
        .Subscribe(tweets => searchResults.ItemsSource = tweets);
                    //通知观察者,绑定书籍集合到 ListBox 控件
}
///<summary>
///处理网络请求的回应
///</summary>
///<param name="webResponse">网络回应</param>
///<returns>返回的字符串</returns>
private string WebResponseToString(WebResponse webResponse)
{
    HttpWebResponse response = (HttpWebResponse)webResponse;
    using (StreamReader reader = new StreamReader(response.GetResponseStream()))
    {
        return reader.ReadToEnd();
    }
}
///<summary>
///解析 XML 文件
///</summary>
///<param name="response">xml 文件字符串</param>
///<returns>返回书籍的枚举集合</returns>
private IEnumerable<Book> ParseSearch(string response)
{
    int i=1;
    var doc = XDocument.Parse(response);
    return doc.Descendants(_entryName)
            .Select(entryElement => new Book()
            {
                Num = i++,
                Title = entryElement.Descendants(_titleName).Single().Value,
                Id = entryElement.Descendants(_idName).Single().Value,
                Author = entryElement.Descendants(_nameName).First().Value
            });
}
}
```

```
    ///<summary>
    ///书的实体类
    ///</summary>
    public class Book
    {
        public int Num { get; set; }
        public string Id { get; set; }
        public string Title { get; set; }
        public string Author { get; set; }
    }
}
```

程序的运行效果如图 20.7 所示。

图 20.7　豆瓣搜书

第21章

C++编程

在 Windows Phone 7 操作系统时代，微软不提供给第三方开发者使用 C++编程，但在 Windows Phone 8 诞生后这种情况得到了改变。Windows Phone 8 支持原生的 C++代码编程，在应用程序开发方面提供了部分的 C++语言的 API，开发者可以通过 Windows 运行时组件来调用；在游戏开发方面，提供了 DirectX 游戏开发框架，完全基于 C++的编程。在 Windows Phone 8 里面的 C++编程将会使用一种新的 C++语法，叫做 C++/CX，同时也支持使用标准 C++编程和调用标准 C++的组件，但是在普通应用程序里面，不可以完全使用 C++/CX 语法来编写，因为大部分基于 UI 的 API 和使用系统特性的 API 不支持 C++的编程，同时也不可以直接创建基于 C++的应用程序项目，在普通应用程序里面只可以通过 Windows 运行时组件来使用 C++和调用 C++的 API。本章将会详细地讲解 Windows Phone 8 的 C++/CX 编程语法，基于 C++的 Windows 运行时组件和如何使用标准的 C++等内容，带领读者进入 Windows Phone 8 C++编程的世界。

21.1 C++/CX 语法

C++/CX 是 Windows 运行时的 C++语法，在 Windows 运行时中提供的 C++编程的 API 就是基于 C++/CX 进行设计的，所以在 Windows 运行时下进行 C++编程将无处不在地使用着这样一种语法。目前 C++/CX 的编程应用在 Windows Phone 8 和 Windows 8 这两个新平台下，开启了微软对 C++扩展的新的路线，它提供了很多新的 C++的语法结构，不过也有很多语法特性与 C#语言很类似。C++/CX 是属于 Native C++，它不使用 CLR 也没有垃圾回收机制，这个和微软之前在.NET 平台下扩展的 C++/CLR 是有本质的区别的，虽然两者在语法方面会比较类似，但是这两者是两种完全不同的东西。C++/CX 是基于 Windows 运行时而诞生的，它运行在 Windows 运行时上，比在.NET 平台上的托管 C++具有更高的效率，是一种新的技术产物。下面来看一些特性。

21.1.1 命名空间

命名空间(Namespace)表示标识符(identifier)的上下文(context)。一个标识符可在多个命名空间中定义，它在不同命名空间中的含义是互不相干的。这样，在一个新的命名空间中可定义任何标识符，它们不会与任何已有的标识符发生冲突，因为已有的定义都处于其他命名空间中。在标准 C++ 里命名空间是为了防止类型冲突，但在 Windows 运行时中，使用 C++ 编程需要给所有的程序类型添加上命名空间，这是 Windows 运行时的一种语法规范。这是 C++/CX 语法的命名空间与标准 C++ 的命名空间最大的区别。在 C++/CX 中命名空间可以嵌套着使用，看下面的例子：

```
namespace Test
{
    public ref class MyClass{};
    public delegate void MyDelegate();
    namespace NestedNamespace
    {
        public ref class MyClass2
        {
            event Test::MyDelegate ^ Notify;
        };
    }
}
```

在 MyClass2 里面来使用 Test 空间下面的类型需要通过 Test::来调用。

Windows 运行时的 API 都在 Windows::* 命名空间里面，在 windows.winmd 文件里面可以找到定义。这些命名空间都是为了 Windows 保留的，其他第三方自定义的类型不能够使用这些命名空间。Windows 运行时 API 的命名空间如表 21.1 所示。

表 21.1　C++/CX 的命名空间

命 名 空 间	描　　述
default	包含了数字和 char16 类型
Platform	包含了 Windows 运行时的一些主要的类型，如 Array<T>，String，Guid 和 Boolean；也包含了一些特别的帮助类型如 Platform::Agile<T> 和 Platform::Box<T>
Platform::Collections	包含了一些从 IVector，Imap 等接口中集成过来的集合类型；这些类型在 collection.h 头文件中定义，并不是在 platform.winmd 里面
Platform::Details	包含了在编译器里面使用的类型并不公开来使用

21.1.2　基本的类型

Windows 运行时的 C++/CX 扩展了标准 C++ 里面的基本的类型，C++/CX 实现了布尔类型，字符类型和数字类型，这些类型都在 default 命名空间里面。另外，C++/CX 还封装了一些在 Windows 运行时环境中独有的类型，这些类型在 Windows 运行时中也很常用。C++/CX 的布尔类型和字符类型如表所示，如果在公共接口中使用标准 C++ 的 bool 和 wchar_t 类型，编译器会自动转换成 C++/CX 的类型。C++/CX 的数字类型如表所示，并不是所有标准C++内置类型都会在Windows运行时中得到支持，比如Windows运行时中不支持标准C++的long类型。Windows运行时类型如表21.2～表21.4所示，这些类型在Platform空间中定义，using namespace Platform。

表21.2 布尔类型和字符类型

命名空间	C++/CX名字	定 义	标准C++名字	取值范围
Platform default	Boolean	一个8-bit布尔类型	bool	true(非零)和false(零)
	char16	一个表示Unicode(UTF-16) 16-bit指针的非数字值	wchar_t或L'c'	标准Unicode字符

表21.3 数字类型

C++/CX名字	定义	标准C++名字	取值范围
int8	有符号8-bit整型	signed char	－128～127
uint8	无符号8-bit整型	unsigned char	0～255
int16	有符号16-bit整型	short	－32 768～32 767
uint16	无符号16-bit整型	unsigned short	0～65 535
int32	有符号32-bit整型	int	－2 147 483 648～2 147 483 647
uint32	无符号32-bit整型	unsigned int	0～4 294 967 295
int64	有符号64-bit整型	long long-or-__int64	－9 223 372 036 854 775 808～9 223 372 036 854 775 807
uint64	无符号64-bit整型	unsigned long long-or-unsigned__int64	0～18 446 744 073 709 551 615
float32	32-bit浮点类型	float	3.4E＋/－38 (7 digits)
float64	64-bit浮点类型	double	1.7E＋/－308 (15 digits)

表21.4 Windows运行时类型

名 字	定 义	名 字	定 义
Object	表示任何一个Windows运行时类型	Guid	表示一个128-bit的唯一字符串类型
String	字符串类型	UIntPtr	一个无符号64位值作为指针使用
Rect	矩形	IntPtr	一个有符号64位值作为指针使用
SizeT	一个表示宽和高的大小类型	Enum	枚举类型
Point	点类型		

21.1.3 类和结构

C++/CX语法定义一个类和创建一个对象所使用的语法和标准C++是有区别的。在C++/CX中定义一个类和结构时需要使用ref关键字，如ref class MyClass{…}，在程序中定义一个该类的对象时需要使用符号^来表示，如MyClass ^ myClass＝ ref new MyClass()。C++/CX中的ref关键字和C＃中的ref关键字是不一样的东西，C＃的ref关键字表示使参数按引用传递，C++/CX表示Windows运行时的引用类。在Windows运行时中需要使用

public ref class 才可以提供给外部访问，也就是需要使用 C++/CX 的语法，也可以在 Windows 运行时中定义和使用标准 C++的类，但是无法提供给外部的程序进行调用和访问。

例如：

```
MyRefClass ^ myClass = ref new MyRefClass();
MyRefClass ^ myClass2 = myClass;
```

myClass2 和 myClass 是指向相同的内存地址。

^符号的变量代表它是引用类型的，表示的是一个对象的指针，但是该指针不需要手动进行销毁，系统会负责他们的引用计数，当引用计数为 0 时，它们会被销毁。

public ref class 的定义形式，一般是为了达到能够将这个对象在其他不同语言中使用的目的。Windows 运行时的 API 的开发模式是对外接口使用 ref class 内部使用标准 C++代码编写，在内部实现标准 C++和 C++/CX 的类型转换，然后通过转型到公共接口上。

下面来看一个 C++/CX 类的示例：

Person.h 头文件

```
#include <map>
using namespace std;    //引入标准 C++库
namespace WFC = Windows::Foundation::Collections; //使用集合命名空间，在程序中定义为 WFC
ref class Person sealed    //使用 ref 关键字定义类
{
  //构造方法
  Person(Platform::String ^ name);
  //添加电话号码方法
  void AddPhoneNumber(Platform::String ^ type, Platform::String ^ number);
  //电话号码集合属性
  property WFC::IMapView<Platform::String ^, Platform::String ^>^ PhoneNumbers
  {
    //通过上面定义的集合空间名字来访问集合空间的 IMapView
    WFC::IMapView<Platform::String ^, Platform::String ^>^ get();
  }
private:
  //名字
  Platform::String ^ m_name;
  //电话号码
  std::map<Platform::String ^, Platform::String ^> m_numbers;
};
```

Person.cpp 源文件

```
using namespace Windows::Foundation::Collections;
using namespace Platform;
using namespace Platform::Collections;
//实现构造方法
Person::Person(String ^ name): m_name(name) { }
//实现添加电话号码方法
```

```
void Person::AddPhoneNumber(String^ type, String^ number)
{
   m_numbers[type] = number;
}
//实现属性的 get 方法
IMapView< String^, String^>^ Person::PhoneNumbers::get()
{
   return ref new MapView< String^, String^>(m_numbers);
}
```

使用自定义的 Person 类

```
using namespace Platform;
//使用 ref new 来新建一个对象
Person^ p = ref new Person("Clark Kent");
//调用 Person 对象的方法
p->AddPhoneNumber("Home", "425-555-4567");
p->AddPhoneNumber("Work", "206-555-9999");
//调用 Person 对象的属性
String^ workphone = p->PhoneNumbers->Lookup("Work");
```

C++/CX 语法的结构体需要使用 value 关键字来进行声名，在结构体里面可以使用枚举、数据类型或者结构体等等的数据，C++/CX 语法的结构体可以通过 Windows 运行时组件来进行传递。

下面来看一个 C++/CX 类的示例：

```
//定义一个枚举
public enum class Continent {  Africa, Asia, Australia, Europe, NorthAmerica, SouthAmerica,
Antarctica };
//定义一个结构体，包含两个数字
value struct GeoCoordinates
{
  double Latitude;
  double Longitude;
};
//定义一个结构体，包含各种的数据结构
value struct City
{
  Platform::String^ Name;
  int Population;
  double AverageTemperature;
  GeoCoordinates Coordinates;
  Continent continent;
};
```

21.1.4 对象和引用计数

Windows 运行时是一个面向对象的库，在 Windows 运行时编程都是以对象的形式进行操作，但是 Windows 运行时中的 C++对象跟 C#里面的对象和标准 C++里面的对象都是不一样的。在 Windows 运行时中，使用引用计数管理对象，关键字 ref new 新建一个对象，返回的是对象的指针，但是在 WinRT 中，使用^符号替代 *，仍然使用一->操作符使用对象的成员方法。另外，由于使用引用计数，不需要使用 delete 去释放，对象会在最后一个引用失效的时候自动释放。Windows 运行时内部是使用 COM 实现的，对于引用计数，在底层，Windows 运行时对象是一个使用智能指针管理的 COM 对象。创建一个 C# 对象，使用完毕的时候不需要释放而是用 GC 垃圾回收期自动管理去回收对象。创建一个 Windows 运行时的 C++对象，使用完毕之后并不需要像标准 C++一样去释放内存，而是由引用计数自动去释放。

在基于引用计数的任何类型的系统中，对类型的引用可以形成循环，即第一个对象引用第二个对象，第二个对象引用第三个对象，依此类推，直到某个最终对象引用回第一个对象。在一个循环中，当一个对象的引用计数变为零时，将无法正确删除该对象。为了帮助解决此问题，C++/CX 提供了 WeakReference 类。WeakReference 对象支持 Resolve() 方法，如果对象不再存在，则返回 Null；或如果对象是活动的但不是类型 T，则将引发 InvalidCastException。

21.1.5 属性

属性的语法是从 C# 的语法里面引进，作用与 C# 的一样，它提供灵活的机制来读取、编写或计算某个私有字段的值。可以像使用公共数据成员一样使用属性，但实际上它们是称作“访问器”的特殊方法。这使得可以轻松访问数据，此外还有助于提高方法的安全性和灵活性。属性里面包含了一个 get 访问器和一个 set 设置器，比其他的变量更加的灵活和强大，比如在属性里面可以定义该变量的访问权限和读取权限，也可以设置该变量的取值和赋值的复杂逻辑。

属性和字段的区别：属性是逻辑字段；属性是字段的扩展，源于字段；属性并不占用实际的内存，字段占内存位置及空间。属性可以被其他类访问，而大部分字段不能直接访问。属性可以对接收的数据范围作限定，而字段不能。最直接的说：属性是被“外部使用”，字段是被“内部使用”。

下面来看一个 C++/CX 属性的示例：

```
public ref class Prescription sealed
{
private:
    //字段
    Platform::String ^ doctor;
    int quantity;
public:
    //属性
    property Platform::String ^ Name;
    //只读属性
    property Platform::String ^ Doctor
    {
        Platform::String ^ get() { return doctor; }
    }
```

```
    //只写属性
    property int Quantity
    {
        int get() { return quantity; }
        void set(int value)
        {
            if (value <= 0) { throw ref new Platform::InvalidArgumentException(); }
            quantity = value;
        }
    }
};
```

21.1.6 接口

接口是一种约束形式，其中只包括成员定义，不包含成员实现的内容。接口的主要目的是为不相关的类提供通用的处理服务，接口是让一个类具有两个以上基类的唯一方式。接口是面向对象中一种非常强大的语法，但是在标准C++面并没有接口的概念，不过在C++/CX语法中，微软引入了接口的语法。C++/CX的接口与C#的接口类似，一个C++/CX的ref类可以继承一个基类和多个接口，一个接口同时也可以继承多个接口。接口里面可以有事件，属性，方法等类型，这些类型必须是公开的，在C++/CX的接口里面不能够实用标准C++的类型，也不能使用静态的类型。

下面来看一个C++/CX接口的示例：

接口的定义：定义一个多媒体播放器的接口

```
//播放状态
public enum class PlayState {Playing, Paused, Stopped, Forward, Reverse};
//播放器的事件参数
public ref struct MediaPlayerEventArgs
{
    property PlayState oldState;
    property PlayState newState;
    };
//播放状态改变事件
public delegate void OnStateChanged(Platform::Object ^ sender, MediaPlayerEventArgs ^ a);
//接口
public interface class IMediaPlayer
{
    //接口的成员
    event OnStateChanged ^ StateChanged;
    property Platform::String ^ CurrentTitle;
    property PlayState CurrentState;
    void Play();
    void Pause();
    void Stop();
    void Forward(float speed);
};
```

接口的类实现：实现一个多媒体播放器的类

```
public ref class MyMediaPlayer sealed : public IMediaPlayer
    {
    public:
        //实现接口的成员
        virtual event OnStateChanged^ StateChanged;
        virtual property Platform::String^ CurrentTitle;
        virtual property PlayState CurrentState;
        virtual void Play()
        {
        // …
            auto args = ref new MediaPlayerEventArgs();
            args->newState = PlayState::Playing;
            args->oldState = PlayState::Stopped;
            StateChanged(this, args);
        }
        virtual void Pause(){/*...*/}
        virtual void Stop(){/*...*/}
        virtual void Forward(float speed){/*...*/}
        virtual void Back(float speed){/*...*/}
    private:
        //...
    };
```

21.1.7　委托

委托是一种 ref 类，相当于标准 C++ 中函数对象的 Windows 运行时。它是封装可执行代码的类型。委托指定其包装函数必须具有的返回类型和参数类型。委托常与事件一起使用。委托的概念最先出现在.NET 框架中，现在在 C++/CX 中引入了这样的语法，目的是解决函数指针的不安全性。函数指针是函数的地址、函数的入口点，函数指针既没有表示这个函数有什么返回类型，也没有指示这个函数有什么形式的参数，更没有指示有几个参数，函数指针是非安全的。

委托使用 delegate 进行定义，可以在程序中定义委托，以便定义事件处理程序。委托的声明类似函数声明，只不过委托是一种类型。通常在命名空间范围声明委托，也可以将委托声明嵌套在类声明中。

下面来看一个 C++/CX 委托的示例：

定义一个联系人信息类

```
public ref class ContactInfo sealed
{
public:
    ContactInfo(){}
    ContactInfo(Platform::String^ saluation, Platform::String^ last, Platform::String^
```

```
first, Platform::String ^ address1);
    property Platform::String ^ Salutation;
    property Platform::String ^ LastName;
    property Platform::String ^ FirstName;
    property Platform::String ^ Address1;
    //其他属性
    Platform::String ^ ToCustomString(CustomStringDelegate ^ func)
    {
        return func(this);
    }
};
```

定义、创建和使用委托

```
//以下委托封装任何将 ContactInfo ^ 作为输入的函数并返回 Platform::String ^
public delegate Platform::String ^ CustomStringDelegate(ContactInfo ^ ci);
//创建一个自定义的委托
CustomStringDelegate ^ func = ref new CustomStringDelegate([] (ContactInfo ^ c)
{
    return c->FirstName + " " + c->LastName;
});
//使用委托
Platform::String ^ name = ci->ToCustomString( func );
```

委托和函数对象一样，包含将在未来某个时刻执行的代码。如果创建和传递委托的代码和接受并执行委托的函数在同一线程上运行，则情况相对简单。如果该线程是 UI 线程，则委托可以直接操作用户界面对象(如 XAML 控件)。

如果客户端应用程序加载在一个线程单元运行的 Windows 运行时组件，并将委托提供给该组件，则默认情况下，在 STA 线程上直接调用该委托。大多数 Windows 运行时组件既可在 STA 中也可在 MTA 中运行。

如果执行委托的代码在不同线程(如在 concurrency::task 对象的上下文中)运行，则将负责同步对共享数据的访问。例如，如果委托中包含对某个向量的引用，而 XAML 控件也具有对这一向量的引用，则必须采取措施，避免因委托和 XAML 控件同时尝试访问该向量而可能发生的死锁或争用现象。还必须注意，在调用委托前，委托不会通过引用来尝试捕获可能超出范围的本地变量。

如果希望在与创建委托相同的线程上回调委托，例如，将它传递给在 MTA 单元中运行的组件，并希望在和创建器相同的线程上调用它，则使用第二个 CallbackContext 参数的委托构造函数重载。只在具有注册代理/存根的委托上使用此重载，并非所有在 Windows.winmd 中定义的委托都会注册。

21.1.8 事件

Windows 运行时类型可以声明事件，同一组件或其他组件中的客户端代码可以订阅这些事件并在发布程序引发事件时执行自定义操作。可以在 ref 类或接口中声明事件，并且可以将其声明为公共、内部(公共/私有)、公共受保护、受保护、私有受保护或私有事件。使用“+=”运算符订阅事件。多个处理程序可以与同一事件关联。事件源按顺序从同一线程调用到所有事件处理程序。如果一个事件接收器在事件处理程序方法内受阻，则它将阻止事件源为该事件调用其他事件处理程序。事件源对事件接收器调用事件处理程序的顺序不能保证，可能因调用而异。

下面的示例演示如何声明和触发事件。请注意，该事件有一个委托类型并用“^”符号声明。

定义事件

```
namespace EventTest
{
    ref class Class1;
    public delegate void SomethingHappenedEventHandler(Class1 ^ sender, Platform::String ^ s);
    public ref class Class1 sealed
    {
    public:
        Class1(){}
        event SomethingHappenedEventHandler ^ OnSomethingHappened;
        void DoSomething()
        {
            //其他一些逻辑处理
            …
            //触发事件
            OnSomethingHappened(this, L"Something happened.");
        }
    };
}
```

调用事件

```
namespace EventClient
{
    using namespace EventTest;
    public ref class Subscriber sealed
    {
    public:
        Subscriber() : eventCount(0)
        {
            //创建拥有事件的类对象
            publisher = ref new EventTest::Class1();
            //订阅事件
            publisher->OnSomethingHappened +=
                ref new EventTest::SomethingHappenedEventHandler(
                this,
                &Subscriber::MyEventHandler);
        }
        //事件的处理方法
        void MyEventHandler(EventTest::Class1 ^ mc, Platform::String ^ msg)
        {
            //处理触发事件要做的事情
```

```
        }
    private:
        EventTest::Class1 ^ publisher;
    };
}
```

21.1.9 自动类型推导 auto

auto 自动类型推导,用于从初始化表达式中推断出变量的数据类型。通过 auto 的自动类型推断,可以大大简化编程工作。在 C++/CX 中的 auto 关键字与 C# 之中的 var 关键字的用法是一样的。auto 自动类型推导可以用到表示任何的一种类型,例如:

```
auto x = 0; //因为 0 是 int 型,所以 x 为 int 类型
auto y = 3.14; //y 为 double 类型
```

auto 也可以让代码更加简单和简洁,例如:

```
vector<int>::const_iterator temp = v.begin();
```

可以写成:

```
auto temp = v.begin();
```

21.1.10 Lambda 表达式

Lambda 表达式是一个匿名函数,它可以包含表达式和语句,并且可用于创建委托或表达式目录树类型。也就是说,一个 lambda 表达式很类似于普通的方法定义的写法,但是它没有函数名,这个方法没有普通方法那样的返回值类型。C++/CX 中 Lambda 表达式的标准形式是:[外部变量](参数)->返回值 {函数体},其中"->返回值"部分可以省略,如果省略则会有返回值类型推导,它的外部变量的传递方式如表 21.5 所示。

表 21.5 外部变量的传递方式

外部变量传递方式	说明
[]	没有定义任何变量
[x, &y]	x 以传值方式传入(默认),y 以引用方式传入
[&]	所有变量都以引用方式导入
[=]	所有变量都以传值方式导入
[&, x]	除 x 以传值方式导入外,其他变量以引用方式导入
[=, &z]	除 z 以引用方式导入外,其他变量以传值方式导入

使用 Lambda 表达式的两个整数相加的语法示例如下:

```
//两个整数相加
auto add = [](int x, int y){ return x + y;};
//调用 Lambda 表达式
int z = add(1, 2);
```

使用 Lambda 表达式的向量求和语法示例如下：

```
//向量
std::vector<int> someList;
int total = 0;
std::for_each(someList.begin(), someList.end(), [&total](int x) {
  total += x;
});
```

21.1.11 集合

在 Windows 运行时里面可以使用 STL (Standard Template Library)集合库，但是不能在对外的接口和方法里面进行传递，所以在 Windows 运行时中需要使用 C++/CX 语法的集合进行信息的传递。Windows 运行时的集合在 collection.h 头文件中定义。

Platform::Collections::Vector 类继承了 Windows::Foundation::Collections::IVector 接口类似于 C++标准库的 std::vector。

Platform::Collections::Map 类继承了 Windows::Foundation::Collections::IMap 接口类似于 C++标准库的 std::map。

VectorView 类继承了 IVectorView 接口和 MapView，类继承了 IMap 接口。

Vector 类继承了 Windows::Foundation::Collections::IObservableVector 接口，Map 类继承了 Windows::Foundation::Collections::IObservableMap 接口，可以通过事件来监视集合的变化。

下面来看一下标准 C++的 Vector 转换成 C++/CX 的 Vector 的示例：

```
#include <collection.h>
#include <vector>
#include <utility>
using namespace Platform::Collections;
using namespace Windows::Foundation::Collections;
using namespace std;
IVector<int>^ Class1::GetInts()
{
    vector<int> vec;
    for(int i = 0; i < 10; i++)
    {
        vec.push_back(i);
    }
    //转换为 Vector
    return ref new Vector<int>(std::move(vec));
}
```

21.2 Windows 运行时组件

Windows 运行时是一个新的框架，目前主要应用在 Windows Phone 8 和 Windows 8 上，正是因为 Windows 运行时才能让 Windows Phone 8 和 Windows 8 实现了代码的共享。

在 Windows Phone 8 中编写普通的 XAML 应用程序不能够像在 Windows 8 里一样使用 C++进行编程，在 Windows Phone 8 里面必须通过 Windows 运行时组件才能够使用 C++编程和使用基于 C++的 API。本节将会基于 Windows 运行时来介绍如何在 Windows Phone 8 上使用 C++编程和调用 C++的 API。

21.2.1　Windows Phone 8 支持的 C++ API

在 Windows Phone 8 中有一部分的 API 支持 C++语言编程，这部分的 API 需要通过 Windows 运行时组件来进行使用，当然这部分的 API 也有对应的 C＃版本，但是 C＃版本不需要使用 Windows 运行时组件来进行调用，直接在项目里面的类库就可以进行调用。Windows Phone 8 支持的 C++编程的 API 主要分为两类：一类是针对 Windows Phone 8 特性的 API；另一类是 Window 运行时公用的 API，也就是在 Windows 8 和 Windows Phone 8 都可以公用的 API。

21.2.2　在项目中使用 Windows 运行时组件

通过 Windows 运行时组件可以让 C++的编程和 C＃的项目无缝地进行整合起来，C＃的 Windows Phone 8 项目调用 C++的 Windows 运行时组件就跟调用 C＃的类库一样方便。下面通过一个实例来介绍如何在项目中使用 Windows 运行时组件。

打开 Visual Studio 2012 新建一个 Visual C++项目，可以看到如图 21.1 所示的 Windows 运行时组件的创建。

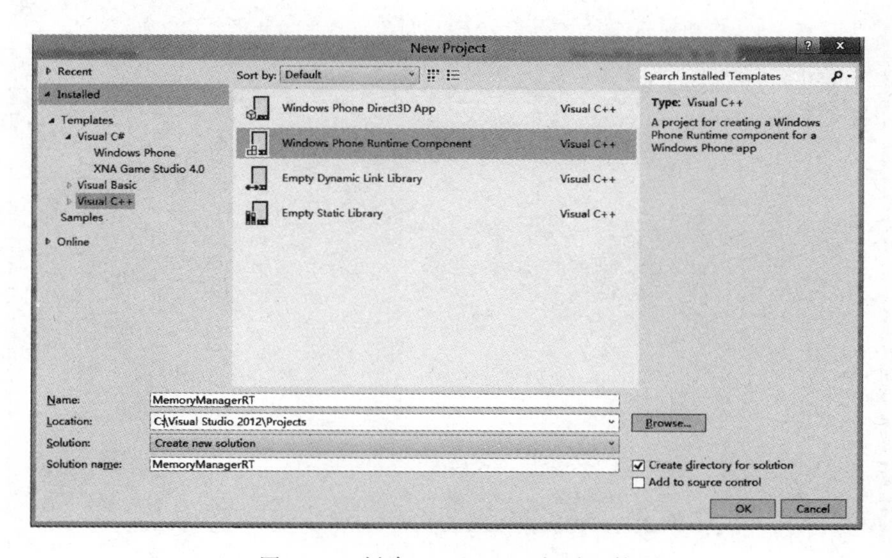

图 21.1　创建 Windows 运行时组件

下面通过一个内存管理的例子（源代码：第 21 章\Examples_21_1）来看一下如何在 Windows 运行时中使用 C++的 API 并且通过 C＃的项目进行调用。

首先创建一个 C#的 Windows Phone 8 应用程序，如图 21.2 所示。

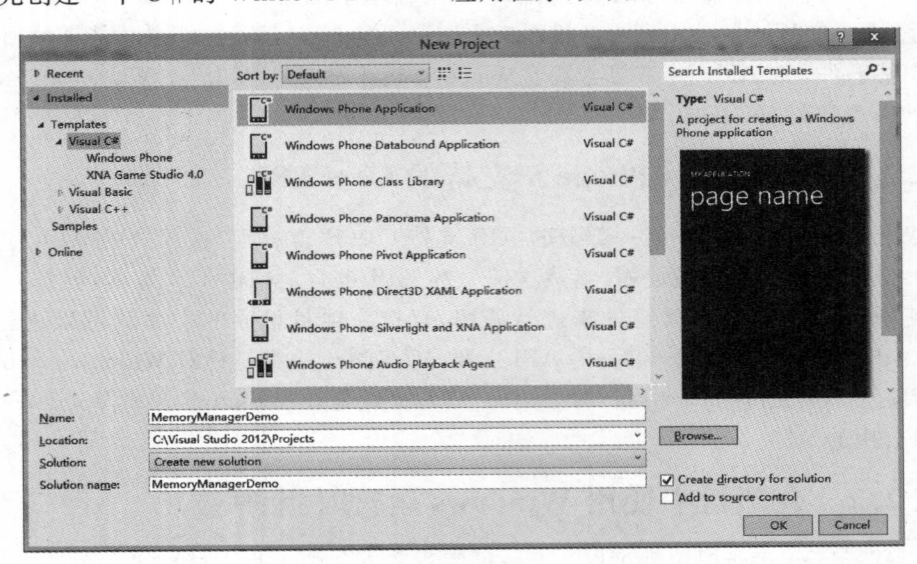

图 21.2 创建一个 Windows Phone 项目

创建一个 C++的 Windows 运行时组件，并在 C#项目中引用该 Windows 运行时组件，如图 21.3 所示。

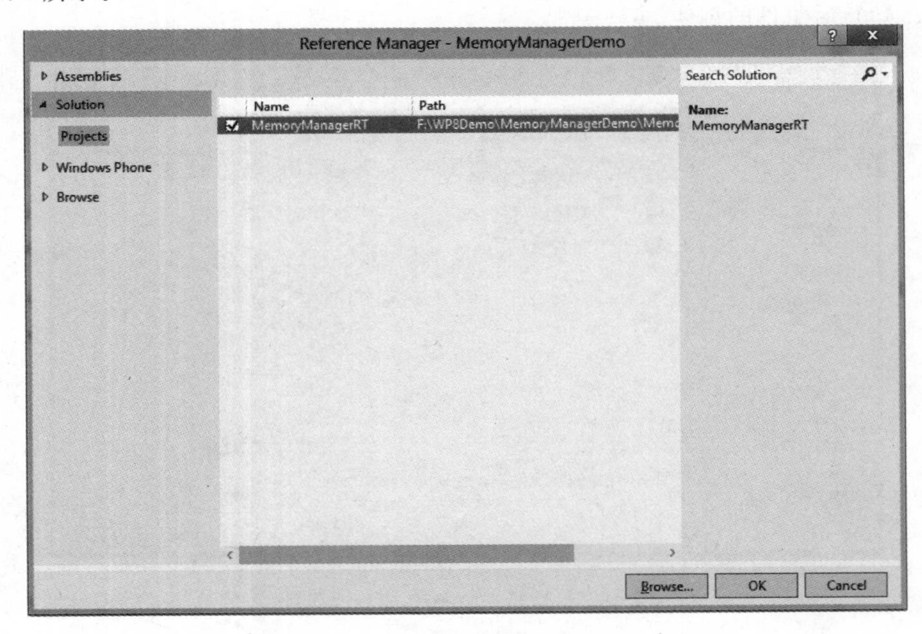

图 21.3 引用 Windows 运行时组件

那么现在整个项目的结构如图 21.4 所示。

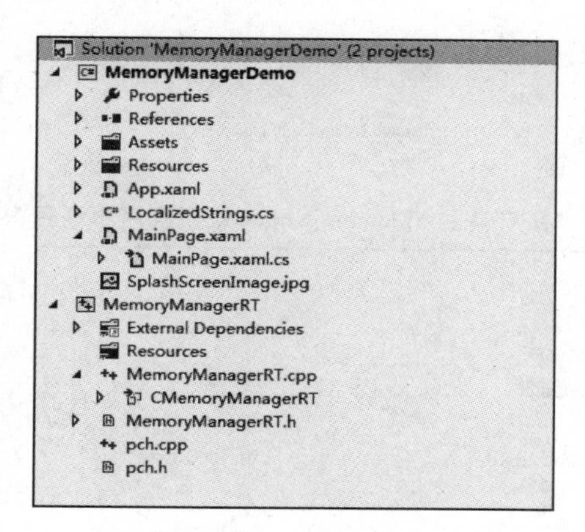

图 21.4 项目结构

在项目结构中，其中 pch.h 和 pch.cpp 是预编译文件，下面来看一下在 Windows 运行时组件中编写的代码。

MemoryManagerRT.h 头文件

```
#pragma once
namespace MemoryManagerRT
{
    public ref class CMemoryManagerRT sealed
    {
    public:
        CMemoryManagerRT();
        static uint64 GetProcessCommittedBytes();//获取程序运行的内存大小
        static uint64 GetProcessCommittedLimit();//获取程序可用内存的大小
    };
}
```

MemoryManagerRT.cpp 源文件

```
#include "pch.h"
#include "MemoryManagerRT.h"
using namespace MemoryManagerRT;
using namespace Platform;
using namespace Windows::Phone::System::Memory;//引入内存管理的 C++空间
CMemoryManagerRT::CMemoryManagerRT()
{
}
//获取程序运行的内存大小
uint64 CMemoryManagerRT::GetProcessCommittedBytes()
{
    return MemoryManager::ProcessCommittedBytes;
```

```
}
//获取程序可用内存的大小
uint64 CMemoryManagerRT::GetProcessCommittedLimit()
{
    return MemoryManager::ProcessCommittedLimit;
}
```

MainPage.xaml.cs 文件：在 C # 的 Windows Phone 应用程序中调用 Windows 运行时的方法

```
using Microsoft.Phone.Controls;
using MemoryManagerRT;

namespace MemoryManagerDemo
{
    public partial class MainPage : PhoneApplicationPage
    {
        public MainPage()
        {
            InitializeComponent();
            tbMemoryUsed.Text = CMemoryManagerRT.GetProcessCommittedBytes().ToString() +
"字节";
            tbMaxMemoryUsed.Text = CMemoryManagerRT.GetProcessCommittedLimit().ToString()
 + "字节";
        }
    }
}
```

程序的运行效果如图 21.5 所示。

图 21.5　查看内存

21.3　使用标准 C++

虽然在 Windows Phone 8 里面的 C++使用 C++/CX 语法，但是对于标准 C++也同样支持。Windows Phone 8 运行时组件支持使用标准 C++的编程，但只是内部支持，在对外调用的方法之中传递的参数必须使用 C++/CX 语法的类型，所以在 Windows 运行时里使用标准 C++编程通常需要做的事情就是将标准 C++的类型与 C++/CX 的类型进行转换。在 Windows 运行时里使用标准 C++通常传入参数的是 C++/CX 的类型，然后在方法里面将 C++/CX 的类型转化为标准 C++的类型，然后进行编写标准 C++的代码，最后将结果再从标准 C++转化为 C++/CX 的类型，然后作为返回值返回。

21.3.1　标准 C++与 C++/CX 的类型自动转换

标准 C++的布尔值、字符和数字基础类型会被自动转换为 C++/CX 对应的类型，所以标准 C++的布尔值、字符和数字基础类型无须再进行转换，可以直接进行赋值和使用。它们的互相对应关系如表 21.2 和表 21.3 所示。

21.3.2　标准 C++与 C++/CX 的字符串的互相转换

标准 C++的字符串和 C++/CX 里面的字符串类型是不一样的类型，在 Windows 运行时中使用的时候需要将两者进行互相转换。在 C++/CX 里是使用 Platform::String 类来表示字符串的类型，在 Windows 运行时的接口和方法中，需要使用 Platform::String 来作为字符串参数的传递。使用标准 C++的字符串类型如 wstring 或者 string 的时候，可以将 Platform::String 与标准的 C++的字符串进行互相的转换。

String 类型表示的是 char16 的字符串，可以直接通过字符串的赋值来进行构造也可以使用标准 C++的 wchar_t * 指针进行构造。

```
//使用 wchar_t * 和 wstring 初始化一个 String
wchar_t msg[] = L"Test";
String^ str8 = ref new String(msg);
std::wstring wstr1(L"Test");
String^ str9 = ref new String(wstr1.c_str());
String^ str10 = ref new String(wstr1.c_str(), wstr1.length());
```

String 提供了相关的方法来操作字符串，其中可以使用 String::Data()方法来返回一个 String^对象的 wchar_t * 指针。下面来看一下 String 类型和标准 C++的 wstring 进行互相的转换。

```
#include <string>
using namespace std;
using namespace Platform;
…
    //创建一个 String^ v 变量
```

```
    String^ str1 = "AAAAAAAA";
    //使用 str1 的值创建一个 wstring 变量
    wstring ws1( str1->Data() );
    //操作 wstring 的值
    wstring replacement( L"BBB" );
    ws1 = ws1.replace ( 1, 3, replacement );
    //这时候 ws1 的值是 L"ABBBAAAA"
    //把 wstring 转化为 String^类型
    str1 = ref new String( ws1.c_str() );
…
```

21.3.3 标准 C++与 C++/CX 的数组的互相转换

标准 C++的数组类型 std::vector 和 C++/CX 中的数组类型 Platform::Array 在使用过程中经常会需要进行转换，不过 Platform::Array 类型既不像 std::vector 那样高效也不像它那样功能强大，因此，一般原则是，应避免在对数组元素执行大量操作的内部代码中使用该类型。下面来看一下这两者之间如何进行转换。

C++/CX 数组转化为标准 C++数组

```
#include <vector>
#include <collection.h>
using namespace Platform;
using namespace std;
using namespace Platform::Collections;
void ArrayConversions(const Array<int>^ arr)
{
    //构建一个标准 C++数组的两种方式
    vector<int> v1(begin(arr), end(arr));
    vector<int> v2(arr->begin(), arr->end());
    //使用循环的方式构建数组
    vector<int> v3;
    for(int i : arr)
    {
        v3.push_back(i);
    }
}
```

标准 C++数组转化为 C++/CX 数组

```
Array<int>^ App::GetNums()
{
    int nums[] = {0,1,2,3,4};
    return ref new Array<int>(nums, 5);
}
```

21.3.4 在 Windows 运行时组件中使用标准 C++

下面给出 MD5 加密的示例：该示例通过 Windows 运行时组件来使用标准 C++编写的

MD5 加密算法，然后在 Windows Phone 的项目里面调用使用这个 MD5 加密算法来加密字符串。

代码清单 21-2：MD5 加密（源代码：第 21 章\Examples_21_2）

md5.h 头文件

```
#ifndef BZF_MD5_H
#define BZF_MD5_H
#include <string>
#include <iostream>
class MD5
{
public:
  typedef unsigned int size_type;
//默认构造方法
  MD5();
//计算 MD5 的字符串
  MD5(const std::string& text);
//MD5 块的更新操作
  void update(const unsigned char * buf, size_type length);
  void update(const char * buf, size_type length);
//结束 MD5 的操作
  MD5& finalize();
  std::string hexdigest() const;
  std::string md5() const;
  friend std::ostream& operator <<(std::ostream&, MD5 md5);
private:
  void init();
  typedef unsigned char uint1;                  //8bit
  typedef unsigned int uint4;                   //32bit
  enum {blocksize = 64};
  //采用 MD5 算法块
  void transform(const uint1 block[blocksize]);
  //解码输入到输出
  static void decode(uint4 output[], const uint1 input[], size_type len);
  //解码输入到输出
  static void encode(uint1 output[], const uint4 input[], size_type len);
  bool finalized;
  uint1 buffer[blocksize];                      //数据
  uint4 count[2];                               //64bit 计算高低位
  uint4 state[4];                               //状态
  uint1 digest[16];                             //结果
  //低位的逻辑操作
  static inline uint4 F(uint4 x, uint4 y, uint4 z);
  static inline uint4 G(uint4 x, uint4 y, uint4 z);
  static inline uint4 H(uint4 x, uint4 y, uint4 z);
  static inline uint4 I(uint4 x, uint4 y, uint4 z);
  static inline uint4 rotate_left(uint4 x, int n);
  static inline void FF(uint4 &a, uint4 b, uint4 c, uint4 d, uint4 x, uint4 s, uint4 ac);
  static inline void GG(uint4 &a, uint4 b, uint4 c, uint4 d, uint4 x, uint4 s, uint4 ac);
```

```
    static inline void HH(uint4 &a, uint4 b, uint4 c, uint4 d, uint4 x, uint4 s, uint4 ac);
    static inline void II(uint4 &a, uint4 b, uint4 c, uint4 d, uint4 x, uint4 s, uint4 ac);
};
#endif
```

创建一个 Windows 运行时组件，命名为 PhoneRuntimeComponent1，在组件里面添加上面的两个 md5 文件的源代码，然后编写 Windows 运行时的对外的方法，代码如下：

PhoneRuntimeComponent1.h 头文件

```
#pragma once
using namespace std;
namespace PhoneRuntimeComponent1
{
    public ref class CPhoneRuntimeComponent1 sealed
    {
    public:
        CPhoneRuntimeComponent1();
        Platform::String^ GetMd5(Platform::String^text);
    };
}
```

PhoneRuntimeComponent1.cpp 头文件

```
#include "pch.h"
#include "PhoneRuntimeComponent1.h"
#include "IntArray.h"
#include "md5.h"
#include <string>
#include <vector>
#include <iostream>
using namespace PhoneRuntimeComponent1;
using namespace Platform;
using namespace std;
CPhoneRuntimeComponent1::CPhoneRuntimeComponent1()
{
}
Platform::String^ CPhoneRuntimeComponent1::GetMd5(Platform::String^text)
{
    ///先把 Platform::String 转化为 std::string
    //转化为一个 std::wstring 对象
    std::wstring strr(text->Data());
    //获取区域设置
    std::locale const loc("");
    wchar_t const* from = strr.c_str();
    std::size_t const len = strr.size();
    std::vector<char> buffer(len + 1);
    //字符转换为类型 char 对应的字符在本机字符集
    std::use_facet<std::ctype<wchar_t>>(loc).narrow(from, from + len, '_', &buffer[0]);
    std::string str =  std::string(&buffer[0], &buffer[len]);
    //调用 MD5 加密算法
```

```
    MD5 md5(str);
    string result = md5.md5();
    //把 std::string 转化为 Platform::String 然后返回
    std::wstring wstr;
    wstr.resize(result.size() + 1);
    size_t charsConverted;
    //将多字节字符序列转换为相应的宽字符序列
    errno_t err = ::mbstowcs_s( &charsConverted, (wchar_t * )wstr.data(), wstr.size(),
result.data(), result.size() );
    //清除最后一个字符
    wstr.pop_back();
    return ref new String( wstr.c_str() );
}
```

在 Windows Phone 项目中，引入编写完成的 Windows 运行时组件，然后通过 Windows Phone 的应用程序用 C# 代码来调用 Windows 运行时组件，从而实现了使用标准 C++的 MD5 加密算法。

MainPage.xaml 文件

```
<Grid x:Name = "ContentPanel" Grid.Row = "1" Margin = "12,0,12,0">
    <StackPanel>
    <TextBlock Text = "请输入字符串:"/>
        <TextBox x:Name = "tbStr"/>
        <Button Content = "获取 MD5 字符串" Click = "Button_Click_1"/>
        <TextBlock x:Name = "tbMd5"/>
        </StackPanel>
</Grid>
```

MainPage.xaml.cs 文件

```
using System.Windows;
using Microsoft.Phone.Controls;
using PhoneRuntimeComponent1;//引入 Windows 运行时组件
namespace CppDLLDemo
{
    public partial class MainPage : PhoneApplicationPage
    {
        public MainPage()
        {
            InitializeComponent();
        }
        private void Button_Click_1(object sender, RoutedEventArgs e)
        {
            //调用 Windows 运行时组件的类
            CPhoneRuntimeComponent1 rt = new CPhoneRuntimeComponent1();
            //调用 Windows 运行时组件的加密方法
            tbMd5.Text = rt.GetMd5(tbStr.Text);
        }
    }
}
```

程序的运行效果如图 21.6 所示。

图 21.6　C++ MD5

21.4　Direct3D

在 Windows Phone 8 里面支持 Direct3D 游戏的开发，Direct3D 游戏是需要完全使用 C++ 来进行开发的，以前在 Windows Phone 7 里面的 XNA 游戏开发框架不能够在 Windows Phone 8 的 SDK 上面进行开发了，当然 Windows Phone 8 的手机依然是会兼容 XNA 的游戏的。

21.4.1　Direct3D 简介

Direct3D 是一种低层图形 API，它能让开发者利用 3D 硬件加速来渲染 3D 世界。开发者可以把 Direct3D 看作是应用程序和图形设备之间的中介。通过利用 Direct3D API 编程，能够屏蔽许多底层实现的技术细节，缩短开发周期。图 21.7 显示了 Direct3D，HAL (Hardware Abstraction Layer，硬件抽象层)及硬件之间的关系。

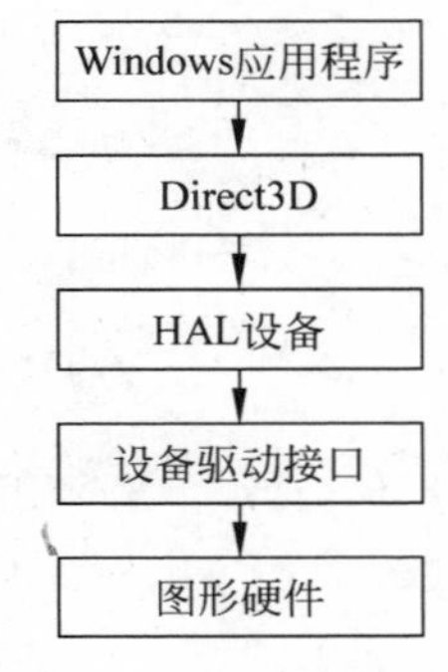

图 21.7　Direct3D 层次

图 21.7 中 Direct3D 表示的是 Direct3D 中已定义的，供程序员使用的 Direct3D 接口和函数的集合。这些接口和函数代表了当前版本的 Direct3D 所支持的全部特性。注意：仅仅因为 Direct3D 支持某种特性，并不意味着所使用的图形硬件(显卡)也能支持它。

如图 21.7 所示，在 Direct3D 和图形设备之间有一层中介——硬件抽象层(Hardware Abstraction Layer，HAL)。Direct3D 不能直接作用于图形设备，Direct3D 要求设备制造商实现 HAL。HAL 是一组指示设备执行某种操作的特殊设备代码的集合。用这种方法，Direct3D 避免了必须去了解某个设备的特殊细节，使它能够独立于硬件设备。

设备制造商在 HAL 中实现他们的产品所支持的所有特性。HAL 将不会实现那些 Direct3D 支持但硬件产品不支持的特性。调用一个 HAL 中没有实现的 Direct3D 的函数将会出错，除非它是顶点处理操作，因为这个功能可以由软件模拟来实现。因此当使用某些仅由市面上少数显卡所支持的高级特性时，必须检测一下设备是否支持。

21.4.2 Direct3D 重要概念

1）近似顶点

一个场景是多个物体或模型的集合。一个物体可以用三角形网格来表示，3D 世界中最基本的图元就是三角形，一个多边形的两边相交的点叫做顶点。为了描述一个三角形，通常指定三个点的位置来对应三角形的三个顶点，这样就能够很明确表示出这个三角形。3D 物体中的三角形经常会有许多共用顶点。为了解决复杂精细场景重复顶点占用更多的渲染带宽，在创建一个顶点列表的同时也创建一个索引列表。顶点列表包含所有不重复的顶点，索引列表中则用顶点列表中定义的值来表示每一个三角形的构造方式。通常每个顶点都包含如下信息：x，y，z 坐标值、颜色值、用于计算灯光的法线和纹理坐标（u，v）。

2）表面

是一个像素点阵，主要用来存储 2D 图形数据。表面数据就像一个矩阵，像素数据实际上是存储在线性数组里面。

3）多重采样（MultiSampling）

由于使用像素来表示图像，在显示时会出现锯齿状。MultiSampling 就是使图像变得平滑的技术。它的最普通的用法就是全屏抗锯齿。D3DMULTISAMPLE_TYPE 枚举类型能使开发者制定全屏抗锯齿的质量等级。

4）交换链和页面交换

Direct3D 通常建立 2～3 个页面组成一个集合，即为交换链，通常由 IDirect3DSwapChain 接口来表示。交换链和页面交换技巧被用在使两帧动画之间过渡更加平滑。通常由 Direct3D 自己去管理。

5）深度缓冲

深度缓冲是一个表面，但它不是用来存储图像数据而是用来记录像素的深度信息。深度缓冲为每一个像素计算深度值，并进行深度测试。通过深度测试可以知道哪个像素离摄像机近从而把它画出来。这样就可以只绘制最靠近摄像机的像素，被遮住的像素就不会被画出来。

6）顶点处理

顶点是 3D 图形学的基础，它能够通过两种不同的方法来处理：一种是软件顶点处理（Software Vertex Processing），另一种是硬件顶点处理（Hardware Vertex Processing）。前者总是被支持而且永久可用，后者要显卡硬件支持顶点处理才可用。使用硬件顶点处理总是首选，因为它比软件方式更快，而且不占用 CPU 资源。如果一块显卡支持硬件顶点处理的话，也就是说它支持硬件几何变换和光照计算。

7）设备能力

Direct3D 支持的每一种特性都对应于 D3DCAPS9 结构的一个数据成员，初始化一个 D3DCAPS9 实例应该以你的设备实际支持的特性为基础。因此，在我们的应用程序里，我们能够通过检测 D3DCAPS9 结构中相对应的某一成员来检测设备是否支持这一特性。

8）ARGB

ARGB 表示一种色彩模式，也就是 RGB 色彩模式附加上 Alpha（透明度）通道，常见于 32 位位图的存储结构。

9）D3DPOOL

D3DPOOL 表示内存池。

21.4.3 创建一个 Direct3D 项目

打开 Visual Studio 新建一个 Visual C++的 Direct3D 项目如图 21.8 所示。Direct3D 的项目结构如图 21.9 所示。

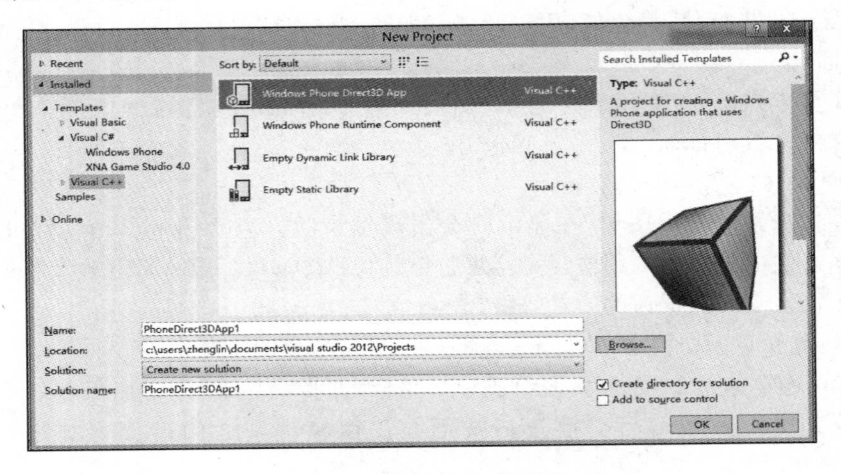

图 21.8 创建 Direct3D 项目

下面来看一下 Direct3D 项目中的文件结构（源代码：第 21 章\Examples_21_3），Direct3DBase 类（Direct3DBase. h，Direct3DBase. cpp）是一个抽象的基类定义了游戏初始化等的流程，Direct3DDemo（Direct3DDemo. h，Direct3DDemo. cpp）类是游戏的主程序类，继承了 Direct3DBase 基类并且实现了里面的一些方法，CubeRenderer（CubeRenderer. h，CubeRenderer. cpp）类是实现 3D 立方体的渲染类，pch. cpp 和 pch. h 是程序的预编译文件，SimplePixelShader. hlsl 和 SimpleVertexShader. hlsl 是 HLSL（High Level Shading

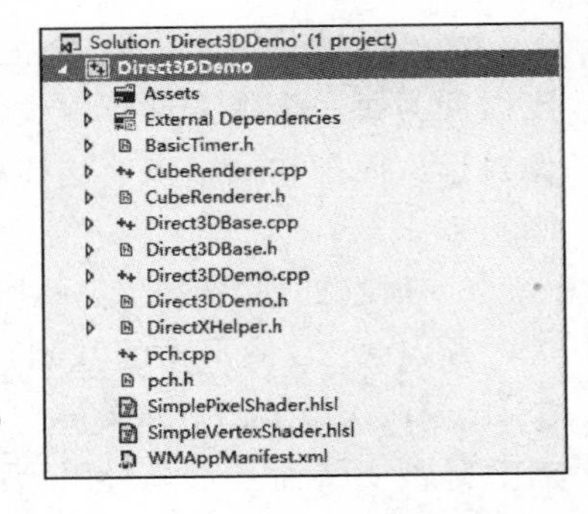

图 21.9 Direct3D 项目结构

Language,高级渲染语言)文件是 Direct3D 里面的着色器。

Direct3DBase.cpp 源文件

```
#include "pch.h"
#include "Direct3DBase.h"
using namespace Microsoft::WRL;
using namespace Windows::UI::Core;
using namespace Windows::Foundation;
//构造方法
Direct3DBase::Direct3DBase()
{
}
//初始化所需要的 Direct3D 资源
void Direct3DBase::Initialize(CoreWindow^ window)
{
    m_window = window;

    CreateDeviceResources();
    CreateWindowSizeDependentResources();
}
//创建依赖于设备的资源
void Direct3DBase::CreateDeviceResources()
{
    //创建的标识符
    UINT creationFlags = D3D11_CREATE_DEVICE_BGRA_SUPPORT;
#if defined(_DEBUG)
    //Debug 测试模式标识符
    creationFlags |= D3D11_CREATE_DEVICE_DEBUG;
#endif
    //支持的 DirectX 硬件数组
    D3D_FEATURE_LEVEL featureLevels[] =
    {
        D3D_FEATURE_LEVEL_11_1,
        D3D_FEATURE_LEVEL_11_0,
        D3D_FEATURE_LEVEL_10_1,
        D3D_FEATURE_LEVEL_10_0,
        D3D_FEATURE_LEVEL_9_3,
        D3D_FEATURE_LEVEL_9_2,
        D3D_FEATURE_LEVEL_9_1
    };
    //创建 DX11 API 的设备对象,并得到相应的内容
    ComPtr<ID3D11Device> device;
    ComPtr<ID3D11DeviceContext> context;
    DX::ThrowIfFailed(
        D3D11CreateDevice(
            nullptr,                        //指定为空来使用默认的适配器
            D3D_DRIVER_TYPE_HARDWARE,
            nullptr,
            creationFlags,                  //Direct2D 和 debug 调试的标识符
```

```
            featureLevels,                      //应用支持的功能列表
            ARRAYSIZE(featureLevels),           //功能列表的数目
            D3D11_SDK_VERSION,                  //设置版本号
            &device,                            //返回创建的 Direct3D 设备
            &m_featureLevel,                    //返回设备创建创建的功能特色
            &context                            //返回设备的直接上下文
            )
        );
    //获取 DirectX11.1 设备
    DX::ThrowIfFailed(
        device.As(&m_d3dDevice)
        );
    //获取上下文
    DX::ThrowIfFailed(
        context.As(&m_d3dContext)
        );
}
//在窗口 SizeChanged 事件,给所有的资源分配内存
void Direct3DBase::CreateWindowSizeDependentResources()
{
    //获取窗口的矩形边界,保存下来,防止 SizeChanged 事件重复地创建
    m_windowBounds = m_window->Bounds;
    //如果已经存在的交换链,改变它的大小
    if(m_swapChain != nullptr)
    {
        DX::ThrowIfFailed(
            m_swapChain->ResizeBuffers(2, 0, 0, DXGI_FORMAT_B8G8R8A8_UNORM, 0)
            );
    }
    //否则创建一个新的
    else
    {
        //创建交换链的一些描述
        DXGI_SWAP_CHAIN_DESC1 swapChainDesc = {0};
        swapChainDesc.Width = 0;                                    //使用自动大小
        swapChainDesc.Height = 0;
        swapChainDesc.Format = DXGI_FORMAT_B8G8R8A8_UNORM;          //交换链格式
        swapChainDesc.Stereo = false;
        swapChainDesc.SampleDesc.Count = 1;                         //不要使用多重采样
        swapChainDesc.SampleDesc.Quality = 0;
        swapChainDesc.BufferUsage = DXGI_USAGE_RENDER_TARGET_OUTPUT;
        swapChainDesc.BufferCount = 1;                              //WP8: 单缓冲
        swapChainDesc.Scaling = DXGI_SCALING_STRETCH;               //WP8: 须具有伸缩性
        swapChainDesc.SwapEffect = DXGI_SWAP_EFFECT_DISCARD;
                                                //WP8: 必须使用和 windows 8 不同的交换效果
        swapChainDesc.Flags = 0;
        //交换链描述配置好之后,必须创建一个相同的适配器作为我们的 D3D 设备
        //首先,检索的基础 DXGI 设备的 D3D 设备
        ComPtr<IDXGIDevice1> dxgiDevice;
        DX::ThrowIfFailed(
```

```
            m_d3dDevice.As(&dxgiDevice)
            );
        //确定此设备上运行的物理适配器(GPU或卡)
        ComPtr<IDXGIAdapter> dxgiAdapter;
        DX::ThrowIfFailed(
            dxgiDevice->GetAdapter(&dxgiAdapter)
            );
        //获得工厂对象并创建它
        ComPtr<IDXGIFactory2> dxgiFactory;
        DX::ThrowIfFailed(
            dxgiAdapter->GetParent(
                __uuidof(IDXGIFactory2),
                &dxgiFactory
                )
            );
        Windows::UI::Core::CoreWindow^ p = m_window.Get();
        //创建这个窗口的交换链
        DX::ThrowIfFailed(
            dxgiFactory->CreateSwapChainForCoreWindow(
                m_d3dDevice.Get(),
                reinterpret_cast<IUnknown*>(p),
                &swapChainDesc,
                nullptr,                                    //允许所有的展现
                &m_swapChain
                )
            );
        //确保DXGI排队不超过一帧的时间,减少了延迟,并确保应用程序将只呈现在每个垂直同
        //步,可减少功耗
        DX::ThrowIfFailed(
            dxgiDevice->SetMaximumFrameLatency(1)
            );
}
//获得这个窗口,这将是最终的3D渲染目标的后备缓冲
ComPtr<ID3D11Texture2D> backBuffer;
DX::ThrowIfFailed(
    m_swapChain->GetBuffer(
        0,
        __uuidof(ID3D11Texture2D),
        &backBuffer
        )
    );
//创建一个视图界面上使用绑定的渲染目标
DX::ThrowIfFailed(
    m_d3dDevice->CreateRenderTargetView(
        backBuffer.Get(),
        nullptr,
        &m_renderTargetView
        )
    );
//在helper类,方便使用高速缓存渲染目标尺寸
```

```
    D3D11_TEXTURE2D_DESC backBufferDesc;
    backBuffer->GetDesc(&backBufferDesc);
    m_renderTargetSize.Width  = static_cast<float>(backBufferDesc.Width);
    m_renderTargetSize.Height = static_cast<float>(backBufferDesc.Height);
    //创建一个深度/模板缓冲的描述
    CD3D11_TEXTURE2D_DESC depthStencilDesc(
        DXGI_FORMAT_D24_UNORM_S8_UINT,
        backBufferDesc.Width,
        backBufferDesc.Height,
        1,
        1,
        D3D11_BIND_DEPTH_STENCIL);
    //分配的 2D 的表面的深度/模板缓冲区
    ComPtr<ID3D11Texture2D> depthStencil;
    DX::ThrowIfFailed(
        m_d3dDevice->CreateTexture2D(
            &depthStencilDesc,
            nullptr,
            &depthStencil
            )
        );
    //在此表面上使用绑定创建一个 DepthStencil 的点
    DX::ThrowIfFailed(
        m_d3dDevice->CreateDepthStencilView(
            depthStencil.Get(),
            &CD3D11_DEPTH_STENCIL_VIEW_DESC(D3D11_DSV_DIMENSION_TEXTURE2D),
            &m_depthStencilView
            )
        );
    //创建一个视口描述符的窗口大小
    CD3D11_VIEWPORT viewPort(
        0.0f,
        0.0f,
        static_cast<float>(backBufferDesc.Width),
        static_cast<float>(backBufferDesc.Height)
        );
    //设置当前视口中使用的描述符
    m_d3dContext->RSSetViewports(1, &viewPort);
}
void Direct3DBase::UpdateForWindowSizeChange()
{
    if (m_window->Bounds.Width  != m_windowBounds.Width ||
        m_window->Bounds.Height != m_windowBounds.Height)
    {
        m_renderTargetView = nullptr;
        m_depthStencilView = nullptr;
        CreateWindowSizeDependentResources();
    }
}
void Direct3DBase::ReleaseResourcesForSuspending()
{
    m_swapChain = nullptr;
```

```
        m_renderTargetView = nullptr;
        m_depthStencilView = nullptr;
}
void Direct3DBase::Present()
{
        //第一个参数指示 DXGI 阻塞,直到垂直同步,将应用程序睡觉,直到下一个垂直同步,这将确保
        //不浪费任何渲染帧的周期,将永远不会被显示在屏幕上
        HRESULT hr = m_swapChain->Present(1, 0);
        //如果设备已删除或者断开连接或升级驱动程序,我们必须完全重新初始化渲染
        if (hr == DXGI_ERROR_DEVICE_REMOVED || hr == DXGI_ERROR_DEVICE_RESET)
        {
            Initialize(m_window.Get());
        }
        else
        {
            DX::ThrowIfFailed(hr);
        }
}
```

Direct3DDemo.cpp 源文件

```
#include "pch.h"
#include "Direct3DDemo.h"
#include "BasicTimer.h"

using namespace Windows::ApplicationModel;
using namespace Windows::ApplicationModel::Core;
using namespace Windows::ApplicationModel::Activation;
using namespace Windows::UI::Core;
using namespace Windows::System;
using namespace Windows::Foundation;
using namespace Windows::Graphics::Display;
//构造方法
Direct3DDemo::Direct3DDemo() :
    m_windowClosed(false)
{
}
//初始化
void Direct3DDemo::Initialize(CoreApplicationView^ applicationView)
{
    applicationView->Activated +=
        ref new TypedEventHandler<CoreApplicationView^, IActivatedEventArgs^>(this,
&Direct3DDemo::OnActivated);
    CoreApplication::Suspending +=
        ref new EventHandler<SuspendingEventArgs^>(this, &Direct3DDemo::OnSuspending);
    CoreApplication::Resuming +=
        ref new EventHandler<Platform::Object^>(this, &Direct3DDemo::OnResuming);
    m_renderer = ref new CubeRenderer();
}
//设置窗口事件
void Direct3DDemo::SetWindow(CoreWindow^ window)
{
    window->Closed +=
```

```
        ref new TypedEventHandler < CoreWindow ^, CoreWindowEventArgs ^ >(this, &Direct3DDemo::
OnWindowClosed);
    window -> PointerPressed +=
        ref new TypedEventHandler < CoreWindow ^, PointerEventArgs ^ >(this, &Direct3DDemo::
OnPointerPressed);
    window -> PointerMoved +=
        ref new TypedEventHandler < CoreWindow ^, PointerEventArgs ^ >(this, &Direct3DDemo::
OnPointerMoved);
        m_renderer -> Initialize(CoreWindow::GetForCurrentThread());
}
//加载
void Direct3DDemo::Load(Platform::String ^ entryPoint)
{
}
//运行
void Direct3DDemo::Run()
{
    BasicTimer ^ timer = ref new BasicTimer();
    while (!m_windowClosed)
    {
        //更新时间
        timer -> Update();
        CoreWindow::GetForCurrentThread() -> Dispatcher -> ProcessEvents(CoreProcessEventsOption
::ProcessAllIfPresent);
        //更新
        m_renderer -> Update(timer -> Total, timer -> Delta);
        //渲染
        m_renderer -> Render();
        //呈现
        m_renderer -> Present();     //此呼叫被同步到显示帧速率。This call is synchronized
                                     //to the display frame rate
    }
}
//反初始化
void Direct3DDemo::Uninitialize()
{
}
//关闭窗口
void Direct3DDemo::OnWindowClosed(CoreWindow ^ sender, CoreWindowEventArgs ^ args)
{
    m_windowClosed = true;
}
//激活
void Direct3DDemo::OnActivated(CoreApplicationView ^ applicationView, IActivatedEventArgs ^ args)
{
    CoreWindow::GetForCurrentThread() -> Activate();
}
//挂起程序
void Direct3DDemo::OnSuspending(Platform::Object ^ sender, SuspendingEventArgs ^ args)
{
    SuspendingDeferral ^ deferral = args -> SuspendingOperation -> GetDeferral();
    m_renderer -> ReleaseResourcesForSuspending();
```

```
    deferral->Complete();
}
//从挂起状态恢复
void Direct3DDemo::OnResuming(Platform::Object^ sender, Platform::Object^ args)
{
    m_renderer->CreateWindowSizeDependentResources();
}
//单击
void Direct3DDemo::OnPointerPressed(CoreWindow^ sender, PointerEventArgs^ args)
{
}
//释放单击
void Direct3DDemo::OnPointerReleased(CoreWindow^ sender, PointerEventArgs^ args)
{
}
//单击移动
void Direct3DDemo::OnPointerMoved(CoreWindow^ sender, PointerEventArgs^ args)
{
}
//创建视图
IFrameworkView^ Direct3DApplicationSource::CreateView()
{
    return ref new Direct3DDemo();
}
//游戏主方法
[Platform::MTAThread]
int main(Platform::Array<Platform::String^>^)
{
    auto direct3DApplicationSource = ref new Direct3DApplicationSource();
    CoreApplication::Run(direct3DApplicationSource);
    return 0;
}
```

程序的运行效果如图 21.10 所示。

图 21.10　3D 立方体

开发实例篇

通过了开发技能篇的学习，读者应该掌握了一定的Windows Phone的开发技术，但是要真正地掌握这些技术必须要通过不断的实例来训练才能熟能生巧，才能够更加深入地去研究这些技术。本篇是以应用实例的形式来介绍Windows Phone的知识，每个实例都会对应到开发技术篇的若干个技术点，本篇实例涉及的知识点不再做详细的讲解，大部分的知识点在开发技术篇都有说明。读者对本篇的学习可以从实际的应用开发的角度去学习；就是准备去开发一个实际应用的时候，要先问清楚自己怎么去实现这些功能，用怎样的技术去实现这些功能，思路是怎样的，然后再去编码实现。对于业务复杂的应用的开发，更加需要开发前期的准备工作，如详细设计文档等。本篇的实例都是比较简单易懂，是Windows Phone技术训练的典型例子，有一定Windows Phone编程基础的人都可以看懂，其中快递100应用是一个在Windows Phone应用市场上发布过的应用程序，这也是本篇一个很好的应用实例的例子。这些实例涉及Windows Phone的多方面的技术知识，是对前面所讲的知识的一个综合的运用，在学习本篇的过程中也是对前面所学的Windows Phone的知识的一个加强训练。

本篇包含了第22章普通应用实例，第23章网络应用实例和第24章记账本应用，在这两章当中都是综合地使用了Windows Phone的各种开发技术，是在掌握了一定的Windows Phone的编程知识后进行训练的好实例。其中，第22章是不需要网络支持就可以正常运行的应用实例，由浅入深地介绍了5个应用实例的开发，可以一步一步地来加深对Windows Phone的编程技术的理解；第23章介绍的是互联网的应用的开发，也就是应用实例必须在手机接入网络的情况下才能够正常地运行和操作，互联网应用是当前智能手机应用的热点。第23章就是从这个热点出发，演示了4个不同角度的互联网手机应用的开发。第24章是通过讲解一个记账本应用的开发来讲解一个相对完善的应用程序的开发。

开发实例篇包括了以下的章节：

第22章　普通应用实例

介绍非网络模式下的应用开发。

第23章　网络应用实例

介绍网络模式下的应用开发。

第24章　记账本应用

介绍记账本的应用开发。

通过本篇的学习，将会更加深刻地理解和运用Windows Phone的技术知识来开发应用，同时也会帮助读者学会综合运用Windows Phone的各种技术，打造出属于自己的Windows Phone应用程序。

普通应用实例

本章通过一些 Windows Phone 的应用实例来综合地应用前文介绍的 Windows Phone 编程的基本知识，主要涉及 Windows Phone 的 Silverlight 编程的各个方面。这些实例简单易懂，并且非常通用，是 Windows Phone 编程入门训练的最佳实践。

22.1 时钟

实例动画时钟：使用动画编程创建一个简易的时钟。

实例说明：本实例主要是使用动画编程的知识，涉及使用 Storyboard 类启动动画和控制动画的运动的知识，时钟的时、分、秒针是由 3 个矩形的图形组成，每个矩形的运动就是按照该矩形表示的时钟的针来定义运动规律，最后使得整个动画的效果反应的就是时钟的运动规律。

代码清单 22-1：动画时钟(源代码：第 22 章\Examples_22_1)

MainPage. xaml 文件主要代码

```
<Grid Width = "400" Height = "400">
    <Ellipse x:Name = "ClockFaceEllipse" Stroke = "Blue" StrokeThickness = "10" >
    </Ellipse>
    <Canvas x:Name = "ClockHandsCanvas">
        <Canvas.Resources>
            <Storyboard x:Name = "ClockStoryboard">
                <DoubleAnimation x:Name = "HourAnimation"
                                 Storyboard.TargetName = "HourHandTransform"
                                 Storyboard.TargetProperty  = "Angle"
                                 Duration = "12:0:0" RepeatBehavior = "Forever" To = "360" />
                <DoubleAnimation x:Name = "MinuteAnimation"
                                 Storyboard.TargetName = "MinuteHandTransform"
                                 Storyboard.TargetProperty = "Angle"
                                 Duration = "1:0:0" RepeatBehavior = "Forever" To = "360" />
                <DoubleAnimation x:Name = "SecondAnimation"
                                 Storyboard.TargetName = "SecondHandTransform"
                                 Storyboard.TargetProperty = "Angle"
                                 Duration = "0:1:0" RepeatBehavior = "Forever" To = "360" />
```

```
            </Storyboard>
        </Canvas.Resources>
        <!-- 表示秒针的矩形 -->
        <Rectangle Width="4" Height="230" RadiusX="2" RadiusY="2"
Canvas.Left="198" Canvas.Top="20" Fill="Red">
            <Rectangle.RenderTransform>
                <TransformGroup>
                    <RotateTransform CenterX="2" CenterY="180" x:Name="SecondHandTransform" />
                </TransformGroup>
            </Rectangle.RenderTransform>
        </Rectangle>
        <!-- 表示分针的矩形 -->
        <Rectangle Width="8" Height="145" RadiusX="3" RadiusY="3"  Canvas.Left="196"
Canvas.Top="60" Fill="White">
            <Rectangle.RenderTransform>
                <TransformGroup>
                    <RotateTransform CenterX="4" CenterY="140" x:Name="MinuteHandTransform" />
                </TransformGroup>
            </Rectangle.RenderTransform>
        </Rectangle>
        <!-- 表示时针的矩形 -->
        <Rectangle Width="10" Height="105" RadiusX="5" RadiusY="5"  Canvas.Left="195"
Canvas.Top="100" Fill="Yellow">
            <Rectangle.RenderTransform>
                <TransformGroup>
                    <RotateTransform CenterX="5" CenterY="100" x:Name="HourHandTransform" />
                </TransformGroup>
            </Rectangle.RenderTransform>
        </Rectangle>
        <Ellipse Width="10" Height="10"  Canvas.Left="195" Canvas.Top="195" />
    </Canvas>
</Grid>
```

MainPage.xaml.cs 文件代码

```
using System;
using System.Windows;
using Microsoft.Phone.Controls;
namespace Clock1
{
    public partial class MainPage : PhoneApplicationPage
    {
        public MainPage()
        {
            InitializeComponent();
        }
        private void PhoneApplicationPage_Loaded(object sender, RoutedEventArgs e)
        {
            var now = DateTime.Now;
            //计算角度
            //时
```

```
            double hourAngle = ((float)now.Hour) / 12 * 360 + now.Minute / 2;
            //分
            double minuteAngle = ((float)now.Minute) / 60 * 360 + now.Second / 10;
            //秒
            double secondAngle = ((float)now.Second) / 60 * 360;
            //时针的角度运动规律
            HourAnimation.From = hourAngle;
            HourAnimation.To = hourAngle + 360;
            //分针的角度运动规律
            MinuteAnimation.From = minuteAngle;
            MinuteAnimation.To = minuteAngle + 360;
            //秒针的角度运动规律
            SecondAnimation.From = secondAngle;
            SecondAnimation.To = secondAngle + 360;
            //启动与此 Storyboard 关联的那组动画
            ClockStoryboard.Begin();
        }
    }
}
```

图 22.1　动画时钟

程序运行的效果如图 22.1 所示。

实例触发器时钟：使用定时触发器(DispatcherTimer)的方式来创建一个简易的时钟。

实例说明：上一个时钟的实例用的是 Silverlight 的动画来实现一个时钟的应用，那么本实例将会用一种完全不一样的方式来实现另外一个时钟应用。本实例使用 TextBlock 控件来表示时钟的时、分、秒针，通过定时触发器(DispatcherTimer)每隔一个很短的时间段不停地来更新三个 TextBlock 控件的位置(即时、分、秒针的位置)，从而实现了时钟的效果。

DispatcherTimer 类是在 System. Windows. Threading 命名空间下的定时器，集成到按指定时间间隔和指定优先级处理的 Dispatcher 队列中的计时器，在每个 Dispatcher 循环的顶端重新计算 DispatcherTimer。不能保证会正好在时间间隔发生时执行计时器，但能够保证不会在时间间隔发生之前执行计时器，这是因为 DispatcherTimer 操作与其他操作一样被放置到 Dispatcher 队列中。何时执行 DispatcherTimer 操作取决于队列中的其他作业及其优先级。每当将对象方法绑定到计时器时，DispatcherTimer 都将使对象保持活动状态。首先，DispatcherTimer 只会创建一个线程，那么它的执行顺序就很明显，要等执行完成之后，才会重头开始执行。每当线程执行完一次，等待 Interval 设置的时之后才会重新开始，也就是说，要执行完→然后再等待时间→然后重复。

代码清单 22-2：触发器时钟(源代码：第 22 章\Examples_22_2)

MainPage.xaml 文件主要代码

```
<StackPanel x:Name="TitlePanel" Grid.Row="0" Margin="12,17,0,28">
    <TextBlock x:Name="ApplicationTitle" Text="利用定时器设计的时钟" Style="{StaticResource PhoneTextNormalStyle}"/>
```

```
</StackPanel>
<Grid x:Name="ContentPanel" Grid.Row="1" Margin="12,0,12,0"
      SizeChanged="OnContentPanelSizeChanged">
    <TextBlock Name="referenceText" Text="参考对象文本控件"/>
    <!-- 时针 -->
    <TextBlock Name="hourHand">
        <!-- 设置 TextBlock 的位置变换为混合变换的模式 -->
        <TextBlock.RenderTransform>
            <CompositeTransform />
        </TextBlock.RenderTransform>
    </TextBlock>
    <!-- 分针 -->
    <TextBlock Name="minuteHand">
        <TextBlock.RenderTransform>
            <CompositeTransform />
        </TextBlock.RenderTransform>
    </TextBlock>
    <!-- 秒针 -->
    <TextBlock Name="secondHand">
        <TextBlock.RenderTransform>
            <CompositeTransform />
        </TextBlock.RenderTransform>
    </TextBlock>
</Grid>
```

MainPage.xaml.cs 文件代码

```
using System;
using System.Windows;
using System.Windows.Controls;
using System.Windows.Media;
using Microsoft.Phone.Controls;
using System.Windows.Threading;
namespace Clock2
{
    public partial class MainPage : PhoneApplicationPage
    {
        Point gridCenter;
        Size textSize;
        double scale;
        public MainPage()
        {
            InitializeComponent();
            //创建定时器
            DispatcherTimer tmr = new DispatcherTimer();
            //设置定时器刻度之间的时间段,每秒触发一次
            tmr.Interval = TimeSpan.FromSeconds(1);
            //DispatcherTimer.Tick 事件,超过定时器间隔时触发
            tmr.Tick += OnTimerTick;
            //定时器开始
            tmr.Start();
```

```
        }
        //这个事件会在刚进入这个 grid 面板的时候触发
        void OnContentPanelSizeChanged(object sender, SizeChangedEventArgs args)
        {
            gridCenter = new Point(args.NewSize.Width / 2,args.NewSize.Height / 2);
            //参考对象的高度和宽度
            textSize = new Size(referenceText.ActualWidth,referenceText.ActualHeight);
            scale = Math.Min(gridCenter.X, gridCenter.Y) / textSize.Width;
            UpdateClock();
        }
        //计时器触发的事件函数
        void OnTimerTick(object sender, EventArgs e)
        {
            UpdateClock();
        }
        void UpdateClock()
        {
            DateTime dt = DateTime.Now;
            double angle = 6 * dt.Second;
            SetupHand(secondHand, "--------------- 秒针 " + dt.Second, angle);
            angle = 6 * dt.Minute + angle / 60;
            SetupHand(minuteHand, "---------- 分针 " + dt.Minute, angle);
            angle = 30 * (dt.Hour % 12) + angle / 12;
            SetupHand(hourHand, "------ 时针 " + (((dt.Hour + 11) % 12) + 1), angle);
        }
        void SetupHand(TextBlock txtblk, string text, double angle)
        {
            txtblk.Text = text;
            //获取 txtblk 的 RenderTransform 属性,并转化为 CompositeTransform
            CompositeTransform xform = txtblk.RenderTransform as CompositeTransform;
            //移动的中心点的 x 坐标
            xform.CenterX = textSize.Height / 2;
            //移动的中心点的 y 坐标
            xform.CenterY = textSize.Height / 2;
            //设置 x 轴的缩放比例
            xform.ScaleX = scale;
            //设置 y 轴的缩放比例
            xform.ScaleY = scale;
            //获取或设置顺时针旋转角度(以度为单位)
            xform.Rotation = angle - 90;
            //设置沿 x 轴平移对象的距离
            xform.TranslateX = gridCenter.X - textSize.Height / 2;
            //设置沿 y 轴平移对象的距离
            xform.TranslateY = gridCenter.Y - textSize.Height / 2;
        }
    }
}
```

程序运行的效果如图 22.2 所示。

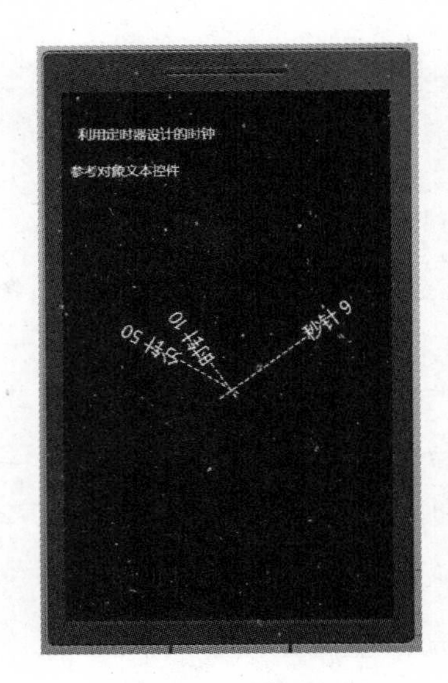

图 22.2　定期器时钟

22.2　日历

实例日历：创建一个简单的日历应用。

实例说明：在 Windows Phone 上实现一个日历的程序有很多种方式，下面将用一种很简单的方法来实现一个日历的应用程序。日历主体是用一个 WrapPanel 面板加上 Button 控件来实现的，每个日期用一个 Button 来表示。WrapPanel 根据其中 Button 元素的尺寸和其自身可能的大小自动地把其中的 Button 元素排列到下一行或下一列。该日历程序实现的功能包括显示当前的日期，可以通过上、下按钮来查看不同月份的日期。

代码清单 22-3：日历（源代码：第 22 章\Examples_22_3）

MainPage.xaml 文件主要代码

```
<phone:PhoneApplicationPage
    …
    xmlns:phone = "clr - namespace:Microsoft.Phone.Controls;assembly = Microsoft.Phone"
    xmlns:toolkit = "clr - namespace:Microsoft.Phone.Controls; assembly = Microsoft.Phone.
Controls.Toolkit">
    <! -- 加载样式文件 -->
    <phone:PhoneApplicationPage.Resources>
        <ResourceDictionary>
            <ResourceDictionary.MergedDictionaries>
                <ResourceDictionary Source = "style.xaml"/>
            </ResourceDictionary.MergedDictionaries>
        </ResourceDictionary>
    </phone:PhoneApplicationPage.Resources>
    …
```

```
            <Grid>
                <Button x: Name = "BackBtn" VerticalAlignment = "Top"
HorizontalAlignment = "Left" Style = "{StaticResource RoundButton}" Height = "72" Width = "72"
Margin = "27,1,0,0" Click = "OnChangeMonth" BorderBrush = "White">
                    <Image Source = "/CalendarControl; component/Images/appbar. back.
png" Height = "42" Width = "42" VerticalAlignment = "Top" HorizontalAlignment = "Left" />
                </Button>
                <TextBlock x: Name = " MonthYear " Text = " November 2010 " Style =
"{StaticResource PhoneTextLargeStyle}" Margin = " 101, 14, 124, 0 " HorizontalAlignment =
"Center" VerticalAlignment = "Top" Width = "231" TextAlignment = "Center" Foreground = "White"/>
                <Button x: Name = "NextBtn" VerticalAlignment = "Top"
HorizontalAlignment = "Right" Style = "{StaticResource RoundButton}" Height = "72" Width = "74"
Margin = "0,0,45,0" Click = "OnChangeMonth" Foreground = "White" BorderBrush = "White">
                    <Image Source = "/CalendarControl; component/Images/appbar. next.
png" Height = "42" Width = "42" VerticalAlignment = "Top" HorizontalAlignment = "Right" />
                </Button>
            </Grid>
            <ListBox x:Name = "CalendarListBox" Margin = "1,78,0,70" Height = "459">
                <ListBoxItem Margin = "12,0,0,0">
                    <toolkit:WrapPanel x:Name = "CalendarWrapPanel" HorizontalAlignment =
"Left" VerticalAlignment = "Top"/>
                </ListBoxItem>
            </ListBox>
        </Grid>
    …
</phone:PhoneApplicationPage>
```

MainPage. xaml. cs 文件代码

```
using System;
using System. Windows;
using System. Windows. Controls;
using Microsoft. Phone. Controls;
using Microsoft. Phone. Shell;
namespace CalendarControl
{
    public partial class MainPage : PhoneApplicationPage
{
    //"?"加上之后表示可以有空值(null)
        DateTime? _entryDate = DateTime. Now;
        public MainPage()
        {
            InitializeComponent();
            //第一次进入日历程序,初始化当前日期的显示
            InitializeCalendar(_entryDate. Value);
        }
        //月份前进后退事件
        private void OnChangeMonth(object sender, RoutedEventArgs e)
        {
            //如果是单击下一个月的按钮
            if (((Button)sender). Name == "NextBtn")
```

```
            _entryDate = _entryDate.Value.AddMonths(1);
        else
            _entryDate = _entryDate.Value.AddMonths(-1);
        CalendarListBox.Visibility = Visibility.Collapsed;
        //初始化该日期的显示
        InitializeCalendar(_entryDate.Value);
    }
    //初始化日历不同月份的日期
    protected void InitializeCalendar(DateTime entryDate)
        MonthYear.Text = String.Format("{0:yyyy 年 MM 月 }", _entryDate.Value);
        DateTime todaysDate = DateTime.Now;
        //获取显示的月份的天数
        int numDays = DateTime.DaysInMonth(entryDate.Year, entryDate.Month);
        //CalendarWrapPanel 面板中检查按钮的数量
        int count = CalendarWrapPanel.Children.Count;
        if (count > numDays)
        {
            //从最后减去多余的日期的按钮,日期用的是按钮控件
            for (int i = 1; i <= count - numDays; i++)
                CalendarWrapPanel.Children.RemoveAt(count - i);
        }
        else
        {
            //从最后加上缺少的日期的按钮
            int start = count + 1;
            for (int i = start; i <= numDays; i++)
            {
                Border border = new Border();
                border.Background = new SolidColorBrush(Color.FromArgb(255, 103, 183, 212));
                border.Margin = new Thickness(0, 0, 5, 5);
                border.Width = 65;
                border.Height = 65;
                border.CornerRadius = new CornerRadius(10);
                Button btn = new Button();
                btn.Name = "Day" + i;
                btn.Content = i.ToString();
                btn.BorderBrush = new SolidColorBrush(Colors.Transparent);
                btn.Width = 60;
                btn.Height = 60;
                btn.FontSize = 25;
                //Button 放进 Border 里面
                border.Child = btn;
                btn.Style = this.Resources["HasDataButtonStyle"] as Style;
                //将按钮添加到面板中
                CalendarWrapPanel.Children.Add(border);
            }
        }
        for (int i = 0; i < numDays; i++)
        {
```

```
                Border border = (Border)CalendarWrapPanel.Children[i];
                if (border != null)
                {
                    Button btn = (Button)border.Child;
                    DateTime currDate = new DateTime(entryDate.Year, entryDate.Month, i + 1);
                    //如果日期是今天则设置日期的按钮为橙色
                    if (currDate.Date.CompareTo(DateTime.Now.Date) == 0)
                    {
                        border.Background = new SolidColorBrush(Color.FromArgb(255, 255, 165, 0));
                        btn.Style = this.Resources["TodayHasDataButtonStyle"] as Style;
                    }
                    else
                    {
                        border.Background = new SolidColorBrush(Color.FromArgb(255, 103,
183, 212));
                    }
                }
            }
            //更新 CalendarWrapPanel 的显示
            CalendarWrapPanel.UpdateLayout();
            //设置为可见
            CalendarListBox.Visibility = Visibility.Visible;
        }
    }
}
```

程序运行的效果如图 22.3 所示。

图 22.3　日历应用

22.3 统计图表

实例统计图表：创建 Windows Phone 上的各种统计图表。

实例说明：利用 Silverlight 的 Chart 组件可以创建出各种常用的图表图形，在 Chart 组件里面有线性图（LineSeries）、饼图（PieSeries）、柱形图（ColumnSeries）、区域图（AreaSeries）、条状图（BarSeries）、散点图（ScatterSeries）和气泡图（BubbleSeries）等 7 种图形。在应用中使用这些图标需要在工程项目引入 Chart 组件的 DLL 文件，在空间 System. Windows. Controls. DataVisualization. Charting 下便可以找到各种图表相关的类，这些图表共用的属性如表 22.1 所示。

表 22.1　Chart 组件的常用属性

名　称	说　明
ActualAxes	获取显示在该组件上的实际轴数
Axes	获取或设置在组件中的轴的序列
ChartAreaStyle	获取或设置 ISeriesHost 的图表区域样式
LegendItems	获取图例项的集合
LegendStyle	获取或设置图例的样式
LegendTitle	获取或设置图例的标题内容
PlotAreaStyle	获取或设置该组件绘图区域的样式
Series	获取或设置显示在该组件中的数据序列集合
StylePalette	获取或设置一个由 ISeriesHost 子项所使用的样式调色板
Title	获取或设置该组件的标题
TitleStyle	获取或设置该组件的标题的样式

代码清单 22-4：统计图表（源代码：第 22 章\Examples_22_4）

MainPage. xaml 文件主要代码

```
<phone:PhoneApplicationPage
    ...
xmlns:charting = "clr - namespace: System. Windows. Controls. DataVisualization. Charting;
assembly = System.Windows.Controls.DataVisualization.Toolkit"
     xmlns:local = "clr - namespace:DataVisualizationOnWindowsPhone"
     xmlns:DataVisualization = "clr - namespace: System. Windows. Controls. DataVisualization;
assembly = System.Windows.Controls.DataVisualization.Toolkit"
     xmlns:controls = "clr - namespace:Microsoft. Phone. Controls; assembly = Microsoft. Phone.
Controls">
    <!-- 加载活动的集合类作为资源,将用于绑定图表的数据 -->
    <phone:PhoneApplicationPage.Resources>
        <local:Activities x:Key = "Activities" />
    </phone:PhoneApplicationPage.Resources>
    <!-- 设置字体的颜色 -->
    <phone:PhoneApplicationPage.FontSize>
```

```
        < StaticResource ResourceKey = "PhoneFontSizeNormal" />
    </phone:PhoneApplicationPage.FontSize >
    < phone:PhoneApplicationPage.Foreground >
        < StaticResource ResourceKey = "PhoneForegroundBrush" />
    </phone:PhoneApplicationPage.Foreground >
    …
        < controls:Panorama Title = "图表控件"  >
            < controls:PanoramaItem Header = "线形图">
              …
            </controls:PanoramaItem >
            < controls:PanoramaItem Header = "饼图">
                < Grid >
                    < charting:Chart x:Name = "pieChart"
                                    Style = "{StaticResource PhoneChartStyle}"
                                    Template = "{StaticResource PhoneChartPortraitTemplate}"
Margin = "0,0, - 7,0">
                        < charting:Chart.Palette >
                            < DataVisualization:ResourceDictionaryCollection >
                                <! -- 设置饼图的样式 -- >
                                < ResourceDictionary >
                                    < Style x:Key = "DataPointStyle" TargetType = "Control">
                                        < Setter Property = "Background">
                                            < Setter.Value >
                                                    < RadialGradientBrush MappingMode =
"Absolute">
                                                    < GradientStop Color = "Blue"
                                                                Offset = "0.9" />
                                                    < GradientStop Color = "DarkBlue"
                                                                Offset = "1.0" />
                                                </RadialGradientBrush >
                                            </Setter.Value >
                                        </Setter >
                                        < Setter Property = "BorderBrush"
                                                Value = "Transparent" />
                                    </Style >
                                </ResourceDictionary >
                                …
                            </DataVisualization:ResourceDictionaryCollection >
                        </charting:Chart.Palette >
                        <! -- 设置饼图绑定的数据源 -- >
                        < charting:PieSeries x:Name = "pieSeries"
                                        ItemsSource = "{StaticResource Activities}"
                                        DependentValuePath = "Count"
                                        IndependentValuePath = "Activity"
                                        AnimationSequence = "FirstToLast" />
                    </charting:Chart >
                </Grid >
            </controls:PanoramaItem >
            < controls:PanoramaItem Header = "柱形图">
             …
```

```
            </controls:PanoramaItem>
            <controls:PanoramaItem Header="区域图">
             ...
            </controls:PanoramaItem>
        </controls:Panorama>
    ...
</phone:PhoneApplicationPage>
```

MainPage.xaml.cs 文件代码

```
using System.Collections.Generic;
using System.Windows;
using System.Windows.Controls;
using Microsoft.Phone.Controls;
using System.Windows.Controls.DataVisualization.Charting;
using System.Windows.Media;
namespace DataVisualizationOnWindowsPhone
{
    public partial class MainPage : PhoneApplicationPage
    {
        public MainPage()
        {
            InitializeComponent();
        }
    }
    //活动信息的类
    public class ActivityInfo
    {
        public string Activity { get; set; }
        public int Count { get; set; }
    }
    //活动的活动列表类,数据源
    public class Activities : List<ActivityInfo>
    {
        public Activities()
        {
            Add(new ActivityInfo { Activity = "上班", Count = 100 });
            Add(new ActivityInfo { Activity = "吃饭", Count = 26 });
            Add(new ActivityInfo { Activity = "聊 QQ", Count = 6 });
            Add(new ActivityInfo { Activity = "陪老婆", Count = 60 });
            Add(new ActivityInfo { Activity = "旅游", Count = 10 });
            Add(new ActivityInfo { Activity = "发呆", Count = 18 });
        }
    }
}
```

程序运行的效果如图 22.4～图 22.7 所示。

图 22.4　饼图

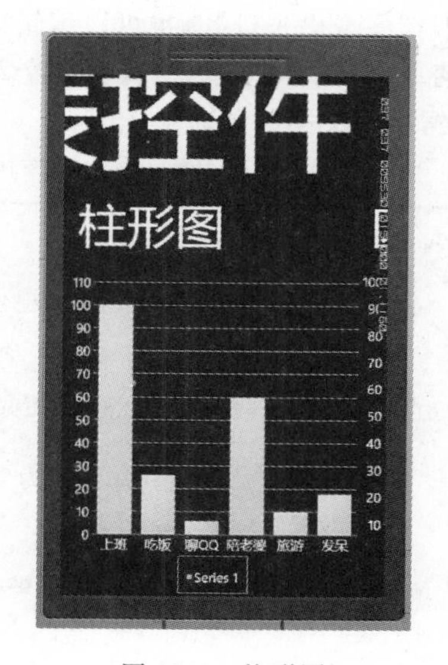

图 22.5　柱形图

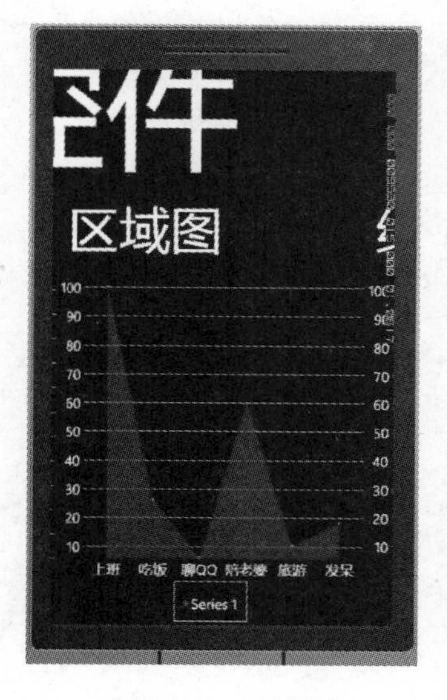

图 22.6　区域图

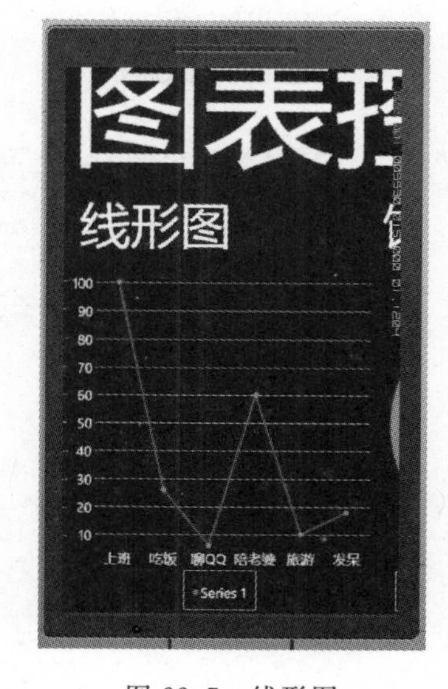

图 22.7　线形图

22.4　记事本

实例记事本：一个记事本应用，使用了 SQL Server 数据库来存储数据。

实例说明：记事本应用的功能是用于记录你的一些事情或者备忘录之类的。这个记事本应用程序使用了 SQL Server 数据库来存储记事的时间、标题和内容，在首页可以将以前所有的记事都展现出来，同时也可以编辑以前的记事内容和新增一条记事本的记录。

代码清单 22-5：记事本（源代码：第 22 章\Examples_22_5）

1. 数据库相关类的创建

NoteTable.cs 文件代码：创建一个记事本的表类

```
using System.Data.Linq.Mapping;
using System.ComponentModel;
namespace Note
{
    [Table]
    public class NoteTable : INotifyPropertyChanged, INotifyPropertyChanging
    {
        //记事本表独立增长 ID,设置为主键
        private int _id;
        [Column(IsPrimaryKey = true, IsDbGenerated = true, DbType = "INT NOT NULL
Identity", CanBeNull = false, AutoSync = AutoSync.OnInsert)]
        public int ID
        {
            get
            {
                return _id;
            }
            set
            {
                if (_id != value)
                {
                    NotifyPropertyChanging("ID");
                    _id = value;
                    NotifyPropertyChanged("ID");
                }
            }
        }
        //定义记事的标题
        private string _title;
        [Column]
        public string Title
        {
            get
            {
                return _title;
            }
            set
            {
                if (_title != value)
                {
                    NotifyPropertyChanging("Title");
                    _title = value;
                    NotifyPropertyChanged("Title");
                }
```

```
        }
    }
    //定义记事的时间
    private string _time;
    [Column]
    public string Time
    {
        get
        {
            return _time;
        }
        set
        {
            if (_time != value)
            {
                NotifyPropertyChanging("Time");
                _time = value;
                NotifyPropertyChanged("Time");
            }
        }
    }
    //记事内容
    private string _content;
    [Column]
    public string Content
    {
        get
        {
            return _content;
        }
        set
        {
            if (_content != value)
            {
                NotifyPropertyChanging("Content");
                _content = value;
                NotifyPropertyChanged("Content");
            }
        }
    }
    public event PropertyChangedEventHandler PropertyChanged;
    //用来通知页面表的字段数据产生了改变
    private void NotifyPropertyChanged(string propertyName)
    {
        if (PropertyChanged != null)
        {
            PropertyChanged(this, new PropertyChangedEventArgs(propertyName));
        }
    }
    public event PropertyChangingEventHandler PropertyChanging;
```

```
        //用来通知数据上下文表的字段数据将要产生改变
        private void NotifyPropertyChanging(string propertyName)
        {
            if (PropertyChanging != null)
            {
                PropertyChanging(this, new PropertyChangingEventArgs(propertyName));
            }
        }
    }
}
```

NoteCollection.cs 文件代码：创建表集合类

```
using System.ComponentModel;
using System.Collections.ObjectModel;
namespace Note
{
    //用于跟页面的数据绑定
    public class NoteCollection : INotifyPropertyChanged
    {
        //定义 ObservableCollection 来绑定页面的数据
        private ObservableCollection<NoteTable> _noteTables;
        //定义集合属性
        public ObservableCollection<NoteTable> NoteTables
        {
            get
            {
                return _noteTables;
            }
            set
            {
                if (_noteTables != value)
                {
                    _noteTables = value;
                    NotifyPropertyChanged("NoteTables");
                }
            }
        }
        public event PropertyChangedEventHandler PropertyChanged;
        //用于通知属性的改变
        private void NotifyPropertyChanged(string propertyName)
        {
            if (PropertyChanged != null)
            {
                PropertyChanged(this, new PropertyChangedEventArgs(propertyName));
            }
        }
    }
}
```

2. 初始化数据库在 App. xaml. cs 中实现

App. xaml. cs 文件主要代码：初始化数据库

```
private void Application_Launching(object sender, LaunchingEventArgs e)
{
    //如果数据库不存在则创建一个数据库
    using (NoteDataContext db = new NoteDataContext(NoteDataContext.DBConnectionString))
    {
        if (db.DatabaseExists() == false)
        {
            //创建一个数据库
            db.CreateDatabase();
        }
    }
}
```

3. 记事本页面的实现

MainPage. xaml 文件主要代码：首页记事本列表

```
<phone:PhoneApplicationPage
    …
    xmlns:phone="clr-namespace:Microsoft.Phone.Controls;assembly=Microsoft.Phone"
    xmlns:shell="clr-namespace:Microsoft.Phone.Shell;assembly=Microsoft.Phone"
    Loaded="PhoneApplicationPage_Loaded">
    …
            <!--用 ListBox 控件绑定记事本的列表-->
            <ListBox Name="noteListBox"  VerticalAlignment="Stretch" HorizontalAlignment=
"Stretch">
                <ListBox.ItemTemplate>
                    <DataTemplate>
                        <StackPanel>
                            <!--单击该超链接按钮将会跳转到记事本的详细内容-->
                            <HyperlinkButton Name="noteLocation" FontSize="40" Content=
"{Binding Title}" HorizontalContentAlignment="Left" Tag="{Binding ID}" Click=
"noteLocation_Click" />
                            <TextBlock Name="noteDateCreated" Text="{Binding Time}"
Margin="10" />
                        </StackPanel>
                    </DataTemplate>
                </ListBox.ItemTemplate>
            </ListBox>
            <!--使用 Canvas 封装记事本的介绍内容,可隐藏和现实-->
            <Canvas Name="helpCanvas" Background="White" VerticalAlignment="Stretch"
HorizontalAlignment="Stretch" Width="400" Height="400" Visibility="Collapsed">
                <!--关闭记事本介绍的按钮,其实是隐藏了 Canvas 控件-->
                <Button Name="helpCloseButton" Canvas.Left="350" Canvas.Top="0" Width=
"50" Height="50" Click="helpCloseButton_Click">
                    <Button.Background>
                        <ImageBrush ImageSource="/Note;component/Images/appbar.close.
```

```
rest.png" Stretch = "None" />
                    </Button.Background>
                </Button>
                <!-- 使用 ScrollViewer 控件可滚动查看记事本的介绍内容 -->
                <ScrollViewer Name = "helpScrollViewer" Canvas.Left = "0" Canvas.Top = "50"
Width = "400" Height = "300">
                    <TextBlock Name = "helpTextBlock" Foreground = "Black" FontSize = "24"
TextWrapping = "Wrap" Height = "500">欢迎使用记事本应用!<LineBreak></LineBreak><LineBreak>
</LineBreak>在这里你可以记录你生活的点点滴滴!<LineBreak></LineBreak> 生活就是一场旅行!
<LineBreak></LineBreak>在这里开始你的记事之旅吧!</TextBlock>
                </ScrollViewer>
            </Canvas>
    ...
<!-- 菜单栏一个添加记事的按钮,一个显示帮助信息的按钮 -->
    <phone:PhoneApplicationPage.ApplicationBar>
        <shell:ApplicationBar IsVisible = "True" IsMenuEnabled = "True">
            <shell:ApplicationBarIconButton IconUri = "/Images/appbar.add.rest.png" Text
= "add" Click = "AppBar_Add_Click" />
            <shell:ApplicationBarIconButton IconUri = "/Images/appbar.questionmark.rest.
png" Text = "help" Click = "AppBar_Help_Click"/>
        </shell:ApplicationBar>
    </phone:PhoneApplicationPage.ApplicationBar>
</phone:PhoneApplicationPage>
```

MainPage.xaml.cs 文件代码

```
using System;
using System.Linq;
using System.Windows;
using System.Windows.Controls;
using Microsoft.Phone.Controls;
using System.IO.IsolatedStorage;
using System.Collections.ObjectModel;
namespace Note
{
    public partial class MainPage : PhoneApplicationPage
    {
        //创建 DataContext 实例用于操作本地的数据库
        private NoteDataContext noteDB;
        private NoteCollection noteCol = new NoteCollection();
        public MainPage()
        {
            InitializeComponent();
        }
        private void PhoneApplicationPage_Loaded(object sender, RoutedEventArgs e)
        {
            //使用独立存储设置来保存当前的记事状态
            IsolatedStorageSettings settings = IsolatedStorageSettings.ApplicationSettings;
            string state = "";
            if (settings.Contains("state"))
            {
```

```
            if (settings.TryGetValue<string>("state", out state))
            {
                //如果是新增状态
                if (state == "add")
                {
                    NavigationService.Navigate(new Uri("/Note;component/Add.xaml",
UriKind.Relative));
                }
                //如果是编辑状态
                else if (state == "edit")
                {
                    NavigationService.Navigate(new Uri("/Note;component/ViewEdit.
xaml", UriKind.Relative));
                }
            }
        }
        bindList();
    }
    //新增一个记事的事件处理
    private void AppBar_Add_Click(object sender, EventArgs e)
    {
        //跳转到新增记事页面 Add.xaml
        NavigationService.Navigate(new Uri("/Note;component/Add.xaml", UriKind.Relative));
    }
    //帮助按钮事件处理
    private void AppBar_Help_Click(object sender, EventArgs e)
    {
        //将含有记事本介绍信息的 Canvas 控件设置为可见状态
        helpCanvas.Visibility = System.Windows.Visibility.Visible;
    }
    //编辑记事事件处理
    private void noteLocation_Click(object sender, RoutedEventArgs e)
    {
        HyperlinkButton clickedLink = (HyperlinkButton)sender;
        string uri = String.Format("/Note;component/ViewEdit.xaml?id={0}", clickedLink.Tag);
        NavigationService.Navigate(new Uri(uri, UriKind.Relative));
    }
    //绑定记事本列表方法
    private void bindList()
    {
        try
        {
            //连接数据库并初始化 DataContext 实例
            noteDB = new NoteDataContext(NoteDataContext.DBConnectionString);
            // 使用 Linq 查询语句查询 EmployeeTable 表的所有数据
            var notesInDB = from NoteTable note in noteDB.Notes select note;
            // 将查询的结果返回到页面数据绑定的集合里面
            noteCol.NoteTables = new ObservableCollection<NoteTable>(notesInDB);
            //赋值给当前页面的 DataContext 用于数据绑定
            this.DataContext = noteCol;
```

```
                noteListBox.ItemsSource = noteCol.NoteTables;
            }
            catch
            {
                MessageBox.Show("数据库查询错误!");
            }
        }
        //关闭记事本介绍按钮
        private void helpCloseButton_Click(object sender, RoutedEventArgs e)
        {
            //将含有记事本介绍信息的Canvas控件设置为不可见状态
            helpCanvas.Visibility = System.Windows.Visibility.Collapsed;
        }
    }
}
```

Add.xaml 文件主要代码：记事新增页面

```
<phone:PhoneApplicationPage
    ...
    xmlns:phone="clr-namespace:Microsoft.Phone.Controls;assembly=Microsoft.Phone"
    xmlns:shell="clr-namespace:Microsoft.Phone.Shell;assembly=Microsoft.Phone"
    Loaded="PhoneApplicationPage_Loaded">
    ...
        <Grid x:Name="ContentPanel" Grid.Row="1" Margin="12,0,12,0">
            <TextBox Name="editTextBox" HorizontalAlignment="Stretch" VerticalAlignment
="Stretch" TextChanged="editTextBox_TextChanged"/>
        </Grid>
    ...
    <!--菜单栏一个取消按钮,一个保存按钮-->
    <phone:PhoneApplicationPage.ApplicationBar>
        <shell:ApplicationBar IsVisible="True" IsMenuEnabled="True">
            <shell:ApplicationBarIconButton IconUri="/Images/appbar.cancel.rest.png" Text
="cancel" Click="AppBar_Cancel_Click"/>
            <shell:ApplicationBarIconButton IconUri="/Images/appbar.save.rest.png" Text
="save" Click="AppBar_Save_Click"/>
        </shell:ApplicationBar>
    </phone:PhoneApplicationPage.ApplicationBar>
</phone:PhoneApplicationPage>
```

Add.xaml.cs 文件代码

```
using System;
using System.Windows;
using System.Windows.Controls;
using Microsoft.Phone.Controls;
using System.IO.IsolatedStorage;
using System.Collections.ObjectModel;
namespace Note
{
    public partial class Add : PhoneApplicationPage
```

```
{
    private IsolatedStorageSettings settings = IsolatedStorageSettings.ApplicationSettings;
    //创建 DataContext 实例用于操作本地的数据库
    private NoteDataContext noteDB;
    private NoteCollection noteCol = new NoteCollection();
    public Add()
    {
        InitializeComponent();
        noteCol.NoteTables = new ObservableCollection<NoteTable>();
    }
    //页面加载事件处理
    private void PhoneApplicationPage_Loaded(object sender, RoutedEventArgs e)
    {
        string state = "";
        if (settings.Contains("state"))
        {
            if (settings.TryGetValue<string>("state", out state))
            {
                if (state == "add")
                {
                    string value = "";
                    if (settings.Contains("value"))
                    {
                        if (settings.TryGetValue<string>("value", out value))
                        {
                            editTextBox.Text = value;
                        }
                    }
                }
            }
        }
        settings["state"] = "add";
        settings["value"] = editTextBox.Text;
        editTextBox.Focus();
        editTextBox.SelectionStart = editTextBox.Text.Length;
    }
    //编辑框的内容改变事件
    private void editTextBox_TextChanged(object sender, TextChangedEventArgs e)
    {
        //将文本框的内容临时存放到独立存储设置里面
        settings["value"] = editTextBox.Text;
    }
    //取消按钮的事件
    private void AppBar_Cancel_Click(object sender, EventArgs e)
    {
        navigateBack();
    }
    //保存按钮事件处理
    private void AppBar_Save_Click(object sender, EventArgs e)
    {
```

```
            if (editTextBox.Text == "")
            {
                MessageBox.Show("朋友,写点东西吧!");
                return;
            }
            try
            {
                //连接数据库并初始化 DataContext 实例
                noteDB = new NoteDataContext(NoteDataContext.DBConnectionString);
                //创建一条表的数据
                NoteTable newNote = new NoteTable { Title = editTextBox.Text.Substring(0, 10).
ToString(),Time = DateTime.Now.ToLongDateString(), Content = editTextBox.Text.ToString() };
                //添加绑定集合的数据
                noteCol.NoteTables.Add(newNote);
                //插入数据库
                noteDB.Notes.InsertOnSubmit(newNote);
                //保存数据库的改变
                noteDB.SubmitChanges();
            }
            catch(Exception)
            {
                MessageBox.Show("保存数据出错!");
            }
            //保存完成,返回主页
            navigateBack();
        }
        //返回主页方法
        private void navigateBack()
        {
            //清空状态和值的标识符
            settings["state"] = "";
            settings["value"] = "";
            NavigationService.Navigate(new Uri("/Note;component/MainPage.xaml", UriKind.Relative));
        }
    }
}
```

ViewEdit.xaml 文件主要代码：记事查看和编辑页面

```
<phone:PhoneApplicationPage
    …
    xmlns:phone = "clr-namespace:Microsoft.Phone.Controls;assembly = Microsoft.Phone"
    xmlns:shell = "clr-namespace:Microsoft.Phone.Shell;assembly = Microsoft.Phone"
    Loaded = "PhoneApplicationPage_Loaded">
    …
        <Grid x:Name = "ContentPanel" Grid.Row = "1" Margin = "12,0,12,0">
            <TextBlock Name = "displayTextBlock" HorizontalAlignment = "Stretch"
VerticalAlignment = "Stretch" TextWrapping = "Wrap" Visibility = "Visible" />
            <TextBox Name = "editTextBox" HorizontalAlignment = "Stretch" VerticalAlignment =
"Stretch" TextWrapping = "Wrap" Visibility = "Collapsed" TextChanged = "editTextBox_
TextChanged"/>
```

```
            <Canvas Name = "confirmDialog" Background = "Red" HorizontalAlignment = "Left"
VerticalAlignment = "Top" Margin = "50, 100, 0, 0" Width = "350" Height = "300" Visibility =
"Collapsed">
                <TextBlock Text = "你确定要删除这个记事本么?" Width = "330" Height = "75"
TextWrapping = "Wrap" Canvas.Left = "10" Canvas.Top = "10" FontSize = "22" />
                <Button Name = "cancelButton" Canvas.Left = "10" Canvas.Top = "150" Content =
"取消" Width = "150" Click = "cancelButton_Click"/>
                <Button Name = "deleteButton" Canvas.Left = "180" Canvas.Top = "150" Width =
"150" Content = "删除" Click = "deleteButton_Click"/>
            </Canvas>
        </Grid>
    …
    <!--菜单栏一个返回按钮、一个编辑按钮、一个保存按钮和一个删除按钮-->
    <phone:PhoneApplicationPage.ApplicationBar>
        <shell:ApplicationBar IsVisible = "True" IsMenuEnabled = "True">
            <shell:ApplicationBarIconButton IconUri = "/Images/appbar.back.rest.png" Text =
"back" Click = "AppBar_Back_Click"/>
            <shell:ApplicationBarIconButton IconUri = "/Images/appbar.edit.rest.png" Text =
"edit" Click = "AppBar_Edit_Click" />
            <shell:ApplicationBarIconButton IconUri = "/Images/appbar.save.rest.png" Text =
"save" Click = "AppBar_Save_Click"/>
            <shell:ApplicationBarIconButton IconUri = "/Images/appbar.delete.rest.png" Text =
"delete" Click = "AppBar_Delete_Click"/>
        </shell:ApplicationBar>
    </phone:PhoneApplicationPage.ApplicationBar>
</phone:PhoneApplicationPage>
```

ViewEdit.xaml.cs 文件代码

```
using System;
using System.Linq;
using System.Windows;
using System.Windows.Controls;
using Microsoft.Phone.Controls;
using System.IO.IsolatedStorage;
using System.Collections.ObjectModel;
namespace Note
{
    public partial class ViewEdit : PhoneApplicationPage
    {
        //获取独立存储设置
        private IsolatedStorageSettings settings = IsolatedStorageSettings.ApplicationSettings;
        //创建 DataContext 实例用于操作本地的数据库
        private NoteDataContext noteDB;
        private NoteCollection noteCol = new NoteCollection();
        private string id = "";
        public ViewEdit()
        {
            InitializeComponent();
            noteCol.NoteTables = new ObservableCollection<NoteTable>();
        }
```

```
private void PhoneApplicationPage_Loaded(object sender, RoutedEventArgs e)
{
    string state = "";
    if (settings.Contains("state"))
    {
        if (settings.TryGetValue<string>("state", out state))
        {
            if (state == "edit")
            {
                string value = "";
                if (settings.Contains("id"))
                {
                    if (settings.TryGetValue<string>("id", out value))
                    {
                        id = value;
                    }
                }
                if (settings.Contains("value"))
                {
                    if (settings.TryGetValue<string>("value", out value))
                    {
                        bindEdit(value);
                    }
                }
            }
            else
            {
                bindView();
            }
        }
    }
    else
    {
        bindView();
    }
}
private void editTextBox_TextChanged(object sender, TextChangedEventArgs e)
{
    settings["value"] = editTextBox.Text;
}
private void AppBar_Back_Click(object sender, EventArgs e)
{
    navigateBack();
}
private void AppBar_Edit_Click(object sender, EventArgs e)
{
    if (displayTextBlock.Visibility == System.Windows.Visibility.Visible)
    {
        bindEdit(displayTextBlock.Text);
    }
```

```
}
private void AppBar_Save_Click(object sender, EventArgs e)
{
    if (editTextBox.Visibility == System.Windows.Visibility.Visible)
    {
        var appStorage = IsolatedStorageFile.GetUserStoreForApplication();
        try
        {
            //获取编辑的 NoteTable 对象
            NoteTable note = (NoteTable)State["note"];
            note.Title = editTextBox.Text.Substring(0,10);
            note.Content = editTextBox.Text;
            //保存数据库的改变
            noteDB.SubmitChanges();
            State["note"] = note;
        }
        catch
        {
            MessageBox.Show("编辑出错!");
        }
        displayTextBlock.Text = editTextBox.Text;
        displayTextBlock.Visibility = System.Windows.Visibility.Visible;
        editTextBox.Visibility = System.Windows.Visibility.Collapsed;
    }
}
private void AppBar_Delete_Click(object sender, EventArgs e)
{
    confirmDialog.Visibility = System.Windows.Visibility.Visible;
}
private void bindView()
{
    id = NavigationContext.QueryString["id"];
    try
    {
        //连接数据库并初始化 DataContext 实例
        noteDB = new NoteDataContext(NoteDataContext.DBConnectionString);
        //使用 Linq 查询语句查询 NoteTable 表的所有数据
        var notesInDB = from NoteTable note in noteDB.Notes
                        where note.ID == Int32.Parse(id)
                        select note;
        //将查询的结果返回到页面数据绑定的集合里面
        noteCol.NoteTables = new ObservableCollection<NoteTable>(notesInDB);
        displayTextBlock.Text = noteCol.NoteTables[0].Content;
        //将需要编辑的表实例存储在 State 里面
        State["note"] = noteCol.NoteTables[0];
    }
    catch
```

```
            {
                MessageBox.Show("记事已经删除了!");
            }
        }
        private void bindEdit(string content)
        {
            editTextBox.Text = content;
            displayTextBlock.Visibility = System.Windows.Visibility.Collapsed;
            editTextBox.Visibility = System.Windows.Visibility.Visible;
            editTextBox.Focus();
            editTextBox.SelectionStart = editTextBox.Text.Length;
            settings["state"] = "edit";
            settings["value"] = editTextBox.Text;
            settings["id"] = id;
        }
        private void navigateBack()
        {
            settings["state"] = "";
            settings["value"] = "";
            settings["id"] = "";
            NavigationService.Navigate(new Uri("/Note;component/MainPage.xaml", UriKind.Relative));
        }
        private void cancelButton_Click(object sender, RoutedEventArgs e)
        {
            confirmDialog.Visibility = System.Windows.Visibility.Collapsed;
        }
        private void deleteButton_Click(object sender, RoutedEventArgs e)
        {
            var appStorage = IsolatedStorageFile.GetUserStoreForApplication();
            try
            {
                //删除的 NoteTable 实例
                NoteTable noteForDelete = State["note"] as NoteTable;
                //移除数据库里面要删除的 NoteTable 记录
                noteDB.Notes.DeleteOnSubmit(noteForDelete);
                //保存数据库的改变
                noteDB.SubmitChanges();
            }
            catch
            {
                MessageBox.Show("删除失败!");
            }
            navigateBack();
        }
    }
}
```

程序运行的效果如图 22.8～图 22.10 所示。

图 22.8　记事列表

图 22.9　你的记事

图 22.10　添加记事

22.5　快速邮件

实例快速邮件：快速邮件发送器，可以将准备要发的邮件添加到快速邮件里面去，等你想要发送的时候就可以快速地调出来进行发送了。

实例说明：上一个记事本实例使用了 SQL Server 数据库实现了数据的增删改的功能，那么快速邮件的实例是使用独立存储来实现了数据的增删改功能，添加一个新的邮件，编辑一个邮件和删除一个邮件。邮件的发送调用了系统的邮件发送器来实现。

代码清单 22-6：快速邮件（源代码：第 22 章\Examples_22_6）

1. 邮件实体类以及独立存储的处理类

QuickMailTemplate.cs 文件代码：邮件实体类

```
using System;
namespace QuickMail.Models
{
    public class QuickMailTemplate
    {
        //唯一 ID
        public Guid Id
        {
            get; set;
        }
        //邮件标题
        public string Subject
```

```
        {
            get; set;
        }
        //邮件内容
        public string Body
        {
            get; set;
        }
    }
}
```

TemplateStorage.cs 文件代码：独立存储的处理类

```
using System;
using System.Collections.Generic;
using System.IO;
using System.IO.IsolatedStorage;
using System.Linq;
using System.Xml.Linq;
namespace QuickMail.Models
{
    public class TemplateStorage
    {
        //用 IList<T>类型来表示邮件类列表信息
        private IList<QuickMailTemplate> templates;
        //独立存储的文件名
        private const string Filename = "template-list.xml";
        protected IList<QuickMailTemplate> Templates
        {
            get
            {
                return templates ?? (templates = LoadTemplates().ToList());
            }
            set
            {
                templates = value;
            }
        }
        //加载独立存储文件存储的邮件列表信息
        protected IEnumerable<QuickMailTemplate> LoadTemplates()
        {
            using(var applicationStorage = IsolatedStorageFile.GetUserStoreForApplication())
            {
                if(!applicationStorage.FileExists(Filename))
                    return Enumerable.Empty<QuickMailTemplate>();
                using(var speedListFile = applicationStorage.OpenFile(Filename, FileMode.
Open, FileAccess.Read))
                {
                    var document = XDocument.Load(speedListFile);
                    //返回所有的邮件信息
                    return from t in document.Root.Elements("template")
```

```
                        select new QuickMailTemplate
                        {
                            Id = new Guid(t.Attribute("id").Value),
                            Subject = t.Attribute("subject").Value,
                            Body = t.Attribute("body").Value
                        };
                }
            }
        }
        //返回邮件列表信息
        public IEnumerable<QuickMailTemplate> GetItems()
        {
            return Templates;
        }
        //添加一封邮件信息
        public void Save(QuickMailTemplate template)
        {
            Templates.Add(template);
        }
        //删除一封邮件信息
        public void Delete(QuickMailTemplate template)
        {
            Templates.Remove(template);
        }
        //保存邮件列表信息的改变
        public void SaveChanges()
        {
            using(var applicationStorage = IsolatedStorageFile.GetUserStoreForApplication())
            using (var speedListFile = applicationStorage.OpenFile (Filename, FileMode.
Create, FileAccess.Write))
            {
                var document = new XDocument(new XDeclaration("1.0", "utf-8", "yes"),
                    new XElement("templates",
                        from t in Templates
                        select new XElement("template",
                            new XAttribute("id", t.Id),
                            new XAttribute("subject", t.Subject),
                            new XAttribute("body", t.Body))));
                document.Save(speedListFile);
            }
        }
    }
}
```

2. 快速邮件的列表页面、新增编辑页面和详情发送页面

MainPage.xaml 文件主要代码：暂存邮件列表页面

```
<!--用 List 控件绑定邮件列表,单击触发 SelectionChanged 事件-->
<ListBox x:Name="Templates" SelectionChanged="OnSelectionChanged">
    <ListBox.ItemTemplate>
```

```
                    <DataTemplate>
                        <TextBlock Text = "{Binding Subject}" Style = "{StaticResource
PhoneTextLargeStyle}" />
                    </DataTemplate>
                </ListBox.ItemTemplate>
            </ListBox>
...
<!--菜单栏添加暂存邮件-->
<phone:PhoneApplicationPage.ApplicationBar>
    <shell:ApplicationBar>
        <shell:ApplicationBarIconButton IconUri="/icons/appbar.add.rest.png" Text="add"
Click="OnAdd"/>
    </shell:ApplicationBar>
</phone:PhoneApplicationPage.ApplicationBar>
```

MainPage.xaml.cs 文件代码

```
using System;
using System.Windows;
using System.Windows.Controls;
using QuickMail.Models;
namespace QuickMail
{
    public partial class MainPage
    {
        //创建一个邮件独立存储信息类
        private readonly TemplateStorage storage = new TemplateStorage();
        public MainPage()
        {
            InitializeComponent();
            //注册页面加载事件
            Loaded += OnLoaded;
        }
        private void OnLoaded(object sender, RoutedEventArgs e)
        {
            //把邮件列表信息绑定到 List 控件中
            Templates.ItemsSource = storage.GetItems();
        }
        //新增暂存邮件事件,跳转到新增页面
        private void OnAdd(object sender, EventArgs e)
        {
            NavigationService.Navigate(new Uri("/EditPage.xaml", UriKind.Relative));
        }
        //单击暂存邮件列表里面的邮件触发的事件
        private void OnSelectionChanged(object sender, SelectionChangedEventArgs e)
        {
            if(Templates.SelectedItem == null)
                return;
            //获取单击中的 List 选项的信息
            var template = (QuickMailTemplate)Templates.SelectedItem;
            //跳转到邮件详细信息页面
```

```
                NavigationService.Navigate(new Uri("/DetailsPage.xaml?Id=" + template.Id,
UriKind.Relative));
                //设置选中的邮件选项为空
                Templates.SelectedItem = null;
            }
        }
}
```

EditPage.xaml 文件主要代码：邮件新增编辑页面

```
        <TextBlock Style="{StaticResource PhoneTextNormalStyle}" Text="标题"/>
        <TextBox x:Name="Subject" Grid.Row="1" InputScope="Text"/>
        <TextBlock Style="{StaticResource PhoneTextNormalStyle}" Text="内容" Grid.Row="2" />
        <TextBox x:Name="Body" Grid.Row="3" InputScope="Text" TextWrapping="Wrap"/>
...
<!-- 菜单栏保存和取消按钮 -->
<phone:PhoneApplicationPage.ApplicationBar>
    <shell:ApplicationBar>
        <shell:ApplicationBarIconButton IconUri="/icons/appbar.save.rest.png" Text=
"save" Click="OnSave"/>
        <shell:ApplicationBarIconButton IconUri="/Icons/appbar.cancel.rest.png" Text=
"cancel" Click="OnCancel"/>
    </shell:ApplicationBar>
</phone:PhoneApplicationPage.ApplicationBar>
```

EditPage.xaml.cs 文件代码

```
using System;
using System.Linq;
using System.Windows;
using QuickMail.Models;
namespace QuickMail
{
    public partial class EditPage
    {
        private readonly TemplateStorage storage = new TemplateStorage();
        public EditPage()
        {
            InitializeComponent();
            Loaded += OnLoaded;
        }
        //获取邮件的唯一 ID
        private Guid? TemplateId
        {
            get
            {
                if(!NavigationContext.QueryString.ContainsKey("Id"))
                    return null;
                return new Guid(NavigationContext.QueryString["Id"]);
            }
        }
```

```
        //页面加载处理
        private void OnLoaded(object sender, RoutedEventArgs e)
        {
            if(TemplateId == null)
                return;
            //获取通过邮件的 ID 从邮件列表里面获取对应的邮件
            var template = storage.GetItems()
                .Single(t => t.Id == TemplateId);
            //赋值到邮件标题的文本框
            Subject.Text = template.Subject;
            //赋值到邮件内容的文本框
            Body.Text = template.Body;
        }
        //保存邮件
        private void OnSave(object sender, EventArgs e)
        {
            QuickMailTemplate template;
            if(TemplateId == null)
            {
                //TemplateId == null 为新增邮件的状态
                template = new QuickMailTemplate
                {
                    Id = Guid.NewGuid(),
                    Subject = Subject.Text,
                    Body = Body.Text
                };
                storage.Save(template);
            }
            else
            {
                //编辑邮件的状态
                template = storage.GetItems()
                    .Single(t => t.Id == TemplateId);
                template.Subject = Subject.Text;
                template.Body = Body.Text;
            }
            storage.SaveChanges();
            //保存成功后跳转到邮件详情和发送页面
            NavigationService.Navigate(new Uri("/DetailsPage.xaml?Id=" + template.Id,
UriKind.Relative));
        }
        //取消邮件编辑将返回邮件列表页面
        private void OnCancel(object sender, EventArgs e)
        {
            NavigationService.Navigate(new Uri("/MainPage.xaml", UriKind.Relative));
        }
    }
}
```

DetailsPage. xaml 文件主要代码：邮件详情以及发送页面

```
        <TextBlock Style="{StaticResource PhoneTextNormalStyle}" Text="标题"/>
            <TextBlock x:Name="Subject" Grid.Row="1" Style="{StaticResource
PhoneTextLargeStyle}"/>
        <TextBlock Style="{StaticResource PhoneTextNormalStyle}" Text="内容" Grid.Row="2"/>
            <TextBlock x:Name="Body" Grid.Row="3" TextWrapping="Wrap" Style=
"{StaticResource PhoneTextLargeStyle}"/>
…
<!--菜单栏的发送按钮、编辑按钮、删除按钮和取消按钮-->
<phone:PhoneApplicationPage.ApplicationBar>
    <shell:ApplicationBar>
        <shell:ApplicationBarIconButton IconUri="/Icons/appbar.feature.email.rest.png"
Text="send" Click="OnSend"/>
            <shell:ApplicationBarIconButton IconUri="/Icons/appbar.edit.rest.png" Text=
"edit" Click="OnEdit"/>
        <shell:ApplicationBarIconButton IconUri="/Icons/appbar.delete.rest.png" Text=
"delete" Click="OnDelete"/>
        <shell:ApplicationBarIconButton IconUri="/Icons/appbar.cancel.rest.png" Text=
"cancel" Click="OnCancel"/>
    </shell:ApplicationBar>
</phone:PhoneApplicationPage.ApplicationBar>
```

DetailsPage. xaml. cs 文件主要代码

```
//发送邮件
private void OnSend(object sender, EventArgs e)
{
    var template = storage.GetItems()
        .Single(t => t.Id == TemplateId);
    var composeTask = new EmailComposeTask
    {
        Subject = template.Subject,
        Body = template.Body
    };
    composeTask.Show();
}
//编辑邮件
private void OnEdit(object sender, EventArgs e)
{
    var template = storage.GetItems()
        .Single(t => t.Id == TemplateId);
    NavigationService.Navigate(new Uri("/EditPage.xaml?Id=" + template.Id, UriKind.
Relative));
}
//删除邮件
private void OnDelete(object sender, EventArgs e)
{
    var template = storage.GetItems()
        .Single(t => t.Id == TemplateId);
    storage.Delete(template);
```

```
        storage.SaveChanges();
        NavigationService.Navigate(new Uri("/MainPage.xaml", UriKind.Relative));
    }
    //取消邮件
    private void OnCancel(object sender, EventArgs e)
    {
        NavigationService.Navigate(new Uri("/MainPage.xaml", UriKind.Relative));
    }
```

程序运行的效果如图 22.11～图 22.13 所示。

图 22.11　邮件列表

图 22.12　新增邮件

图 22.13　邮件详情

网络应用实例

移动互联网是当前互联网发展的新的方向，而在智能手机客户端软件应用的热潮当中，将移动互联网和客户端应用整合起来是目前大部分互联网产品在手机端开发的理念。那么本章将会介绍一些 Windows Phone 上的网络应用实例，通过这些实例来加深对 Windows Phone 互联网编程的认识，以及学会如何去开发一个网络应用。本章从以下多个不同的角度来实现 Windows Phone 上的网络应用：单纯的获取网页信息的角度，如 RSS 阅读器实例；WebBrowser 控件操作网页的角度，如博客园主页实例；手机客户端与主机服务器端结合的角度，如网络留言板实例；利用第三方开发的网络接口的角度，如快递 100 实例。

23.1 RSS 阅读器

实例 RSS 阅读器：实现一个 RSS 阅读器，通过输入 RSS 地址来获取 RSS 的信息列表和查看 RSS 文章中的详细内容。

实例说明：RSS 阅读器是使用了 WebClient 类来获取网络上的 RSS 的信息，然后再转化为自己定义好的 RSS 实体类对象的列表，最后绑定到页面上。

代码清单 23-1：RSS 阅读器（源代码：第 23 章\Examples_23_1）

1. RSS 实体类和 RSS 服务类

RssItem.cs 文件代码：RSS 实体类

```
using System.Net;
using System.Text.RegularExpressions;
namespace WindowsPhone.Helpers
{
    ///<summary>
    ///RSS 对象类
    ///</summary>
    public class RssItem
    {
        ///<summary>
        ///初始化一个 RSS 目录
        ///</summary>
```

```
        ///< param name = "title">标题</param >
        ///< param name = "summary">内容</param >
        ///< param name = "publishedDate">发表事件</param >
        ///< param name = "url">文章地址</param >
        public RssItem(string title, string summary, string publishedDate, string url)
        {
            Title = title;
            Summary = summary;
            PublishedDate = publishedDate;
            Url = url;
            //解析 html
            PlainSummary = HttpUtility.HtmlDecode(Regex.Replace(summary, "<[^>]+?>",""));
        }
        //标题
        public string Title { get; set; }
        //内容
        public string Summary { get; set; }
        //发表时间
        public string PublishedDate { get; set; }
        //文章地址
        public string Url { get; set; }
        //解析的文本内容
        public string PlainSummary { get; set; }
    }
}
```

RssService.cs 文件代码：RSS 服务类

```
using System;
using System.Collections.Generic;
using System.IO;
using System.Net;
using System.ServiceModel.Syndication;
using System.Xml;
namespace WindowsPhone.Helpers
{
    //获取网络 RSS 服务类
    public static class RssService
    {

        ///< summary>
        ///获取 RSS 目录列表
        ///</summary>
        ///< param name = "rssFeed"> RSS 的网络地址</param >
        ///< param name = "onGetRssItemsCompleted">获取完成事件</param >
        public static void GetRssItems ( string rssFeed, Action < IEnumerable < RssItem >>
onGetRssItemsCompleted = null, Action<Exception> onError = null, Action onFinally = null)
        {
            WebClient webClient = new WebClient();
            //注册 webClient 读取完成事件
            webClient.OpenReadCompleted += delegate(object sender, OpenReadCompletedEventArgs e)
```

```
                {
                    try
                    {
                        if (e.Error != null)
                        {
                            if (onError != null)
                            {
                                onError(e.Error);
                            }
                            return;
                        }
                        //将网络获取的信息转化成 RSS 实体类
                        List<RssItem> rssItems = new List<RssItem>();
                        Stream stream = e.Result;
                        XmlReader response = XmlReader.Create(stream);
                        SyndicationFeed feeds = SyndicationFeed.Load(response);
                        foreach (SyndicationItem f in feeds.Items)
                        {
                            RssItem rssItem = new RssItem(f.Title.Text, f.Summary.Text,
f.PublishDate.ToString(), f.Links[0].Uri.AbsoluteUri);
                            rssItems.Add(rssItem);
                        }
                        //通知完成返回事件执行
                        if (onGetRssItemsCompleted != null)
                        {
                            onGetRssItemsCompleted(rssItems);
                        }
                    }
                    finally
                    {
                        if (onFinally != null)
                        {
                            onFinally();
                        }
                    }
                };
                webClient.OpenReadAsync(new Uri(rssFeed));
            }
        }
}
```

2. RSS 的页面处理

MainPage.xaml 文件主要代码

```
<TextBlock FontSize="30" Grid.Row="1" Height="49" HorizontalAlignment="Left" Margin=
"0,6,0,0" Name="textBlock1" Text="RSS 地址" VerticalAlignment="Top" Width="116" />
  <TextBox Grid.Row="1" Height="72" HorizontalAlignment="Left" Margin="107,0,0,0" Name=
"rssURL" Text="http://www.cnblogs.com/rss" VerticalAlignment="Top" Width="349" />
  <Button Content="加载 RSS" Click="Button_Click" Margin="-6,72,6,552" Grid.Row="1" />
  <ListBox x:Name="listbox" Grid.Row="1" SelectionChanged="OnSelectionChanged" Margin=
```

```
"0,150,6, - 11">
        <ListBox.ItemTemplate>
            <DataTemplate>
                <Grid>
                    <Grid.RowDefinitions>
                        <RowDefinition Height = "Auto" />
                        <RowDefinition Height = "Auto" />
                        <RowDefinition Height = "60" />
                    </Grid.RowDefinitions>
                    <TextBlock Grid.Row = "0" Text = "{Binding Title}" Foreground = "Blue" />
                    <TextBlock Grid.Row = "1" Text = "{Binding PublishedDate}" Foreground = "Green" />
                    <TextBlock Grid.Row = "2" TextWrapping = "Wrap" Text = "{Binding PlainSummary}" />
                </Grid>
            </DataTemplate>
        </ListBox.ItemTemplate>
    </ListBox>
```

MainPage.xaml.cs 文件代码

```
using System.Windows;
using System.Windows.Controls;
using Microsoft.Phone.Controls;
using WindowsPhone.Helpers;
namespace ReadRssItemsSample
{
    public partial class MainPage : PhoneApplicationPage
    {
        private  string WindowsPhoneBlogPosts = "";
        public MainPage()
        {
            InitializeComponent();
        }
        private void Button_Click(object sender, RoutedEventArgs e)
        {
            if (rssURL.Text != "")
            {
                WindowsPhoneBlogPosts = rssURL.Text;
            }
            else
            {
                MessageBox.Show("请输入 RSS 地址!");
                return;
            }
            //加载 RSS 列表
            RssService.GetRssItems(
                WindowsPhoneBlogPosts,
                (items) => { listbox.ItemsSource = items; },
                (exception) => { MessageBox.Show(exception.Message); },
                null
                );
        }
```

```
        //查看文章的详细内容
        private void OnSelectionChanged(object sender, SelectionChangedEventArgs e)
        {
            if (listbox.SelectedItem == null)
                return;
            var template = (RssItem)listbox.SelectedItem;
            MessageBox.Show(template.PlainSummary);
            listbox.SelectedItem = null;
        }
    }
}
```

程序运行的效果如图 23.1 和图 23.2 所示。

图 23.1　RSS 列表

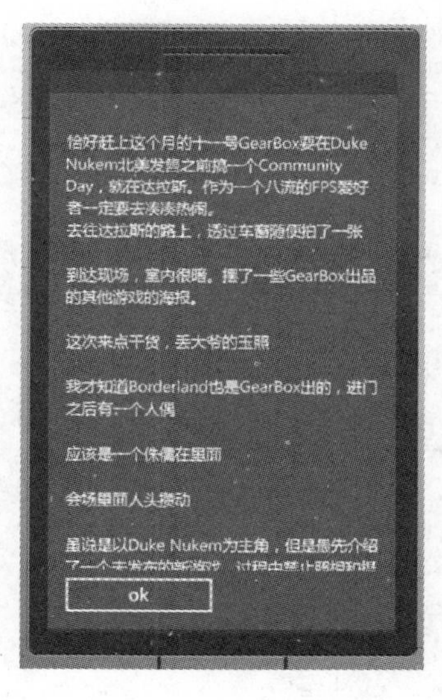

图 23.2　文章内容详情

23.2　博客园主页

实例博客园主页：在 WebBrowser 控件中操作博客园的网站。

实例说明：该实例演示了查看 WebBrowser 控件网页的 HTML 源代码以及在 WebBrowser 控件中执行页面的 JavaScript 方法。通过查看博客园(http://www.cnblogs.com)网页的源代码可以发现关闭博客园主页右侧遍览的 JavaScript 代码如下：

```
<script type="text/javascript">
    $("#span_ing").css("display", "inline");
    function close_side_right() {
        $("#side_right").css("display", "none");
        $("#main").css("margin-right", "10px");
        $("#rss_block").css("right", "10px");
    }
```

```
</script>
```

在实例中将会调用这一个 JavaScript 方法来关闭博客园主页右侧的边栏，实现了在 Windows Phone 客户端应用上面操作网页的脚本事件。

代码清单 23-2：博客园主页(源代码：第 23 章\Examples_23_2)

MainPage.xaml 文件主要代码

```
<Grid x:Name="ContentPanel" Grid.Row="1" Margin="12,0,12,0">
    <phone:WebBrowser HorizontalAlignment="Left" Name="webBrowser1" VerticalAlignment=
"Top" Height="431" Width="456" Margin="0,64,0,0"
                      Navigating="webBrowser1_Navigating"
                      NavigationFailed="webBrowser1_NavigationFailed"
                      LoadCompleted="webBrowser1_LoadCompleted" />
    <Button Content = " Go " Height = " 72 " HorizontalAlignment = " Left " Name = " go "
VerticalAlignment="Top" Width="160" Click="go_Click" />
    <Button Content="查看 HTML 源码" Height="72" HorizontalAlignment="Left" Margin="-
12,515,0,0" Name = "html" VerticalAlignment = "Top" Width = "226" Visibility = "Collapsed
" Click="html_Click" />
    <Button Content="关闭博客园右侧" Height="72" HorizontalAlignment="Left" Margin=
"220,515,0,0" Name="script" VerticalAlignment="Top" Width="230" Visibility="Collapsed"
Click="script_Click" />
</Grid>
```

MainPage.cs 文件代码

```
using System;
using System.Windows;
using Microsoft.Phone.Controls;
namespace WebBrowserScript
{
    public partial class MainPage : PhoneApplicationPage
    {
        public MainPage()
        {
            InitializeComponent();
        }
        //在 WebBrowser 控件中打开博客园网站
        private void go_Click(object sender, RoutedEventArgs e)
        {
            webBrowser1.IsScriptEnabled = true;
            webBrowser1.Navigate(new Uri("http://www.cnblogs.com/", UriKind.Absolute));
        }
        //查看博客园的网页 HTML 源码
        private void html_Click(object sender, RoutedEventArgs e)
        {
            MessageBox.Show(webBrowser1.SaveToString());
        }
        //执行 script,关闭博客园网页右侧
```

```
        private void script_Click(object sender, RoutedEventArgs e)
        {
            webBrowser1.InvokeScript("close_side_right");
        }
        //正在加载网页,这时候隐藏,查看 HTML 源码按钮和关闭博客园右侧按钮
        private void webBrowser1_Navigating(object sender, NavigatingEventArgs e)
        {
            html.Visibility = Visibility.Collapsed;
            script.Visibility = Visibility.Collapsed;
        }
        //导航失败
        private void webBrowser1_NavigationFailed(object sender, System.Windows.Navigation.
NavigationFailedEventArgs e)
        {
            MessageBox.Show("请检查网络!");
        }
        //网页加载完成,显示查看 HTML 源码按钮和关闭博客园右侧按钮
        private void webBrowser1_LoadCompleted(object sender, System.Windows.Navigation.
NavigationEventArgs e)
        {
            html.Visibility = Visibility.Visible;
            script.Visibility = Visibility.Visible;
        }
    }
}
```

程序运行的效果如图 23.3 和图 23.4 所示。

图 23.3　打开博客园的效果

图 23.4　查看博客园的源码

23.3 网络留言板

实例网络留言板：通过手机客户端结合云端主机的服务器实现一个留言板的功能。

实例说明：本实例使用 Windows Phone 客户端应用程序结合 WCF 实现一个网络留言板的实例，数据库存储使用 Access 关系型数据库，通过 WCF 服务来查询数据以及插入数据。数据库设计使用 Access 数据库创建一个表命名为 about，表的字段如表 23.1 所示。在 Windows Phone 客户端程序上会调用创建好的 WCF 服务器端的接口进行查询留言的记录和插入留言的记录。

表 23.1 information 表结构

字段名称	数据类型	说明
id	自动编号	自增长 id
name	文本	网名
description	文本	留言内容

代码清单 23-3：网络留言板(源代码：第 23 章\Examples_23_3)

1. 服务器端 WCF Service 的实现

Service1.svc 文件代码

```
using System;
using System.Collections.Generic;
using System.Linq;
using System.Runtime.Serialization;
using System.ServiceModel;
using System.ServiceModel.Web;
using System.Text;
using System.Data.OleDb;
using System.Data;
using System.Collections.ObjectModel;
namespace WcfService1
{
    public class Service1 : IService1
    {
        //获取留言板数据
        public List<string> GetMessage()
        {
            List<string> Messages = new List<string>();
            OleDbCommand cmd = new OleDbCommand();
            SQLExcute("SELECT * from about order by id desc", cmd);
            OleDbDataAdapter da = new OleDbDataAdapter();
            da.SelectCommand = cmd;
            DataSet ds = new DataSet();
            da.Fill(ds);
```

```
            if (ds.Tables[0] != null)
            {
                for (int i = 0; i < ds.Tables[0].Rows.Count; i++)
                {
                    Messages.Add(ds.Tables[0].Rows[i]["name"].ToString() + "|" + ds.Tables
[0].Rows[i]["description"].ToString());
                }
            }
            return Messages;
        }
        //往留言板插入数据
        public string InsertMessage(string name, string description)
        {
            try
            {
                string sql = "insert into about(name,description) values('" + name + "','"
+ description + "')";
                SQLExcute(sql);
                return "ok";
            }
            catch (Exception)
            {
                return "no";
            }
        }
        //SQL 的操作
        private void SQLExcute(string SQLCmd)
        {
            //请改成你的路径
            string ConnectionString = "PROVIDER = Microsoft.Jet.OLEDB.4.0;DATA SOURCE = D:\
code\WcfService1\WcfService1\App_Data\information.mdb";
            OleDbConnection conn = new OleDbConnection(ConnectionString);
            conn.Open();
            OleDbCommand cmd = new OleDbCommand();
            cmd.Connection = conn;
            cmd.CommandTimeout = 15;
            cmd.CommandType = CommandType.Text;
            cmd.CommandText = SQLCmd;
            cmd.ExecuteNonQuery();
            conn.Close();
        }
        //SQL 的操作是 SQLExcute 的重构
        private void SQLExcute(string SQLCmd, OleDbCommand Cmd)
        {
            //请改成你的路径
            string ConnectionString = "PROVIDER = Microsoft.Jet.OLEDB.4.0;DATA SOURCE = D:\
code\WcfService1\WcfService1\App_Data\information.mdb";
            OleDbCommand cmd = new OleDbCommand();
            OleDbConnection Conn = new OleDbConnection(ConnectionString);
            Conn.Open();
```

```
                Cmd.Connection = Conn;
                Cmd.CommandTimeout = 15;
                Cmd.CommandType = CommandType.Text;
                Cmd.CommandText = SQLCmd;
                Cmd.ExecuteNonQuery();
                Conn.Close();
            }
        }
    }
```

IService1.cs 文件代码

```
using System.Data;
using System.Collections.ObjectModel;
namespace WcfService1
{
    [ServiceContract]
    public interface IService1
    {
        [OperationContract]
        List<string> GetMessage();
        [OperationContract]
        string InsertMessage(string name, string description);
    }
}
```

2. Windows Phone 客户端的实现

MainPage.xaml 文件主要代码：留言列表页面

```
        <ListBox x:Name="MessagesListBox" Grid.Row="1"   >
            <ListBox.ItemTemplate>
                <DataTemplate>
                    <StackPanel x:Name="DataTemplateStackPanel" Orientation="Horizontal">
                        <StackPanel>
                            <StackPanel Orientation="Horizontal">
                                <TextBlock x:Name="name" Text="{Binding Name}" Margin=
"0,0,0,0" Style="{StaticResource PhoneTextExtraLargeStyle}"/>
                                <TextBlock x:Name="say" Text="留言：" Margin="10,0,0,
5" VerticalAlignment="Bottom"  Style="{StaticResource PhoneTextNormalStyle}"/>
                                <TextBlock x:Name="description" Text="{Binding
Description}" Margin="70,0,0,0" VerticalAlignment="Bottom"  Style="{StaticResource
PhoneTextAccentStyle}"/>
                            </StackPanel>
                        </StackPanel>
                    </StackPanel>
                </DataTemplate>
            </ListBox.ItemTemplate>
        </ListBox>
 ...
<phone:PhoneApplicationPage.ApplicationBar>
    <shell:ApplicationBar IsVisible="True" IsMenuEnabled="True">
```

```
        <shell:ApplicationBar.MenuItems>
            <shell:ApplicationBarMenuItem Text="发表留言" Click="New_Click"/>
        </shell:ApplicationBar.MenuItems>
    </shell:ApplicationBar>
</phone:PhoneApplicationPage.ApplicationBar>
```

MainPage.xaml.cs 文件代码：从 WCF Service 中获取留言列表的信息

```
using System;
using System.Collections.Generic;
using System.Linq;
using System.Net;
using System.Windows;
using System.Collections.ObjectModel;
using System.ComponentModel;
namespace WCFDemo
```

MainPage.xaml.cs 文件代码：从 WCF Service 中获取留言列表的信息

```
{
    public partial class MainPage : PhoneApplicationPage
    {
        ObservableCollection<message> Messages = new ObservableCollection<message>();
        public MainPage()
        {
            InitializeComponent();
            //实例化留言板 WCF Service 代理的对象
            WCFDemo.WCFService.Service1Client proxy = new WCFService.Service1Client();
            //注册 GetMessage 方法调用结束之后触发的事件
            proxy.GetMessageCompleted += new EventHandler<WCFService.
GetMessageCompletedEventArgs>(proxy_GetMessageCompleted);
            //异步调用 WCF Service 的 GetMessage 方法
            proxy.GetMessageAsync();
        }
        //获取消息列表
        void proxy_GetMessageCompleted(object sender, WCFService.GetMessageCompletedEventArgs e)
        {
            if (e.Error == null)
            {
                //获取 WCF Service 服务器端返回来的留言字符串,并进行分解
                List<string> list = e.Result.ToList<string>();
                for (int i = 0;i < list.Count; i++)
                {
                    string[] messges = list[i].ToString().Split('|');
                    Messages.Add(new message(messges[0], messges[1]));
                }
                //将留言列表的内容绑定到 ListBox 控件中
                this.MessagesListBox.ItemsSource = Messages;
            }
            else
            {
```

```
                MessageBox.Show("网络错误,请检查是否部署好 WCF service");
            }
        }
        //跳转到发表留言的页面
        private void New_Click(object sender, EventArgs e)
        {
            NavigationService.Navigate(new Uri("/addMessage.xaml", UriKind.Relative));
        }
    }
    //创建一个留言信息类
    public class message : INotifyPropertyChanged
    {
        public string name;
        public string description;
        //初始化留言信息类的构造方法
        public message(string myname, string mydescription)
        {
            name = myname;
            description = mydescription;
        }
        //留言者的名字
        public string Name
        {
            get { return name; }
            set
            {
                name = value;
                InvokePropertyChanged(new PropertyChangedEventArgs("Name"));
            }
        }
        //留言的内容
        public string Description
        {
            get { return description; }
            set
            {
                description = value;
                InvokePropertyChanged(new PropertyChangedEventArgs("Description"));
            }
        }
        //定义属性改变事件
        public event PropertyChangedEventHandler PropertyChanged;
        //实现属性改变事件
        private void InvokePropertyChanged(PropertyChangedEventArgs e)
        {
            var handler = PropertyChanged;
            if (handler != null) handler(this, e);
        }
    }
}
```

addMessage. xaml 文件主要代码：发表留言的页面

```
<Grid x:Name = "ContentPanel" Grid.Row = "1" Margin = "12,0,12,0">
    <TextBlock FontSize = "40" Height = "59" HorizontalAlignment = "Left" Margin = "9,23,0,0"
Name = "name1" Text = "你的网名：" VerticalAlignment = "Top" Width = "219" />
    <TextBox Height = "72" HorizontalAlignment = "Left" Margin = "-4,88,0,0" Name =
"yourName" Text = "" VerticalAlignment = "Top" Width = "460" />
    <TextBlock FontSize = "40" Height = "59" HorizontalAlignment = "Left" Margin = "9,166,0,0"
Name = "textBlock1" Text = "留言的内容：" VerticalAlignment = "Top" Width = "219" />
    <TextBox Height = "243" HorizontalAlignment = "Left" Margin = "0,231,0,0" Name =
"yourMessage" Text = "" VerticalAlignment = "Top" Width = "460" />
    <Button Content = "发表留言" Height = "72" HorizontalAlignment = "Left" Margin = "83,516,
0,0" Name = "save" VerticalAlignment = "Top" Width = "231" Click = "save_Click" />
</Grid>
```

addMessage. xaml. cs 文件主要代码：调用 WCF Service 提交内容到 Access 数据库中去

```
//发表留言
private void save_Click(object sender, RoutedEventArgs e)
{
    if (yourName.Text == "")
    {
        MessageBox.Show("请输入你的网名!");
    }
    else if (yourMessage.Text == "")
    {
        MessageBox.Show("请输入你的留言!");
    }
    else
    {
        //实例化留言板 WCF Service 代理的对象
        WCFDemo.WCFService.Service1Client proxy = new WCFService.Service1Client();
        //注册 InsertMessage 方法调用结束之后触发的事件
        proxy.InsertMessageCompleted += new EventHandler<WCFService.
InsertMessageCompletedEventArgs>(proxy_InsertMessageCompleted);
        //异步调用 WCF Service 的 InsertMessage 方法
        proxy.InsertMessageAsync(yourName.Text, yourMessage.Text);
    }
}
void proxy_InsertMessageCompleted(object sender, WCFService.InsertMessageCompletedEventArgs e)
{
    if (e.Error == null)
    {
        //成功后返回留言板列表页面
        NavigationService.Navigate(new Uri("/MainPage.xaml", UriKind.Relative));
    }
    else
    {
        MessageBox.Show("网络错误,请检查是否部署好 WCF service");
    }
}
```

程序运行的效果如图 23.5 和图 23.6 所示。

图 23.5　网络留言表列表运行的效果

图 23.6　发表留言页面

23.4　快递 100

实例快递 100：实现中国大部分快递公司的快递跟踪查询。

实例说明：本实例是通过调用快递 100(http://www.kuaidi100.com)网站的快递查询接口来获取快递的配送信息的，使用快递 100 的接口需要向快递 100 官方网站申请的开发者 AppKey，只有成功申请到开发者的 AppKey 才可以使用其网站提供的快递查询的接口。该应用在成功查询了快递的配送信息后会自动把快递的信息保存到查询的历史记录中去，这样可以免去重复输入快递单号进行多次查询，可以直接刷新历史记录就可以跟踪到快递单的最新配送信息，快递信息的历史记录使用了 Windows Phone 的第三方数据组件 Perst 嵌入式数据库来保存数据。本实例已经发布到 Windows Phone 的 App Market 上面，可以免费下载来查看应用的运行效果。

代码清单 23-4：快递 100(源代码：第 23 章\Examples_23_4)

1. 创建 Perst 数据库的相关处理类和代码

App.xaml.cs 文件主要代码：处理初始化数据库和关闭数据库

```
using System.Windows;
using System.Windows.Navigation;
using Microsoft.Phone.Controls;
using Microsoft.Phone.Shell;
using Perst;
using System.IO.IsolatedStorage;
using Perst.FullText;
namespace 快递 100
{
```

```
public partial class App : Application
{
    …
    //定义一个数据库对象
    public Database Database { get; internal set; }
    //打开数据库方法
    internal void OpenPerstDatabase()
    {
        using (var stor = IsolatedStorageFile.GetUserStoreForApplication())
        {
            InitializePerstStorage();
        }
    }
    //初始化数据库方法
    internal void InitializePerstStorage()
    {
        //创建 Perst 存储 Storage 实例
        Storage storage = StorageFactory.Instance.CreateStorage();
        //初始化存储大小为 512KB
        storage.SetProperty("perst.file.extension.quantum", 512 * 1024);
        //每次递增的存储大小为 256KB
        storage.SetProperty("perst.extension.quantum", 256 * 1024);
        //打开 Storage
        storage.Open("PerstDemoDB.dbs", 0);
        //使用上面初始化的 Storage 实例创建数据库
        Database = new Database(storage, false, true, new FullTextSearchHelper(storage));
        //关闭自动索引,使用人工索引
        Database.EnableAutoIndices = false;
    }
    //关闭数据库方法
    internal void ClosePerstDatabase()
    {
        if (Database != null && Database.Storage != null)
            //关闭数据库存储
            Database.Storage.Close();
    }
    //应用初始化事件处理
    public App()
    {
        OpenPerstDatabase();
        …
    }
    …
    //应用关闭事件处理
    private void Application_Closing(object sender, ClosingEventArgs e)
    {
        //关闭数据库
        ClosePerstDatabase();
    }
    …
```

```
        }
    }
```

Base.cs 文件代码：创建一个数据库存储的面向对象的类的基类

```
using System.Windows;
using Perst;
namespace 快递100
{
    public class Base : Persistent
    {
        //获取数据库对象
        protected static Database Database
        {
            get { return ((App)Application.Current).Database; }
        }
        //删除记录
        public override void Deallocate()
        {
            Database.DeleteRecord(this);
        }
        //保存,相当于保存表
        public void Save()
        {
            Store();
            //更新索引
            Database.UpdateFullTextIndex(this);
        }
    }
}
```

History.cs 文件代码：快递查询历史记录的数据表类

```
using Perst;
using System.ComponentModel;
using Perst.FullText;
namespace 快递100
{
    public class History : Persistent, INotifyPropertyChanged
    {
        [FullTextIndexable]
        //快递公司名称
        public string companyName;
        [FullTextIndexable]
        //快递公司代码
        public string companyNO;
        [FullTextIndexable]
        //快递单号
        public string expressNO;
        [FullTextIndexable]
        //最新的时间
```

```
public string time;
[FullTextIndexable]
//最新的状态信息
public string context;
[FullTextIndexable]
//查询的地址
public string url;
//重载 Persistent 类的 OnLoad 方法
public override void OnLoad()
{
    base.OnLoad();
}
public string CompanyName
{
    get { return companyName; }
    set
    {
        companyName = value;
        InvokePropertyChanged(new PropertyChangedEventArgs("CompanyName"));
    }
}
public string CompanyNO
{
    get { return companyNO; }
    set
    {
        companyNO = value;
        InvokePropertyChanged(new PropertyChangedEventArgs("CompanyNO"));
    }
}
public string ExpressNO
{
    get { return expressNO; }
    set
    {
        expressNO = value;
        InvokePropertyChanged(new PropertyChangedEventArgs("ExpressNO"));
    }
}
public string Time
{
    get { return time; }
    set
    {
        time = value;
        InvokePropertyChanged(new PropertyChangedEventArgs("Time"));
    }
}
public string Context
{
```

```
                get { return context; }
                set
                {
                    context = value;
                    InvokePropertyChanged(new PropertyChangedEventArgs("Context"));
                }
            }
            public string Url
            {
                get { return url; }
                set
                {
                    this.url = value;
                    InvokePropertyChanged(new PropertyChangedEventArgs("Url"));
                }
            }
            public event PropertyChangedEventHandler PropertyChanged;
            //删除对象的数据
            public override void Deallocate()
            {
                base.Deallocate();
            }
            //属性改变事件
            private void InvokePropertyChanged(PropertyChangedEventArgs e)
            {
                var handler = PropertyChanged;
                if (handler != null) handler(this, e);
            }
        }
    }
```

2. 数据绑定的面向对象实体类和历史记录数据操作的类

Company.cs 文件代码：快递公司实体类

```
namespace 快递100
{
    public class Company
{
    //公司名称
        public string Name { get; set; }
        //公司编码
        public string NO { get; set;}
    }
}
```

Common.cs 文件代码：快递公司信息公共类

```
namespace 快递100
{
    public class Common
{
```

```
        //通过公司名获取公司的编码
        static public string GetExpressCompanyNO(string name)
        {
            string typeCom = name;
            if (typeCom == "AAE全球专递")
            {
                typeCom = "aae";
            }
            …
        //获取快递查询所支持的所有公司名称和编码
        static public string[,] GetExpressCompany()
        {
            string typeCom = "";
            typeCom += "AAE全球专递:aae";
            typeCom += "|AAE安捷快递:anjiekuaidi";
            typeCom += "|安信达快递:anxindakuaixi";
            …
            string[] typeCom1 = typeCom.Split('|');
            string[,] typeCom11 = new string[typeCom1.Length,2];
            for(int i = 0;i < typeCom1.Length;i++)
            {
                typeCom11[i,0] = typeCom1[i].Split(':')[0];
                typeCom11[i,1] = typeCom1[i].Split(':')[1];
            }
            return typeCom11;
        }
    }
}
```

Data.cs 文件代码：快递查询的查询信息类

```
using System.ComponentModel;
namespace 快递100
{
    public class Data: INotifyPropertyChanged
    {
        //时间
        private string time;
        //内容
        private string context;
        public string Time
        {
            get
            {
                return time;
            }
            set
            {
                if (value != time)
                {
                    time = value;
```

```
                    NotifyPropertyChanged("Time");
                }
            }
        }
        public string Context
        {
            get
            {
                return context;
            }
            set
            {
                if (value != context)
                {
                    context = value;
                    NotifyPropertyChanged("Context");
                }
            }
        }
        public event PropertyChangedEventHandler PropertyChanged;
        //构造
        public Data(string time, string context)
        {
            Time = time;
            Context = context;
        }
        //构造
        public Data()
        {
        }
        //用于绑定属性值改变触发的事件,动态改变
        private void NotifyPropertyChanged(string property)
        {
            if (PropertyChanged != null)
            {
                PropertyChanged(this, new PropertyChangedEventArgs(property));
            }
        }
    }
}
```

Express.cs 文件代码：快递查询的查询信息集合类,用于绑定查询出来的快递信息

```
using System;
using System.ComponentModel;
using System.Collections.ObjectModel;
namespace 快递100
{
    public class Express : INotifyPropertyChanged
    {
        //返回的消息
```

```
private string message;
//返回的状态
private string status;
//快递的信息集合
public ObservableCollection<Data> DataList
{
    get;
    set;
}
public String Message
{
    get
    {
        return message;
    }
    set
    {
        if (value != message)
        {
            message = value;
            NotifyPropertyChanged("Message");
        }
    }
}
public String Status
{
    get
    {
        return status;
    }
    set
    {
        if (value != status)
        {
            status = value;
            NotifyPropertyChanged("Status");
        }
    }
}
//初始化快递查询信息集合类
public Express()
{
    DataList = new ObservableCollection<Data>();
}
public event PropertyChangedEventHandler PropertyChanged;
private void NotifyPropertyChanged(string property)
{
    if (PropertyChanged != null)
    {
        PropertyChanged(this, new PropertyChangedEventArgs(property));
```

```
            }
        }
    }
}
```

HistorysViewModel. cs 文件代码：历史记录数据绑定的集合类

```
using System;
using System.Windows;
using System.ComponentModel;
using System.Collections.ObjectModel;
using Perst;
namespace 快递100
{
    public class HistorysViewModel : INotifyPropertyChanged
    {
        public HistorysViewModel()
        {
            Historys = new ObservableCollection<History>();
            //从数据库中获取所有的 Account 记录
            if (Database != null)
            {
                //数据库查询,查询出 Account 类(相当于表)的所有对象,通过时间进行排序
                Historys = Database.Select<History>("order by Time").ToObservableCollection();
            }
        }
        public ObservableCollection<History> Historys { get; private set; }
        private static Database Database
        {
            get { return ((App)Application.Current).Database; }
        }
        private static Storage Storage
        {
            get { return Database.Storage; }
        }
        public event PropertyChangedEventHandler PropertyChanged;
        private void NotifyPropertyChanged(String propertyName)
        {
            if (null != PropertyChanged)
            {
                PropertyChanged(this, new PropertyChangedEventArgs(propertyName));
            }
        }
    }
}
```

3. 快递查询的页面

MainPage. xaml 文件主要代码：使用 Panorama 控件来布局快递 100 应用的界面

```
<phone:PhoneApplicationPage
    …
```

```
    xmlns:phone = "clr - namespace:Microsoft.Phone.Controls;assembly = Microsoft.Phone"
    xmlns:shell = "clr - namespace:Microsoft.Phone.Shell;assembly = Microsoft.Phone"
    xmlns:controls = "clr - namespace: Microsoft.Phone.Controls; assembly = Microsoft.Phone.
Controls"
    xmlns: watermark = " clr - namespace: WatermarkedTextBoxControl; assembly =
WatermarkedTextBoxControl"
    xmlns:toolkit = "clr - namespace: Microsoft.Phone.Controls; assembly = Microsoft.Phone.
Controls.Toolkit">
    <!-- 添加按钮的样式到应用的资源中 -->
    <phone:PhoneApplicationPage.Resources>
        <ResourceDictionary>
            <ResourceDictionary.MergedDictionaries>
                <ResourceDictionary Source = "ButtonStyle.xaml"/>
            </ResourceDictionary.MergedDictionaries>
        </ResourceDictionary>
    </phone:PhoneApplicationPage.Resources>
    <Grid x:Name = "LayoutRoot" Background = "AntiqueWhite">
        <!-- 添加列表采集器的样式到 Grid 控件的资源中 -->
        <Grid.Resources>
            <ResourceDictionary>
                <ResourceDictionary.MergedDictionaries>
                    <ResourceDictionary Source = "ListPickerStyle.xaml"/>
                </ResourceDictionary.MergedDictionaries>
            </ResourceDictionary>
        </Grid.Resources>
        …
        <Grid x:Name = "ContentPanel" Grid.Row = "1" Margin = "12,0,12,0">
            <controls:Panorama Title = "">
                <controls:PanoramaItem Header = "快递查询" Foreground = "Blue">
                        <StackPanel>
                        <toolkit: ListPicker x: Name = " listPicker " toolkit: TiltEffect.
IsTiltEnabled = "true" Style = "{StaticResource style}"
                                ItemTemplate = " { StaticResource PickerItemTemplate }"
FullModeItemTemplate = "{StaticResource PickerFullModeItemTemplate}"
                                FullModeHeader = "快递公司"
                                Header = " Cities " Height = " 86 " HeaderTemplate =
"{StaticResource PickerHeadTemplate}"/>
                        <watermark:WatermarkedTextBox BorderBrush = "Blue" Name = "No" Watermark
= "请输入您要查询的单号" WatermarkStyle = "{StaticResource styleBlue}"Height = "73" />
                        <Button Click = " Button _ Click " Content = " 查 询 " Style =
"{StaticResource CustomButton}" Height = "80"BorderThickness = "0" />
                        <ListBox Name = "expressList"Height = "383" >
                            <ListBox.ItemTemplate>
                                <DataTemplate>
                                    <StackPanel>
                                        <TextBlock Text = "{Binding Time}" TextWrapping
= "Wrap"Foreground = "Blue" />
                                        <TextBlock Text = "{Binding Context}" TextWrapping
= "Wrap" FontWeight = "Bold"Foreground = "Blue"/>
                                    </StackPanel>
```

```
                                </DataTemplate>
                            </ListBox.ItemTemplate>
                        </ListBox>
                    </StackPanel>
                </controls:PanoramaItem>
                <controls:PanoramaItem Header="历史记录"  Foreground="Red">
                    <ListBox x:Name="HistorysListBox"toolkit:TiltEffect.IsTiltEnabled=
"true">
                        <ListBox.ItemTemplate>
                            <DataTemplate>
                                <StackPanel x:Name="DataTemplateStackPanel" Orientation
="Horizontal" >
                                    <Button Width="120" Height="120" Click="update_
Click" Name="update" Background="Red" Foreground="White" BorderThickness="0" Content=
"刷新"></Button>
                                    <StackPanel>
                                        <TextBlock x:Name="companyName" Text="{Binding
CompanyName}" Margin="0,0,0,0" Foreground="Red"/>
                                        <TextBlock x:Name="expressNO" Text="{Binding
ExpressNO}" Margin="10,0,0,5"Foreground="Red"/>
                                        <TextBlock x:Name="time" Text="{Binding Time}"
Margin="0,-6,0,3" Foreground="Red" />
                                        <TextBlock x:Name="context" Text="{Binding
Context}" Margin="0,-6,0,3" Foreground="Red"FontWeight="Bold"/>
                                    </StackPanel>
                                    <toolkit:ContextMenuService.ContextMenu>
                                        <toolkit:ContextMenu>
                                                <toolkit:MenuItem Header="删除" Click=
"MenuItem_Click"/>
                                        </toolkit:ContextMenu>
                                    </toolkit:ContextMenuService.ContextMenu>
                                </StackPanel>
                            </DataTemplate>
                        </ListBox.ItemTemplate>
                    </ListBox>
                </controls:PanoramaItem>
                <controls:PanoramaItem Header="关于"Foreground="YellowGreen">
                    <ListBox>
                        <StackPanel Orientation="Horizontal" Margin="0,0,0,0" >
                            <Button toolkit:TiltEffect.IsTiltEnabled="True" Style=
"{StaticResource myButton}" Name="send" Height="160" Width="160" Background=
"YellowGreen" BorderThickness="0" Content="反馈问题" Margin="12,0,0,0" Click="send_
Click"></Button>
                            <Button toolkit:TiltEffect.IsTiltEnabled="True"  Style=
"{StaticResource myButton}" Name="home" Height="160" Width="160" Background=
"YellowGreen" BorderThickness="0" Content="作者主页" Margin="12,0,0,0" Click="home_
Click"></Button>
                        </StackPanel>
                        <StackPanel Orientation="Horizontal" Margin="0,0,0,0" >
                            <TextBlock Foreground="YellowGreen" Width="400" Text="使用
```

```
帮助：每次查询会自动保存查询的成功的历史记录，可以通过长按历史记录带出菜单进行删除，单击刷新可以更新快递的最新情况." FontWeight = "Bold" FontSize = "23" TextWrapping = "Wrap"Margin = "0,12,0,0"></TextBlock>
                        </StackPanel>
                        <StackPanel Orientation = "Horizontal" Margin = "0,0,0,0" >
                            <TextBlock Foreground = "YellowGreen" Width = "400"
Text = "http://www.cnblogs.com/linzheng" FontSize = "20" TextWrapping = "Wrap" FontWeight = "Bold" Margin = "0,12,0,0"></TextBlock>
                        </StackPanel>
                        <StackPanel Orientation = "Horizontal" Margin = "0,0,0,0" >
                            <TextBlock Foreground = "YellowGreen" Width = "400" Text = "作者的书籍" FontSize = "20" FontWeight = "Bold" TextWrapping = "Wrap" Margin = "0, 12, 0, 0">
</TextBlock>
                        </StackPanel>
                        <StackPanel Orientation = "Horizontal" Margin = "0,0,0,0" >
                            <TextBlock Foreground = "YellowGreen" Width = "400" Text = "《深入浅出 Windows Phone 7 应用开发》" FontSize = "20" FontWeight = "Bold" TextWrapping = "Wrap"
Margin = "0,12,0,0"></TextBlock>
                        </StackPanel>
                    </ListBox>
                </controls:PanoramaItem>
            </controls:Panorama>
        </Grid>
    </Grid>
</phone:PhoneApplicationPage>
```

MainPage.xaml.cs 文件主要代码

```
using System;
using System.Collections.Generic;
using System.Linq;
using System.Net;
using System.Windows;
using System.Windows.Controls;
using Microsoft.Phone.Controls;
using System.IO;
using System.Collections.ObjectModel;
using System.Xml.Linq;
using Perst;
using Microsoft.Phone.Tasks;
namespace 快递100
{
    public partial class MainPage : PhoneApplicationPage
    {
        //快递查询信息的集合
        Express express;
        //公司名称
        public string companyName = "";
        //公司编码
        public string companyNO = "";
        //快递单号
```

```
        public string expressNO = "";
        //快递信息的时间
        public string time = "";
        //快递信息的内容
        public string context = "";
        //快递查询的接口地址
        public string url = "";
        public History edit = null;
        //手机当前网络的状态
        public bool networkIsAvailable;
        //进度条
        ProgressBar pro = new ProgressBar();
        public MainPage()
        {
            InitializeComponent();
            //获取快递公司的信息
            string[,] company = Common.GetExpressCompany();
            List<Company> source = new List<Company>();
            //将快递公司的信息用List<T>类型存储起来
            for (int i = 0; i < company.Length/2; i++)
            {
                source.Add(new Company() { Name = company[i, 0].ToString(), NO = company
[i, 1].ToString() });
            }
            //将快递公司的信息集合绑定到listPicker控件的数据源
            this.listPicker.ItemsSource = source;
            //将快递查询的历史记录信息集合绑定到HistorysListBox控件的数据源
            this.HistorysListBox.ItemsSource = new HistorysViewModel().Historys;
            //判断当前手机网络是否可用
            networkIsAvailable = Microsoft.Phone.Net.NetworkInformation.NetworkInterface.
GetIsNetworkAvailable();
        }
        //快递查询的事件处理
        private void Button_Click(object sender, RoutedEventArgs e)
        {
            if (!networkIsAvailable)
            {
                MessageBox.Show("手机网络不可用!");
                return;
            }
            express = new Express();
            //获取listPicker控件选中的公司的编码
            string company = ((Company)listPicker.SelectedItem).NO;
            string no = No.Text;
            if (no == "")
            {
                MessageBox.Show("请输入单号!");
                return;
            }
            //查询快递信息
```

```
            GetExpress(company, no);
        }
        //获取快递信息
        public void GetExpress(string company,string no)
        {
            try
            {
                //设置进度条为重复模式进度条
                pro.IsIndeterminate = true;
                pro.Height = 20;
                //将进度条添加到页面上
                ContentPanel.Children.Add(pro);
                time = "";
                context = "";
                //拼接快递查询接口的地址和传递参数
                UriBuilder fullUri = new UriBuilder("http://api.kuaidi100.com/api?id=你
申请的AppKey&com=" + company + "&nu=" + no + "&show=1&muti=1&order=desc");
                companyName = ((Company)listPicker.SelectedItem).Name;
                companyNO = company;
                expressNO = no;
                url = "http://api.kuaidi100.com/api?id=你申请的AppKey&com=" + company
+ "&nu=" + no + "&show=1&muti=1&order=desc";
                //创建WebRequest类
                WebRequest request = HttpWebRequest.Create(fullUri.Uri);
                //返回异步操作的状态
                IAsyncResult result = (IAsyncResult)request.BeginGetResponse
(ResponseCallback, request);
            }
            catch (FormatException)
            {
                MessageBox.Show("请检查手机网络!");
                //设置进度条为不可见
                ContentPanel.Children.Remove(pro);
                return;
            }
        }
        //异步获取信息
        private void ResponseCallback(IAsyncResult asyncResult)
        {
            try
            {
                //获取异步操作返回的信息
                HttpWebRequest request = (HttpWebRequest)asyncResult.AsyncState;
                //结束对网络资源的异步请求
                WebResponse response = request.EndGetResponse(asyncResult);
                Stream streamResult;
                string message = "";
                string status = "";
                //创建一个临时的快递信息的集合
                ObservableCollection<Data> newDataList =
```

```
        new ObservableCollection<Data>();
    //获取接口返回的数据流
    streamResult = response.GetResponseStream();
    //加载XML
    XElement xmlData = XElement.Load(streamResult);
    //找到message节点下的信息,表示快递的信息
    message = xmlData.Descendants("message").First().Value.ToString();
    //找到status节点下的信息,表示查询的状态
    status = xmlData.Descendants("status").First().Value.ToString();
    //status != "1"表示接口失败
    if (status != "1")
    {
        Deployment.Current.Dispatcher.BeginInvoke(() =>
        {
            MessageBox.Show(message);
            //设置进度条为不可见
            ContentPanel.Children.Remove(pro);
        });
        return;
    }
    Data newData;
    //解析XML里面的快递信息并绑定到ListBox控件的数据源显示出来
    foreach (XElement curElement in xmlData.Descendants("data"))
    {
        try
        {
            newData = new Data();
            newData.Time = (string)(curElement.Element("time").Value);
            newData.Context = (string)(curElement.Element("context").Value);
            newDataList.Add(newData);
            if (time == "")
            {
                time = (string)(curElement.Element("time").Value);
            }
            if (context == "")
            {
                context = (string)(curElement.Element("context").Value);
            }
        }
        catch (FormatException)
        {
        }
    }
    //获取在App中定义的数据库对象
    Database Database = ((App)App.Current).Database;
    if (edit != null)
    {
        Deployment.Current.Dispatcher.BeginInvoke(() =>
```

```
                {
                    edit.Time = time;
                    edit.Context = context;
                    Database.AddRecord(edit);
                    //关闭数据库存储
                    Database.Storage.Commit();
                    edit = null;
                    time = "";
                    context = "";
                    ContentPanel.Children.Remove(pro);
                });
            }
            else
            {
                //初始化一个表对象
                History tem1 = new History { CompanyName = companyName, CompanyNO =
companyNO, ExpressNO = expressNO, Time = time, Context = context, Url = url };
                //添加数据库记录
                Database.AddRecord(tem1);
                //关闭数据库存储
                Database.Storage.Commit();
                Deployment.Current.Dispatcher.BeginInvoke(() =>
                {
                    express.Message = message;
                    express.Status = status;
                    express.DataList.Clear();
                    foreach (Data data in newDataList)
                    {
                        express.DataList.Add(data);
                    }
                    DataContext = express;
                    expressList.ItemsSource = express.DataList;
                    this.HistorysListBox.ItemsSource = new HistorysViewModel().Historys;
                    ContentPanel.Children.Remove(pro);
                });
            }
        }
        catch (FormatException)
        {
            Deployment.Current.Dispatcher.BeginInvoke(() =>
            {
                MessageBox.Show("请检查手机网络!");
                ContentPanel.Children.Remove(pro);
            });
            return;
        }
    }
```

```
        …
        //刷新历史记录
        private void update_Click(object sender, RoutedEventArgs e)
        {
            if (!networkIsAvailable)
            {
                MessageBox.Show("手机网络不可用!");
                return;
            }
            //获取单击的按钮实例
            var button = sender as Button;
            if (button != null)
            {
                //获取当前按钮绑定的DataContext,即当前的编辑的EmployeeTable实例
                edit = button.DataContext as History;
                GetExpress(edit.CompanyNO, edit.ExpressNO);
            }
        }
        //这个是MenuItem的单击事件
        private void MenuItem_Click(object sender, RoutedEventArgs e)
        {
            //获取选中的menuItem对象
            MenuItem menuItem = (MenuItem)sender;
            //获取对象的标题头的内容
            string header = (sender as MenuItem).Header.ToString();
            //获取选中的ListBoxItem
            ListBoxItem selectedListBoxItem = this.HistorysListBox.ItemContainerGenerator.
ContainerFromItem((sender as MenuItem).DataContext) as ListBoxItem;
            //如果没有选中则返回
            if (selectedListBoxItem == null)
            {
                return;
            }
            if (menuItem.Header.ToString() == "删除")
            {
                History his = menuItem.DataContext as History;
                //获取在App中定义的数据库对象
                Database Database = ((App)App.Current).Database;
                Database.DeleteRecord(his);
                //关闭数据库存储
                Database.Storage.Commit();
                this.HistorysListBox.ItemsSource = new HistorysViewModel().Historys;
            }
        }
    }
}
```

程序运行的效果如图 23.7～图 23.10 所示。

图 23.7 快递公司选择

图 23.8 快递查询页面

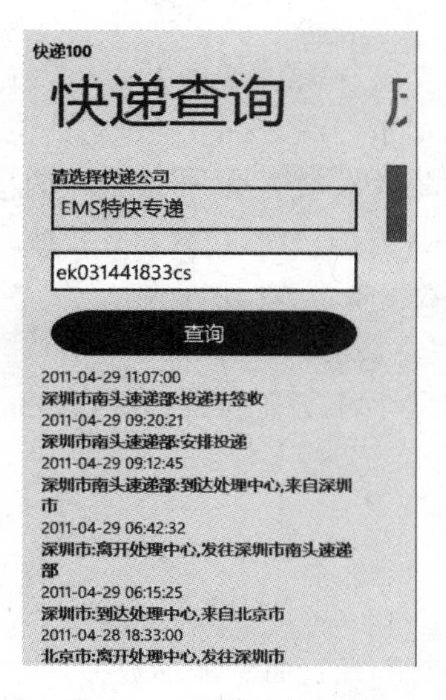

图 23.9 快递查询结果

图 23.10 历史记录

第24章

记账本应用

本章介绍一个记账本应用的开发，讲解在 Windows Phone 平台下如何开发一个记账本的应用软件。每一个软件都是从刚开始的第一个版本以及基本的功能开始，然后逐步地进行完善和改进，这是软件工程项目管理的过程。那么这个记账本的应用就是第一个版本，完成了记账的基本功能，读者可以在这个基础上继续改进和开发。

24.1 记账本简介

记账是理财的一部分，随着智能手机和网络的发展，越来越多的人选择了使用智能手机记账软件来进行记账，因为这种记账方式既方便又简单。那么一个手机上的记账软件需要具备哪些功能呢？首先，最基本的功能就是添加一笔收入和支出，新增记账的记录，查看记账记录，查看分类报表收支报表等；其次就是一些智能的功能，包括超支提醒、智能理财计划分析等；最后还会包括一些安全类的功能，比如数据备份数据恢复等。

本实例的记账本应用实现了记账应用软件的基本功能，包括添加一笔收入，添加一笔支出、月报表、年报表、查询记录、分类图表。其中，理财计划、添加类别和设置的相关模块留着读者发挥想象的空间可以继续进行开发和完善。

24.2 对象序列化存储

记账本应用采用了独立存储作为数据存储的机制，通过将记账的数据对象序列化成一个 XML 文件，然后将文件保存到独立存储里面，读取数据的时候在打开独立存储文件，把数据通过反序列化转化为具体的对象数据。这种数据存储的机制对于存储的数据比较灵活，把各种的数据结构体交给序列化类 DataContractSerializer 类来进行处理就可以了，不过这种机制对于大数据量的效率不高，有兴趣的读者可以使用 SQL Server CE 数据库存储来编写数据存储模块，SQL Server CE 数据库的知识请参考前面章节的知识。

下面来看一下数据存储的公共类。

IsolatedStorageHelper.cs 文件代码：IsolatedStorageHelper 类是独立存储的公共类，负责数据文件的读写操作

```
using System;
using System.IO;
using System.IO.IsolatedStorage;
using System.Runtime.Serialization;
namespace AccountBook
{
    public class IsolatedStorageHelper
    {
        //获取应用独立存储文件
        private static IsolatedStorageFile iso = IsolatedStorageFile.GetUserStoreForApplication();
        ///<summary>
        ///创建一个独立存储文件
        ///</summary>
        ///<param name="path">文件路径</param>
        ///<returns>独立存储文件流</returns>
        public static IsolatedStorageFileStream CreateFile(string path)
        {
            try
            {
                //如果文件存在需要先将文件删除
                if (isFileExist(path))
                {
                    DeleteFile(path);
                }
                return iso.CreateFile(path);
            }
            catch (Exception)
            {
                return null;
            }
        }
        ///<summary>
        ///除一个独立存储文件
        ///</summary>
        ///<param name="path">文件路径</param>
        ///<returns>是否删除成功</returns>
        public static bool DeleteFile(string path)
        {
            try
            {
                iso.DeleteFile(path);
                return true;
            }
            catch (Exception)
            {
                return false;
            }
```

```
}
///<summary>
///判断独立存储文件是否存在
///</summary>
///<param name="path">文件路径</param>
///<returns>是否存在</returns>
public static bool isFileExist(string path)
{
    return iso.FileExists(path);
}
///<summary>
///打开一个独立存储文件
///</summary>
///<param name="path">文件路径</param>
///<param name="mode">文件模型</param>
///<returns>独立存储文件流</returns>
public static IsolatedStorageFileStream OpenFile(string path, FileMode mode)
{
    if (isFileExist(path))
    {
        return iso.OpenFile(path, mode);
    }
    return null;
}
///<summary>
///读取独立存储文件流的内容
///</summary>
///<param name="fs">独立存储文件流</param>
///<returns>文件内容字符串</returns>
public static string ReadAllText(IsolatedStorageFileStream fs)
{
    try
    {
        //转化为刻度流
        StreamReader reader = new StreamReader(fs);
        //把流转化为字符串
        string str = reader.ReadToEnd();
        //关闭流
        reader.Close();
        return str;
    }
    catch (Exception)
    {
        return "";
    }
}
///<summary>
///读取独立存储文件流的内容
///</summary>
///<param name="FileFullName">文件路径</param>
```

```
///< returns >文件内容字符串</returns >
public static string ReadAllText(string FileFullName)
{
    try
    {
        IsolatedStorageFileStream fs = OpenFile(FileFullName, FileMode.Open);
        if (fs == null)
        {
            return "";
        }
        string str = ReadAllText(fs);
        fs.Close();
        return str;
    }
    catch (Exception)
    {
        return null;
    }
}
///< summary >
///读取文件里面的
///</summary >
///< param name = "path">文件路径</param >
///< param name = "type">对象类型</param >
///< returns > XML 序列化对象</returns >
public static object ReadObjectFromFile(string path, Type type)
{
    try
    {
        FileStream stream = OpenFile(path, FileMode.Open);
        object obj2 = new DataContractSerializer(type).ReadObject(stream);
        stream.Close();
        return obj2;
    }
    catch (Exception)
    {
        return null;
    }
}
///< summary >
///把字符串写入独立存储文件
///</summary >
///< param name = "fs">独立存储文件流</param >
///< param name = "Text">字符串</param >
///< returns >是否成功</returns >
public static bool WriteAllText(IsolatedStorageFileStream fs, string Text)
{
    try
    {
        StreamWriter writer = new StreamWriter(fs);
```

```
                writer.Write(Text);
                writer.Close();
                return true;
            }
            catch (Exception)
            {
                return false;
            }
        }
        ///<summary>
        ///把字符串写入独立存储文件
        ///</summary>
        ///<param name="FileFullName">文件名字</param>
        ///<param name="Text">字符串</param>
        ///<returns>是否成功</returns>
        public static bool WriteAllText(string FileFullName, string Text)
        {
            IsolatedStorageFileStream fs = CreateFile(FileFullName);
            if (WriteAllText(fs, Text))
            {
                fs.Close();
                return true;
            }
            return false;
        }
        ///<summary>
        ///把可序列化对象写入独立存储文件
        ///</summary>
        ///<param name="path">文件路径</param>
        ///<param name="type">对象类型</param>
        ///<param name="obj">可序列化对象</param>
        ///<returns></returns>
        public static bool WriteObjectToFile(string path, Type type, object obj)
        {
            try
            {
                FileStream stream = CreateFile(path);
                new DataContractSerializer(type).WriteObject(stream, obj);
                stream.Close();
                return true;
            }
            catch (Exception)
            {
                return false;
            }
        }
    }
}
```

24.3 记账本首页磁贴设计

记账本应用的设计风格采用了 Windows Phone 的磁贴设计风格，通过单击相关功能的磁贴进入功能的页面，磁铁也可以动态地展示相关信息，比如收入的磁贴，在磁贴的右上角会显示总收入的数据。磁贴按钮的控件使用了开源控件 Coding4Fun 控件库里面的磁铁按钮控件来实现，Coding4Fun 控件库里面封装了很多种类的 Windows Phone 的开发控件，是一个非常优秀的控件库，有兴趣的读者可以深入地了解一下。

记账本应用的首页使用了 Panorama 控件进行布局，左边是功能的磁贴按钮，右边是今日的记账记录，如图 24.1 和图 24.2 所示。

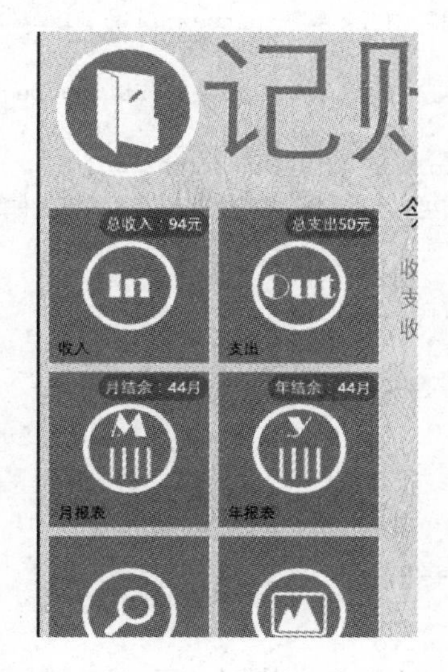

图 24.1 首页左边

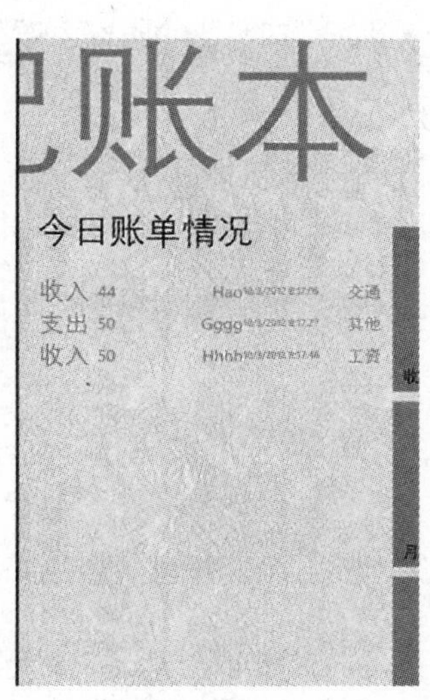

图 24.2 首页右边

首页的相关代码如下：

MainPage.xaml 文件主要代码

```
<controls:Panorama>
        <controls:Panorama.Background>
            <ImageBrush ImageSource = "Skin/PanoramaBG.jpg" />
        </controls:Panorama.Background>
        <controls:Panorama.Title>
            <StackPanel Orientation = "Horizontal">
                <Image Margin = "12,80,0,0" MaxHeight = "180" Source = "Skin/logo.png" />
                <TextBlock Margin = "0,30,0,0" Foreground = "#FFEB6416" Text = "记账本" />
            </StackPanel>
        </controls:Panorama.Title>
        <controls:PanoramaItem x:Name = "_columnItem" Foreground = "Black" >
            <Grid>
```

```
                    <ScrollViewer>
                        <StackPanel>
                            <!-- 收入、支出 -->
                            <StackPanel Orientation="Horizontal">
                                <c4fToolkit:Tile Margin=" 0, 0, 12, 0" Background=
"#FFEB6416" Width="203" Height="203" Title="收入"  Click="Income_Tile_Click">
                                    <Grid>
                                        <Image Source="Images/inlogo.png" Stretch=
"None" />
                                        <c4fToolkit:TileNotification x:Name="SummaryIncome"
  Content="总收入: 0" Background="{StaticResource PhoneAccentBrush}" Margin="0,2,0,0" />
                                    </Grid>
                                </c4fToolkit:Tile>
                                <c4fToolkit:Tile Margin="0, 0, 12, 0" Background="#
FFEB6416" Width="203" Height="203" Title="支出" Click="Expenses_Tile_Click">
                                    <Grid>
                                        <Image Source="Images/outlogo.png" Stretch=
"None" />
                                        <c4fToolkit: TileNotification  x: Name =
"SummaryExpenses" Content = "总支出: 0" Background = "{StaticResource PhoneAccentBrush}"
Margin="0,2,0,0" />
                                    </Grid>
                                </c4fToolkit:Tile>
                            </StackPanel>
                            …
                        </StackPanel>
                    </ScrollViewer>
                </Grid>
            </controls:PanoramaItem>
            <controls:PanoramaItem x:Name="_historyItem" Foreground="Black">
                <controls:PanoramaItem.Header>
                    <StackPanel Orientation="Horizontal">
                        <TextBlock Text = "今日账单情况" FontSize = "{StaticResource
PhoneFontSizeExtraLarge}" Margin="0,0,5,0" />
                    </StackPanel>
                </controls:PanoramaItem.Header>
                <Grid>
                    <ListBox x:Name="listToday" >
                        <ListBox.ItemTemplate>
                            <DataTemplate>
                                <StackPanel Orientation="Horizontal" Margin="12,0,0,0">
                                    <TextBlock Foreground = "#FFEB6416" FontSize = "30"
Text = "{Binding Type, Converter = {StaticResource VoucherTypeConverter}}" TextWrapping =
"NoWrap" HorizontalAlignment="Left" VerticalAlignment="Center" Width="69" />
                                    <TextBlock Foreground = "#FFEB6416" Text = "{Binding
Money}" TextWrapping = "NoWrap" HorizontalAlignment = "Left" VerticalAlignment = "Center"
Width="55" />
                                    <TextBlockForeground = "#FFEB6416" Text = "{Binding
Desc, Converter = { StaticResource VoucherDescConverter }}" TextWrapping = " NoWrap "
HorizontalAlignment="Left" TextAlignment="Right" VerticalAlignment="Center" Width="118" />
```

```
                                        <TextBlock Foreground = " #FFEB6416" FontSize = "12"
Text = "{Binding DT}" TextWrapping = "NoWrap" HorizontalAlignment = "Left" VerticalAlignment
 = "Center" Width = "94"/>
                                        <TextBlock Foreground = " #FFEB6416" FontSize = "20"
Text = " { Binding Category }" TextWrapping = " NoWrap " TextAlignment = " Right "
HorizontalAlignment = "Right" VerticalAlignment = "Center" Width = "70" />
                                    </StackPanel>
                                </DataTemplate>
                            </ListBox.ItemTemplate>
                        </ListBox>
                    </Grid>
                </controls:PanoramaItem>
            </controls:Panorama>
```

MainPage.xaml.cs 文件主要代码

```
//页面加载处理
private void MainPage_Loaded(object sender, RoutedEventArgs e)
{
    trexStoryboard.Begin();
    //设置收入 Tile 的总收入金额
    SummaryIncome.Content = "总收入:" + Common.GetSummaryIncome().ToString() + "元";
    //设置支出 Tile 的总支出金额
    SummaryExpenses.Content = "总支出" + Common.GetSummaryExpenses().ToString() + "元";
    //计算月结余
    double mouthIncome = Common.GetThisMouthSummaryIncome();
    double mouthExpenses = Common.GetThisMouthSummaryExpenses();
    MouthBalance.Content = "月结余:" + (mouthIncome - mouthExpenses).ToString() + "月";
    //计算年结余
    double yearIncome = Common.GetThisYearSummaryIncome();
    double yearExpenses = Common.GetThisYearSummaryExpenses();
    YearBalance.Content = "年结余:" + (yearIncome - yearExpenses).ToString() + "月";
    //获取今日的账单记录,并绑定到首页的 ListBox 控件进行显示
    listToday.ItemsSource = Common.GetThisDayAllRecords(DateTime.Now.Day, DateTime.Now.
Month, DateTime.Now.Year);
}
```

24.4 添加一笔收入

收入和支出的记账数据我们使用了同一个数据对象来保存，然后通过一个字段来区分，记账的数据对象如下：

Voucher.cs 文件代码：Voucher 类表示账单实体类

```
using System;
namespace AccountBook
{
    public class Voucher
    {
```

```
        //金额
        public double Money { get; set; }
        //账单类型 0 表示收入 1 表示支出
        public short Type { get; set; }
        //说明
        public string Desc { get; set; }
        //时间
        public DateTime DT { get; set; }
        //唯一 id
        public Guid ID { get; set; }
        //图片
        public byte[] Picture { get; set; }
        //图片高度
        public int PictureHeight { get; set; }
        //图片宽度
        public int PictureWidth { get; set; }
        //类别
        public string Category { get; set; }
    }
}
```

记账的数据需要通过独立存储文件来存储，所以还需要一个记账数据存储的帮助类，记账数据存储的帮助类如下：

VoucherHelpr.cs 文件代码：VoucherHelpr 类表示账单操作帮助类

```
using System;
using System.Collections.Generic;
namespace AccountBook
{
    public class VoucherHelpr
    {
        //记账列表
        private List<Voucher> _data;
        //初始化
        public VoucherHelpr()
        {
            if (!this.LoadFromFile())
            {
                this._data = new List<Voucher>();
            }
        }
        //添加一条记账记录
        public void AddNew(Voucher item)
        {
            item.ID = Guid.NewGuid();
            this.data.Add(item);
        }
        //读取记账列表
        public bool LoadFromFile()
        {
```

```
            this._data = IsolatedStorageHelper.ReadObjectFromFile("Voucher.dat", typeof
(List<Voucher>)) as List<Voucher>;
            return (this._data != null);
        }
        //移除一条记录
        public void Remove(Voucher item)
        {
            this.data.Remove(item);
        }
        //保存记账列表
        public bool SaveToFile()
        {
            return IsolatedStorageHelper.WriteObjectToFile("Voucher.dat", typeof(List
<Voucher>), this._data);
        }
        //获取记账列表
        public List<Voucher> data
        {
            get
            {
                if (this._data == null)
                {
                    this.LoadFromFile();
                }
                if (this._data == null)
                {
                    this._data = new List<Voucher>();
                }
                return this._data;
            }
        }
    }
}
```

数据保存的代码在 AddAccount.xaml.cs 文件下，代码如下：

AddAccount.xaml.cs 文件保存收入数据的处理代码

```
//一条记账记录的对象
Voucher voucher = new Voucher
{
    Money = double.Parse(this.textBox_Income.Text),
    Desc = this.textBox_IncomeDesc.Text,
    DT = DateTime.Parse(this.DatePickerIncome.Value.Value.ToString("yyyy/MM/dd") + " " +
this.TimePickerIncome.Value.Value.ToString("HH:mm:ss")),
    Category = listPickerIncome.SelectedItem.ToString(),
    Type = 0
};
//添加一条记录
App.voucherHelper.AddNew(voucher);
```

添加一笔收入的页面设计如图 24.3 所示。

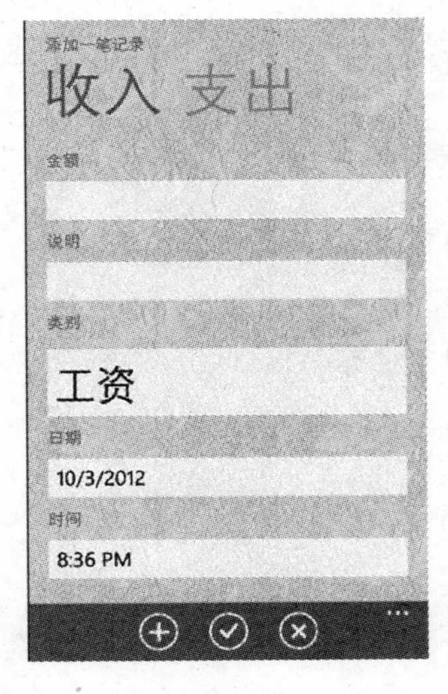

图 24.3　收入

24.5　添加一笔支出

支出的数据结构与收入的数据结构一样，添加支出支持添加图片的记录，通过 ImageHelper 类的相关方法来处理图片的保存。ImageHelper 类的代码如下：

ImageHelper.cs 文件代码：ImageHelper 类表示图片处理类

```
using System;
using System.IO;
using System.Windows;
using System.Windows.Controls;
using System.Windows.Media.Imaging;
namespace AccountBook
{
    ///<summary>
    ///图片处理类
    ///</summary>
    public class ImageHelper
    {
        ///<summary>
        ///把字节数组转化为位图
        ///</summary>
        ///<param name="imageBytes">字节数组</param>
        ///<returns>位图</returns>
        public static BitmapImage ByteToImageSource(byte[] imageBytes)
        {
            BitmapImage image = new BitmapImage();
            MemoryStream streamSource = new MemoryStream(imageBytes);
            image.SetSource(streamSource);
```

```
            return image;
        }
        ///<summary>
        ///从图片的目录获取位图
        ///</summary>
        ///<param name="path">目录路径</param>
        ///<returns>位图</returns>
        public static BitmapImage GetImageFromContentPath(string path)
        {
            Uri uriResource = new Uri(path, UriKind.Relative);
            using (BinaryReader reader = new BinaryReader(Application.GetResourceStream
(uriResource).Stream))
            {
                return ByteToImageSource(reader.ReadBytes((int)reader.BaseStream.Length));
            }
        }
        ///<summary>
        ///从图片的资源的路径获取位图
        ///</summary>
        ///<param name="path">图片路径</param>
        ///<returns>位图</returns>
        public static BitmapImage GetImageFromResourcePath(string path)
        {
            return new BitmapImage(new Uri(path, UriKind.Relative));
        }
        ///<summary>
        ///把图片转化成字节数组
        ///</summary>
        ///<param name="img">图片控件</param>
        ///<returns>字节数组</returns>
        public static byte[] ToByteArray(Image img)
        {
            //把图片转化为可写位图
            WriteableBitmap bitmap = new WriteableBitmap(img, null);
            MemoryStream streamSource = new MemoryStream();
            System.Windows.Media.Imaging.Extensions.SaveJpeg(bitmap, streamSource, bitmap.
PixelWidth, bitmap.PixelHeight, 0, 100);
            //像素宽度
            int pixelWidth = bitmap.PixelWidth;
            //像素高度
            int pixelHeight = bitmap.PixelHeight;
            //像素 int 数组
            int[] pixels = bitmap.Pixels;
            //图片像素 int 数组的长度
            int length = pixels.Length;
            //初始化 int 数组对应的 byte 数组
            byte[] buffer = new byte[(4 * pixelWidth) * pixelHeight];
            //把 int 数组转化成 byte 数组
            int index = 0;
            for (int i = 0; index < length; i += 4)
            {
                int num6 = pixels[index];
                buffer[i] = (byte)(num6 >> 0x18);
                buffer[i + 1] = (byte)(num6 >> 0x10);
```

```
                buffer[i + 2] = (byte)(num6 >> 8);
                buffer[i + 3] = (byte)num6;
                index++;
            }
            new BitmapImage().SetSource(streamSource);
            return buffer;
        }
    }
}
```

支出数据保存的代码在 AddAccount. xaml. cs 文件下，代码如下：

AddAccount. xaml. cs 文件保存支出数据的处理代码

```
WriteableBitmap bmp = new WriteableBitmap(this.ImagePic, null);
//一条记账记录的对象
Voucher voucher = new Voucher
{
    Money = double.Parse(this.textBox_Expenses.Text),
    Desc = this.textBox_ExpensesDesc.Text,
    DT = DateTime.Parse(this.DatePickerExpenses.Value.Value.ToString("yyyy/MM/dd") + " " +
this.TimePickerExpenses.Value.Value.ToString("HH:mm:ss")),
    Picture = ImageHelper.ToByteArray(this.ImagePic),
    PictureHeight = bmp.PixelHeight,
    PictureWidth = bmp.PixelWidth,
    Category = listPickerExpenses.SelectedItem.ToString(),
    Type = 1
};
//添加一条记录
App.voucherHelper.AddNew(voucher);
```

添加一笔支出的页面设计如图 24.4 所示。

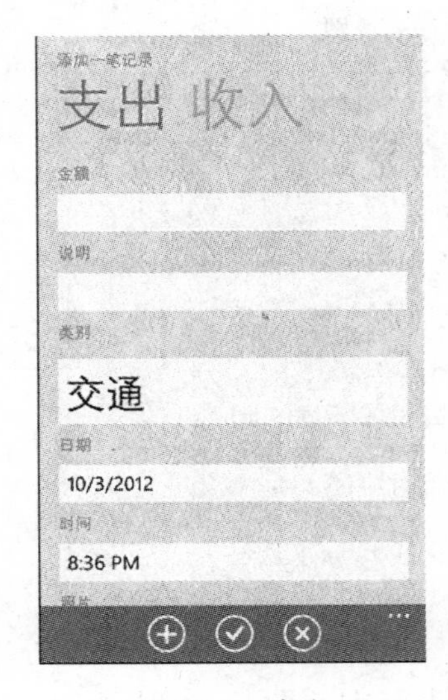

图 24.4　支出

24.6 月报表

月报表是指按照月份来分类查询记账的数据记录。下面来看一下月报表的界面设计和代码实现。

MouthReport.xaml 文件主要代码：月报表的界面设计

```
<Grid x:Name = "ContentPanel" Grid.Row = "1" Margin = "12,0,12,0">
        <!-- 记账记录列表头项目 -->
        <StackPanel Height = "31" Orientation = "Horizontal" Grid.Row = "0" VerticalAlignment
= "Top">
            <Border BorderThickness = "0,0,5,0"Width = "90" Background = "#FFEB6416">
                <TextBlock TextWrapping = " Wrap " Text = " 收 支 " Foreground = " White "
FontWeight = "Bold" HorizontalAlignment = "Center" VerticalAlignment = "Center"/>
            </Border>
            <Border BorderThickness = "0,0,5,0"Width = "90" Background = "#FFEB6416">
                <TextBlock TextWrapping = " Wrap " Text = " 金 额 " Foreground = " White "
FontWeight = "Bold" HorizontalAlignment = "Center" VerticalAlignment = "Center"/>
            </Border>
            <Border BorderThickness = "0,0,5,0"Width = "190" Background = "#FFEB6416">
                <TextBlock TextWrapping = " Wrap " Text = " 备 注 " Foreground = " White "
FontWeight = "Bold" HorizontalAlignment = "Center" VerticalAlignment = "Center"/>
            </Border>
            <Border BorderThickness = "0,0,5,0"Width = "90" Background = "#FFEB6416">
                <TextBlock TextWrapping = " Wrap " Text = " 类 别 " Foreground = " White "
FontWeight = "Bold" HorizontalAlignment = "Center" VerticalAlignment = "Center"/>
            </Border>
        </StackPanel>
        <!-- 记账记录数据绑定列表 -->
        <ListBox x:Name = "listMouthReport" Margin = "0,35,0,80"Grid.Row = "1" >
            <ListBox.ItemTemplate>
                <DataTemplate>
                    <StackPanel Orientation = "Vertical">
                        <StackPanel Orientation = "Horizontal">
                            <TextBlock Foreground = " #FFEB6416 " FontSize = " 30 " Text =
"{Binding Type, Converter = {StaticResource VoucherTypeConverter}}" TextWrapping = " NoWrap"
HorizontalAlignment = "Center"Width = "90" />
                            <TextBlock Foreground = " #FFEB6416 " FontSize = " 30 " Text =
"{Binding Money}" TextWrapping = "NoWrap" HorizontalAlignment = "Center" Width = "90" />
                            <TextBlock Foreground = " #FFEB6416 " Text = " {Binding Desc,
Converter = { StaticResource VoucherDescConverter }}" TextWrapping = "NoWrap"
HorizontalAlignment = "Center"Width = "190" />
                            <TextBlock Foreground = " #FFEB6416 " FontSize = " 20 " Text =
"{Binding Category}" TextWrapping = "NoWrap" HorizontalAlignment = "Center" Width = "93" />
                        </StackPanel>
                        <TextBlock Foreground = " # FFEB6416 " FontSize = " 20 " Text =
"{Binding DT}" TextWrapping = "NoWrap" HorizontalAlignment = "Right" Width = "200"/>
                    </StackPanel>
```

```
                </DataTemplate>
            </ListBox.ItemTemplate>
        </ListBox>
        <!--显示本月收入、支出和结余-->
        <StackPanel Height="57"Orientation="Horizontal" Grid.Row="2" VerticalAlignment
="Bottom" Margin="0,0,0,10">
            <Border BorderThickness="0,0,0,0"Width="155" Background="#FFEB6416">
                <TextBlock TextWrapping="Wrap" x:Name="inTB" Text="" Foreground=
"White" FontWeight="Bold" HorizontalAlignment="Center" VerticalAlignment="Center"/>
            </Border>
            <Border BorderThickness="0,0,0,0"Width="155" Background="#FFEB6416">
                <TextBlock TextWrapping="Wrap" x:Name="exTB" Text="" Foreground=
"White" FontWeight="Bold" HorizontalAlignment="Center" VerticalAlignment="Center"/>
            </Border>
            <Border BorderThickness="0,0,0,0"Width="155" Background="#FFEB6416">
                <TextBlock TextWrapping="Wrap" x:Name="balanceTB" Text="" Foreground=
"White" FontWeight="Bold" HorizontalAlignment="Center" VerticalAlignment="Center"/>
            </Border>
        </StackPanel>
    </Grid>
</Grid>
<!--菜单栏-->
<phone:PhoneApplicationPage.ApplicationBar>
    <shell:ApplicationBar IsVisible="True" BackgroundColor="#FFEB6416" IsMenuEnabled="True">
        <shell:ApplicationBarIconButton IconUri="/Images/appbar.First.rest.png" Text="上
一月" Click="ApplicationBarIconButton_Click"/>
        <shell:ApplicationBarIconButton IconUri="/Images/appbar.Last.rest.png" Text="下
一月" Click="ApplicationBarIconButton_Click"/>
    </shell:ApplicationBar>
</phone:PhoneApplicationPage.ApplicationBar>
```

MouthReport.xaml.cs 文件主要代码

```
using System;
using System.Windows;
using Microsoft.Phone.Controls;
using Microsoft.Phone.Shell;
namespace AccountBook
{
    public partial class MouthReport : PhoneApplicationPage
    {
        private int mouth;
        private int year;
        public MouthReport()
        {
            InitializeComponent();
        }
        //处理页面加载事件
        private void PhoneApplicationPage_Loaded(object sender, RoutedEventArgs e)
```

```
{
    mouth = DateTime.Now.Month;
    year = DateTime.Now.Year;
    DisplayVoucherData();
}
//处理菜单栏单击事件
private void ApplicationBarIconButton_Click(object sender, EventArgs e)
{
    try
    {
        switch ((sender as ApplicationBarIconButton).Text)
        {
            case "上一月":
                this.mouth--;
                if (this.mouth <= 0)
                {
                    this.year--;
                    this.mouth = 12;
                }
                break;
            case "下一月":
                this.mouth++;
                if (this.mouth >= 12)
                {
                    this.year++;
                    this.mouth = 1;
                }
                break;
        }
        DisplayVoucherData();
    }
    catch
    {
    }
}
//展现记账的数据
private void DisplayVoucherData()
{
    //本月的收入
    double inSum = Common.GetMouthSummaryIncome(mouth, year);
    //本月的支出
    double exSum = Common.GetMouthSummaryExpenses(mouth, year);
    //显示本月收入
    inTB.Text = "收入:" + inSum;
    //显示本月支出
    exTB.Text = "支出:" + exSum;
    //显示本月结余
```

```
            balanceTB.Text = "结余:" + (inSum - exSum);
            //绑定当前月份的记账记录
            listMouthReport.ItemsSource = Common.GetThisMonthAllRecords(mouth,year);
            PageTitle.Text = year + "年" + mouth + "月";
        }
    }
}
```

月报表的页面设计如图 24.5 所示。

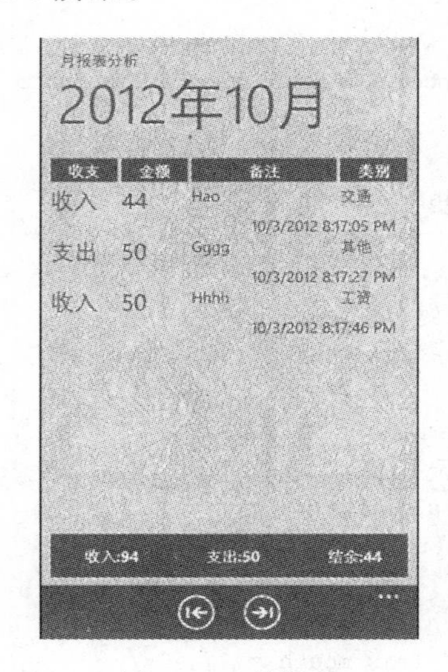

图 24.5 月报表

24.7 年报表

年报表是指按照年份来分类查询记账的数据记录。年报表的页面设计与月报表差不多，下面来看一下年报表的代码处理。

YearReport.xaml.cs 文件主要代码：年报表的实现处理

```
using System;
using System.Windows;
using Microsoft.Phone.Controls;
using Microsoft.Phone.Shell;
namespace AccountBook
{
    public partial class YearReport : PhoneApplicationPage
    {
        private int year;
        public YearReport()
        {
            InitializeComponent();
```

```
        }
        //处理页面加载事件
        private void PhoneApplicationPage_Loaded(object sender, RoutedEventArgs e)
        {
            year = DateTime.Now.Year;
            DisplayVoucherData();
        }
        //处理菜单栏单击事件
        private void ApplicationBarIconButton_Click(object sender, EventArgs e)
        {
            try
            {
                switch ((sender as ApplicationBarIconButton).Text)
                {
                    case "上一年":
                        this.year--;
                        break;
                    case "下一年":
                        this.year++;
                        break;
                }
                DisplayVoucherData();
            }
            catch
            {
            }
        }
        //展现记账的数据
        private void DisplayVoucherData()
        {
            //本年的收入
            double inSum = Common.GetYearSummaryIncome(year);
            //本年的支出
            double exSum = Common.GetYearSummaryExpenses(year);
            //显示本年收入
            inTB.Text = "收入:" + inSum;
            //显示本年支出
            exTB.Text = "支出:" + exSum;
            //显示本年结余
            balanceTB.Text = "结余:" + (inSum - exSum);
            //绑定当前年份的记账记录
            listYearReport.ItemsSource = Common.GetThisYearAllRecords(year);
            PageTitle.Text = year + "年";
        }
    }
}
```

年报表的页面设计如图 24.6 所示。

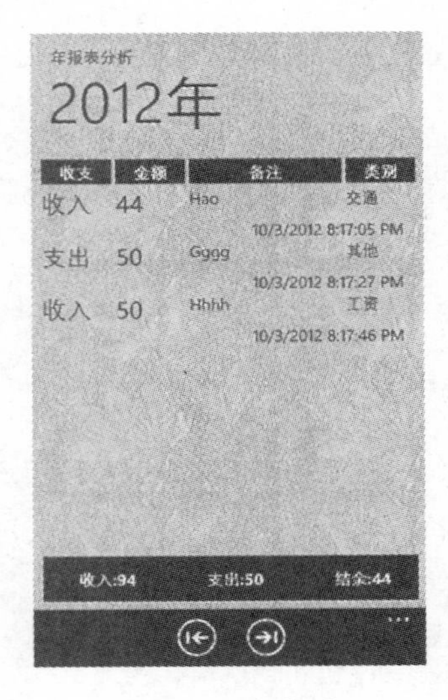

图 24.6 年报表

24.8 查询记录

查询记录的功能是指可以通过时间或者关键字来查询记账的记录。对于关键字查询，可以使用 LINQ 语句的查询语句来实现。下面来看一下查询记录的代码实现。

Search. xaml. cs 文件主要代码

```
//处理菜单栏单击事件,查询记账记录
private void ApplicationBarIconButton_Click(object sender, EventArgs e)
{
    DateTime? begin = DatePickerBegin.Value;
    DateTime? end = DatePickerEnd.Value;
    listReport.ItemsSource = Common.Search(begin, end, keyWords.Text);
}
```

Common. cs 文件查询记账记录的 LINQ 语句代码

```
///<summary>
///查询记账记录
///</summary>
///<param name = "begin">开始日期</param>
///<param name = "end">结束日期</param>
///<param name = "keyWords">关键字</param>
///<returns>记账记录</returns>
public static IEnumerable<Voucher> Search(DateTime? begin, DateTime? end, string keyWords)
{
    if (keyWords == "")
```

```
    {
        return (from c in App.voucherHelper.data
                where c.DT >= begin && c.DT <= end select c);
    }
    else
    {
        return (from c in App.voucherHelper.data
                where c.DT >= begin && c.DT <= end && c.Desc.IndexOf(keyWords) >= 0 select c);
    }
}
```

查询记录的页面设计如图 24.7 所示。

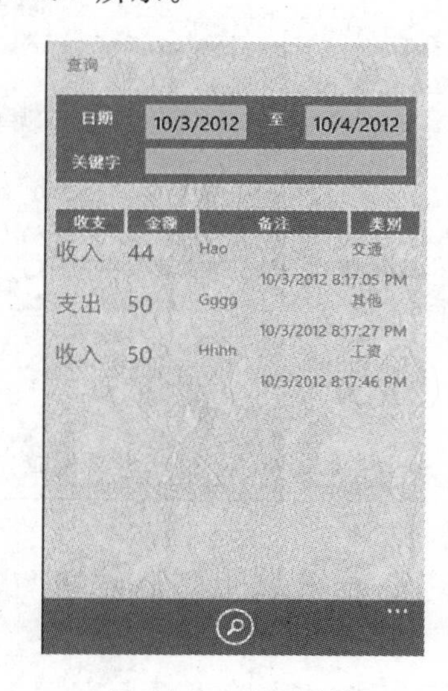

图 24.7　查询记录

24.9　分类图表

图表是很直观的数据展示，同时也很美观，对于记账类软件图表的展示是一个不可缺少的一部分。在记账本里面实现了按照类别进行图表统计，使用了饼状图和柱形图的表现形式，展现了每个类别的记账数目大小和所占的比例。图表控件使用的是 Toolkit 控件库，下面来看一下分类图表的实现。

首先定义一个图表数据展现的实体类。

ChartData.cs 文件代码：ChartData 类表示图表数据类

```
public class ChartData
{
    public double Sum { get; set; }//数值
    public string TypeName { get; set; }//类型
}
```

ChartPage.xaml 文件代码：图表的 UI 绑定设计

```
<controls:PivotItem Header="圆饼图">
    <Grid>
        <charting:Chart HorizontalAlignment="Left" Name="chart1" VerticalAlignment=
"Top" Height="460" Width="455" FontSize="16" RenderTransformOrigin="0.5,0.5">
            <charting:PieSeries x:Name="PieChart1" />
        </charting:Chart>
    </Grid>
</controls:PivotItem>
<controls:PivotItem Header="条形图">
    <Grid>
        <charting:Chart HorizontalAlignment="Left" x:Name="chart2" VerticalAlignment=
"Top" Height="460" Width="455" FontSize="16" RenderTransformOrigin="0.5,0.5">
            <charting:Chart.LegendStyle>
                <Style TargetType="DataVisualization:Legend">
                    <Setter Property="Width" Value="0" />
                    <Setter Property="Height" Value="0" />
                </Style>
            </charting:Chart.LegendStyle>
            <charting:ColumnSeries x:Name="ColumnSeries1" />
        </charting:Chart>
    </Grid>
</controls:PivotItem>
```

ChartPage.xaml.cs 文件主要代码

```
//页面加载事件处理
private void Report_Loaded(object sender, RoutedEventArgs e)
{
    //创建图表的数据源对象
    ObservableCollection<ChartData> collecion = new ObservableCollection<ChartData>();
    //获取所有的记账记录
    IEnumerable<Voucher> allRecords = Common.GetAllRecords();
    //获取所有记账记录里面的类别
    IEnumerable<string> enumerable2 = (from c in allRecords select c.Category).Distinct
<string>();
    //按照类别来统计记账的数目
    foreach (var item in enumerable2)
    {
        //获取该类别下的钱的枚举集合
        IEnumerable<double> enumerable3 = from c in allRecords.Where<Voucher>(c =>
c.Category == item) select c.Money;
        //添加一条图表的数据
        ChartData data = new ChartData
        {
            Sum = enumerable3.Sum(),
            TypeName = item
        };
        collecion.Add(data);
    }
```

```
    //新建一个饼状图表的控件对象
    PieSeries series = new PieSeries();
    //新建一个柱形图表的控件对象
    ColumnSeries series2 = new ColumnSeries();
    //绑定数据源
    series.ItemsSource = collecion;
    series.DependentValueBinding = new Binding("Sum");
    series.IndependentValueBinding = new Binding("TypeName");
    series2.ItemsSource = collecion;
    series2.DependentValueBinding = new Binding("Sum");
    series2.IndependentValueBinding = new Binding("TypeName");
    //添加到图表里面
    this.chart1.Series.Clear();
    this.chart1.Series.Add(series);
    this.chart2.Series.Clear();
    this.chart2.Series.Add(series2);
}
```

分类图表的页面设计如图 24.8 和图 24.9 所示。

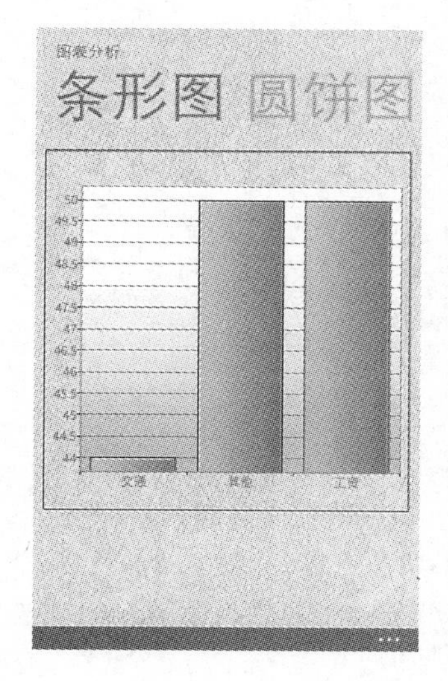

图 24.8　柱形图

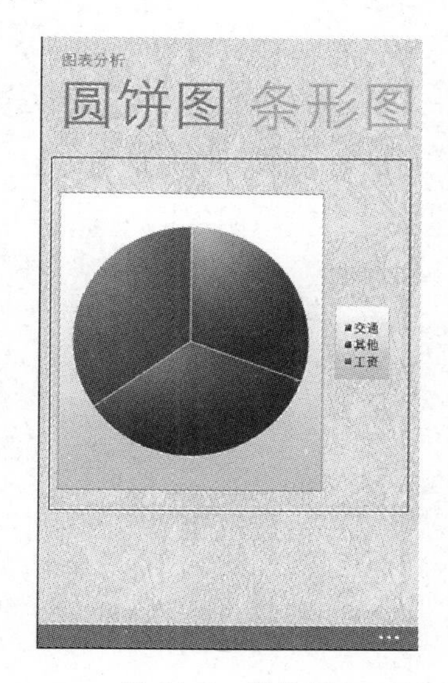

图 24.9　饼状图